KENDALL'S

ADVANCED
Theory of
STATISTICS

Volume 2B

BAYESIAN INFERENCE

KENDALL'S ADVANCED THEORY OF STATISTICS

Volume 1: Distribution Theory
Alan Stuart and Keith Ord
ISBN 0 340 61430 7

Volume 2A: Classical Inference and the Linear Model
Alan Stuart, Keith Ord and Steven Arnold
ISBN 0 340 66230 1

Volume 2B: Bayesian Inference
Anthony O'Hagan
ISBN 0 340 52922 9

KENDALL'S LIBRARY OF STATISTICS

1. Multivariate Analysis
   Part 1: Distributions, Ordination and Influence
   WJ Krzanowski (University of Exeter) and FHC Marriott (University of Oxford)
   1994, ISBN 0 340 59326 1

2. Multivariate Analysis
   Part 2: Classification, Covariance Structures and Repeated Measurements
   WJ Krzanowski (University of Exeter) and FHC Marriott (University of Oxford)
   1995, ISBN 0 340 59325 3

3. Multilevel Statistical Models
   Second edition
   H Goldstein (University of London)
   1995, ISBN 0 340 59529 9

4. The Analysis of Proximity Data
   BS Everitt (Institute of Psychiatry) and S Rabe-Hesketh (Institute of Psychiatry)
   1997, ISBN 0 340 67776 7

5. Robust Nonparametric Statistical Methods
   Thomas P Hettmansperger (Penn State University) and Joseph W McKean (Western Michigan University)
   1998, ISBN 0 340 54937 8

6. Statistical Regression with Measurement Error
   Chi-Lun Cheng (Academia Sinica, Republic of China) and John Van Ness (University of Texas at Dallas)
   1999, ISBN 0 340 61461 7

7. Latent Variable Models and Factor Analysis
   Second edition
   DJ Bartholomew (London School of Economics) and M Knott (London School of Economics)
   1999, ISBN 0 340 69243 X

8. Statistical Inference for Diffusion Type Processes
   BLS Prakasa Rao (Indian Statistical Institute)
   1999, ISBN 0 340 74149 X

KENDALL'S

ADVANCED *Theory of* STATISTICS

Volume 2B

BAYESIAN INFERENCE

Anthony O'Hagan
Professor of Statistics
University of Nottingham, UK

ARNOLD

A member of the Hodder Headline Group
LONDON • SYDNEY • AUCKLAND
Co-published in the USA by Oxford University Press Inc., New York

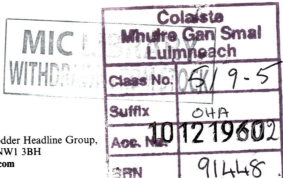

First published in 1994 by
Arnold, a member of the Hodder Headline Group,
338 Euston Road, London NW1 3BH
http://www.arnoldpublishers.com

Co-published in the United States of America by
Oxford University Press Inc.,
198 Madison Avenue,
New York, NY 10016

Publishing History

Volume 1	Distribution Theory		
		First edition	1943
		New editions and reprints	1945, 1947, 1948, 1952, 1958, 1963, 1969, 1977, 1987, 1994
Volume 2	Classical Inference and Relationships		
		First edition	1946
		New editions and reprints	1947, 1951, 1961, 1967, 1973, 1979, 1991
		As Volume 2A	1999
Volume 2B	Bayesian Inference		
		First edition	1994
		reprint	1999
Volume 3	Design and Analysis and Time Series		
		First edition	1966
		New editions	1968, 1976, 1983
	Now replaced by *Kendall's Library of Statistics*		

British Library Cataloguing in Publication Data
A catalogue record for this book is available from the British Library

Library of Congress Cataloging-in-Publication Data
A catalog record for this book is available from the Library of Congress

ISBN 0 340 52922 9

Printed and bound in Great Britain by St Edmundsbury Press, Bury St Edmunds, Suffolk

To Dennis V Lindley

Agh! Macintosh! He was Goldwasser's closest friend, and the very thought of him filled Goldwasser with a warm, familiar irritation. He irritated Goldwasser in two ways—sometimes by seeming to be too stupid to talk to, and sometimes by seeming to be possibly cleverer than Goldwasser. Even more irritating than either state was the ambiguity of fluctuating between the two.

Was Macintosh cleverer than him or wasn't he? There must have been some objective way of telling. Once, he was sure, he had been indubitably cleverer than Macintosh. But he was slowing down. At least, he was afraid he was slowing down. He was pretty certain his brain was of the genus *Cerebrum Dialectici*—a logician's or child prodigy's brain, an early-flowering plant already past its best by the age of thirty. His anxiety on the point had developed into a sort of cerebral hypochondria. He was for ever checking his mental performance for symptoms of decline. He borrowed sets of IQ tests from his colleagues, and timed his performances, plotting the results on graphs. When he produced a graph with a curve that went downwards, he assured himself that it was merely a misleading technique; and when he produced one with a curve that went upwards, he told himself sceptically that it must be the result of experimental error.

Michael Frayn, *The Tin Men*, 1965. Collins: London.

CONTENTS

PREFACE

The Bayesian approach to statistics has developed rapidly, and at an ever-increasing rate, over the last thirty or forty years. This volume is a sign, if one were needed, that it has become accepted as a sound and viable method. With this addition to the other two volumes, *Kendall's Advanced Theory of Statistics* can claim still to fulfil Kendall's original intention, fifty years ago, 'to develop a systematic treatment of [statistical] theory as it exists at the present time'.

It was a privilege to be asked to write this volume, and a great challenge. '*Kendall*' is legendary for its authoritative coverage of its subject, and for the clarity of its exposition. I have tried to reach that high standard in producing a thorough but readable text on Bayesian inference. However, I knew that its coverage could not be as complete and authoritative as the other volumes. I have not had decades in which to build that kind of depth. The coverage and the list of references inevitably have many gaps, particularly in those areas which have not formed part of my own research interests. Readers will, I hope, help me to fill those gaps in future by drawing my attention to work and references that I have failed to cite.

Cross-references within this volume follow the established practice of *Kendall's Advanced Theory of Statistics*. For instance, **2.13** refers to section **2.13** within Chapter 2, while (2.13) refers to equation (2.13), which will also be in Chapter 2 of this volume. However, I have also tried to make use of the excellent material in the other two volumes wherever they present ideas or results of relevance to Bayesian inference. Cross-references to the other '*Kendall*' volumes are indicated by a prefix as in A**2.13**, (A2.13) or Chapter A2.

Although this volume carries just my name, and of course I alone am responsible for any errors, it could not have been written without the help and encouragement of many other people. Chief among these is Dennis Lindley, who has encouraged and stimulated my own development in Bayesian statistics enormously, as he has done for so many others. His influence on the subject has been tremendous, both indirectly by enthusing new generations with Bayesian thinking but also directly through his many theoretical contributions. His comments on early drafts of this book have been as perceptive and valuable as always. This volume is dedicated to Dennis Lindley.

In addition to the many other colleagues who have contributed their ideas and knowledge, I should acknowledge help from the Universities of Warwick and Nottingham in each giving me several months free from teaching duties to work on this book. Purdue University gave me generous hospitality for some of that time. The editorial staff at Edward Arnold have always been helpful and supportive, too. And last, but far from least, my ever patient and loving wife Anne typed the manuscript, put up with my long hours of work, and kept me going. Without her, this volume would have been completed much later, or not at all.

Nottingham, October 1993

GLOSSARY OF ABBREVIATIONS

The following abbreviations and symbols are sometimes used:

$B(a,b)$	the beta function with arguments a and b
$D(a,b)$	the Dirichlet distribution with parameters a and b
d.f.	distribution function
$DP(a,b)$	the Dirichlet process with parameters a and b
f.f.	frequency function
$GP(a,b)$	the (multivariate) Gaussian process with mean (vector) a and variance (covariance matrix) b
$IG(a,b)$	the inverse-gamma distribution with parameters a and b
iid	independently and identically distributed
$IW(a,b)$	the inverse-Wishart distribution with parameters a and b
LP	Likelihood Principle
MCMC	Markov Chain Monte Carlo
MLE	Maximum Likelihood Estimator
m.s.e.	mean-square-error
$N(a,b)$	the (multivariate) normal distribution with mean (vector) a and variance (covariance matrix) b
$NIG(a,b,c,d)$	the normal-inverse-gamma distribution with parameters a, b, c and d
$NIW(a,b,c,d)$	the normal-inverse-Wishart distribution with parameters a, b, c and d
$t_d(a,b)$	the (multivariate) t distribution with degrees of freedom d and parameters a and b
$\Gamma(a)$	the gamma function with argument a

CHAPTER 1

THE BAYESIAN METHOD

1.1 The fundamental problem towards which the study of statistics is addressed, is that of inference. Some data are observed and we wish to make statements, *inferences*, about one or more unknown features of the physical system which gave rise to these data.

The last sentence is deliberately couched in broad terms because the circumstances in which the inference problem arises are very varied. For example, an opinion poll produces data concerning which people in a sample answer 'Yes' to a certain question. The system giving rise to these data consists of a population of individuals, a mechanism for selecting a sample from that population, and a mechanism which produces a 'Yes' or 'No' response from each individual in the sample. Typically, the answering mechanism is a direct question to which the individual answers 'Yes' if and only if he or she possesses a certain characteristic, and the sampling mechanism is some probabilistic scheme which is known. The population is some well-defined group of people, but the unknown feature of the system is the proportion, θ, of individuals in that population who possess the characteristic. The object of inference then is to use the observed responses in the sample to make statements about θ. Sometimes, however, the system will have other unknown features of interest. Direct questioning about a sensitive characteristic will not always elicit an honest answer, and the proportion of people who give false responses will generally be another unknown feature. Some features of the sampling mechanism may also be unknown. Exceptionally, such features could be the main object of interest, and θ may even be known.

1.2 The problem of inference has been the subject of considerable attention since the systematic study of probability theory began in the eighteenth century. Many different theories of inference have been proposed, and there has hardly been a time when inference was not a matter of real controversy. Throughout most of Volumes 1 and 2A of this book, particularly in Volume 2A, the development has followed firmly the theory of inference which was almost uncontested in the statistical community during the middle of the twentieth century. This theory has been given a number of names, but it is most often called *classical* or *frequentist* inference. Since about 1960 there has been a steady revival of interest in *Bayesian inference*, to the extent that the Bayesian approach is now a well-established alternative to classical inference.

Bayes' theorem

1.3 Consider two events, A and B. From the identity

$$P(A)\,P(B \mid A) = P(A, B) = P(B)\,P(A \mid B) \tag{1.1}$$

we have the simplest form of Bayes' theorem,

$$P(B \mid A) = P(B)\,P(A \mid B)/P(A). \tag{1.2}$$

We can interpret (1.2) in the following way. We are interested in the event B, and begin with an initial, *prior* probability $P(B)$ for its occurrence. We then observe the occurrence of A. The proper description of how likely B is when A is known to have occurred is the *posterior* probability $P(B \mid A)$. Bayes' theorem can be understood as a formula for updating from prior to posterior probability, the updating consisting of multiplying by the ratio $P(A \mid B)/P(A)$. It therefore describes how a probability changes as we learn new information.

Observing the occurrence of A will increase the probability of B if $P(A \mid B) > P(A)$. Using the law of total probability (A7.7),

$$P(A) = P(A \mid B)\,P(B) + P(A \mid B^c)\,P(B^c),$$

where B^c denotes the complement or negation of B, and remembering that $P(B^c) = 1 - P(B)$, we have

$$P(A \mid B) - P(A) = \{P(A \mid B) - P(A \mid B^c)\}\,P(B^c). \tag{1.3}$$

Assuming that $P(B^c) > 0$ (otherwise B is a certain event, and its probability would not be of interest), $P(A \mid B) > P(A)$ if and only if $P(A \mid B) > P(A \mid B^c)$.

1.4 It is more usual to present Bayes' theorem in a more general form. Let B_1, B_2, ..., B_n be a set of mutually exclusive and exhaustive events. Then

$$\begin{aligned} P(B_r \mid A) &= P(B_r)\,P(A \mid B_r)/P(A) \\ &= \frac{P(B_r)\,P(A \mid B_r)}{\sum_r P(B_r)\,P(A \mid B_r)}. \end{aligned} \tag{1.4}$$

(1.4) is a simple generalization of (1.2), which may be proved directly (or see A**8.2**). We can think of the B_rs as a set of hypotheses, one and only one of which is true. (If hypothesis r is true we will say that the event B_r occurs.) Observing event A changes the prior probabilities $P(B_r)$ to posterior probabilities $P(B_r \mid A)$. Notice that the posterior probabilities sum to one, since we still know that one and only one hypothesis is true.

The denominator $P(A)$ in (1.4) is a weighted average of the probabilities $P(A \mid B_r)$, the weights being the $P(B_r)$s (which sum to one). The occurrence of A increases the probability of B_r if $P(A \mid B_r)$ is greater than this average of all the $P(A \mid B_r)$s. The hypothesis whose probability is increased the most by A (in the sense of being multiplied by the largest factor) is the one for which $P(A \mid B_r)$ is highest.

Likelihood

1.5 The probabilities $P(A \mid B_r)$ in (1.4) are known as *likelihoods*. Specifically, $P(A \mid B_r)$ is the likelihood given to B_r by A, or 'the likelihood of B_r given A'. The notion of likelihood is used extensively in Volumes 1 and 2A. In particular, the method of maximum likelihood is first mentioned in A**8.7** and developed fully as a method of estimation in Chapter A18.

The primitive notion, that hypotheses given greater likelihood by A should somehow have higher probability when A is observed to occur, has a compelling logic which is clearly understood from examples.

Example 1.1
I observe through my window a tall, branched thing with green blobs covering its branches: why do I think it is a tree? Because that is what trees generally look like. I do not think it is a man because men rarely look like that. Converting into formal notation, A is the event that I see a tall, branched thing partially covered in small green things, B_1 is the event that it is a tree, B_2 that it is a man, and B_3 that it is something else. The statement that 'trees generally look like that' implies that $P(A \mid B_1)$ is close to one, whereas 'men rarely look like that' means that $P(A \mid B_2)$ is close to zero. I convert these facts into a belief that the object is far more likely to be, i.e. has a much higher posterior probability of being, a tree than a man.

Example 1.2
A less extreme example occurs when I hear a piece of music but do not know its composer. I decide that the music is more probably by Beethoven than Bach because, to me, it sounds more like the music that Beethoven typically composed. That is, Beethoven has higher likelihood because a Beethoven composition is more likely to sound like this, and this suggests a higher probability for Beethoven.

1.6 It is clear that this primitive notion of likelihood underlies one of our most natural thought processes. However, likelihood is not the only consideration in this reasoning.

Example 1.3
Continuing Example 1.1, there are other things that might look like a tree, particularly at a distance. I might be seeing a cardboard replica of a tree. This hypothesis would have essentially the same likelihood as the hypothesis that it is a tree, but it is not a hypothesis that I seriously entertain because it has a very much lower prior probability.

1.7 Bayes' theorem, (1.4), is in complete accord with this natural reasoning. The posterior probabilities of the various hypotheses are in proportion to the products of their prior probabilities and their likelihoods. Bayes' theorem thus combines two sources of information: the prior information is represented by the prior probabilities, the new information A is represented by the likelihoods, and the posterior probabilities represent the totality of this information.

Posterior distributions

1.8 Whereas sometimes the 'unknown features of the physical system' may be expressed in terms of a discrete set of hypotheses $B_1, B_2, \ldots, B_n$, by far the more common situation is that those 'unknown features' are unknown *quantities*. Any such unknown quantity in a statistical problem is called a *parameter*, and inference can most conveniently be thought of as concerned with statements about the unknown values of parameters.

For instance, **1.1** discussed a sampling problem in which the unknown parameter was the population proportion having a certain characteristic, denoted by θ. The value of θ might be anywhere between 0 and 1, and we might attempt various kinds of inferences concerning this value. If we wish to assert a specific value for θ, this is a problem of point estimation, the classical theory of which is given in Chapters A17 to A19. If instead we wish to assert a range of values within which θ is, in some sense, most likely to lie then this is a question of interval estimation, for which Chapter A20 gives the classical theory. The other kind of inference traditionally addressed in classical theory is hypothesis testing, dealt with in Chapters A21 to A23, where we wish to decide whether a particular statement about θ, such as that $\theta > 0.5$, is true. Chapter 2 demonstrates that Bayesian inference can address all these classical forms of inference and many others, within a very general framework for investigating parameters.

Statistical problems are also typically concerned with quantitative data rather than observations of discrete events. We therefore need to restate Bayes' theorem in terms of random variables instead of events.

1.9 Consider a general problem in which we have data x and require inference about a parameter θ. The identity corresponding to (1.1) is

$$dF(x)\,dF(\theta\,|\,x) = dF(x,\,\theta) = dF(\theta)\,dF(x\,|\,\theta)$$

$$\therefore dF(\theta\,|\,x) = dF(\theta)\,dF(x\,|\,\theta)/dF(x). \tag{1.5}$$

We shall generally assume that every distribution possesses a corresponding density function, and therefore rewrite (1.5) as

$$f(\theta\,|\,x) = f(\theta)\,f(x\,|\,\theta)/f(x)$$
$$= \frac{f(\theta)\,f(x\,|\,\theta)}{\int f(\theta)\,f(x\,|\,\theta)\,d\theta}. \tag{1.6}$$

Most commonly both the parameter θ and the data x are continuous valued, and all the frequency functions in (1.6) are continuous. There are, however, many instances where the data are discrete, so that both $f(x\,|\,\theta)$ and $f(x)$ would be discrete frequency functions, giving probabilities of individual values of x. Exceptionally, θ can be discrete. Then both the prior and posterior 'densities' are discrete frequency functions, and in this case we replace the integral in the denominator by a sum. In fact we should rewrite the denominator in the more general Stieltjes form $\int f(x\,|\,\theta)\,dF(\theta)$ to cover both cases. However, since it really is exceptional for us to be concerned with a discrete parameter, we find it most useful to express Bayes' theorem in the form (1.6).

This represents a shift in notation from Volumes 1 and 2, where densities and discrete frequency functions are commonly denoted by dF, and where the Stieltjes form of integral is generally used. That approach produces expressions having strictly greater generality. However, the use of f in this volume as a density function results in clearer formulae, which the reader may easily adapt to discrete or more general cases as appropriate.

1.10 This version of Bayes' theorem is interpreted in the same way as for the simpler case of events. The prior density $f(\theta)$ represents the prior information about θ. We then observe the data x and Bayes' theorem constructs the posterior density $f(\theta \mid x)$ as proportional to the product of the prior density and the likelihood $f(x \mid \theta)$, thereby combining the two sources of information.

Posterior density for particular values of θ will be low if they have low prior density or low likelihood, so that they are essentially discounted by one or other source of information. Appreciable posterior density will exist at values of θ for which neither prior density nor likelihood is very low. If there are values that are well supported by both information sources, i.e. having high prior density and high likelihood, then these values will also have high posterior density.

Example 1.4

Let x be the number of successes in a series of n trials with probability θ of success in each. That is, x has the (discrete) binomial f.f., given θ,

$$f(x \mid \theta) = \binom{n}{x} \theta^x (1-\theta)^{n-x}, \qquad x = 0, 1, \ldots, n. \tag{1.7}$$

This is the likelihood for θ. Now let θ have the prior beta distribution with density

$$f(\theta) = \frac{1}{B(p, q)} \theta^{p-1}(1-\theta)^{q-1}, \qquad 0 \leqslant \theta \leqslant 1, \tag{1.8}$$

for some $p, q > 0$. We now apply Bayes' theorem. First,

$$\begin{aligned} f(x) &= \int f(\theta) f(x \mid \theta)\, \mathrm{d}\theta \\ &= \binom{n}{x} \frac{1}{B(p,q)} \int_0^1 \theta^{p+x-1}(1-\theta)^{q+n-x-1}\, \mathrm{d}\theta \\ &= \binom{n}{x} \frac{B(p+x, q+n-x)}{B(p,q)}. \end{aligned}$$

Now (1.6) gives

$$f(\theta \mid x) = \frac{1}{B(p+x, q+n-x)} \theta^{p+x-1}(1-\theta)^{q+n-x-1}. \tag{1.9}$$

The posterior distribution is therefore another beta distribution.

The likelihood favours values of θ near the maximum of (1.7). Since

$$\frac{\partial}{\partial \theta} \log f(x \mid \theta) = \frac{x}{\theta} - \frac{n-x}{1-\theta} = \frac{x - n\theta}{\theta(1-\theta)},$$

the maximum is at $x - n\theta = 0$, i.e. $\theta = x/n$. The prior distribution, on the other hand, favours values of θ near to the maximum of (1.8). We suppose now that $p, q > 1$ so that (1.8) has a maximum within (0,1). This is then at $\theta = (p-1)/(p+q-2)$. Taking both sources of information, we find that the posterior density (1.9) is maximized at

$$\frac{x+p-1}{n+p+q-2} = a\left(\frac{x}{n}\right) + (1-a)\left(\frac{p-1}{p+q-2}\right), \tag{1.10}$$

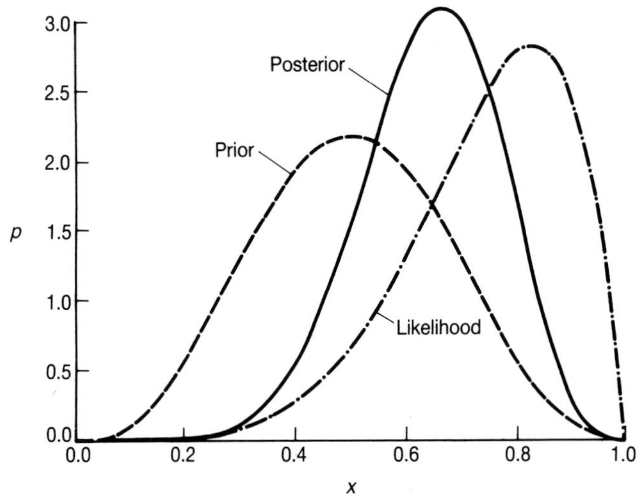

Fig. 1.1 Prior to posterior for binomial trials model

where $a = n/(n+p+q-2)$. Since $0 < a < 1$, (1.10) shows that the posterior distribution synthesises and compromises by favouring values *between* the maxima of prior density and likelihood.

Figure 1.1 demonstrates this visually by showing all three functions in the case $p = q = 4$, $n = 6$, $x = 5$. Notice that the posterior density is taller and narrower than the prior density. It therefore favours strongly a smaller range of θ values, reflecting the fact that we now have more information.

Posterior inference

1.11 Having obtained the posterior density $f(\theta \mid x)$, the final step of the Bayesian method is to derive from it some suitable inference statements. The most usual inference question is this: After seeing the data x, what do we now know about the parameter θ? The only answer to this question is to present the entire posterior distribution. The posterior distribution encapsulates all that is known about θ following the observation of the data x, and can be thought of as comprising an all-embracing inference statement about θ. All posterior degrees of belief concerning θ are expressed therein. Unfortunately, the posterior distribution codes this information in a very compact form, and to respond to the question simply by stating a mathematical formula for the posterior density fails to convey the information in a useful form. It should be supplemented by a range of helpful *summaries*. Perhaps the most useful summary is to plot the density function. Quantitative summaries like measures of location and dispersion are also useful.

If, for instance, the posterior distribution is $N(m, v)$ then a plot of the density shows the familiar, unimodal curve centred on m. The most 'typical', or indicative, single value of θ is m. It is the 'most probable' value of θ, in the sense that the $N(m, v)$ density is maximized at m. It is the posterior expected value of θ and also has the property that θ has equal probabilities of being less than m or greater than m. In other words, m is the posterior mode, mean and median. A dispersion measure may be interpreted as an indicator of how weak the posterior information is. If our information is such that θ has high dispersion, then the range of likely values of θ is large and our information is consequently weak. Conversely, low dispersion indicates strong information. The reciprocal of variance is known as *precision*, and is therefore an indicator of *strength* of information.

We may be interested in some specific hypothesis H, asserting that θ lies in some set Θ. The posterior probability that H is true is simply $P(H \mid x) = P(\theta \in \Theta \mid x)$. If for instance H asserts that $\theta \leqslant \theta_0$ then $P(H \mid x) = P(\theta \leqslant \theta_0 \mid x) = F(\theta_0 \mid x)$, where $F(\theta \mid x)$ is the posterior d.f. The basic Bayesian inference about a hypothesis is its posterior probability, in contrast with classical inference where hypotheses are either rejected or accepted (or 'not rejected'). Accepting or rejecting a hypothesis is not so much an inference as an action or *decision*. The posterior probability will be one factor in making that decision, but we will also wish to take account of the consequences of right or wrong decisions.

We may formulate Bayesian analogues of the other classical inferences. A measure of location like the posterior mean, median or mode, for instance, is an obvious point estimate. There is an equally natural way of producing interval estimates, which is to assert an interval such that the posterior probability that θ lies in the interval has some suitably high value, say 0.9 or 0.99. Thus, if $F(\theta \mid x)$ is the posterior d.f. and if $F(\theta_1 \mid x) = p_1$, $F(\theta_2 \mid x) = p_2 > p_1$, then the interval $(\theta_1, \theta_2]$ is a posterior interval estimate of θ with *coverage* probability $p_2 - p_1$.

There are therefore clear parallels between certain kinds of summary which could be used for general inference, and the natural answers to the three specific inference questions considered by classical theory.

Example 1.5

Let x have a normal distribution $N(\theta, v)$ with unknown mean θ and known variance v, and let the prior distribution for θ be $N(m, w)$. Applying (1.6) the posterior distribution is $N(m_1, w_1)$, where

$$m_1 = \frac{wx + vm}{w + v}, \qquad w_1 = \frac{vw}{w + v}.$$

(See Example A8.6.) Notice that the posterior mean is a weighted average of prior mean m and observation x. Therefore the posterior mean (also median and mode) is a compromise between the prior mean and the observation. Writing

$$m_1 = \frac{v^{-1}x + w^{-1}m}{v^{-1} + w^{-1}},$$

it is clear that each source of information is weighted proportionately to its own precision. Therefore the posterior mean will lie closer to whichever source has the stronger information. If, for instance, prior information is very weak, expressed by w being very large, then m_1 will be close to x.

The posterior precision w_1^{-1} is the sum of the prior and data precisions, reflecting the combination of information from the two sources. The posterior information is stronger than either source of information alone.

1.12 Either x or θ or both may be vector random variables, in which case the appropriate densities in (1.6) are joint densities. Indeed, in most statistical problems the data comprise more than one observation. Then the likelihood function is the joint density $f(x\,|\,\theta)$ of all the observations $x = (x_1, x_2, \ldots, x_n)$ given the parameter(s) θ. Thus, if the data are independent given θ, as in the case of a random sample,

$$f(x\,|\,\theta) = f(x_1\,|\,\theta)\,f(x_2\,|\,\theta)\,\ldots\,f(x_n\,|\,\theta).$$

Also typically, there is more than one unknown parameter, so that θ is also a vector $\theta = (\theta_1, \theta_2, \ldots, \theta_p)$. Then $f(\theta)$ is the joint prior density and $f(\theta\,|\,x)$ is the joint posterior density. The higher the number of parameters, the more important it becomes to summarize the posterior distribution thoroughly and carefully.

To see some of the complexities involved, consider a specific inference. A point estimate for the whole vector θ would be a vector $t = (t_1, t_2, \ldots, t_p)$. It is natural to interpret this as a vector of point estimates, and clearly we could let each t_i be an individual point estimate of the corresponding θ_i. However, there are other possibilities. For instance, the mode of the joint posterior density $f(\theta\,|\,x)$ is a vector whose components are not in general the modes of the individual, marginal posterior densities $f(\theta_i\,|\,x)$.

The same is true in general of all inference. We can distinguish between joint inferences or summaries (such as a joint mode, a covariance matrix or a contour plot of a bivariate density) and individual inferences or summaries. Consider in general inference about a subvector ϕ comprising from one or more of the elements of θ. Any such inference derives from the posterior density $f(\phi\,|\,x)$, obtained by integrating the full joint density $f(\theta|x)$ with respect to all elements of θ which are not elements of ϕ.

Example 1.6

Let $x = (x_1, x_2, \ldots, x_n)$ be a sample from the distribution $N(\mu, \sigma^2)$, where both the mean μ and variance σ^2 are unknown. Therefore $\theta = (\mu, \sigma^2)$. The likelihood function is

$$\begin{aligned} f(x\,|\,\theta) &= \prod_{i=1}^{n} \frac{1}{\surd(2\pi)\,\sigma}\,\exp\{-\frac{1}{2\sigma^2}(x_i-\mu)^2\} \\ &= (2\pi\sigma^2)^{-n/2}\exp\left[-\frac{1}{2\sigma^2}\{s^2 + n(\bar{x}-\mu)^2\}\right], \end{aligned}$$

where $\bar{x}$ is the sample mean and $s^2 = \sum_{i=1}^{n}(x_i - \bar{x})^2$. Let the prior density be

$$f(\theta) = k(a, b, w)(\sigma^2)^{-(b+3)/2}\exp\left[-\frac{1}{2\sigma^2}\{a + (\mu - m)^2/w\}\right],$$

where

$$k(a,\, b,\, w) = a^{b/2}\,2^{-(b+1)/2}\,(\pi w)^{-1/2}\{\Gamma(\tfrac{1}{2}b)\}^{-1}$$

is a constant. Applying Bayes' theorem, the posterior density is

$$f(\theta\,|\,x) = k(a_1,\, b_1,\, w_1)\,(\sigma^2)^{-(b_1+3)/2}\,\exp\left[-\frac{1}{2\sigma^2}\{a_1 + (\mu - m_1)^2/w_1\}\right], \qquad (1.11)$$

where

$$w_1 = \frac{w}{1+nw}, \qquad m_1 = \frac{m+nw\bar{x}}{1+nw},$$
$$b_1 = b+n, \qquad a_1 = a + s^2 + \frac{n(\bar{x}-m)^2}{1+nw}.$$

Suppose now that we require inference about μ alone. We therefore calculate the marginal posterior density

$$f(\mu \mid x) = \int_0^\infty f(\theta \mid x)\, d\sigma^2.$$

This integral is easily done with the substitution $y = \{a_1 + (\mu - m_1)^2/w_1\}/(2\sigma^2)$, and we find that the posterior density of μ is

$$f(\mu \mid x) = k_\mu(t_1, b_1)\{1 + (\mu - m_1)^2/(b_1 t_1)\}^{-(b_1+1)/2},$$

where $t_1 = w_1 a_1 / b_1$ and

$$k_\mu(t_1, b_1) = (t_1 b_1)^{-1/2}\{B(\tfrac{1}{2}, \tfrac{1}{2}b_1)\}^{-1}.$$

Therefore $(\mu - m_1)/\sqrt{t_1}$ has a Student t-distribution with b_1 degrees of freedom. It is now straightforward to make inferences about μ. For instance, we can plot its density, which is symmetric and, for t_1 not too small, very like a normal density. m_1 is the mean, median and mode (provided $t_1 > 1$).

For inference about σ^2,

$$f(\sigma^2 \mid x) = \int_{-\infty}^{\infty} f(\theta \mid x)\, d\mu$$
$$= k_{\sigma^2}(a_1, b_1)(\sigma^2)^{-(b_1+2)/2} \exp\left(-\frac{a_1}{2\sigma^2}\right),$$

where

$$k_{\sigma^2}(a_1, b_1) = (\tfrac{1}{2}a_1)^{b_1/2}\{\Gamma(\tfrac{1}{2}b_1)\}^{-1}.$$

This is a kind of inverse chi-squared distribution, because the distribution of a_1/σ^2 is chi-squared with b_1 degrees of freedom. A plot of its density shows it to be quite like an ordinary chi-squared distribution – unimodal and skewed to the right. Its mode may be found by

$$\frac{\partial}{\partial \sigma^2} \log f(\sigma^2 \mid x) = -(b_1+2)\frac{1}{2\sigma^2} + \frac{a_1}{2\sigma^4},$$

so that the mode is at $a_1/(b_1+2)$. However, the mean is at $a_1/(b_1-2)$. Whereas for a symmetric distribution we can summarize location by a single quantity, in this case neither mean nor mode is uniquely appropriate. Both can be quoted, and their difference is also indicative of the degree of skewness. The median, which lies between these two values, may only be found numerically.

Finally, we can find the joint mode by equating both partial derivatives $\partial f(\theta \mid x)/\partial \mu$ and $\partial f(\theta \mid x)/\partial \sigma^2$ to zero, obtaining $(m_1, a_1/(b_1+3))$.

Example 1.7
Let a random sample of n be drawn from a population with a proportion θ_i of individuals of type i, $i = 1, 2, \ldots, p$. If we have x_i individuals of type i in the sample, the LF is

$$f(x \mid \theta) = \frac{n!}{x_1! \, x_2! \ldots x_p!} \theta_1^{x_1} \theta_2^{x_2} \ldots \theta_p^{x_p},$$

where $\theta_i \geqslant 0$ and $\sum_{i=1}^{p} \theta_i = 1$. Let the prior density be

$$f(\theta) = \frac{\Gamma(a_1 + a_2 + \ldots + a_p)}{\Gamma(a_1)\Gamma(a_2) \ldots \Gamma(a_p)} \theta_1^{a_1 - 1} \theta_2^{a_2 - 1} \ldots \theta_p^{a_p - 1}.$$

Then the posterior density is

$$f(\theta \mid x) = \frac{\Gamma(b_1 + b_2 + \ldots + b_p)}{\Gamma(b_1)\Gamma(b_2) \ldots \Gamma(b_p)} \theta_1^{b_1 - 1} \theta_2^{b_2 - 1} \ldots \theta_p^{b_p - 1}. \tag{1.12}$$

To find the joint mode we need to recognize that $\sum_{i=1}^{p} \theta_i = 1$ and we will use the method of Lagrange multipliers. Define

$$g(\theta) = \log f(\theta \mid x) + \lambda(\theta_1 + \theta_2 + \ldots + \theta_p - 1).$$

Equating $\partial g(\theta)/\partial \theta_i$ and $\partial g(\theta)/\partial \lambda$ to zero we obtain $\theta_i = (b_i - 1)/(b_1 + b_2 + \ldots + b_p - p)$. These are the components of the joint mode. Similar care is needed in finding marginal distributions and inferences. See Exercise 1.7.

The Bayesian method

1.13 In the preceding sections we have sketched the essentials of the Bayesian method as it is most commonly applied. We have deliberately done so in terms which have left many questions unasked. In the remainder of this chapter we will pose and briefly discuss these questions, before proceeding to study them in detail in the remainder of this volume.

1.14 The Bayesian method briefly comprises the following principal steps.

Likelihood. Obtain the likelihood function, i.e. $f(x \mid \theta)$. This step simply describes the process giving rise to the data x in terms of the unknown parameters θ.

Prior. Obtain the prior density $f(\theta)$. The prior distribution expresses what is known about θ prior to observing the data.

Posterior. Apply Bayes' theorem to derive the posterior density $f(\theta \mid x)$. This will now express what is known about θ after observing the data.

Inference. Derive appropriate inference statements from the posterior distribution. These will generally be designed to bring out the information expressed in the posterior distribution, and may include specific inferences such as point estimates, interval estimates or probabilities of hypotheses.

The prior distribution

1.15 Perhaps the most important questions concern the prior distribution. What do prior probabilities represent, and how can they be derived? We cannot answer these without first raising the even more fundamental question of what a probability represents. The theory of classical inference is based on the frequency interpretation of probability, which asserts that the probability of an event is the limiting relative frequency of occurrence of the event in an infinite sequence of trials. Classical rules of inference are judged on their long-run behaviour in repeated sampling. For instance, a point estimator has a sampling distribution and is said to be unbiased if the mean of that distribution equals the parameter to be estimated. The sampling distribution is composed of frequency probabilities, where the infinite sequence of trials is a sequence of hypothetical repetitions of the sampling process which generated the data.

In most inference problems, the prior distribution cannot realistically be thought of as comprising frequency probabilities. Consider for instance the opinion poll example of **1.1**. The parameter θ is the proportion of individuals in a population who possess some characteristic. If the probability that $\theta < 0.5$ is to be a frequency probability, then we require an infinite sequence of trials, but the population of interest will usually be unique. It is unrealistic, or at least artificial, to contemplate a sequence of populations so that $P(\theta < 0.5)$ can be defined as a limiting frequency, of populations in which the proportion of individuals having the characteristic is less than one-half. In contrast, there is no difficulty in contemplating repeated sampling from the unique population in order to derive the sampling distributions required by classical inference.

Many proponents of classical inference are unwilling to admit any notion of probability other than frequency probability. To them, unless prior probabilities can be formulated as frequency probabilities they do not exist. They therefore deny that the Bayesian method is a viable approach to inference. Classical inference is often called frequentist inference because of its strong association with frequency probability.

The fundamental distinction between classical and Bayesian inference is that in Bayesian inference the parameters are random variables, and therefore have both prior and posterior distributions. In classical inference the parameters take unique values, although these values are unknown, and it is not permitted to treat them as random or to give them probabilities.

1.16 In order to follow the Bayesian method it is necessary to adopt a less restrictive interpretation of probability. In A**7.12** to A**7.16** the principal alternative views to frequency probability are described, namely *logical* probability and *subjective* probability. In both of these theories, probability represents a degree of belief in a proposition, based on all the available information. A subjective probability is a measure of one person's degree of belief. Another person may have a different degree of belief in the same proposition, and so have a different probability. The only constraint is that a single person's probabilities should not be inconsistent, and therefore they should obey all the axioms of probability.

We would expect two people's degrees of belief in a proposition to differ if they have different information, and in practice two people will never have exactly the same information. The question does not arise, therefore, whether two people with identical information might have different subjective probabilities, or whether there is a unique degree of belief implied by a given body of information. The latter view is taken in the theory of logical probability, which is then concerned with trying to identify logical degrees of belief. Proponents of logical probabilities see them as extending the theory of logic. In logic, a body of information may imply either the truth or falsehood of a given proposition, or may be insufficient to establish either truth or falsehood. Logical probability is a measure of degree of implication when the information does not suffice to prove a proposition true or false.

Subjective and logical probabilities exist for any proposition, given any information. This difference from frequency probabilities is reinforced by using the word 'proposition'. Frequency probabilities are usually applied to 'events', which can be considered as repeatable. A proposition may assert that a certain event occurs, but the notion of a proposition is much more general than that of an event.

1.17 One difference between logical and subjective probabilities lies in propositions which are theoretically provable, such as the proposition that the four-hundredth digit in the decimal expansion of π is a zero. The logical probability of this proposition must be zero or one, and cannot be determined without computing the decimal expansion as far as the four-hundredth digit. Most adherents of subjective probability would allow the individual to have a probability strictly between zero and one, if the decimal expansion of π is not immediately available. A common reaction would be to assign equal degrees of belief in the four-hundredth digit being 0, 1, 2, 3, 4, 5, 6, 7, 8 or 9, and hence to give the proposition of it being a zero a probability of one-tenth.

1.18 The Bayesian method requires prior probabilities to be given explicit values. A logical prior probability must be the unique value logically implied by the available prior information. Unfortunately, for almost every kind of prior information this value cannot be found; the necessary theory simply does not exist.

The exception is the case where prior information is non-existent or, more accurately, where there is no prior information which is relevant to the parameter θ. For this case there exists some theory which can be used to construct prior distributions. However, it is a matter of contention whether such a state of complete prior ignorance of θ can exist. Sufficient prior information always exists to identify θ as a meaningful quantity in the data-generating process, and to have some purpose in requiring inference about it. Nevertheless, many proponents of subjective probability, even if they do not accept a true state of complete ignorance, will adopt these logical probabilities as approximations whenever prior information is very weak.

1.19 Almost all proponents of Bayesian inference adopt a subjective interpretation of probability. Even though logical probability is attractive to some, at least as an ideal, it cannot be used in practice. In contrast there is a well-developed and practically applicable theory of subjective probability. Chapter 4 develops subjective probability at a philosophical level, and Chapter 5 discusses characteristics of both frequentist and logical probability in detail. Practical evaluation of subjective probabilities is the subject of Chapter 6.

The status of likelihood and posterior

1.20 The likelihood function is $f(x \mid \theta)$, regarded as a function of θ for fixed x. However, individual values of this function are obtained from the probability distributions of x for various θ. It is natural now to ask what kinds of probabilities comprise the likelihood function. In classical inference, $f(x \mid \theta)$ represents, and is a consequence of, the investigator's statistical *model* of the process generating the data. The classical statistician usually regards that process as one of random sampling from a (usually infinite) population of data, and so interprets $f(x \mid \theta)$ as a frequency probability.

Good (1959) points out that in most cases such probabilities would more properly be called *tautological*. The model hypothesizes that the data have this distribution given the unknown parameters. In most models it cannot be verified as a true frequency probability distribution without performing an infinite number of observations. Therefore its real status is as a tautological consequence of the model, and all inference is conditional on the correctness of the model.

From the subjective point of view, $f(x \mid \theta)$ measures the investigator's degrees of belief in the data taking certain values given the hypothetical information that the parameters take certain values, in addition to all the other available (prior) information. The statistical model is then not so much a hypothesis about the data generating process as a statement of the investigator's subjective views about that process. This statement is still conditioned on the values of the unknown parameters, but even this is not necessary. In a fully subjectivist framework, parameters can be thought of as arising from *exchangeabilities* in the investigator's beliefs – see Chapter 4.

1.21 To the subjectivist, $f(\theta)$, $f(x \mid \theta)$ and $f(\theta \mid x)$ are all subjective probability distributions. Any subjective probability distribution may be determined by the investigator assessing his or her degrees of belief in the random variable in question taking particular values. This is the process used to determine the prior distribution, but it may also be used for the posterior distribution. That is, $f(\theta \mid x)$ may be assessed directly in terms of the investigator's degrees of belief in θ taking certain values, based on all the available information, which comprises the observed values x of the data in addition to the prior information. This seems to be in direct opposition to the Bayesian method, which requires the posterior distribution to be derived via Bayes' theorem. How do advocates of subjective probability reconcile this conflict?

1.22 To answer this question, consider an even more basic criticism of subjective probabilities, that they merely reflect a person's prejudices and have no scientific basis at all. On the contrary, a subjective probability should be the result of very careful weighing of all the available information. In doing this, the individual should take account of the logical implications of probability theory. For instance, if a subjective probability has been determined for an event A, say $P(A) = p$, then the complementary probability is implied for its negation. It clearly would not constitute careful weighing of the available information to assign $P(A^c) = q \neq 1 - p$.

Now suppose that A and B are mutually exclusive events. It follows that $P(A \text{ or } B) = P(A) + P(B)$. In trying to assign personal probabilities, it will often be more difficult to think about one's degree of belief in 'A or B' than about one's degrees of belief in the simpler events A and B. It would therefore be natural to use the formula to give a value to $P(A \text{ or } B)$, i.e. to assign values first to $P(A)$ and $P(B)$, and then to let $P(A \text{ or } B)$ be their sum. The theory of subjective probability is very much concerned with methods like this for constructing probabilities.

1.23 The Bayesian method has, to the subjectivist, precisely this status. It is usually difficult to measure posterior probabilities accurately by weighing up prior and data information together. Bayes' theorem allows this judgement to be decomposed into the simpler, separate tasks of assigning prior probabilities and likelihood. Wherever the totality of available information includes some well-defined data x, it is generally good practice to think about the data separately from the other information, and to combine them by Bayes' theorem. Statistical analysis traditionally deals with exactly this context, i.e. inference *after* observing some specific data, and it is in this context that the Bayesian method is the accepted technique for assigning subjective posterior probabilities.

What is a good inference?

1.24 Classical inference theory is very concerned with constructing *good* inference rules. For instance, since any function $t(x)$ of the data can formally serve as a classical estimator of a parameter θ, it is necessary to identify criteria for comparing estimators, and for saying that one estimator is somehow better than another. The primary concern of Bayesian inference, as outlined in **1.11** and **1.12**, is entirely different. The objective is to extract information concerning θ from its posterior distribution, and to present it helpfully via effective use of summaries. There are two criteria in this process. The first is to identify interesting features of the posterior distribution. If, for instance, the posterior density is bimodal, or if a bivariate distribution has highly nonlinear regressions, it would be important to identify these facts.

The second criterion is good communication. Summaries should be chosen to convey clearly and succintly all the features of interest. In this framework it is not possible to construct a formal mathematical structure to measure how good inference is. We cannot even say what we mean by 'interesting features of the posterior distribution'. Interest is very dependent on context. We can give examples of the sorts of features that are likely to be of interest, develop a strategy to identify them, and construct good summaries to display them, but interest often resides in the unusual or the unexpected. When we have a complex, multi-dimensional posterior distribution, we can never be sure that we have summarized it exhaustively. This is one aspect of the statistician's work that relies heavily on experience.

In Bayesian terms, therefore, a good inference is one which contributes effectively to appreciating the information about θ which is conveyed by the posterior distribution.

The first half of Chapter 2 is concerned with identifying a range of useful summaries. It is possible, however, to develop ideas analogous to classical concerns about goodness of inferences, through the concepts of decision theory.

1.25 The classical theory of estimation regards an estimator as good if it is unbiased and has small variance, or more generally if its mean-square-error (m.s.e.) is small (see **A17.30**). The m.s.e. is an average squared error, where the *error* is the difference between θ and the estimate t. In accord with classical theory, the average is taken with respect to the sampling distribution of the estimator.

In Bayesian inference, θ is a random variable, and it is therefore appropriate to average the squared error with respect to the (posterior) distribution of θ. Consider

$$\begin{aligned} E\{(t-\theta)^2 \mid x\} &= E(t^2 \mid x) - E(2t\theta \mid x) + E(\theta^2 \mid x) \\ &= t^2 - 2t\,E(\theta \mid x) + E(\theta^2 \mid x) \\ &= \{t - E(\theta \mid x)\}^2 + \text{var}\,(\theta \mid x) \end{aligned}$$

Therefore the estimate t which minimizes posterior expected squared error is $t = E(\theta \mid x)$, the posterior mean. The posterior mean can therefore be seen as an estimate of θ which is best in the sense of minimizing expected squared error. This is distinct from, but clearly related to, its more natural role as a useful summary of location of the posterior distribution.

1.26 There is nothing sacred about squared error as a measure of how close the estimate is to θ. We could use absolute error $|t-\theta|$ or some other function. In this approach, the choice of an estimate t is seen as a *decision* the result of which is to incur an error $t-\theta$. The choice should depend on how seriously we regard different magnitudes of error. Squared error implicitly says that large errors are relatively much more serious than they would be under absolute error. Given any measure $\ell(t-\theta)$ of 'seriousness', we can find an estimate to minimize the posterior expected 'seriousness' $E\{\ell(t-\theta) \mid x\}$. The choice of estimate will depend on ℓ.

1.27 If the investigator genuinely must take a decision, to 'estimate' θ by t, and the consequence of the error $t-\theta$ can be represented by a specific, unique *loss* function l, then the decision theory approach identifies a unique best decision. Although this is rarely the case in practice, a study of decision theory has two potential benefits. First, it provides a link to classical inference. It thereby shows to what extent classical estimators, confidence intervals and hypotheses tests can be given a Bayesian interpretation or motivation. Second, it helps to identify suitable summaries to give Bayesian answers to the stylized inference questions which classical theory addresses. These ideas are pursued in the second half of Chapter 2.

Implementation

1.28 There are several potential difficulties in any practical implementation of the Bayesian method. We have already discussed the key issue of specifying the prior distribution, which is dealt with in Chapter 6. Chapter 2 considers procedures for inference.

However, extra difficulties arise in actually calculating the various quantities required. First, in applying Bayes' theorem we need to compute the integral in the denominator of (1.6). Second, the process of inference may require the calculation of further integrals or other operations on the posterior distribution. These calculations may be difficult to perform in practice.

Example 1.8

Let x be normally distributed with mean θ and known unit variance:

$$f(x \mid \theta) = \frac{1}{\sqrt{2\pi}} \exp\{-\tfrac{1}{2}(x-\theta)^2\}.$$

In Example 1.5 we analysed this problem assuming a normal prior distribution for θ, but now suppose that θ has a Student t-distribution:

$$f(\theta) = \frac{1}{\sqrt{d}B(\tfrac{1}{2}, \tfrac{1}{2}d)} (1 + \theta^2/d)^{-(d+1)/2}.$$

If we now attempt to apply Bayes' theorem, the denominator is the integral of the product of these two expressions. This integral cannot be performed in closed form. Therefore we cannot find the posterior density in an explicit closed form, either. Following on from this, there are many summaries that cannot be calculated, such as the posterior mean or the posterior probability of any hypothesis.

1.29 In general, if an arbitrary likelihood $f(x \mid \theta)$ is combined with an arbitrary prior density $f(\theta)$, then the product will almost certainly be mathematically intractable. It will not be possible to integrate it with respect to θ in order to express the posterior density in closed form. One 'solution' to this problem is to use numerical integration, also known as quadrature. That is, we apply a numerical technique for approximating the value of an integral. There are a great many quadrature rules available, and given sufficient computing power arbitrarily accurate approximations can be obtained, but this approach has its limitations.

Finite and realistic computing power and time can only achieve a finite accuracy. The more difficult the integration, and in particular the more dimensions over which it is necessary to integrate numerically, the less the achievable accuracy. Sophisticated techniques may be necessary. In high dimensions, even the best methods may break down.

1.30 Numerical integration is a process of approximation, but entirely different approximations are possible. Instead of using $f(x \mid \theta)$ and $f(\theta)$ to express the investigator's beliefs about the data generating process and prior knowledge about θ as accurately as possible, alternative, perhaps simpler, forms could be used as approximations. In particular, functions can be chosen that combine together to produce an easily integrable product. The likelihood-prior combinations in Examples 1.4 to 1.7 are of this form. They were chosen here for convenience and simplicity but in practice such choices are made as pragmatic approximations. Specification of subjective probability distributions always entails an element of approximation, because it is not possible to make an infinite number of careful probability judgements. Therefore, the question is really one of degree of approximation.

1.31 Although it may not be possible to integrate $f(x \mid \theta) f(\theta)$ analytically, it is almost always possible to differentiate. Various summaries, such as the posterior mode, depend only on differentiation and do not require the denominator of Bayes' theorem to be known. Integrals may even be approximated using derivatives, by integrating a suitable expansion of $f(x \mid \theta) f(\theta)$ up to a finite number of terms. All of these techniques are considered in Chapter 8.

1.32 Wherever an approximation is used, it is prudent to try to assess its accuracy. In particular, we have mentioned the degree of approximation inherent in specifying $f(x \mid \theta)$ and $f(\theta)$. The effect of such approximations on posterior inference can be assessed by deriving the posterior distribution under a range of similar specifications. The general question of assessing sensitivity or robustness of posterior inferences is considered briefly in Chapter 4 and then more fully in the first half of Chapter 7.

Comparison with classical inference

1.33 There is one major outstanding question: Why should one use Bayesian inference, as opposed to classical inference? There are various answers. Broadly speaking, some of the arguments in favour of the Bayesian approach are that it is fundamentally sound, very flexible, produces clear and direct inferences and makes use of all the available information. In contrast, the classical approach suffers from some philosphical flaws, has a restrictive range of inferences with rather indirect meanings and ignores prior information. However, the arguments are not entirely one-sided. Although the Bayesian approach is unassailable on the fundamental level, there are more practical difficulties in implementing it than are found with classical inference.

It is not surprising that a method which is fundamentally superior, and in particular which makes use of more information, requires more effort to implement. One purpose of the present volume is to describe the techniques that have evolved, many of them quite recently, to combat the difficulties. The other purpose is to present the theory, both general and particular, which underlies those techniques and which demonstrates the advantages of the Bayesian approach.

1.34 The first key argument in favour of the Bayesian approach can be called the axiomatic argument. We can formulate systems of axioms of good inference, and under some persuasive axiom systems it can be proved that Bayesian inference is a consequence of adopting any of these systems. Chapter 4 refers to one such system based on axioms of coherent decision making. A similar result, that only Bayesian inference generally yields decision rules that are *admissible*, i.e. cannot be uniformly improved upon by another decision rule, is proved in Chapter 5. Chapter A31 presents another argument. If one adopts two principles known as the ancillarity and sufficiency principles, then under some statements of these principles it follows that one must also adopt another known as the likelihood principle. Bayesian inference conforms to the likelihood principle whereas classical inference does not. Classical procedures regularly violate the likelihood principle or one or more of the other axioms of good inference. There are no such axiomatic arguments in favour of classical inference.

1.35 Pursuing this approach, we can try to detect poor behaviour of one or other inference method, via *counter-examples*. There are several specific counter-examples showing unacceptable behaviour of classical inference, which appear at various places in this volume. A fully subjective Bayesian approach seems to be immune to any such counter-examples. It should be said, however, that counter-examples occur in some variants of the Bayesian approach, notably those concerned with attempts to formulate prior ignorance. As we have said earlier, the main problems with the Bayesian method are practical. One argument for Bayesian inference is that it is better to wrestle with the practicalities of a method that is fundamentally sound, than to work with one having fundamental flaws.

Counter-examples are important because axiomatic arguments can be deceptive. An axiom system is like a tiny seed, which has the capacity to grow into a great tree. (Probability theory itself is an example of how a few axioms can imply a huge variety of theorems.) It can also grow into an unpleasant weed, and on the whole it is much easier to say whether one likes the look of the plant rather than its seed. Accordingly, most advocates of Bayesian inference do so not entirely because they are convinced by the axiomatic arguments but also because they find Bayesian inference more natural, more pleasing and more powerful than classical inference.

The latter reaction is as much aesthetic as intellectual, and results from experience in applying the Bayesian approach through the whole range of statistical problems. It embraces not just an absence of counter-examples but also the more positive fact that Bayesian methods are repeatedly found to accord extremely well with practical intuition. In Chapter 4 we examine the general properties of Bayesian inference. The behaviour of Bayesian inference in particular problems is obviously a much wider field, but Chapters 9 and 10 provide a detailed study of the most important statistical models.

1.36 Another important issue concerns the meaningfulness of inference. The Bayesian response to a question of whether θ is less than 0.5 is to compute $P(\theta < 0.5 \mid x)$, giving a probability that the statement is true. This is a natural and meaningful reply to the question. In contrast, a classical hypothesis test simply does not answer the question. To reject the hypothesis that $\theta < 0.5$ at, say, the 5% level seems to suggest that the probability that $\theta < 0.5$ is 0.05 or less, but in fact says nothing of the kind. It asserts only that *if* $\theta < 0.5$ then the probability of the event that x lies in some critical set C is less than or equal to 0.05 (and x was indeed observed to lie in C). It is a probability statement about the data, not about θ. *All* classical inference statements are of this kind. They are probability statements about x given θ, phrased so as to appear to be probability statements about θ.

Similar remarks apply to classical confidence intervals. A statement that one should have 90% confidence in a parameter θ lying in a given interval is a probability statement about the random interval, not about θ. In contrast, a Bayesian posterior probability interval states the probability that θ lies in that fixed interval. So although there are strong parallels between the two theories, as suggested in **1.11**, there are deep differences between them in regard to the interpretation that can be placed on the inferences they provide.

Probability statements about the parameters are simply not possible in classical statistics, because parameters are not allowed to be random variables. Yet statements about parameters are what the investigator actually wants in practice. It is hard to think that the phrase 'statistical inference about θ' could mean anything else. Because of the way classical inference statements are phrased, they may seem to give this kind of inference, when in fact they do not. Only Bayesian inference makes probability statements about θ, and so only Bayesian inference actually answers the investigator's questions.

1.37 The purpose of this volume is not to criticize classical statistics, nor simply to extol the virtues of the Bayesian approach. The primary purpose is to set out in detail the theory and methodology of Bayesian statistics. However, it is necessary to explain why the Bayesian approach is formulated as it is, and in this chapter particularly that has meant criticizing classical ideas. Subsequent chapters will spend less time contrasting classical and Bayesian methods, but will concentrate on developing Bayesian ideas. For this reason, the reader should now take careful note of the key features of Bayesian methods as listed in the next section. They are fundamental to the Bayesian approach, and a reader who has been brought up on classical statistics is likely to misunderstand much of this volume unless these fundamentals are clearly appreciated.

Basic concepts

1.38 The following features characterize the Bayesian approach and distinguish it from the classical approach.

1. *Prior information.* The prior distribution is an integral part of Bayesian statistics, but is absent from classical methods. It represents the investigator's knowledge about θ before seeing the data. Classical statistics uses only the likelihood, and this fact has another important implication. Classical statistics may assert that if x is binomially distributed with unknown parameter θ then the best estimator of θ, according to some criterion, is x/n. This is supposed to apply to all problems in which x is binomially distributed, but to a Bayesian every such problem is unique. Every problem has a real context in which θ is some meaningful quantity, about which the investigator has beliefs expressed in the prior distribution. The likelihood may be the same in each case, but the prior distribution changes, and so the Bayesian analysis is different in each case.

   The importance of the practical context causes difficulties for a book such as this, which attempts to give the general theory and methodology rather than practical case studies. That distinction is even more artificial in Bayesian statistics than it is in the classical approach.

2. *Subjective probability*. All probabilities and distributions are subjective, or personalistic. Probabilities represent the investigator's degrees of belief. Because the investigator does not know θ, it is a random variable (whereas in classical statistics it is treated as a fixed value, even though it is unknown). Once the data are observed, they cease to be random (whereas all classical inference relies on treating x as a random variable). This is recognized in the posterior distribution $f(\theta \mid x)$, which is a distribution for the random θ, conditional on the observed x.
3. *Self-consistency*. The Bayesian method relies exclusively on probability operations. The prior distribution and likelihood are combined by Bayes' theorem, and all inference derives directly from the resulting posterior distribution. The inference process will typically involve further probability operations, such as finding the marginal posterior distribution of a single parameter in a multiparameter problem, or deriving posterior mean and variance. Probability theory is a completely self-consistent system. Any question of probabilities has one and only one answer, although there may be many ways to derive it.
4. *No adhockery*. Faced with a new problem, a classical statistician is free to invent new estimators, confidence intervals or hypothesis tests. There is an obligation to consider the properties of any new procedure, such as bias or power, but it is rare that any single inference rule can be identified as unequivocally the best. This freedom to make up statistical methods is called 'adhockery' by De Finetti (1974, paragraph 11.1.4), who points out that the proliferation of classical methods for a given problem leads to confusion and disagreement. In contrast, there is a unique Bayesian solution to any problem. That is the posterior distribution, which expresses the investigator's knowledge about θ after observing x. The Bayesian statistician's task is to identify the posterior distribution as accurately as possible, which usually entails identifying the prior distribution and the likelihood and then applying Bayes' theorem. There is no room for adhockery in Bayesian statistics.

EXERCISES

1.1 Data x have the Poisson distribution with unknown mean θ, i.e.

$$f(x \mid \theta) = \frac{\theta^x}{x!} \mathrm{e}^{-\theta}, \qquad x = 0, 1, \ldots$$

The prior density of θ has the gamma form

$$f(\theta) = \frac{a^b \, \theta^{b-1}}{\Gamma(b)} \mathrm{e}^{-a\theta}, \qquad \theta > 0.$$

Show that the posterior distribution is another gamma distribution, with parameters $a_1 = a + 1$ and $b_1 = b + x$.

1.2 In Exercise 1.1, suppose that we have a sample $(x_1, x_2, \ldots, x_n)$ from the same Poisson distribution instead of a single observation. The prior distribution is the same. Show that the posterior distribution is again a gamma distribution. Express the posterior mean as a weighted average of prior mean and sample mean $\bar{x}$, and interpret the weights in terms of precisions.

1.3 As in Example 1.5, x has a normal distribution $N(\theta, v)$ with unknown mean θ and known variance v. The prior density is

$$f(\theta) = \frac{c}{\sqrt{2\pi w}} \exp\{-\tfrac{1}{2w}(\theta - m)^2\} + \frac{1-c}{\sqrt{2\pi W}} \exp\{-\tfrac{1}{2W}(\theta - M)^2\}.$$

Show that the posterior density has the same form, with m, w, M, W and c respectively replaced with

$$m_1 = \frac{vm + wx}{v + w}, \qquad w_1 = \frac{vw}{v + w},$$

$$M_1 = \frac{vM + Wx}{v + W}, \qquad W_1 = \frac{vW}{v + W},$$

$$c_1 = \left(1 + (c^{-1} - 1)\left(\frac{v + w}{v + W}\right)^{1/2} \exp\left[\frac{1}{2}\left\{\frac{(m - x)^2}{v + w} - \frac{(M - x)^2}{v + W}\right\}\right]\right)^{-1}.$$

1.4 Following Exercise 1.3, show that the posterior density can be bimodal.

1.5 A series of n trials is performed with unknown probability θ of success, so that the number of successes x_1 has the binomial distribution with f.f.

$$f(x \mid \theta) = \binom{n}{x_1} \theta^{x_1} (1 - \theta)^{n - x_1}, \qquad x = 0, 1, \ldots, n.$$

An extra trial is performed, independent of the others given θ, but with probability $\frac{\theta}{2}$ of success. The data comprise $x = (x_1, x_2)$, where $x_2 = 1$ if the extra trial is a success, otherwise $x_2 = 0$. θ has the uniform prior density

$$f(\theta) = 1, \qquad 0 \leqslant \theta \leqslant 1.$$

Show that for $x_2 = 0$ the posterior mean of θ is

$$E(\theta \mid x_1, x_2 = 0) = \frac{(x_1 + 1)(2n - x_1 + 4)}{(n + 3)(2n - x_1 + 3)},$$

and that this is less than $E(\theta \mid x_1, x_2 = 1)$. Show also that $f(\theta \mid x)$ is unimodal for all x.

1.6 Data x comprise n_1+n_2 independent, normally distributed observations with known variance v. The first n_1 observations have mean μ and the other n_2 have mean $\mu+\delta$. The parameters $\theta = (\mu, \delta)$ have independent normal prior distributions. The prior distribution of μ is $N(m, w)$ and the prior distribution of δ is $N(d, g)$. Obtain the posterior joint density $f(\theta \mid x)$, and show that the posterior distribution of δ is $N(d_1, g_1)$, where

$$d_1 = \frac{c_1 g\,(\bar{x}_2 - \bar{x}_1) + c_2 g\,(\bar{x}_2 - m) + c_3 v\, d}{(c_1 + c_2)g + c_3 v},$$

$$g_1 = \frac{c_3 v\, g}{(c_1 + c_2)g + c_3 v},$$

and where $c_1 = n_1 n_2 w$, $c_2 = n_2 v$, $c_3 = (n_1 + n_2)w + v$, $\bar{x}_1$ is the sample mean of the first n_1 observations and $\bar{x}_2$ is the sample mean of the last n_2 observations.

1.7 Following Example 1.7, we wish to derive marginal distributions from (1.12). Let $\theta^{\star}_{p-1} = \theta_{p-1} + \theta_p$. Show that the distribution of $\phi = (\theta_1, \theta_2, \ldots, \theta_{p-2}, \theta^{\star}_{p-1})$ is

$$f(\phi \mid x) = \frac{\Gamma(b_1 + b_2 + \ldots + b_p)}{\Gamma(b_1)\Gamma(b_2) \ldots \Gamma(b_{p-2})\Gamma(b_{p-1} + b_p)}\, \theta_1^{b_1-1}\, \theta_2^{b_2-1} \ldots \theta_{p-2}^{b_{p-2}-1}\, \theta^{\star\;\; b_{p-1}+b_p-1}_{p-1}.$$

Hence show by induction that the marginal posterior distribution of θ_1 is a beta distribution with parameters b_1 and $\sum_{i=2}^{p} b_i$. Show that the marginal posterior mode of θ_i is greater then the θ_i component of the joint mode found in Example 1.7.

1.8 From (1.11) derive the conditional distributions of μ given σ^2 and σ^2 given μ. Show that μ and σ^2 are uncorrelated but not independent.

CHAPTER 2

INFERENCE AND DECISIONS

2.1 This chapter considers techniques for deriving inferences about θ from its posterior distribution. It begins with a study of inference through summarization, followed by a development based on statistical decision theory. We shall develop the main ideas of decision theory in a general framework, because such ideas may be useful at various times throughout this volume.

We shall save tedious repetition throughout this chapter by referring to the posterior density as simply $f(\theta)$ rather than $f(\theta \mid x)$. This may seem to create unnecessary confusion between prior and posterior densities, but the ambiguity is deliberate. In later chapters we shall have occasion to apply inference techniques to different kinds of distribution. For instance, it is often of interest to separate the roles of prior information and data in so far as they affect specific inferences. This may be done by comparing the corresponding inferences derived from prior and posterior distributions. Therefore one may genuinely wish to derive inference from the prior density $f(\theta)$. As another example, Chapter 3 considers predictive inference, in which inference about future data y is derived from its *predictive* density $f(y \mid x)$.

In this chapter, therefore, $f(\theta)$ will denote the density function of whatever is the current distribution of interest, be it a prior, posterior or predictive distribution.

Summarizing univariate distributions

2.2 In **1.11** we identified inference with the process of summarization. The posterior density formally expresses all the information that we have about θ having seen the data x, but it does so in the form of a mathematical formula. In order to express the posterior information in forms that are clearly and easily understood, we need to derive suitable summaries of the posterior distribution.

As stated in **1.24**, there are two criteria of good summarization. The first is that the summaries should identify every useful or interesting feature of the posterior distribution. The second is that they should do so clearly and succinctly.

2.3 Consider first the case of scalar θ. In fact, summarizing a univariate distribution is quite a simple matter. Nevertheless we shall consider this case in detail because many higher-dimensional summaries are more readily developed as generalizations of one-dimensional cases. Summaries that are very similar to one another in one dimension may have important distinctions in higher dimensions.

The most useful summary of all is a plot of the density function. This will show the general *shape* of the density and gives very clear inference of a qualitative nature. Such a plot shows primarily which ranges of θ values are the most probable. If the density has two or more distinct modes then the most probable θ values are grouped into two or more separate ranges. If the plot also shows skewness, it implies that θ values on one side of the 'most probable range' are more probable than those on the other. And so on.

Quantitative summaries of shape are needed to supplement this qualitative view. The first task is to identify turning points of the density, i.e. solutions of $f'(\theta) = 0$. Such points include local maxima and minima of $f(\theta)$, which we call modes and antimodes, respectively. A point θ_0 is characterized as a mode if $f'(\theta_0) = 0$ and $f''(\theta_0) < 0$, whereas it is an antimode if $f'(\theta_0) = 0$ and $f''(\theta_0) > 0$. Any point θ_0 for which $f''(\theta_0) = 0$ is a point of inflexion of the density (whether or not $f'(\theta_0) = 0$). Locating such points adds quantitative detail to the density plot. In practice, by far the most common case is a unimodal density: $f(\theta)$ falls away on either side of the single mode, usually tending to zero at both ends of the range of possible θ values. There are no antimodes, but there are typically two points of inflexion, one on each side of the mode.

Example 2.1

Consider the gamma density

$$f(\theta) = \frac{a^b}{\Gamma(b)}\theta^{b-1}e^{-a\theta}, \qquad \theta > 0,$$

where a and b are positive constants. Derivatives are easily found.

$$f'(\theta) = \frac{a^b}{\Gamma(b)}\{(b-1) - a\theta\}\,\theta^{b-2}e^{-a\theta},$$

$$f''(\theta) = \frac{a^b}{\Gamma(b)}\{a^2\theta^2 - 2a(b-1)\theta + (b-1)(b-2)\}\,\theta^{b-3}e^{-a\theta}.$$

From the first derivative, there are turning points at $\theta = 0$ (for $b > 2$) and $\theta = (b-1)/a$ (for $b > 1$). For $b \leqslant 1$, $f'(\theta) < 0$ for all $\theta \geqslant 0$, so $f(\theta)$ is monotone decreasing and the mode is at $\theta = 0$. For $b > 1$, $f(\theta) \to 0$ as $\theta \to 0$, so the turning point at $\theta = 0$ is not a mode. In this case, $f'(\theta) > 0$ for $\theta < (b-1)/a$, and $f'(\theta) < 0$ for $\theta > (b-1)/a$. Therefore $\theta = (b-1)/a$ is the mode.

Looking at $f''(\theta)$, the quadratic expression in braces has roots at

$$\theta = \frac{b-1}{a} \pm \frac{(b-1)^{1/2}}{a}.$$

Therefore, for $b > 1$ these are the points of inflexion (except that unless $b > 2$ the lower point is negative). Notice that they are equidistant from the mode, although this is not a symmetric density. Indeed, the gamma distribution with small b is quite highly skewed. That the same is true of all distributions in the Pearson system is shown by Johnson and Kotz (1970a, chapter 12). One cannot in general infer skewness or symmetry from the positions of points of inflexion.

Example 2.2

In this example and the next we use mixtures of two normal densities to obtain some interesting shapes. Let

$$f(\theta) = \frac{0.8}{(2\pi)^{1/2}}\exp(-\tfrac{1}{2}\theta^2) + \frac{0.2}{(2\pi)^{1/2}}\exp\{-\tfrac{1}{2}(\theta-4)^2\}, \tag{2.1}$$

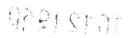

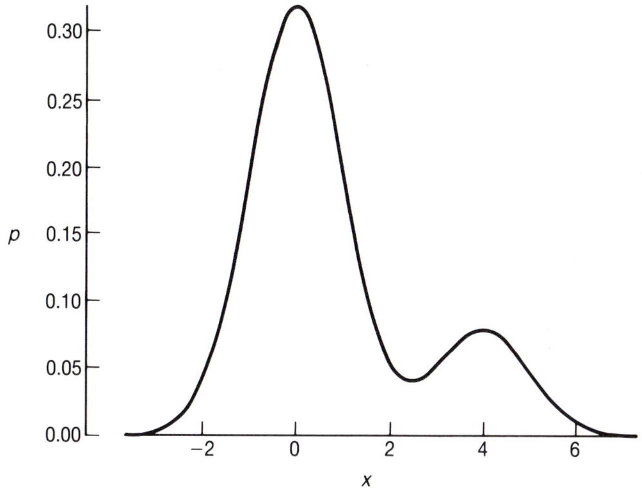

Fig. 2.1 Mixture of $N(0,1)$ and $N(4,1)$, weights 4:1

a mixture of the $N(0,1)$ and $N(4,1)$ densities with weights 0.8 and 0.2 respectively. Figure 2.1 is a plot of (2.1), showing it to be bimodal.

$$f'(\theta) = -\frac{0.8\theta}{(2\pi)^{1/2}}\exp(-\tfrac{1}{2}\theta^2) - \frac{0.2(\theta-4)}{(2\pi)^{1/2}}\exp\{-\tfrac{1}{2}(\theta-4)^2\},$$

which is clearly positive for $\theta \leqslant 0$ and negative for $\theta \geqslant 4$. A little computation locates turning points at $\theta = 0.00034$, 2.46498 and 3.99452. Now

$$f''(\theta) = \frac{0.8}{(2\pi)^{1/2}}(\theta^2 - 1)\exp(-\tfrac{1}{2}\theta^2) + \frac{0.2}{(2\pi)^{1/2}}(\theta^2 - 8\theta + 15)\exp\{-\tfrac{1}{2}(\theta-4)^2\}.$$

This is easily seen to be positive for $\theta \leqslant -1$, for $1 \leqslant \theta \leqslant 3$ and for $\theta \geqslant 5$, confirming that the middle turning point is an antimode. Calculating $f''(\theta)$ at the other two points confirms them to be modes. Calculations also reveal points of inflexion at $\theta = -0.999\,98$, $0.982\,54$, $3.179\,03$ and $4.999\,71$.

Example 2.3
Now let

$$f(\theta) = \frac{0.75}{(2\pi)^{1/2}}\exp(-\tfrac{1}{2}\theta^2) + \frac{0.25}{(2\pi)^{1/2}}\exp\{-\tfrac{1}{2}(\theta-2.5)^2\},$$

which is plotted in Figure 2.2. This density is unimodal but has a hump, known as a shoulder, for θ around 2 or 3. $f'(\theta)$ is found to be zero at $\theta = 0.0398$, which is the only turning point, the mode. However, we find four points of inflexion, at $\theta = -0.993, 0.886, 2.248$ and 3.428. The existence of the shoulder is therefore revealed by a pair of inflexion points with no mode or antimode between them.

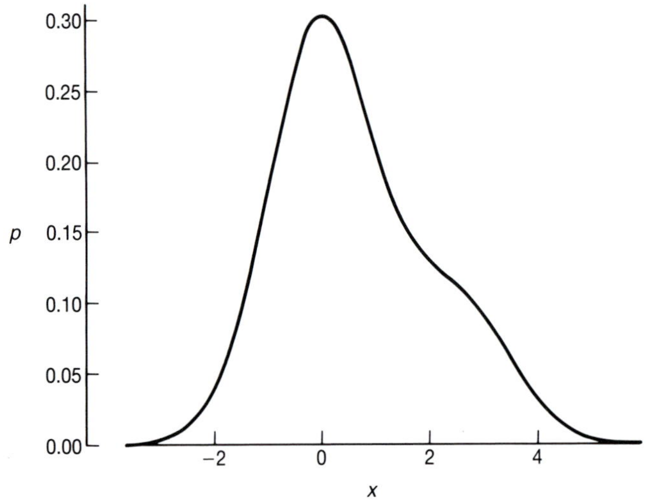

Fig. 2.2 Mixture of $N(0, 1)$ and $N(2.5, 1)$, weights 3:1

2.4 There are several other things that could be categorized as summaries of shape. For a bimodal or multimodal density we might try to describe the relative importance of two modes θ_1 and θ_2 by quoting the ratio of their heights, $f(\theta_1)/f(\theta_2)$, or to describe how well separated they are by the minimum of $f(\theta_1)/f(\theta_3)$ and $f(\theta_2)/f(\theta_3)$, where θ_3 is the antimode between θ_1 and θ_2.

However, a glance at a density like Figure 2.3 will show that the 'importance' of a mode does not simply equate to its height. If we think of a mode as not just the point of a local maximum of $f(\theta)$, which we can call a mode point, but as the whole hump of density around that mode point, then a mode also has width. In Figure 2.3, the mode around the mode point θ_1 is lower but 'wider' than the one around θ_2. It might be considered more important because its area is larger, so it accounts for more probability.

There are several ways of describing a mode's width. In a bimodal density with well separated modes, as in Figures 2.1 and 2.3, we can represent relative width by the ratio $|\theta_1 - \theta_3|/|\theta_2 - \theta_3|$ of distances to the antimode point, but this measure is less help for a trimodal or unimodal density. Alternatively, since there is generally a point of inflexion on either side of each mode point, we could describe a mode's width by the distance between its points of inflexion. This has the advantage of also providing a measure of the width of a shoulder.

A third width summary is to use points of equal density on either side of a mode point. Its width may be described by the distance between points whose density is some constant $c < 1$ times the density at the mode point. Of course, for multimodal densities there is a lower bound on c given by the relative heights of antimodes. Notice that if c is close to 1 this summary is really describing the sharpness of the mode at the mode point, and may be computed more easily in terms of the second derivative of $f(\theta)$ at the mode

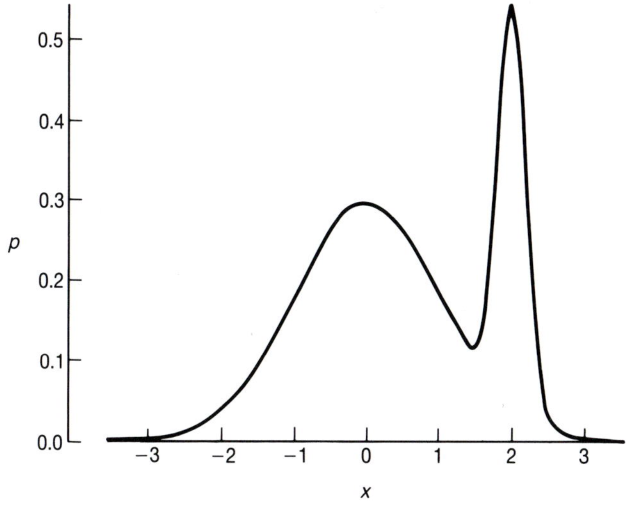

Fig. 2.3 Mixture of $N(0, 1)$ and $N(2, 0.04)$, weights 3:1

point. For, suppose the mode point is at θ_1. We have the expansion

$$f(\theta) = f(\theta_1) + f''(\theta_1)(\theta - \theta_1)^2/2 + o(\theta - \theta_1)^2, \tag{2.2}$$

where $f''(\theta)$ is the second derivative. Equating to $c\,f(\theta_1)$ gives the approximation

$$\theta - \theta_1 \approx \pm\{-2(1 - c)f(\theta_1)/f''(\theta_1)\}^{1/2}, \tag{2.3}$$

which is asymptotically exact as $c \to 1$. This is an increasing function of $-f(\theta_1)/f''(\theta_1)$, which can therefore be used to summarize the width of a mode locally around the mode point θ_1. The quantity $-f(\theta_1)/f''(\theta_1)$ will be referred to as the modal dispersion summary.

Example 2.4

For the density (2.1) we found the mode points $\theta_1 = 0.000\,34$ and $\theta_2 = 3.994\,52$ in Example 2.2, with an antimode at $\theta_3 = 2.464\,98$. The modes have heights

$$f(\theta_1) = 0.319\,18, \qquad f(\theta_2) = 0.079\,90$$

and the antimode has height $f(\theta_3) = 0.039\,86$. Therefore the ratio of heights of the modes is $0.319\,18/0.0799 = 4.004$, and the measure of separation is $.0799/.039\,86 = 2.005$.

Now consider the widths of the two modes. First, $|\theta_1 - \theta_3|/|\theta_2 - \theta_3| = 1.611$, suggesting that the mode around θ_1 is not only four times as high but also rather wider than that around θ_2. This may, however, be considered rather a misleading summary. The density (2.1) is composed of a mixture of two normal densities. The ratio of modal heights accurately reflects the mixing proportions 0.8 : 0.2, and the mode points closely represent the means, 0 and 4, of the two normal densities, but since the two components have equal variances we might prefer a measure of width that gave roughly equal widths in this case.

Consider now the distance between inflexion points. From Example 2.2 we find for the first mode $0.982\,54 + 0.999\,98 = 1.983$, and for the second mode $4.999\,71 - 3.179\,03 = 1.821$. The ratio is 1.09, indicating similar widths. If we now compute

$$f''(\theta_1) = -0.318\,75, \qquad f''(\theta_2) = -0.078\,15$$

the ratios $-f(\theta_i)/f''(\theta_i)$ are both close to unity, again indicating similar widths. The same conclusion would be given by looking at distances between points of equal density.

2.5 Further indication of how the various modal width summaries are related can be obtained by applying them to the simple case of a $N(m, v)$ density. Then the mode is at $\theta = m$ and its height is $(2\pi v)^{-1/2}$. The inflexion points are $m \pm v^{1/2}$, with a distance between them of twice the standard deviation. Points of density c times the modal density are

$$m \pm (-2v \ln c)^{1/2}.$$

In particular, at $c = \exp(-\frac{1}{2}) = 0.606\,53$, the distance is again twice the standard deviation, and at $c = \exp(-\frac{1}{8}) = 0.8825$ it is one standard deviation. Finally, $-f(m)/f''(m) = v$, the variance.

We can also derive indicators of the probability content of each mode in a multimodal density. Since in the normal case there is a single mode, this mode represents the entire distribution and so may be thought of as having a probability content of one. In a multimodal density, to indicate the probability content of a mode whose mode point is θ_1 we can use $(2\pi)^{\frac{1}{2}} f(\theta_1)$ multiplied by some measure of width transformed so that it corresponds to standard deviation. One possibility is

$$\{-2\pi f(\theta_1)/f''(\theta_1)\}^{1/2} f(\theta_1). \tag{2.4}$$

Example 2.5

The density plotted in Figure 2.3 is another mixture of two normals. The components are the $N(0, 1)$ and $N(2, 0.04)$ densities with weights 0.75 and 0.25 respectively. The mode points are $\theta_1 = 0$ and $\theta_2 = 1.993\,44$ with heights $f(\theta_1) = 0.299\,21$, $f(\theta_2) = 0.539\,44$. We find $-f(\theta_1)/f''(\theta_1) = 1$ and $-f(\theta_2)/f''(\theta_2) = 0.043\,77$, which closely approximate the component variances 1 and 0.04. Finally, multiplying the square roots of these by $(2\pi)^{\frac{1}{2}} f(\theta_i)$ give values of 0.75 and 0.283 for the probability contents, close to the true weights of 0.75 and 0.25.

It is important to recognize that bimodal densities often arise without the explicit mixture form of these examples. It is nevertheless useful to examine each mode separately and summarize height, width and probability content. The implied mixture can then often be a useful, simple approximation to $f(\theta)$.

2.6 Having considered a mode's location, height and width, its next most important feature is skewness, which is generally the property of $f(\theta)$ falling away faster on one side of the mode point than on the other. We have seen in Example 2.1 that the locations of inflexion points may be bad indicators of skewness. The most obvious skewness summary is based on points of equal density. If θ_1 is a mode point and for given $c \in (0, 1)$ we have $\theta_{11} < \theta_1 < \theta_{12}$ such that $f(\theta_{11}) = c\,f(\theta_1) = f(\theta_{12})$, then skewness is indicated by the difference between $\theta_{12} - \theta_1$ and $\theta_1 - \theta_{11}$. Specifically, we can divide their difference by their sum to obtain the summary

$$\frac{\theta_{11} + \theta_{12} - 2\theta_1}{\theta_{12} - \theta_{11}}. \tag{2.5}$$

Of course, this will vary with the choice of c. Nor is it feasible to recommend a good, general choice of c, since, in a multimodal density, neighbouring antimodes will affect the available values of c.

It is useful to examine (2.5) as c tends to one. If we take the expansion of $f(\theta)$ in (2.2) to the term in $(\theta - \theta_1)^3$ we can extend the solution (2.3) to its next term (regarding it as an expansion in powers of $(1 - c)^{1/2}$). Then

$$\theta \approx \theta_1 \pm \left(\frac{2(1 - c)f(\theta_1)}{-f''(\theta_1)}\right)^{1/2} + \frac{(1 - c)f(\theta_1)f'''(\theta_1)}{3\{f''(\theta_1)\}^2}.$$

So as c tends to one, whereas the numerator of (2.5) is of order $(1 - c)$ its denominator is of order $(1 - c)^{1/2}$. This emphasizes the dependence of (2.5) on c: it tends to zero as c tends to one. To reduce the dependence on c we should multiply (2.5) always by $(1 - c)^{-1/2}$.

Now taking the limit as $c \to 1$, and dropping a constant factor $(3/\sqrt{2})^{-1}$, gives the following summary of modal skewness in the locality of the mode point θ_1,

$$\frac{f'''(\theta_1)}{-f''(\theta_1)}\left(\frac{f(\theta_1)}{-f''(\theta_1)}\right)^{1/2} \tag{2.6}$$

Example 2.6

As in Example 2.1, consider the gamma density

$$f(\theta) = \frac{a^b}{\Gamma(b)}\theta^{b-1}e^{-a\theta}, \qquad \theta > 0.$$

For $b > 1$, we found its unique mode at $\theta_1 = (b - 1)/a$. We find

$$f(\theta_1) = \frac{a}{\Gamma(b)}(b - 1)^{b-1}e^{1-b},$$

$$f''(\theta_1) = -\frac{a^3}{\Gamma(b)}(b - 1)^{b-2}e^{1-b},$$

$$f'''(\theta_1) = 2\frac{a^4}{\Gamma(b)}(b - 1)^{b-3}e^{1-b}.$$

Considering first the measure of probability content (2.4) we find

$$(2\pi)^{1/2}(b - 1)^{b-1/2}e^{1-b}/\Gamma(b), \tag{2.7}$$

which, by Stirling's expansion of $\Gamma(b)$, tends to one (the correct probability content) as $b \to \infty$, and is close to one even for small b. For example, at $b = 3$, (2.7) equals 0.9595.

The skewness measure (2.6) is found to be $2(b - 1)^{-1/2}$, confirming that the skewness of the gamma distribution decreases as the shape parameter b increases. The table below

Table 2.1

c	0.999	0.9	0.8	0.5	0.2
$b=3$	0.3334	0.3408	0.3491	0.3821	0.4445
$b=10$	0.1572	0.1611	0.1657	0.1839	0.2198

shows the alternative measure (2.5), multiplied by $(1-c)^{-1/2}$ as c varies for two values of b. (The figures are independent of a.) The measure is quite stable for values of c of 0.8 and above. In general, the asymptotic measure (2.6) is a good indication of skewness of a mode over the region inside its points of inflexion.

2.7 One further feature of the shape of a density is sometimes important. Consideration of the mode(s) of $f(\theta)$ is fundamental because the probability of θ lying outside this region, i.e. in the tails of the density, is small. Nevertheless, tails of distributions with otherwise similar shapes can differ dramatically, and inference may occasionally be sensitive to such differences.

2.8 After summaries of shape, the two most important classes of summaries concern location and dispersion. Summaries of location include mean, median and mode. Summaries of dispersion include standard deviation (or, equivalently, variance) and mean deviation. We can also include as dispersion summaries the various indicators of modal width in **2.4**.

Location and dispersion should always be interpreted with reference to shape. For a unimodal distribution, location and dispersion may be taken in broad terms as providing, respectively, a 'typical' value of θ and an indication of how far from that 'typical' value θ is likely to be. Even so, the interpretation of a location summary must allow for skewness, and in fact differences between mean, median and mode can be used as indicators of skewness. Standard deviation and mean deviation are sensitive, to varying extents, to tail thickness, whereas modal width summaries are unaffected by tail behaviour. Conversely, differences in the various dispersion summaries may be used to indicate tail thickness. The more traditional summaries of skewness and tail thickness are based on third and fourth moments, as described in A**3.31** and A**3.32**.

In the case of a distribution with more than one mode, shape summaries are crucial. A single summary of location, such as the mean or median, will now be 'typical' of θ only in a mathematical sense. It is quite possible that either mean or median could fall close to an antimode, and such a value would in a practical sense be rather untypical. A single summary of dispersion is far less informative in such a case than summaries of the various modes. Indeed, the overall dispersion is then composed of dispersion within modes and between modes.

If $f(\theta)$ has k modes, partition the line into segments $C_1 = (-\infty, t_1)$, $C_2 = [t_1, t_2)$, $C_3 = [t_2, t_3), \ldots, C_{k-1} = [t_{k-2}, t_{k-1})$, $C_k = [t_{k-1}, \infty)$, where the t_is are the antimode points, so that each segment contains one mode. Write

$$f_i(\theta) = \begin{cases} c_i^{-1} f(\theta) & \text{if } \theta \in C_i \\ 0 & \text{otherwise} \end{cases}$$

for $i = 1, 2, \ldots, k$, where

$$c_i = \int_{C_i} f(\theta)\, d\theta \,,$$

so that $\sum_{i=1}^{k} c_i = 1$. Let ϕ be a random variable taking values $1, 2, \ldots, k$ with $P(\phi = i) = c_i$, and let the conditional density of θ given $\phi = i$ be $f_i(\theta)$. This provides a representation of the marginal density $f(\theta)$ of θ. The general formula

$$\operatorname{var}(\theta) = E\{\operatorname{var}(\theta \mid \phi)\} + \operatorname{var}\{E(\theta \mid \phi)\} \tag{2.8}$$

applies and we can identify the first component of (2.8) as within mode variance and the second as between mode variance. The summaries of the different modes of $f(\theta)$ now provide analogues of c_i (the probability contained in mode i), $E(\theta \mid \phi = i)$ (the location of mode i) and $\operatorname{var}(\theta \mid \phi = i)$ (the width of mode i), with which to describe this decomposition of the overall variance.

In general, when $f(\theta)$ is multimodal, single overall summaries such as location, dispersion and skewness have very limited value on their own and provide much less information than shape summaries describing the separate modes.

Density plots for bivariate distributions

2.9 In two dimensions we cannot produce a graphical summary of the whole density that is as simple and as powerful as the plot of a univariate density. Two basic forms of graph are available. First, we can try to draw the whole joint density function as a two-dimensional surface in three-dimensional space. This suffers from the usual difficulties of representing a three-dimensional object on the two-dimensional page. Any such graph is essentially a view of the surface from a particular point. A sense of the height of a mode cannot be obtained without obscuring the view of the surface on the other side. Distances and heights cannot be accurately judged without perspective, yet if we view the surface from two different points we lose the simplicity of having a single graph, and it is still quite difficult to judge distances. In particular it is hard to visualize the shape of a mode clearly. In principle a hologram would allow a clear view of a density surface from a wide range of positions, but this technology is not yet sufficiently well developed.

2.10 Another kind of graphical summary is a contour plot. Figure 2.4 is a simple example. The contour lines join points of equal joint density. A mode is identified as the centre of a system of nested contour lines, joining points of increasing density as we move inwards. Figure 2.4 shows a single mode around the point (1.2, 0.2). The shape of a mode is quite clearly shown in such a diagram by the shape of the surrounding contours.

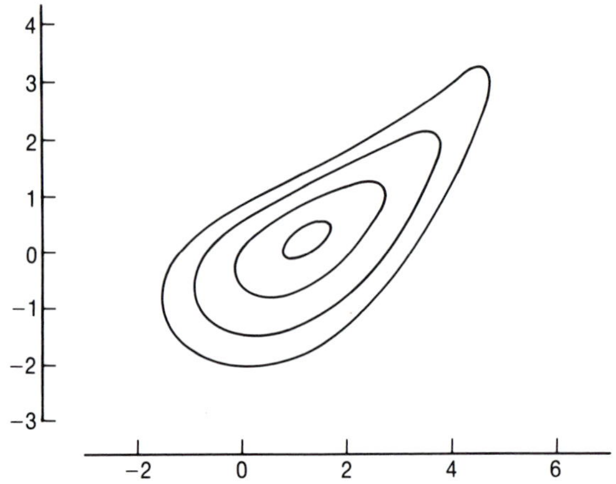

Fig. 2.4 Example of a contour plot

Figure 2.4 shows a mode that is reasonably elliptical, but elongation of the contours indicates a ridge stretching out beyond (4,4).

Other aspects of shape are less easy to see in a contour plot. Essentially, it is a three-dimensional plot in which all perspective of height has been lost by viewing directly from above. Heights are given not visually but numerically by the heights of the contours. Information about heights between contour lines is lost.

In short, no single diagram shows the shape of the joint density fully and clearly. Three-dimensional plots and contour plots bring out different aspects of shape, and so complement each other.

Visualizing multivariate densities

2.11 Beyond two dimensions we cannot hope to produce a graphical summary that represents the whole density. One- and two-dimensional density plots have an important part to play in summarization, particularly in showing marginal distributions, but it is no longer possible to describe the joint distribution fully by such diagrams. The main feature of a joint distribution that is not represented in its margins is the extent and form of associations between the various random variables. There may also be significant shape features in a joint density, such as bimodality, that are not shown in the marginal densities.

Although there are severe problems in visualizing, and so in summarizing, a multivariate distribution, we can construct many useful summaries as generalizations of ideas developed in **2.2** to **2.8** to the multivariate case. It is still helpful to know of the existence and locations of modes and other turning points. The width of a mode is no longer described by a single value, such as the distance between two points of equal density, but the analogous idea of looking at contours will aid in describing the size and shape of a mode. Modal shape can then be used in describing association. We shall also develop some new kinds of summary specifically for the multivariate case. Our discussion will concentrate on measures of shape, location and dispersion; measures of skewness and kurtosis are considered by Mardia (1970).

Turning points

2.12 First consider the locations of turning points. This is simply a question of examining partial derivatives. Consider first a bivariate density $f(\theta, \phi)$. At any point (θ, ϕ), the gradient in the θ direction is the partial derivative

$$\frac{\partial f(\theta, \phi)}{\partial \theta} = \lim_{\delta \to 0} \{f(\theta + \delta, \phi) - f(\theta, \phi)\}/\delta$$

and in the ϕ direction it is $\partial f(\theta, \phi)/\partial \phi$. At a mode, or at a turning point generally, the gradient is zero in all directions. This reduces to the two partial derivatives being zero, for if we consider another direction at an angle α to the θ axis, the gradient is

$$\begin{aligned}
&\lim_{\delta \to 0} \{f(\theta + \delta \sin \alpha, \phi + \delta \cos \alpha) - f(\theta, \phi)\}/\delta \\
&\quad = \sin \alpha \lim_{\delta \to 0} \{f(\theta + \delta \sin \alpha, \ \phi + \delta \cos \alpha) - f(\theta, \ \phi + \delta \cos \alpha\}/(\delta \sin \alpha) \\
&\qquad\qquad + \cos \alpha \lim_{\delta \to 0} \{f(\theta, \ \phi + \delta \cos \alpha) - f(\theta, \ \phi)\}/(\delta \cos \alpha) \\
&\quad = \sin \alpha \frac{\partial f(\theta, \phi)}{\partial \theta} + \cos \alpha \frac{\partial f(\theta, \phi)}{\partial \phi}. \qquad (2.9)
\end{aligned}$$

Therefore, turning points are solutions of the simultaneous equations $0 = \partial f(\theta, \phi)/\partial \theta = \partial f(\theta, \phi)/\partial \phi$.

If we now consider a joint density function $f(\boldsymbol{\theta})$ for a $p \times 1$ vector random variable $\boldsymbol{\theta}$, we have a vector of partial derivatives

$$\boldsymbol{f}'(\boldsymbol{\theta}) = \partial f(\boldsymbol{\theta})/\partial \boldsymbol{\theta},$$

whose ith element is $\partial f(\boldsymbol{\theta})/\partial \theta_i$. We can represent any direction by a vector $\mathbf{t}$ of direction cosines, satisfying $\mathbf{t}'\mathbf{t} = 1$. The generalization of (2.9) is that the derivative of $f(\boldsymbol{\theta})$ in direction $\mathbf{t}$ is

$$\lim_{\delta \to 0} \{f(\boldsymbol{\theta} + \delta \mathbf{t}) - f(\boldsymbol{\theta})\}/\delta = \mathbf{t}'\boldsymbol{f}'(\boldsymbol{\theta}). \qquad (2.10)$$

Turning points are solutions of the system of simultaneous equations

$$\boldsymbol{f}'(\boldsymbol{\theta}) = \mathbf{0}.$$

Turning points may be classified by examining the symmetric matrix $\mathbf{F}''(\boldsymbol{\theta})$ of second-order partial derivatives. For instance, for a bivariate density $f(\theta, \phi)$,

$$\mathbf{F}''(\boldsymbol{\theta}) = \begin{pmatrix} \partial^2 f(\theta, \phi)/\partial \theta^2 & \partial^2 f(\theta, \phi)/\partial \theta \partial \phi \\ \partial^2 f(\theta, \phi)/\partial \theta \partial \phi & \partial^2 f(\theta, \phi)/\partial \phi^2 \end{pmatrix}.$$

$\mathbf{F}''(\boldsymbol{\theta})$ is known as the *hessian* matrix. The second derivative of $f(\boldsymbol{\theta})$ in a direction $\mathbf{t}$ is, after differentiating (2.10), $\mathbf{t}'\mathbf{F}''(\boldsymbol{\theta})\mathbf{t}$. At a mode, this must be negative in all directions, so that the hessian matrix is negative definite. Similarly, at an antimode it is positive definite. In the intermediate case, where $\mathbf{F}''(\boldsymbol{\theta})$ is indefinite, i.e. has both positive and negative eigenvalues, we have a *saddlepoint*.

Notice that at *every* point (θ, ϕ) in the bivariate case there is a direction α in which the gradient (2.9) is zero. This direction is the tangent to the contour line that passes through (θ, ϕ). Another way of characterizing a saddlepoint is as a point where a contour line crosses itself, yielding two different directions of zero gradient without the contour degenerating to a point (as it does at a mode or antimode). In the multivariate case, the gradient is zero in all directions $\mathbf{t}$ satisfying $\mathbf{t}'\boldsymbol{f}'(\boldsymbol{\theta}) = 0$. This defines the tangent plane to the contour surface at $\boldsymbol{\theta}$. The direction given by $\boldsymbol{f}'(\boldsymbol{\theta})$ itself (scaled to unit length by dividing by $\{\boldsymbol{f}'(\boldsymbol{\theta})'\boldsymbol{f}'(\boldsymbol{\theta})\}^{1/2}$) is the normal to the tangent plane, and is the direction of *steepest ascent* of $f(\boldsymbol{\theta})$ at $\boldsymbol{\theta}$.

2.13 A point of inflexion corresponds to the second derivative being zero in some direction $\mathbf{t}$, therefore inflexion points are characterized by $F''(\boldsymbol{\theta})$ being singular. The $\boldsymbol{\theta}$ space can be divided into three kinds of region:

(a) $\mathbf{F}''(\boldsymbol{\theta})$ is positive definite,
(b) $\mathbf{F}''(\boldsymbol{\theta})$ is negative definite,
(c) $\mathbf{F}''(\boldsymbol{\theta})$ is indefinite,

which we could call regions of concave, convex and indefinite curvature respectively. On the boundaries between these regions, one eigenvalue of $\mathbf{F}''(\boldsymbol{\theta})$ is zero, so all points on such boundaries are inflexion points.

In fact, in more than two dimensions we can further subdivide the region of indefinite curvature according to how many positive eigenvalues $\mathbf{F}''(\boldsymbol{\theta})$ has, and all these subregions are also separated by inflexion boundaries.

Example 2.7

The density whose contours are plotted in Figure 2.5 is a mixture of two bivariate normal distributions of the form

$$0.4 \times N\left(\begin{pmatrix} 0 \\ 0 \end{pmatrix}, \begin{pmatrix} 0.5 & 0 \\ 0 & 0.5 \end{pmatrix}\right) + 0.6 \times N\left(\begin{pmatrix} 1.5 \\ 2 \end{pmatrix}, \begin{pmatrix} 0.5 & 0.3 \\ 0.3 & 0.5 \end{pmatrix}\right). \qquad (2.11)$$

Three turning points are found. At (0.0187, 0.0615) there is a mode, with density 0.1316 and hessian matrix

$$\begin{pmatrix} -0.2636 & 0.0250 \\ 0.0250 & -0.2154 \end{pmatrix}.$$

There is another mode at (1.4972, 1.9970), with density 0.2390 and hessian matrix

$$\begin{pmatrix} -0.7443 & 0.4506 \\ 0.4506 & -0.7425 \end{pmatrix},$$

and at (0.4598, 0.8068) there is a saddlepoint with density 0.1025 and hessian matrix

$$\begin{pmatrix} -0.1644 & 0.2591 \\ 0.2591 & 0.0343 \end{pmatrix}.$$

Figure 2.5 also shows the inflexion boundaries as dotted lines. Around each mode is a region of negative definite curvature. Crossing the inflexion boundary then takes us

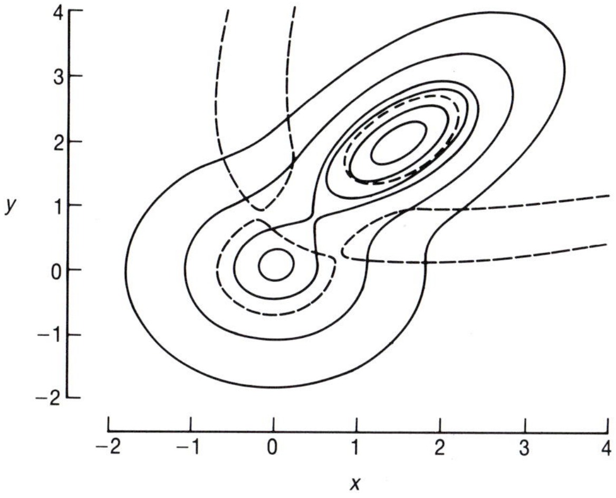

Fig. 2.5 Contour plot with inflexion boundaries

into a region of indefinite curvature. We also see that the density has two regions of positive definite curvature on either side of the saddlepoint. These regions do not contain antimodes, and may be called hollows. As in the one-dimensional case, a region of negative curvature that does not contain a mode is called a shoulder.

Shape and size of modes

2.14 Having located the turning-points of $f(\boldsymbol{\theta})$ the next step in describing its shape is to consider the 'widths' of modes. This is made difficult by the fact that a mode may have a different width in each direction **t**, under any of the one-dimensional summaries of width suggested in **2.4**.

A simplification is offered if we examine the widths of contours in the limiting sense as $c \to 1$, as discussed in **2.4**. Suppose that there is a mode at $\boldsymbol{\theta} = \boldsymbol{\theta}_1$. Expanding $f(\boldsymbol{\theta})$ around the mode to the second order gives

$$f(\boldsymbol{\theta}) \approx f(\boldsymbol{\theta}_1) + \frac{1}{2}(\boldsymbol{\theta} - \boldsymbol{\theta}_1)'\mathbf{F}''(\boldsymbol{\theta}_1)(\boldsymbol{\theta} - \boldsymbol{\theta}_1), \tag{2.12}$$

the multivariate generalization of (2.2). Equating $f(\boldsymbol{\theta})$ to $c\,f(\boldsymbol{\theta}_1)$,

$$-(\boldsymbol{\theta} - \boldsymbol{\theta}_1)'\mathbf{F}''(\boldsymbol{\theta}_1)(\boldsymbol{\theta} - \boldsymbol{\theta}_1) \approx 2(1 - c)f(\boldsymbol{\theta}_1). \tag{2.13}$$

Since $\mathbf{F}''(\boldsymbol{\theta}_1)$ is negative definite, (2.13) says that for c close to 1 the contour is approximately an ellipsoid centred at the mode and with shape described by $-\mathbf{F}''(\boldsymbol{\theta}_1)$.

Example 2.8

Figure 2.6 is a contour map of another mixture of bivariate normal densities. In this case it has the form

$$0.5 \times N\left(\begin{pmatrix}0\\0\end{pmatrix}, \begin{pmatrix}1 & 0\\0 & 10\end{pmatrix}\right) + 0.5 \times N\left(\begin{pmatrix}0\\0\end{pmatrix}, \begin{pmatrix}10 & 0\\0 & 1\end{pmatrix}\right).$$

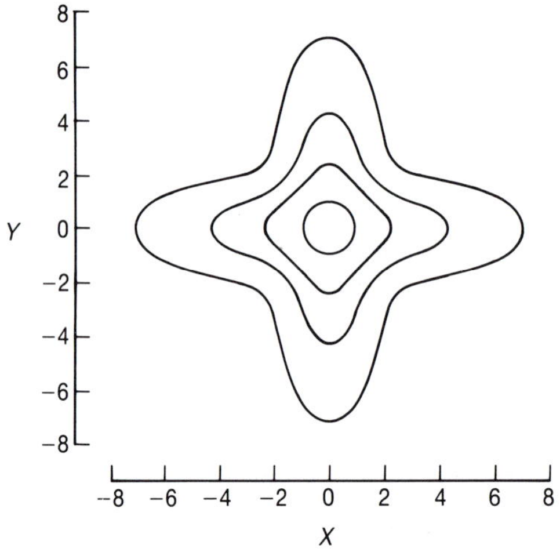

Fig. 2.6 Contour plot for Example 2.8

The mode has height $f(0,0) = 0.0503$ and the innermost contour of Figure 2.6, at height about 80% of the modal height, is approximately circular, reflecting the fact that

$$\mathbf{H}\left(\begin{pmatrix}0\\0\end{pmatrix}\right) = \begin{pmatrix}-0.0277 & 0\\ 0 & -0.0277\end{pmatrix},$$

a multiple of the identity matrix. Yet the other contours, corresponding to smaller values of c, become steadily less circular. The asymptotic argument is valid but the shape of this mode is not well described by its shape in the immediate neighbourhood of the mode point.

2.15 Define the *modal dispersion* matrix $\mathbf{D}$ for a mode at $\boldsymbol{\theta} = \boldsymbol{\theta}_1$ as

$$\mathbf{D} = -f(\boldsymbol{\theta}_1)\mathbf{F}''(\boldsymbol{\theta}_1)^{-1}. \tag{2.14}$$

Notice that $-\mathbf{D}^{-1}$ is the hessian matrix of the logarithm of $f(\boldsymbol{\theta})$ at the mode. $\mathbf{D}$ is positive definite and can therefore be written

$$\mathbf{D} = \mathbf{Q}\Lambda\mathbf{Q}', \tag{2.15}$$

with $\mathbf{Q}$ an orthogonal matrix of eigenvectors of $\mathbf{D}$, and Λ the diagonal matrix of eigenvalues

$$\Lambda = diag(\lambda_1, \lambda_2, \ldots, \lambda_p), \qquad \lambda_1 \geqslant \lambda_2 \geqslant \ldots \geqslant \lambda_p > 0$$

Now a contour ellipsoid (2.13) sets $(\boldsymbol{\theta}-\boldsymbol{\theta}_1)'\mathbf{D}^{-1}(\boldsymbol{\theta}-\boldsymbol{\theta}_1)$ to a constant, and the decomposition (2.15) defines its principal axes. Thus if

$$\mathbf{Q} = (\mathbf{q}_1, \mathbf{q}_2, \ldots, \mathbf{q}_p),$$

the first column vector $\mathbf{q}_1$ is the direction of the longest principal axis and $\mathbf{q}_p$ is the direction of the shortest principal axis. The principal axis lengths are proportional to the square roots of the eigenvalues λ_i.

2.16 Consider the multivariate normal density

$$f(\boldsymbol{\theta}) = (2\pi)^{-p/2} \, |\mathbf{V}|^{-1/2} \exp\{-(\boldsymbol{\theta} - \mathbf{m})'\mathbf{V}^{-1}(\boldsymbol{\theta} - \mathbf{m})/2\},$$

then we find that the hessian matrix of $\log f(\boldsymbol{\theta})$ is $-\mathbf{V}^{-1}$, so that the modal dispersion matrix $\mathbf{D}$ equals the covariance matrix $\mathbf{V}$. The decomposition (2.15) is then identical to the principal components decomposition of $\mathbf{V}$; see A**28.49**. Notice also that in the multivariate normal case the contours are exact ellipsoids. This case therefore provides us with an interpretation of $\mathbf{D}$ as a summary of dispersion in the same way as the covariance matrix. Both are positive definite, and in both cases the eigenvalue decomposition identifies the directions of greatest and least dispersion.

The class of multivariate distributions whose contours are concentric ellipsoids is considered by Johnson and Kotz (1972, chapter 42), Dawid (1977) and Dickey and Chen (1985). This and other forms of symmetry are studied by Fang et al. (1990).

2.17 The volume of an ellipsoid (2.13) is proportional to $|\mathbf{D}|^{1/2}$, which can therefore be used as a single, overall summary of width. This is analogous to the use of the generalized variance $|\mathbf{V}|^{1/2}$ (see A**18.28**) as an overall measure of variance for a random vector with covariance matrix $\mathbf{V}$.

Furthermore, continuing the analogy with the multivariate normal distribution, an indicator of the amount of probability contained within a mode at $\boldsymbol{\theta}_1$ is given by

$$s(\boldsymbol{\theta}_1) = (2\pi)^{p/2} f(\boldsymbol{\theta}_1) \, |\mathbf{D}|^{1/2}. \tag{2.16}$$

This is the multivariate generalization of (2.4). We shall call it the mode size summary.

Example 2.9

Continuing Example 2.7, the $\mathbf{D}$ matrix for the first mode is

$$-0.1316 \begin{pmatrix} -0.2636 & 0.0250 \\ 0.0250 & -0.2154 \end{pmatrix}^{-1} = \begin{pmatrix} 0.505 & 0.059 \\ 0.059 & 0.618 \end{pmatrix},$$

which serves as an approximation to the covariance matrix of the first component of (2.11). The increased 'variance' of ϕ and the positive 'correlation' in the matrix reflect the influence of the second component of (2.11) in drawing out a ridge between the two modes. As Figure 2.5 shows, this effect is more pronounced in contours with values of c not close to one.

Looking at the second mode, we obtain a $\mathbf{D}$ matrix

$$-0.239 \begin{pmatrix} -0.7443 & 0.4506 \\ 0.4506 & -0.7425 \end{pmatrix}^{-1} = \begin{pmatrix} 0.508 & 0.308 \\ 0.308 & 0.509 \end{pmatrix},$$

which is a good approximation to the covariance matrix of the second component of (2.11). Applying (2.16) gives the following indications of size of each mode: for the first mode, 0.46 and for the second, 0.61, close to the corresponding probabilities which are the weights 0.4 and 0.6 in (2.11).

Example 2.10
Consider the general p-dimensional multivariate Student t-density with d degrees of freedom,

$$f(\boldsymbol{\theta}) = \frac{d^{d/2}\,\Gamma\{(d+p)/2\}}{\pi^{p/2}\,\Gamma(d/2)\,|\mathbf{V}|^{1/2}}\{d + (\boldsymbol{\theta}-\mathbf{m})'\mathbf{V}^{-1}(\boldsymbol{\theta}-\mathbf{m})\}^{-(d+p)/2}.$$

Then

$$d\log f(\boldsymbol{\theta})/d\boldsymbol{\theta} = -(d+p)\{d + (\boldsymbol{\theta}-\mathbf{m})'\mathbf{V}^{-1}(\boldsymbol{\theta}-\mathbf{m})\}^{-1}\mathbf{V}^{-1}(\boldsymbol{\theta}-\mathbf{m}).$$

Equating to zero yields $\boldsymbol{\theta} = \mathbf{m}$, as the unique mode. Contours are ellipsoids centred at $\mathbf{m}$ with shape given by $\mathbf{V}$. Also, the hessian matrix of $\log f(\boldsymbol{\theta})$ is

$$\begin{aligned}\frac{d^2\log f(\boldsymbol{\theta})}{d\boldsymbol{\theta}\,d\boldsymbol{\theta}'} &= -(d+p)\{d + (\boldsymbol{\theta}-\mathbf{m})'\mathbf{V}^{-1}(\boldsymbol{\theta}-\mathbf{m})\}^{-1} \\ &\quad + (d+p)\{d + (\boldsymbol{\theta}-\mathbf{m})'\mathbf{V}^{-1}(\boldsymbol{\theta}-\mathbf{m})\}^{-2}\mathbf{V}^{-1}(\boldsymbol{\theta}-\mathbf{m})(\boldsymbol{\theta}-\mathbf{m})'\mathbf{V}^{-1}\end{aligned}.$$

Evaluating at $\boldsymbol{\theta} = \mathbf{m}$ gives $-\mathbf{V}^{-1}(d+p)/d$, therefore

$$\mathbf{D} = \mathbf{V}d/(d+p). \tag{2.17}$$

The covariance matrix of this density is $\mathbf{V}\,d/(d-2)$. Therefore the modal dispersion summary indicates the same correlation structure as the covariance matrix but rather smaller variances. The mode size summary (2.16), representing total probability, is

$$\Gamma\{(d+p)/2\}/[\Gamma(d/2)\{(d+p)/2\}^{p/2}] \tag{2.18}$$

which is less then one for all d and p. It tends to one as $d \to \infty$, for any p, but for large p it can be a very poor indicator of actual probability content. For example, at $d = 10$ and $p = 12$ it gives the value 0.085. This summary is sensitive to $\mathbf{D}$ in the sense that multiplying $\mathbf{D}$ by a constant a multiplies $|\mathbf{D}|^{\frac{1}{2}}$ by $a^{p/2}$. For large p, even a small overall change in $\mathbf{D}$ produces a large change in the summary of mode size. If for instance we changed the expression $(d+p)$ in (2.17), and therefore also the last term of (2.18), to $d + p/2$, (2.18) would generally be much closer to unity.

Regression summaries

2.18 Study of the covariance matrix or modal dispersion matrix can provide information about association between components of $\boldsymbol{\theta}$, mainly in the form of correlations. It is often more useful to employ summaries of regression. Just as the familiar regression line of X on Y plots the expectation of X given $Y = y$ against y, (see A**16.23**), regression summaries represent other features of conditional distributions as functions of the conditioning variables. For instance, a plot of the conditional mode(s) of θ given ϕ and of ϕ given θ is analogous to plotting the two regression lines, but particularly in the case of bimodality can be more informative.

Any kind of location summary may be plotted in this way. We can also plot dispersion summaries. A useful combined summary is to plot a location summary with bounds on either side defined by a dispersion summary. The most obvious case is to plot conditional means, the regression lines, plus and minus one conditional standard deviation. Figure 2.7 shows this plot for the density (2.11) of Example 2.7, whose contours are plotted in

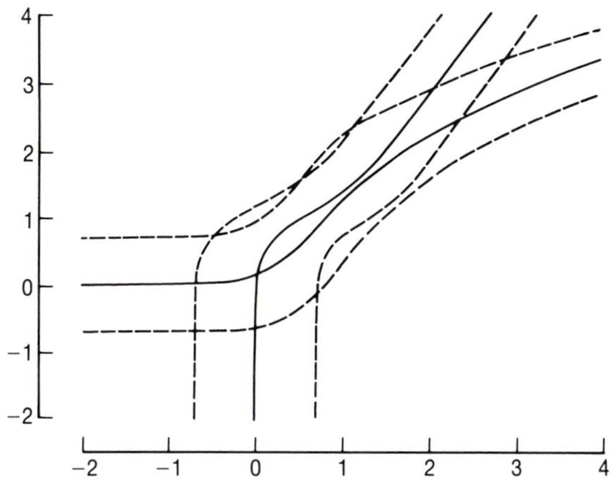

Fig. 2.7 Regression summaries for Example 2.7

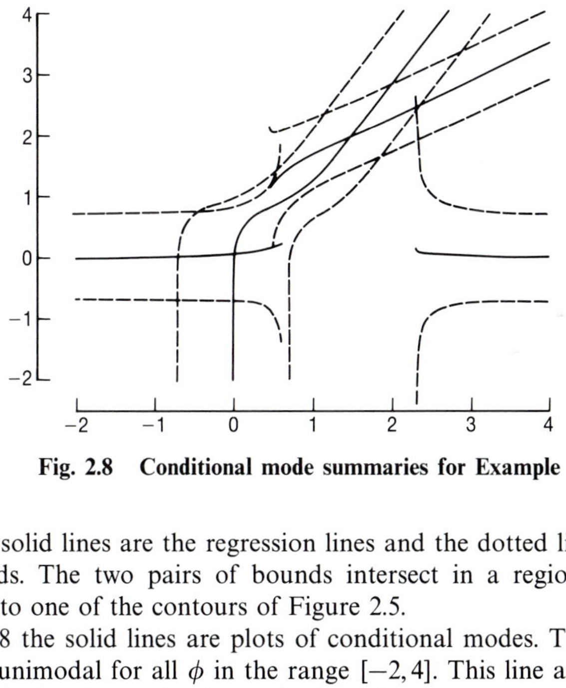

Fig. 2.8 Conditional mode summaries for Example 2.7

Figure 2.5. The solid lines are the regression lines and the dotted lines give one standard deviation bounds. The two pairs of bounds intersect in a region whose boundary is roughly similar to one of the contours of Figure 2.5.

In Figure 2.8 the solid lines are plots of conditional modes. The conditional density of θ given ϕ is unimodal for all ϕ in the range $[-2, 4]$. This line and the dotted lines on either side, which plot the mode plus and minus the square root of the modal dispersion summary (2.14), are very similar to the corresponding lines in Figure 2.7. The conditional densities of ϕ given θ have a more complex shape. The conditional modes are plotted as three separate line segments. The conditional density is bimodal in a narrow interval around $\theta = 0.5$, and again for all θ greater than about 2.4. At the end of each line segment, the mode disappears and becomes a shoulder. At the transition it is a point of inflexion,

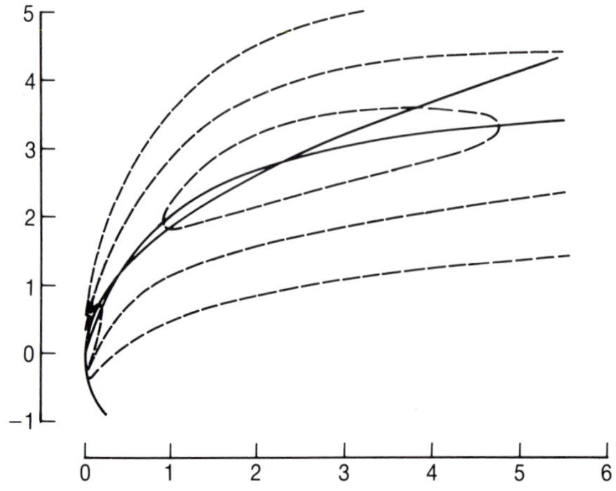

Fig. 2.9 Contours and conditional modes for Example 2.11

so that the modal dispersion becomes infinite. This is why the dotted-line bounds widen out rapidly at these points. Strictly, the bounds become vertical at the point where the mode disappears.

Turning points of the joint density occur at intersections of the lines of conditional turning points. There are three such intersections in Figure 2.8, at the two modes and the saddlepoint identified in Example 2.7. In this example, the intersections of conditional modes produce a joint mode, but such an intersection is not necessarily a joint mode. The fact that the conditional turning points are modes means that the diagonal elements of the hessian matrix are negative, but this is not sufficient to imply that the matrix is negative definite. The intersection of conditional modes may be a joint mode or a saddlepoint. The intersection of a conditional mode with a conditional antimode is necessarily a saddlepoint.

Figure 2.8 lacks any idea of how high the conditional modes are, which is a limitation to such a diagram. For instance, the conditional modes of ϕ given θ along the line marked C are small in comparison with the corresponding points on line B. Nevertheless, their existence, as a continuation of the line marked A, helps to bring out the mixture form of this density. There is a similar bimodality, continuing line B, out of the range of Figure 2.8 for $\theta < -13$. The conditional densities of θ given ϕ are also bimodal for values of ϕ outside the range of Figure 2.8.

Example 2.11

Figure 2.9 shows a contour plot of the density

$$f(\theta, \phi) = 1.1488\,\theta^{-1.7}\exp\{-0.5(\phi - 4)^2 - (0.02 + 0.5\phi^2)/\theta\},$$

where $\theta > 0$. All conditional densities of θ given ϕ and ϕ given θ are unimodal, and the conditional modes are also plotted in Figure 2.9. These lines intersect three times, at two modes and a saddlepoint.

2.19 All of the ideas presented in **2.18** for two dimensions generalize readily to higher dimensions, but we can no longer present the regression summaries graphically.

Summarization strategy

2.20 Thorough summarization of a multivariate distribution should begin by summarizing the one-dimensional marginal densities, then build up through two-dimensional marginals and higher dimensions to end with summarizing the full joint density.

At each stage, when looking at a density function we begin by summarizing its shape. The density may be plotted in one dimension, or in two dimensions a contour map or three-dimensional density plot may be produced. The locations of modes, antimodes and saddlepoints are found. In one or two dimensions it is valuable to locate shoulders and/or hollows. For each mode, its height and modal dispersion matrix may be found, and these are combined in the mode size summary.

In two dimensions, regression summaries may be plotted. In higher dimensions it is valuable to identify conditional distributions whose shape, location or dispersion vary markedly with the value of the conditioning variable.

2.21 Measures of location, dispersion and association, including correlation and regression, may be based on moments or shape summaries. The moment-based summaries are well-known and described fully elsewhere. The mean vector is the location summary and the covariance matrix $\mathbf{V}$ is the dispersion summary. The determinant of $\mathbf{V}$, the generalized variance, provides a single summary of overall dispersion. Its principal components summarize dispersion and association. Its off-diagonal elements, converted to correlations, also summarize association. Other standard calculations on $\mathbf{V}$ yield linear regression coefficients, which provide another form of association summary. Finally, conditional expectations are the regression functions.

2.22 For a unimodal density, analogous summaries may be derived from summaries of the mode. The mode point summarizes location. The modal dispersion summary $\mathbf{D}$ indicates dispersion in the same way as $\mathbf{V}$. The same calculations that produce generalized variance, principal components, correlations and linear regression coefficients may be performed on $\mathbf{D}$ to achieve analogous summaries. Alternatively, contours or the inflexion boundary around the mode, if ellipsoidal (or nearly so), could be used to obtain dispersion matrices analogous to $\mathbf{D}$ and $\mathbf{V}$ and may be manipulated in the same ways. All these summaries, applied to conditional densities and regarded as functions of the conditioning variable, yield regression summaries.

2.23 For a multimodal density, any overall summaries of location, dispersion and association must be interpreted in terms of shape. We can construct overall summaries also from modal summaries. Suppose that there are k modes, at $\boldsymbol{\theta}_1, \boldsymbol{\theta}_2, \ldots, \boldsymbol{\theta}_k$. Let the mode at $\boldsymbol{\theta}_i$ have modal dispersion matrix $\mathbf{D}_i$ and mode size $s(\boldsymbol{\theta}_i)$. Then we can think of the density as a mixture of unimodal densities with weights proportional to $s(\boldsymbol{\theta}_i)$. Define

$$s_i = s(\boldsymbol{\theta}_i) / \sum_{j=1}^{k} s(\boldsymbol{\theta}_j),$$

then an overall summary of location is

$$\mathbf{m} = \sum_{i=1}^{k} s_i \, \boldsymbol{\theta}_i.$$

As in **2.8**, overall dispersion has two components – within and between modes. We can define

$$\mathbf{D} = \sum_{i=1}^{k} s_i \, \mathbf{D}_i + \sum_{i=1}^{k} s_i \, (\boldsymbol{\theta}_i - \mathbf{m})(\boldsymbol{\theta}_i - \mathbf{m})'.$$

This overall dispersion matrix may be used in the same way as the covariance matrix **V**.

Computation

2.24 In **2.20** to **2.23** we have outlined an ideal strategy for summarization which entails, for a high-dimensional density, very substantial computation. It is not practical to implement this approach fully, for two reasons. First, with a p-dimensional density there are 2^{p-1} marginal densities (one-, two-, three-, ..., dimensional), plus the full joint density to summarize. The regression summaries for any q-dimensional marginal density could involve looking at 2^{q-1} sets of conditional densities. In practice this amount of work must be reduced to a manageable level. Certainly, the one-dimensional densities may be plotted and summarized fully. The rest of the exercise is largely concerned with association. An analysis of the covariance or dispersion matrix of the full joint density will point to groups of variables whose association merits further study, through summarizing appropriate densities, perhaps including some regression summaries.

The second practical difficulty lies in calculating many of the summaries. Whenever any but the simplest of statistical models is used, or whenever there is substantive prior information that has also been carefully modelled, the posterior density is typically mathematically intractable. The first line of summarization, i.e. studying the one-dimensional marginal densities, may be closed because we cannot integrate the posterior joint density. Where mathematical analysis fails we must resort to numerical methods. Chapter 8 addresses this very substantial practical problem of computing summaries of complex multi-dimensional densities.

Many of the summaries presented in examples in this chapter, particularly in Examples 2.7 to 2.10, required substantial computation. The necessary techniques will be introduced in Chapter 8. The few exercises on summarization at the end of this chapter may, however, be done with little or no computational skills.

Inference as a decision problem

2.25 We now begin a study of inferences considered as decisions about θ, rather than as partial answers to the general question 'what do we now know about θ?' Any location summary, as discussed in **2.8**, can be thought of as an estimate of θ, and as such it is one useful, concise part of general inference about θ. An estimate is sometimes required in a more specific sense than this. The estimate will be acted upon, using this 'best guess' at θ as a surrogate for θ itself. Then to use a poor estimate, one that is in some sense far from the true value, will inevitably be a worse action than to use a more accurate estimate.

We shall assume that it is possible to give a measurement of how good or bad a particular decision is. This measurement, called a *utility*, will depend upon the context. In some circumstances it may be much worse to overestimate a parameter θ than to underestimate it by the same amount, whereas in other circumstances over and underestimation may be equally serious. The first situation favours underestimation more than the second situation does, so it is clear that the optimal decision will also depend on context. For instance, when estimation is formulated as a decision problem, the choice of estimate will depend on the purpose for which it is required. This is natural, and not very different from thinking of a location summary as an estimate. For there are many possible location summaries, conveying various aspects of the posterior distribution, which would be useful for various purposes. The difference is that in the decision formulation the purpose enters explicitly through the expression of utilities, and this allows an optimal decision to be identified for that purpose.

Utility

2.26 In greatest generality, the investigator's utility will be a function of every aspect of the circumstances in which he finds himself. His utility will change if there is any change in circumstances, in particular if he gains any information or takes any action.

Consider a baker. In the morning he bakes b loaves of bread, and during the day he sells d loaves. If both b and d are large, he will have made a substantial profit and his circumstances at the end of the day will have a high utility for him. At the time when he has just baked b loaves and does not yet know how many he will sell, his utility will depend on his evaluation of how many he is likely to sell. If, later, d turned out to be lower than his expectations, his utility would fall. His utility will also depend on b. If he bakes many more loaves than he could hope to sell, or if he bakes too few so that he cannot take full advantage of the expected demand, his utility will be lower than it would be if he baked a more realistic number.

We will call the investigator's whole complex of circumstances at any time his *status*, and will let $U(S)$ denote his utility for status S.

2.27 To define $U(S)$ we first suppose that there exist statuses S_0 and S_1 such that S_1 is preferable to S and S is preferable to S_0. We further suppose that the utilities of S_0 and S_1 have already been defined as $U(S_0) = u_0$ and $U(S_1) = u_1$. We now introduce a random event E with $P(E) = p$. Consider the status S_p where if E occurs the investigator will be placed in status S_1, but if E fails to occur he will be put in status S_0. Clearly, if $p > q$ then S_p is preferable to S_q. We know that S_1 is preferred to S but S_0 is not. There will be a $p \in (0, 1)$ such that S_q is preferred to S for all $q > p$, and S is preferred to S_q for all $q < p$. Define the utility of S to be

$$U(S) = pu_1 + (1 - p)u_0. \qquad (2.19)$$

Once $U(S_0)$ and $U(S_1)$ are defined a utility value can be given to any S that lies between S_0 and S_1 in the sense that it is preferred to S_0 but not to S_1.

2.28 If we could identify two statuses $S_0^\star$ and $S_1^\star$, such that $S_0^\star$ was the worst possible status in the sense that *every* S was preferable to $S_0^\star$, and $S_1^\star$ was preferred to every S, then we could define $U(S)$ for every S. The choice of values to give to $U(S_0^\star)$ and $U(S_1^\star)$ would be arbitrary, and would serve only to fix the scale of measurement for utility. This is the device used to define the probability scale, where the certain event is given probability one and the impossible event is given probability zero. Unfortunately, $S_0^\star$ and $S_1^\star$ cannot be identified. There are no absolutes. The investigator's death might be thought to be a candidate for $S_0^\star$, yet quite rational people will commit suicide rather than face some kinds of extreme suffering, whether or not they believe in any form of afterlife. Similarly, even those who believe in a heaven do not act as if going to heaven were $S_1^\star$, because they find the extreme way of life that might guarantee that they would achieve $S_1^\star$ unacceptable in practice.

2.29 It is neither possible nor necessary to have a unique, absolutely defined, utility scale for all statuses. It is only necessary that for the purposes of any problem we can identify statuses S_0 and S_1 such that S_1 is preferable to every status that we need to consider in that problem, and every such status is preferable to S_0. We can then assign $U(S_0) = u_0$ and $U(S_1) = u_1$ arbitrarily, to fix an appropriate utility scale for the problem.

Optimal decisions

2.30 Taking a decision implies taking some action that will change the decision-maker's status. Each possible action produces a different status, so a decision problem consists of choosing from a set $\mathscr{S}$ of possible statuses. By definition, the status in $\mathscr{S}$ with highest utility will be the optimal choice. Let S be the decision-maker's status before taking this decision. Status S includes the option to choose from $\mathscr{S}$, which we will write as $S \mapsto \mathscr{S}$. Since that choice will be to maximize utility we have the immediate result that if $S \mapsto \mathscr{S}$ then

$$U(S) = \sup_{T \in \mathscr{S}} U(T). \tag{2.20}$$

2.31 Often, a decision problem includes as one of the options a decision to take no action. It is important to realize that the status reached by taking no action is different from the status before that decision. One aspect of this is obvious from (2.20): if $S \mapsto \mathscr{S}$ and $S \in \mathscr{S}$ then S = 'no action' would have to be an optimal decision, but this is by no means necessarily true. Making a decision should always be recognized as changing status. Even no action invariably results in a loss of future choice, or a loss of utility due to loss of time, or has some other effect.

2.32 The rule represented by (2.18), to choose that decision which maximizes utility, is not much help in practical decision making if utilities must always be specified by the definition of **2.27**. Those occasions when it is difficult to say which of a set of possible decisions is best are precisely those in which it is difficult to assess utilities of the resulting statuses. It is necessary to have other ways of assigning utilities in practice.

Utility of money

2.33 Business decisions are typically concerned with financial gains and losses. In these circumstances, it seems natural to relate utility to money. If two statuses differ only in that under S_1 the decision-maker receives more money than under S_0, then $U(S_1) > U(S_0)$. Suppose that in a given problem the maximum possible financial gain is £m_1, which is achieved in status S_1, and the minimum gain is £m_0, achieved in status S_0. The 'gain' m_0 (or even m_1) might be negative, denoting a financial loss. Now with S_1 and S_0 as defined in (2.19), setting $u_1 = m_1$ and $u_0 = m_0$ is the natural choice. It is reasonable now to ask whether, if a status S yields a gain of £m, the utility of S will be m for all $m_0 \leqslant m \leqslant m_1$. However, the relationship between utility and money is less simple than this.

Let $m_0 = -m_1$, and consider the status S in which the decision maker receives no gain and incurs no loss. Now compare the status $S_{0.5}$ which depends on the event E as defined in **2.27**. This status results in S_1, a gain of £m_1, or S_0, a loss of £m_1, with equal probabilities. Now most decision-makers in practice will prefer the *status quo*, as represented by S, to the gamble $S_{0.5}$. Since by definition $U(S_{0.5}) = 0.5(m_0 + m_1) = 0$, the utility of S is greater than its monetary value of zero.

2.34 In general, almost every rational person will prefer a certain gain to a gamble having the same expected return, and this phenomenon increases as we increase the sums of money at risk in the gamble. Using the above scenario, the preference of S over $S_{0.5}$ will be only marginal if m_1 is small. If m_1 is in the thousands, so that a great deal of money will be lost if $S_{0.5}$ results in S_0 rather than S_1, then S will be strongly preferred to $S_{0.5}$. $U(S_{0.5}) = 0$ always by definition, so $U(S)$ will depend on m_1. Similarly, people will pay more than £m to avoid a gamble with an expected loss of £m; this is the underlying basis of insurance. See Exercise 2.7.

This general behaviour is known as risk aversion. Most rational people are risk averse to some extent. Risk aversion manifests itself in decision problems by a nonlinear relationship between utility and money. Figure 2.10 shows a typical decision-maker's utility function for the business decision problem discussed in **2.33**. The amount of curvature in the relationship, and in particular the difference in utility between the *status quo* S and the gamble $S_{0.5}$, will vary from person to person. The main reason for this is that the loss of £m_1, is more serious for some people than others. We can make some allowance for this by relating utility not to monetary gain or loss but to overall wealth. Figure 2.11 shows a typical utility function for wealth. The curvature decreases as wealth increases. If the sum of £m_1 is small relative to the decision-maker's wealth, then the curvature shown in Figure 2.10 will be slight.

2.35 Therefore, although the relationship between money and utility is not strictly linear, provided the sums of money involved are small compared to the decision-maker's wealth, we can safely ignore the nonlinearity. Then if status S is characterized by a gain of m units of money for the decision-maker we can set $U(S) = m$.

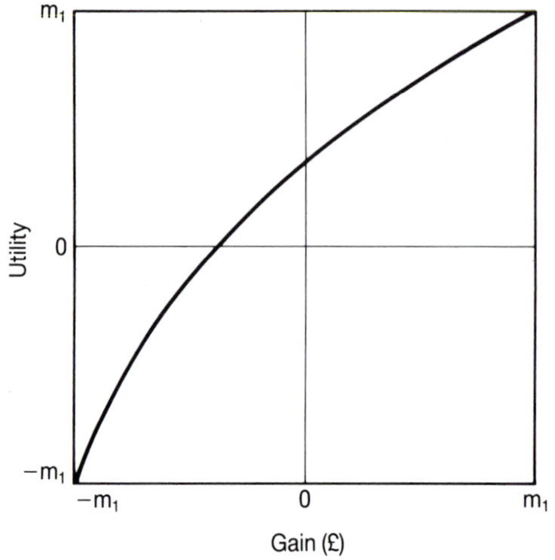

Fig. 2.10 Utility function for a gamble

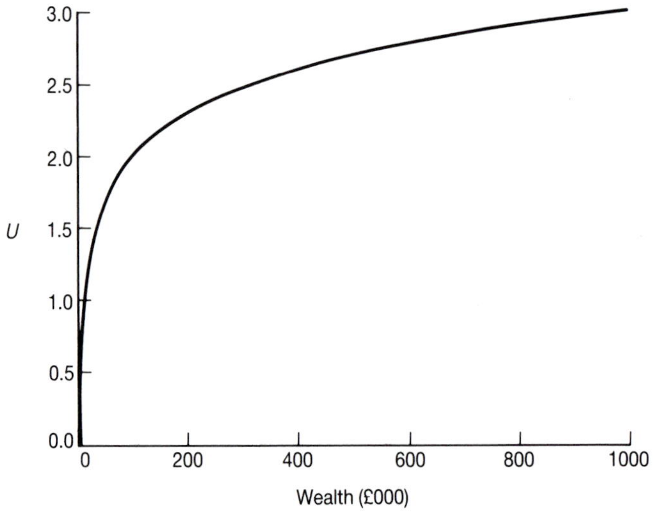

Fig. 2.11 Typical utility for wealth

Extending the argument

2.36 Suppose that in status S the value of a random variable X is unknown, but has a probability distribution with frequency function $f(x)$. Let $S(x)$ be the status identical to S except that the value of X is known to be x, and let $U(S(x)) = u(x) = p(x)u_1 + (1 - p(x))u_0$. Then $S(x)$ is equivalent to the status in which S_1 is achieved with probability $p(x)$ and S_0 is obtained with probability $1 - p(x)$. Now status S will change to $S(x)$ if $X = x$, which has probability $f(x)$. We can therefore consider S as equivalent to the status S_p in which S_1 is achieved with probability $p = \sum_x f(x)p(x)$, and S_0 is given with probability $1 - p = \sum_x f(x)(1 - p(x))$. Therefore

$$\begin{aligned} U(S) &= U(S_p) = p\,u_1 + (1-p)\,u_0 \\ &= \sum_x f(x)\{p(x)\,u_1 + (1-p(x))\,u_0\} \\ &= \sum_x f(x)U(S(x)) = E\{U(S(X))\}\,. \end{aligned} \tag{2.21}$$

We have presented this argument for discrete X, but it clearly applies also for continuous X. This result is known as extending the argument, or the law of expected utility.

Example 2.12

A baker must decide how many loaves of bread to bake on a given day. Let S be his status before baking. Then $S \mapsto \mathscr{S} = \{S(b) : b = 0, 1, 2, \ldots\}$ where b is the number of loaves. In order to make the best decision, the baker must evaluate $U(S(b))$ for each b, but the utility of baking b loaves would be much easier to state if he knew what the demand for bread would be on that day. Let D be the number of loaves that customers will demand, and let $S(b, d)$ be his status at the end of the day after baking b and experiencing demand $D = d$. Suppose that the cost of baking one loaf is c and that the loaf sells for $s > c$. The sums of money involved in one day's trading will be small compared with the baker's wealth, so we equate utility to money. He cannot sell more than he bakes, so

$$U(S(b,d)) = \begin{cases} ds - bc & \text{if } b \geqslant d \\ bs - bc & \text{if } b \leqslant d \end{cases} \tag{2.22}$$

Let the baker's probability distribution for D be $f(d)$, then applying (2.21) to (2.22) gives

$$\begin{aligned} U(S(b)) &= E\{U(S(b,D))\} \\ &= \sum_{d=0}^{b} f(d)\,(ds - bc) + \sum_{d=b+1}^{\infty} f(d)\,(bs - bc) \\ &= s\sum_{d=0}^{b} d\,f(d) + sb\sum_{d=b+1}^{\infty} f(d) - bc\,. \end{aligned}$$

$$\therefore U(S(b)) - U(S(b-1)) = s\sum_{d=b}^{\infty} f(d) - c\,. \tag{2.23}$$

Therefore, the optimal decision is to bake b_0 loaves, where b_0 is the largest integer b for which (2.23) is positive, and from (2.20)

$$U(S) = U(S(b_0)).$$

Value of information

2.37 Suppose that a decision may be made now, or else may be delayed until some further information is available. Choosing to delay is one of the options available now, so the current status has the form

$$S \mapsto \mathscr{S}_+ = \{S_0\} \cup \mathscr{S}. \tag{2.24}$$

where S_0 is the status obtained by delaying the decision, and $\mathscr{S}$ is the set of statuses available by taking the decision now. Write

$$\mathscr{S} = \{S(d) \,:\, d \in \mathscr{D}\}$$

so that $\mathscr{D}$ represents the set of decisions, each decision $d \in \mathscr{D}$ leading to a status $S(d)$.

Now suppose that delaying the decision allows us to observe the value x of a random variable X. Then write $S_0(x)$ for the status reached after delaying and then observing $X = x$. For every x,

$$S_0(x) \mapsto \mathscr{S}(x) = \{S_0(x, d) \,:\, d \in \mathscr{D}\}$$

because after observing $X = x$ we can take any decision $d \in \mathscr{D}$ leading to a new status $S_0(x, d)$.

Let

$$U(S_0(x, d)) = u_0(x, d), \tag{2.25}$$

then

$$U(S_0(x)) = \sup_{d \in \mathscr{D}} u_0(x, d) = u_0(x). \tag{2.26}$$

by (2.20), and

$$U(S_0) = E(u_0(X)) \tag{2.27}$$

by (2.21). We can now decide either to delay or to take a decision $d \in \mathscr{D}$ immediately by comparing $U(S_0)$ with $\sup_{d \in \mathscr{D}} U(S(d))$.

2.38 An important case is represented by the idea that delaying the decision does not incur any cost. This corresponds to the assertion that for all possible x and for all $d \in \mathscr{D}$,

$$U(S_0(x, d)) = U(S(d, x)) = u_0(x, d). \tag{2.28}$$

where $S(d, x)$ is the status reached when we first choose decision d, so reaching status $S(d)$, and then subsequently observe $X = x$. Now (2.28) gives

$$U(S(d)) = E(U(S(d, X)) = E(u_0(X, d))$$

$$\therefore \sup_{d \in \mathscr{D}} U(S(d)) = \sup_{d \in \mathscr{D}} E(u_0(X, d)), \tag{2.29}$$

whereas (2.25) to (2.27) give

$$\begin{aligned} U(S_0) &= E(\sup_{d \in \mathscr{D}} u_0(X, d)) \\ &\geqslant \sup_{d \in \mathscr{D}} U(S(d)) \end{aligned} \tag{2.30}$$

with equality if and only if $u_0(x) = \sup_{d \in \mathscr{D}} u_0(x, d)$ is a constant, independent of x. Otherwise it is always preferable to delay the decision if (2.26) holds.

The difference between (2.30) and (2.29),

$$V(X, \mathscr{D}, S) = E(\sup_{d \in \mathscr{D}} u_0(X, d)) - \sup_{d \in \mathscr{D}} E(u_0(X, d)), \tag{2.31}$$

is called the *value of information* X to the decision $\mathscr{D}$ under current status S. The information only has zero value if knowing X does not affect the optimal decision.

2.39 In the light of **2.31** it is more realistic to recognize that delaying a decision generally incurs some cost, so that $U(S_0(x, d)) < U(S(d, x))$. If there were a constant c such that, for all x and d,

$$u_0(x, d) = U(S(d, x)) = c + U(S_0(x, d)), \tag{2.32}$$

then (2.30) would be replaced by

$$U(S_0) = E(\sup_{d \in \mathscr{D}} u_0(X, d)) - c.$$

Now the option of delaying the decision is preferable only if the value (2.29) exceeds the cost c.

For decisions governed primarily by financial gains and losses, the cost of acquiring information X may be an actual sum of money. Then (2.32) is a realistic assumption if linear utility for money may also be assumed.

2.40 Now suppose that the information X is in two parts, $X = (Y, Z)$. The expectations in (2.31) can then also be taken in two stages, as expectations first over Z given Y, which we denote by $E_{Z \mid Y}$, and then over the marginal distribution of Y, denoted by E_Y. Then

$$\begin{aligned} V(X, \mathscr{D}, S) &= [E_Y \, E_{Z \mid Y} \sup u_0((Y, Z), d) - E_Y \sup E_{Z \mid Y} u_0((Y, Z), d)] \\ &\quad + [E_Y \sup E_{Z \mid Y} u_0((Y, Z), d) - \sup E_Y E_{Z \mid Y} u_0((Y, Z), d)] \\ &= E_Y(V(Z, \mathscr{D}, S_0(Y))) + V(Y, \mathscr{D}, S), \end{aligned} \tag{2.33}$$

where the suprema are all with respect to $d \in \mathscr{D}$. Notice that

$$V(Z, \mathscr{D}, S_0(y)) = E_{Z \mid y} \sup u_0((y, Z), d) - \sup E_{Z \mid y} u_0((y, Z), d) \tag{2.34}$$

is the value of information Z to the decision $\mathscr{D}$ at the time when the decision has first been delayed leading to status S_0, and then $Y = y$ has been observed giving status $S_0(y)$. (2.33) says that the value of $X = (Y, Z)$ is composed of the value of Y plus the expectation of the further value of Z.

If

$$V((Y, Z), \mathscr{D}, S) = V(Y, \mathscr{D}, S) \tag{2.35}$$

for all possible random variables Z, then Y is called *perfect information* for $\mathscr{D}$ at status S. Now (2.33) shows that this occurs if and only if (2.34) is zero for all y and for all Z. In turn this means that for each y, the utility $u_0((y, z), d)$ is maximized at the same $d(y) \in \mathscr{D}$ for all possible values z of all possible random variables Z. So Y represents perfect information if after observing $Y = y$ for any y, no other observation would be relevant to the decision problem.

If Y is perfect information for $\mathscr{D}$ at status S then for any random variable X

$$V(X, \mathscr{D}, S) \leqslant V(Y, \mathscr{D}, S), \tag{2.36}$$

so that the value of perfect information is an upper bound on the value of any information. To prove (2.36) formally, let $(X, X') = (Y, Y')$ so that X' represents information contained in Y but not in X, and Y' represents information contained in X but not in Y. Then

$$\begin{aligned} V(Y, \mathscr{D}, S) &= V((Y, Y'), \mathscr{D}, S) = V((X, X'), \mathscr{D}, S) \\ &= V(X, \mathscr{D}, S) + E_{X' \mid X}(V(X', \mathscr{D}, S_0(X))) \\ &\geqslant V(X, \mathscr{D}, S). \end{aligned}$$

2.41 Decision theory is a large field with a substantial body of literature. We have presented a rather simplified formulation which is adequate for a study of statistical decision theory as it relates to Bayesian inference. In doing so, we have ignored philosophical questions of the precise meaning and properties of utilities, and practical questions about how utilities may be measured in real problems. On the philosophical side, Eatwell et al. (1990) present a wealth of detail and numerous further references. Techniques and tools of Bayesian decision making are described in French (1988) and Smith (1988)

Statistical decision theory

2.42 Some general features will be found in any formulation of a decision problem as an 'inference' about a parameter θ. The parameter θ represents perfect information, because when θ is known any further information is irrelevant. We cannot delay until θ is known, but we can delay until we observe the data x.

If the investigator must make the 'inference' with current status S then

$$S \mapsto \{S(d) : d \in \mathscr{D}\}$$

where the set $\mathscr{D}$ represents the available 'inferences'. Let $S(d, \theta)$ be his status after taking decision d and then learning that the true parameter value is θ, and let $U(S(d, \theta)) = u(d, \theta)$. Then the optimal decision is to choose d to maximize

$$U(S(d)) = E(u(d, \theta)), \tag{2.37}$$

the expectation being taken with respect to the investigator's current distribution of θ. In the statistical context, inference occurs after the data x have been observed, and this is reflected in the current status S. Therefore, the required distribution of θ is the posterior distribution $f(\theta \mid x)$.

The utility function $u(d, \theta)$ represents how good it would be to have made decision d if the parameter value were θ. The optimal decision is to maximize (2.37), the (posterior) expected utility.

2.43 In statistical decision theory, it is usual to formulate utility in terms of a *loss function* $L(d, \theta)$, which expresses how *bad* it would be to make decision d if the parameter value were θ. One might simply let $L(d, \theta) = -u(d, \theta)$ or define it relative to the utility of the optimal decision $d^\star$ by $L(d, \theta) = u(d^\star, \theta) - u(d, \theta)$. As with utilities, the scale of measurement for losses is arbitrary. Instead of maximizing expected utility, we can minimize expected loss.

As in the earlier part of this chapter, we will at this stage consider inference based on information about θ represented by a current distribution $f(\theta)$. This may be a prior or posterior distribution. For predictive inference, θ can be replaced by some future observation, and $f(\theta)$ represents the appropriate predictive distribution.

2.44 Suppose that $\theta = (\phi, \psi)$ and $L(d, \theta) = L(d, \phi)$, a function only of ϕ. Then we would say that for this decision or inference problem ψ is a *nuisance parameter*, since it has no effect on utility. Then, $E(L(d, \theta)) = E(L(d, \phi))$ is evaluated from the marginal distribution of ϕ after 'integrating out' ψ; see **3.13**. This agrees with our discussion of inference through summarization, where inference about ϕ alone is achieved by summarizing its marginal distribution.

Point estimation

2.45 First consider the problem of point estimation, where the decision consists of asserting a single value as an *estimate* of θ. We therefore identify the decision set $\mathscr{D}$ as the set of possible values of θ. The ideal estimate is $d = \theta$, and it is natural to define $L(\theta, \theta) = 0$ for all θ. Then it should also be the case that if θ is further from d_1 than from d_2, according to some appropriate measure of distance, $L(d_1, \theta) \geqslant L(d_2, \theta)$. Beyond these basic remarks, the formulation of $L(d, \theta)$ should depend on context. We return to this in **2.49** but begin by looking at some specific examples of $L(d, \theta)$.

2.46 Suppose that θ is scalar, and therefore so is d. The *quadratic loss* function is $L(d, \theta) = (d - \theta)^2$, also known as squared-error loss.

$$E(L(d, \theta)) = d^2 - 2d\,E(\theta) + E(\theta^2) = (d - E(\theta))^2 + \mathrm{var}(\theta)\,. \tag{2.38}$$

This is minimized at $d = E(\theta)$, which is therefore the optimal estimate.

The *bilinear loss* function takes the form

$$L(d, \theta) = \begin{cases} a(\theta - d) & \text{if } d \leqslant \theta\,, \\ b(d - \theta) & \text{if } d \geqslant \theta\,, \end{cases} \tag{2.39}$$

where a and b are positive constants. Suppose that θ is continuous with density function $f(\theta)$ and distribution function $F(\theta)$. Then

$$-E(L(d, \theta)) = ad\{1 - F(d)\} - bdF(d) - a\int_d^\infty \theta f(\theta)\,d\theta + b\int_{-\infty}^d \theta f(\theta)\,d\theta\,.$$

Differentiating with respect to d gives

$$a\{1 - F(d)\} - ad\,f(d) - b\,F(d) - bd\,f(d) + ad\,f(d) + bd\,f(d) = a - (a + b)F(d)\,.$$

Equating to zero gives $F(d) = a/(a+b)$, so that the optimal d is the $100a/(a+b)\%$ point of the current distribution of θ. Essentially the same answer is obtained for discrete θ. If $a = b$ we have $L(d, \theta) = a|d - \theta|$, which is called *absolute loss*, and the optimal estimate is the median of the distribution of θ.

The *zero-one* loss function takes the form

$$L(d, \theta) = \begin{cases} 0 & \text{if } |d - \theta| \leqslant b, \\ 1 & \text{if } |d - \theta| > b, \end{cases}$$

where b is a constant. Then

$$E(L(d, \theta)) = P(|d - \theta| > b), \tag{2.40}$$

and the optimal d is the centre of the interval of width $2b$ having maximum probability. For continuous θ, note that if we differentiate (2.40) with respect to d and equate to zero we have $f(d-b) = f(d+b)$, so the end points of the optimal interval will necessarily have equal density. In the limit as $b \to 0$, the optimal d tends to the mode of the distribution of θ.

2.47 These examples show how the form of $L(d, \theta)$ influences the choice of estimate. Compared with absolute loss, quadratic loss penalizes large deviations of the estimate d from the true value θ much more heavily. It gives more importance to the tails of the distribution and so leads to the choice of mean rather than median. Zero-one loss penalizes large deviations less than absolute loss. In its limiting form it leads to the mode as an estimate, giving no importance to the tails of the distribution. Looking at the bilinear loss functions with $a \neq b$ shows the effect of asymmetric loss. If $a < b$, overestimation is penalized more than underestimation, so the estimate has a probability $a/(a+b) < 0.5$ of being an overestimate. Conversely, when $a > b$ the optimal choice is more likely to be an overestimate.

There is a clear parallel between this use of the mean, median and mode as formal point estimates and their use as summaries of location in general inference.

2.48 For a vector parameter $\boldsymbol{\theta}$, quadratic loss generalizes readily to the quadratic form

$$L(\mathbf{d}, \boldsymbol{\theta}) = (\mathbf{d} - \boldsymbol{\theta})'\mathbf{A}(\mathbf{d} - \boldsymbol{\theta}), \tag{2.41}$$

where $\mathbf{A}$ is a non-negative definite matrix. If $\mathbf{A}$ is singular there exists a $\mathbf{c}$ such that $\mathbf{Ac} = \mathbf{0}$ and $L(\mathbf{d}, \boldsymbol{\theta})$ does not depend at all on the difference between $\mathbf{c}'\boldsymbol{\theta}$ and its estimate $\mathbf{c}'\mathbf{d}$. Then $\psi = \mathbf{c}'\boldsymbol{\theta}$ is a nuisance parameter as discussed in **2.44**. Now

$$E(L(\mathbf{d}, \boldsymbol{\theta})) = E[\{(\mathbf{d} - E(\boldsymbol{\theta})) - (\boldsymbol{\theta} - E(\boldsymbol{\theta}))\}'\mathbf{A}\{(\mathbf{d} - E(\boldsymbol{\theta})) - (\boldsymbol{\theta} - E(\boldsymbol{\theta}))\}],$$

where $E(\boldsymbol{\theta})$ is the vector of expectations of elements of $\boldsymbol{\theta}$. Expanding this, and noting that

$$E\{(\mathbf{d} - E(\theta))'\mathbf{A}(\boldsymbol{\theta} - E(\boldsymbol{\theta})\} = (\mathbf{d} - E(\boldsymbol{\theta}))'\mathbf{A}E(\boldsymbol{\theta} - E(\boldsymbol{\theta})) = 0,$$

we have

$$E(L(\mathbf{d}, \boldsymbol{\theta})) = (\mathbf{d} - E(\boldsymbol{\theta}))'\mathbf{A}(\mathbf{d} - E(\boldsymbol{\theta})) + E\{(\boldsymbol{\theta} - E(\boldsymbol{\theta}))'\mathbf{A}(\boldsymbol{\theta} - E(\boldsymbol{\theta}))\}. \tag{2.42}$$

Since $\mathbf{A}$ is non-negative definite, the first term in (2.42) is non-negative. The second term does not depend on $\mathbf{d}$. Therefore (2.42) is minimized by making the first term zero, which occurs at $\mathbf{d} = E(\boldsymbol{\theta})$. The mean vector $E(\boldsymbol{\theta})$ is therefore an optimal estimate under quadratic loss for any $\mathbf{A}$. If $\mathbf{A}$ is nonsingular, and so strictly positive definite, this is the unique optimal estimate. Otherwise, as we have remarked, there exists a $\mathbf{c}$ such that $\mathbf{Ac} = \mathbf{0}$ and loss does not depend on the nuisance parameter $\psi = \mathbf{c}'\boldsymbol{\theta}$. Then any $\mathbf{d}$ which gives $\mathbf{a}'\mathbf{d} = \mathbf{a}'E(\boldsymbol{\theta})$ for all $\mathbf{a}$ such that $\mathbf{Aa} \neq \mathbf{0}$ but gives arbitrary $\mathbf{c}'\mathbf{d}$ when $\mathbf{Ac} = \mathbf{0}$ is optimal.

Zero-one loss also generalizes readily to

$$L(\mathbf{d}, \boldsymbol{\theta}) = \begin{cases} 0 & \text{if } \mathbf{d} - \boldsymbol{\theta} \in B, \\ 1 & \text{if } \mathbf{d} - \boldsymbol{\theta} \notin B, \end{cases}$$

where B is some compact set containing the origin. Then the optimal $\mathbf{d}$ maximizes $P(\mathbf{d} - \boldsymbol{\theta} \in B)$. Letting B shrink to an arbitrarily small neighbourhood of the origin yields the mode of $f(\boldsymbol{\theta})$ as the optimal estimate. If B is infinitely wide in some directions, and remains so as we shrink it in other directions, the limiting estimate is the mode of the appropriate marginal distribution, combined with arbitrary values in other directions.

There is no natural definition of the median of a vector random variable. In one dimension, the probability that θ is less than its median is 0.5, but the notion of 'less than' does not have a natural generalization to vector θ. There are similarly various possible generalizations of $L(d, \theta) = a|d - \theta|$. One is the square root of (2.41). Unfortunately, the estimate $\mathbf{d}$ minimizing expected loss is then difficult to compute and does not preserve the sense of $\boldsymbol{\theta}$ being equally likely to be 'less than' or 'greater than' $\mathbf{d}$. If instead we let

$$L(\mathbf{d}, \boldsymbol{\theta}) = \sum a_i \, |d_i - \theta_i|,$$

then $E(L(\mathbf{d}, \boldsymbol{\theta}))$ is also a sum, whose components are minimized independently by letting each d_i be the median of the marginal distribution of θ_i.

2.49 It is clear that different forms of loss function lead to widely differing estimates. There is no unique best estimate of θ. In order to employ decision theory to specify an estimate of θ there needs to be a genuine decision context, in which it is possible to specify an appropriate loss function.

Example 2.13

Consider again the baker's problem of Example 2.12. Deciding how many loaves to bake is in a sense a problem of estimating demand. In the terminology of statistical decision theory, the decision set is $\mathscr{B} = \{0, 1, 2, \ldots\}$ and the parameter is the demand d. We define the baker's loss if he bakes b loaves and demand is d by the difference between his actual profit and the profit he would have made from demand d by making $b = d$ loaves. Thus, if he bakes $b > d$ his loss is $c(b - d)$, the cost of baking surplus loaves, whereas if $b < d$ his loss is $(s - c)(d - b)$, the lost profit from not baking enough. This fits the bilinear loss specification (2.39) with $a = s - c$ and $b = c$. The optimal decision derived in Example 2.12 agrees with the bilinear loss estimate in **2.47**. Notice that the loss function specified here is not just minus the utility function (2.28) from Example 2.12. To make this into an estimation problem we needed $L(b.d) = 0$ when $b = d$. The justification for the adjustment applied here is given in Exercise 2.7.

Interval estimation

2.50 In the classical inference problem of interval estimation the objective is not to assert a single value for θ but rather a range of values. Therefore if the set of possible values for θ is Ω the decision set $\mathscr{D}$ is not Ω but a set of subsets of Ω. The loss associated in making the inference of $d \subset \Omega$ when the parameter value is θ has two components. First, we simply ask whether $\theta \in d$. If so, the inference is right and there is no loss, otherwise it is wrong and there is a non-zero loss. If the loss function were just this, then the optimal inference would always be $d = \Omega$, since then the inference would always be right. There is clearly also a loss associated with d itself. If d is in some sense large, then asserting this as a range of values for θ is not very helpful. It would be more informative to give a 'smaller' d, thereby reducing the range of values given for θ. Thus, the loss function might be

$$L(d, \theta) = \delta(d, \theta) + w(d), \tag{2.43}$$

where $\delta(d, \theta) = 0$ if $\theta \in d$, otherwise $\delta(d, \theta) = 1$, and $w(d)$ is a loss associated with the 'width' of d. Then

$$E(L(d, \theta)) = P(\theta \notin d) + w(d), \tag{2.44}$$

and the optimal d will be a balance between high coverage, $P(\theta \in d)$, and low width.

Suppose that $\boldsymbol{\theta}$ is a p-vector, then it is natural to specify $w(d) = g(v(d))$, where g is an increasing function of the p-dimensional volume $v(d)$ of d. (Formally, $v(d)$ is the Lebesgue measure of d. We will require $\mathscr{D}$ to be restricted to the set of Lebesgue measurable subsets of Ω, but this includes all subsets of practical interest.) In order to minimize (2.44), first consider maximizing $P(\boldsymbol{\theta} \in d)$ over all d having a fixed volume v (and therefore a fixed width loss $g(v)$). It is easy to see that the optimal d is then bounded by a contour of the density $f(\theta)$, being specifically that contour whose volume is v. For suppose that d_0 is this set and let d_1 be some other choice. Let $d_2 = d_0 \cap d_1$, $d_3 = d_0 - d_2 = d_0 \cap d_1^c$ and $d_4 = d_1 - d_2 = d_0^c \cap d_1$. Then $P(\boldsymbol{\theta} \in d_0) = P(\boldsymbol{\theta} \in d_2) + P(\boldsymbol{\theta} \in d_3)$, and $P(\boldsymbol{\theta} \in d_1) = P(\boldsymbol{\theta} \in d_2) + P(\boldsymbol{\theta} \in d_4)$. But since d_0 is bounded by a contour of $f(\boldsymbol{\theta})$, every point $\boldsymbol{\theta}$ in d_0 has higher density than every point not in d_0. Therefore every point in d_3 has higher density than every point in d_4. Since $v = v(d_0) = v(d_2) + v(d_3)$ and $v = v(d_1) = v(d_2) + v(d_4)$, both d_3 and d_4 have the same volume. Therefore $P(\boldsymbol{\theta} \in d_3) > P(\boldsymbol{\theta} \in d_4)$ and so $P(\boldsymbol{\theta} \in d_0) > P(\boldsymbol{\theta} \in d_1)$.

Now let $p(v)$ be $P(\boldsymbol{\theta} \in d_v)$, where d_v is such a contour-bounded set of volume v. Then (2.44) becomes

$$E(L(d_v, \theta)) = 1 - p(v) + g(v). \tag{2.45}$$

Now $f(\boldsymbol{\theta})$ has a constant value, say f_v, all over the contour which bounds d_v. Therefore increasing the volume by an infinitessimal δ increases $p(v)$ by $f_v\delta$, so $dp(v)/dv = f_v$. Therefore if we differentiate (2.45) with respect to v, and equate to zero, the optimal inference is d_v where v satisfies

$$f_v = g'(v) \tag{2.46}$$

and $g'(v) = dg(v)/dv$.

Meeden and Vardeman (1985) point out that (2.46) has no solution if $g(v) = kv$ and $k > f(\boldsymbol{\theta})$ for all $\boldsymbol{\theta}$. Then the optimal set d is empty. That is because the penalty for increasing the volume is too high. A solution can be guaranteed if we let $g(v) = kv^2$. Then $g'(0) = 0$ and $g'(v)$ is increasing, while f_v is decreasing to $f_\infty = 0$, so a unique solution will exist. Using $g(v) = kv^2$ may be seen as analogous to using squared error loss for point estimation, but the constant k will be important in determining the balance between volume and probability.

2.51 In order to apply this solution, we need to specify the function g, which will depend upon the decision context. In practice, it is usually both easier and realistic to specify the coverage $p = P(\boldsymbol{\theta} \in d)$, that is, the set $\mathscr{D}$ comprises only those subsets of Ω having coverage p. Minimizing (2.44) now reduces to finding the set d of mimimum volume having given coverage p. It is easy to see that this is again a set bounded by a contour of $f(\boldsymbol{\theta})$, specifically the contour-bounded set of coverage p.

A set bounded by a contour of $f(\theta)$ is called a *highest-density set*. The optimal interval estimate under any formulation will be a highest-density set. The choice of the bounding density may be by (2.46), or to satisfy a fixed volume or fixed coverage.

Example 2.14

Figure 2.12 shows examples of highest-density sets for the univariate density shown in Figure 2.3 and discussed in Example 2.5. The interval between 1.936 and 2.050 is a highest-density set of volume $2.05 - 1.936 = 0.114$, bounded by density $f(1.936) = f(2.05) = 0.52$. It has coverage probability 0.605. Increasing the size of the set (or decreasing the bounding density) eventually begins to include the second mode. The union of the intervals $[-0.484, 0.486]$ and $[1.729, 2.241]$ is a highest-density set of volume $0.968 + 0.512 = 1.48$, bounded by density 0.2662, and with coverage $0.2785 + 0.2215 = 0.5$. Increasing the coverage further, the two intervals eventually recombine; $[-1.629, 2.407]$ is the highest-density set with coverage 0.95.

Example 2.15

The contours shown in Figures 2.4, 2.5 and 2.6 all define highest-density sets. The contour in Figure 2.5 bounded by density 0.125 comprises two regions with combined coverage probability 0.321. The contour bounded by density 0.005 is a single closed region, with coverage 0.98.

Hypothesis tests

2.52 Now consider a hypothesis H, that $\theta \in \Omega_0 \subset \Omega$. Inference consists of accepting or rejecting H. A correct inference will result in zero loss. The loss associated with an incorrect inference may depend on the kind of error. Let d_0 be the inference to accept H and d_1 be the inference to reject H. $\mathscr{D} = \{d_0, d_1\}$. Let $\Omega_1 = \Omega_0^c$. Then we let

$$L(d_i, \theta) = \begin{cases} 0 & \text{if } \theta \in \Omega_i\,, \\ a_i & \text{if } \theta \notin \Omega_i\,, \end{cases} \tag{2.47}$$

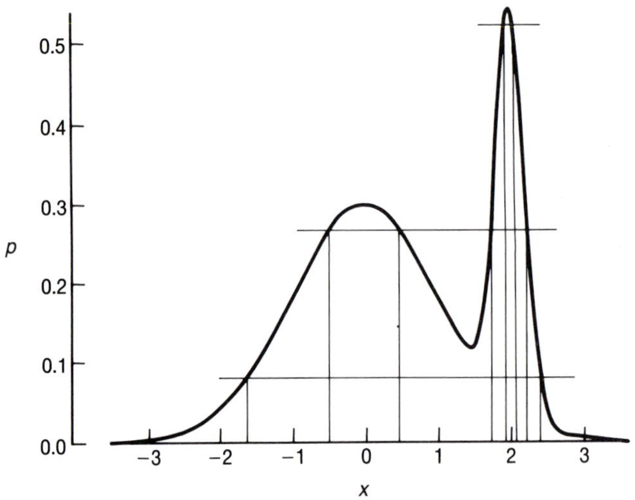

Fig. 2.12 Highest-density sets, Example 2.14

for $i = 0, 1$. Therefore $E(L(d_i, \theta)) = a_i P(\theta \notin \Omega_i)$ and the optimal inference is to reject H if $E(L(d_1, \theta)) < E(L(d_0, \theta))$, i.e. if $a_1 P(\theta \in \Omega_0) < a_0 \{1 - P(\theta \in \Omega_0)\}$, or $P(\theta \in \Omega_0) < a_0/(a_0 + a_1)$. Therefore, we reject H_0 if its probability is sufficiently low. The critical value, $a_0/(a_0 + a_1)$, depends on the relative seriousness of the two kinds of error, as measured by a_0/a_1.

2.53 This formulation of the hypothesis testing problem is very simple because we have expressed the loss function in very stark terms. If we choose d_0 and it turns out that $\theta \in \Omega_1$ then the inference is wrong and incurs a loss of a_0 regardless of where θ lies in Ω_1. In a practical decision problem it may be more appropriate to make a_0 and a_1 functions of θ. Then $E(L(d_i, \theta))$ will not have a simple formula, but may be computed for $i = 0, 1$ and an optimal inference thereby selected. Similarly, in formulating the interval estimation problem through the loss function (2.43) the possibility that if $\theta \notin d$ the loss might depend on how far outside d it was, was ignored. To recognize such a distinction would be possible, although it would complicate the analysis. Considerations such as these demonstrate very clearly how flexible the Bayesian approach is. Even when deliberately attempting to formulate versions of the three standard forms of classical inference, nuances and variations suggest themselves that are very difficult if not impossible to tackle within the classical framework.

Scoring rules

2.54 Rather than a formal hypothesis test, for which there are just two possible inferences, 'accept' or 'reject', the more natural Bayesian inference is simply to state the probability $P(\theta \in \Omega_0)$ that H is true. This may also be formulated as a decision problem. The decision is to assert a number $d \in \mathscr{D} = [0, 1]$, to be interpreted as a probability for H. Consider a loss function

$$L(d,\theta)=\begin{cases}\ell_0(d) & \text{if } \theta\in\Omega_0\,,\\ \ell_1(d) & \text{if } \theta\in\Omega_1\,.\end{cases}$$

Loss functions of this kind are called *scoring rules* and have been studied in the context of asking an expert for his opinion on an uncertain hypothesis. The expert is required to give a probability, but the recipient, who is paying for the information, wishes to ensure that the expert gives a carefully considered judgement. So a penalty arrangement is agreed, wherein the fee is reduced by the amount $\ell_0(d)$ if the stated 'probability' is d and then H is found to be true, but it is reduced by $\ell_1(d)$ if H is found to be false.

This context motivates us setting $\ell_0(1)=0$ and making $\ell_0(d)$ a decreasing function of d, but setting $\ell_1(0)=0$ and making $\ell_1(d)$ an increasing function of d. Since the objective is to encourage a carefully considered probability, the scoring rule should also have the property that the optimal value of d is the expert's actual probabilty $p=P(\theta\in\Omega_0)$. Now

$$E(L(d,\theta))=p\,\ell_0(d)+(1-p)\ell_1(d)\,. \tag{2.48}$$

If we further require that ℓ_0 and ℓ_1 be differentiable, we may differentiate (2.48) to give

$$p\,\ell_0'(d)+(1-p)\ell_1'(d).$$

If the optimal d is the actual probability p, then this will be zero at $d=p$, giving the following condition

$$d\ell_0'(d)+(1-d)\ell_1'(d)=0\,. \tag{2.49}$$

A scoring rule that satisfies (2.49) for all $d\in[0,1]$ is called a proper scoring rule.

Example 2.16

The quadratic scoring rule $\ell_0(d)=(1-d)^2$, $\ell_1(d)=d^2$, is easily seen to satisfy (2.49), and so is proper.

2.55 Instead of asking an expert for the probability of H, we could ask for the probability of its complement, i.e. $\theta\in\Omega_1$. Another desirable property of a scoring rule is that it is the same whether we focus on H or not-H. This means that

$$\ell_0(d)=\ell_1(1-d)\,. \tag{2.50}$$

If this is true for all $d\in[0,1]$ the scoring rule is said to be symmetric.

It is easy to construct scoring rules that are both proper and symmetric. Given any increasing differentiable function $t(d)$ for $d\in[0,\,0.5]$ with $t(0)=0$, define

$$\ell_0(d)=\begin{cases} t(0.5)+\int_d^{0.5}x^{-1}(1-x)\,t'(x)\,dx & \text{if } 0<d\leqslant 0.5\,,\\ t(1-d) & \text{if } 0.5\leqslant d\leqslant 1\,,\end{cases}$$

and define $\ell_1(d)$ by (2.50). This will be seen to be a scoring rule and to satisfy (2.49).

2.56 If the parameter θ can only take two possible values, then to assert the probability that it takes one of those values is to give a complete Bayesian inference statement about θ. A proper scoring rule is a way of formulating this inference as a decision problem. Furthermore, we can generalize the scoring rule concept to give loss functions appropriate to the general inference which is to state the distribution of θ, rather than dealing with the specific classical inferences of point estimation, interval estimation and hypothesis testing.

First suppose that θ is discrete, taking possible values $\Omega = \{\theta_1, \theta_2, \ldots, \theta_k\}$. The inference is to state $\mathbf{d} = (d_1, d_2, \ldots, d_k)$ with loss function $L(\mathbf{d}, \theta_i) = \ell_i(\mathbf{d})$, and $\mathscr{D} = \{\mathbf{d} : \sum_{i=1}^k d_i = 1\}$. The rule is symmetric if the loss $\ell_i(\mathbf{d})$, for inference $\mathbf{d}$ when the true value of θ is θ_i, equals $\ell_j(\mathbf{d}^\star)$ when $d_j^\star = d_i$ and the other elements of $\mathbf{d}^\star$ are any permutation of the remaining elements of $\mathbf{d}$. (We include the case $j = i$.) A symmetric rule has the form

$$L(\mathbf{d}, \theta_i) = m(\mathbf{d}) + \ell(d_i), \tag{2.51}$$

for some symmetric function $m(\mathbf{d})$ and some decreasing function $\ell(d_i)$.

A proper rule is such that the optimal inference is $\mathbf{d} = \mathbf{p} = (p_1, p_2, \ldots, p_k)$, where $p_i = P(\theta = \theta_i)$. Again, there are many proper, symmetric scoring rules. For instance, the quadratic scoring rule with $m(\mathbf{d}) = 1 + \mathbf{d}'\mathbf{d}$ and $\ell(d) = -2d$ is used extensively by De Finetti (1974) in his development of subjective probability theory.

2.57 For $k > 2$, a further constraint may be considered desirable. Under (2.51), the loss when $\theta = \theta_i$ and the inference is $\mathbf{d}$ may depend not just on d_i, the 'probability' asserted for $\theta = \theta_i$, but also on values asserted for other θ_js. We will call a rule *impartial* if $L(\mathbf{d}, \theta_i)$ depends only on d_i. It is now easy to demonstrate that for $k > 2$ there is a unique impartial, symmetric, proper scoring rule.

For, the impartiality and symmetry imply that $L(\mathbf{d}, \theta_i) = \ell(d_i)$, so that $E(L(\mathbf{d}, \theta)) = \sum_{i=1}^k p_i \ell(d_i)$. The optimal decision will minimize $E(L(\mathbf{d}, \theta))$ subject to $\sum_{i=1}^k d_i = 1$. Using the method of Lagrange multiplier define

$$F = \sum_{i=1}^k p_i \ell(d_i) + \lambda \sum_{i=1}^k d_i$$

$$\therefore \frac{\partial F}{\partial d_i} = p_i \ell'(d_i) + \lambda .$$

The rule is proper if this equals zero at $d_i = p_i$, i.e. $d_i \ell'(d_i)$ is the same for each i. When $k = 2$, this reduces to (2.49) and (2.50), so impartiality adds no further constraint. Now suppose that $k = 3$. We now have

$$d_1 \ell'(d_1) = d_2 \ell'(d_2)$$

for all $d_1 + d_2 \leqslant 1$, which implies that $d\,\ell'(d)$ is a constant, and therefore that $\ell(d)$ is a linear function of $\log(d)$. Therefore apart from an arbitrary linear transformation, which simply reflects the arbitrary scale of utilities, the unique impartial, symmetric, proper scoring rule is the logarithmic rule

$$L(\mathbf{d}, \theta_i) = -\log(d_i) . \tag{2.52}$$

2.58 Now if θ is a continuous random variable with density $f(\theta)$, we can consider $\mathscr{D}$ to be the set of all density functions $d(\theta)$ and define

$$L(d, \theta) = -\log d(\theta).$$

The expected loss is

$$E(L(d, \theta)) = -\int_{-\infty}^{\infty} f(\theta) \log d(\theta)\, d\theta, \tag{2.53}$$

and is minimized at $d(\theta) = f(\theta)$. The minimum

$$H(f) = -\int_{-\infty}^{\infty} f(\theta) \log f(\theta)\, d\theta \tag{2.54}$$

is known as the *entropy* of the density $f(\theta)$. The difference between the expected loss and its minimum value is therefore

$$D(f, d) = \int_{-\infty}^{\infty} \log \frac{f(\theta)}{d(\theta)} f(\theta)\, d\theta, \tag{2.55}$$

which is known as the *Kullback–Liebler divergence* between the densities $f(\theta)$ and $d(\theta)$.

See Winkler (1969) and Savage (1975) for further development of scoring rules. Bernardo (1979b) presents the relationship between entropy, the logarithmic scoring rule and an information measure (see **3.48**).

Approximation

2.59 These results find application in several areas of Bayesian statistics. One immediate use in inference is as follows. The process of summarization is one of describing the information contained in $f(\theta)$ in simple terms. One possible simplification is to say that $f(\theta)$ is approximately equal to some standard distribution, such as a normal distribution. If we restrict $\mathscr{D}$ to be the set of all distributions in some standard family, then the best approximating distribution in that family could be defined as the one that minimizes (2.53), or equivalently the Kullback–Liebler divergence (2.55).

Example 2.17

Let $\mathscr{D}$ be the set of normal distributions, so that $d(\theta)$ is the $N(m, v)$ density. Then

$$\begin{aligned} E(L(d, \theta)) &= -\log(2\pi v)^{1/2} - \frac{1}{2v}\int_{-\infty}^{\infty} (\theta - m)^2 f(\theta)\, d\theta \\ &= -\log(2\pi v)^{1/2} - \{\mathrm{var}(\theta) + (E(\theta) - m)^2\}/(2v). \end{aligned}$$

This is obviously minimized for m by $m = E(\theta)$. Differentiating with respect to v, the optimal v satisfies

$$-(1/2v) + \mathrm{var}(\theta)/(2v^2) = 0 \qquad \therefore v = \mathrm{var}(\theta).$$

The best approximating normal distribution is therefore the one whose mean and variance equal the mean and variance of $f(\theta)$.

Overview of inference

2.60 Summarization is the most natural mode of inference in a Bayesian analysis. It consists of examining the posterior distribution and presenting a range of expressions of the information contained therein. This chapter has, however, treated summarization more thoroughly than seems to be usual, and certainly proposes more extensive summarization procedures than are commonly carried out in practical Bayesian analysis. It is nevertheless this author's firm view that careful summarization should not be neglected, particularly the summarization of shape, since an understanding of shape is crucial to interpreting other summaries. Shape is also important in determining appropriate numerical procedures for computing other summaries; see Chapter 8. Many examples of good summarization are to be seen in Box and Tiao (1973), although their work long predates usage of the word 'summarization'.

Statistical decision theory has two uses. It is potentially useful in practice, if a clear decision problem can be identified with a natural utility or loss function, but such applications seem to be rare. Perhaps its greater value to Bayesian statistics is as a theoretical tool, giving insight into other Bayesian processes. It is used in this way in discussing scoring rules in this chapter, and also in approaches to experimental design (Chapter 3). It is also in this spirit that we presented analogues of classical inference procedures in **2.45** to **2.52**, because this can be informative in choosing useful summaries. It is important to recognize that Bayesian inference is by no means limited to those few, stylised, classical forms. On the contrary, summarization allows Bayesian analysis to answer *any* inference question of interest.

EXERCISES

2.1 Let θ have the F distribution with density

$$f(\theta) = \frac{m^{m/2} n^{n/2}}{B(m/2,\, n/2)} \theta^{m/2-1} (n + m\theta)^{-(n+m)/2}.$$

Prove that there is a unique mode at $\theta_1 = \{n(m-2)\}/\{m(n+2)\}$ and points of inflexion at a distance $n\{(m-2)(n+m)\}^{1/2}/\{m(n+2)\}$ on either side of θ_1. Show that the modal dispersion summary is $\{-f(\theta_1)\}/f''(\theta_1) = 2n^2(m-2)(n+m)/\{m^2(n+2)^3\}$.

2.2 Consider the bivariate density

$$f(\theta,\, \phi) = 1.1488\, \theta^{-1.7} \exp\{-0.5(\phi - 4)^2 - (0.02 + 0.5\phi^2)/\theta\}$$

of Example 2.11. Verify that the turning points of this distribution are at (0.0125, 0.0493), (0.411, 1.145) and (2.295, 2.786). Show that although the first mode is six times as high as the second mode, the mode size summary (2.16) suggests that the second mode accounts for almost all of the probability.

2.3 Let the $(p + 1)$-vector $\boldsymbol{\theta} = (\boldsymbol{\mu},\, \phi)$ have the normal-inverse-gamma density

$$f(\boldsymbol{\mu},\, \phi) = k\, \phi^{-(b+p+2)/2} \exp[-\{a + (\boldsymbol{\mu} - \mathbf{m})'\mathbf{V}^{-1}(\boldsymbol{\mu} - \mathbf{m})\}/(2\phi)]$$

(see **9.4**), where

$$k = 2^{-(b+p)/2} a^{b/2} \pi^{-p/2} |\mathbf{V}|^{-1/2} \Gamma(b/2)^{-1},$$

a and b are positive constants, $\mathbf{m}$ is a p-vector and $\mathbf{V}$ is a $p \times p$ positive definite symmetric matrix. This is characterized by $\boldsymbol{\mu}$ having a multivariate normal distribution $N(\mathbf{m},\, \phi\mathbf{V})$ conditional on ϕ, and $a\phi^{-1}$ having the χ^2_b distribution. Show that if p is large relative to b then the ϕ component of the mode of $f(\boldsymbol{\mu},\, \phi)$ can give a misleading summary of location of ϕ.

2.4 Let $\boldsymbol{\theta}$ have density $f(\boldsymbol{\theta})$ and let $\boldsymbol{\psi} = \mathbf{A}\boldsymbol{\theta} + \mathbf{b}$, where $\mathbf{A}$ is a $p \times p$ nonsingular matrix and $\mathbf{b}$ is a p-vector. Show (a) that if $f(\boldsymbol{\theta})$ has a turning point at $\boldsymbol{\theta} = \boldsymbol{\theta}_0$ then the density of $\boldsymbol{\psi}$ has a turning point at $\boldsymbol{\psi} = \boldsymbol{\psi}_0 = \mathbf{A}\boldsymbol{\theta}_0 + \mathbf{b}$, (b) that if $\boldsymbol{\theta}_0$ is a mode then $\boldsymbol{\psi}_0$ is a mode of the density of $\boldsymbol{\psi}$, (c) that if $\mathbf{D}$ is the modal dispersion matrix of $\boldsymbol{\theta}$ at $\boldsymbol{\theta}_0$ then the modal dispersion matrix of $\boldsymbol{\psi}$ at $\boldsymbol{\psi}_0$ is $\mathbf{ADA}'$, and (d) that the mode size summary for $\boldsymbol{\theta}$ at $\boldsymbol{\theta}_0$ is the same as the mode size summary for $\boldsymbol{\psi}$ at $\boldsymbol{\psi}_0$.

2.5 Prove that the multivariate normal distribution $N(\mathbf{m},\, \mathbf{V})$ has a single inflexion boundary given by the ellipsoid $(\boldsymbol{\theta} - \mathbf{m})'\mathbf{V}^{-1}(\boldsymbol{\theta} - \mathbf{m}) = 1$.

2.6 An individual faces the risk of losing £m if an event E occurs. His wealth is less than that of an insurance company. Both the individual and the insurance company have the same utility function $u(w)$ for wealth w, where $u(w)$ has the same shape as in 2.11, characterized by $u'(w) > 0$, $u''(w) < 0$ and $u'''(w) > 0$ for all w. Both the individual and the insurance company assign the same probability to the occurrence of E. Prove that there exists a premium £k such that (a) the individual is prepared to pay £k to have the risk of losing £m if E occurs removed, and (b) the company is prepared to accept that risk in return for £k.

2.7 In a statistical decision problem with loss function $L(d,\, \theta)$, consider the alternative loss function $L^\star(d,\, \theta) = a\, L(d,\, \theta) + g(\theta)$, where a is a positive constant and $g(\theta)$ is any bounded function of θ. Prove that $L(d,\, \theta)$ and $L^\star(d,\, \theta)$ lead to the same optimal decision.

2.8 If $L(d,\, \theta) = (d - \theta)^2$, show that no information can have value greater than $\text{var}(\theta)$.

2.9 In an estimation problem with $\theta > 0$ and loss function $L(d,\, \theta) = (d - \theta)^2/\theta$, prove that the optimal estimate is $E(\theta^{-1})^{-1}$. Given instead $L(d,\, \theta) = (d - \theta)^2/d$, prove that the optimal estimate is $E(\theta^2)^{1/2}$.

2.10 $\boldsymbol{\theta}$ has the p-dimensional multivariate normal distribution $N(\mathbf{m}, \boldsymbol{\theta})$. Show that the highest-density set with probability content α is the ellipsoid

$$(\boldsymbol{\theta} - \mathbf{m})'\mathbf{V}^{-1}(\boldsymbol{\theta} - \mathbf{m}) \leqslant t,$$

where t is the $100\alpha\%$ point of the χ_p^2 distribution.

2.11 In **2.55** we motivated the use of scoring rules in the context of eliciting expert opinion. Explain why the logarithmic scoring rule cannot be used in practice in this context.

2.12 A joint density $f(\theta, \phi)$ is to be approximated by a density $d(\theta, \phi) = d_1(\theta)d_2(\phi)$, in which θ and ϕ are independent. Show that the choice of $d(\theta, \phi)$ that minimizes the Kullback–Liebler divergence is given by equating $d_1(\theta)$ and $d_2(\phi)$ to the marginal densities $\int f(\theta, \phi)\, d\phi$ and $\int f(\theta, \phi)\, d\theta$.

2.13 Generalize Example 2.17 to show that the best multivariate normal approximation to an arbitrary $f(\boldsymbol{\theta})$, in the Kullback–Liebler sense, is given by setting the mean vector and variance matrix of the normal distribution to those of $f(\boldsymbol{\theta})$.

CHAPTER 3

GENERAL PRINCIPLES AND THEORY

3.1 This chapter deals with ideas and theory that are generally applicable to Bayesian inference. After emphasizing the paramount role of likelihood, we turn to considerations that arise when we imagine splitting the data into two parts, $x = (x_1, x_2)$, or splitting the parameters into $\theta = (\theta_1, \theta_2)$. These include the sequential processing of data, sufficiency, ancillarity, nuisance parameters, marginal likelihoods, nonidentifiability, indirect information and constrained parameters.

After a discussion of zero probabilities and Cromwell's dictum, we move to some asymptotic results. The first of these links with earlier ideas by letting $x = (x_1, x_2, \ldots, x_n)$ and then examining the posterior distribution as n tends to infinity. Asymptotic analysis of the prior distribution leads to consideration of improper priors. Other asymptotic questions arise when the data conflict with the prior information or with each other, as with outlying observations.

Classical inference is based on the behaviour of inference rules regarded as functions of the random data, for fixed θ. In Bayesian analysis generally, the data are regarded as fixed and θ is random, as in the posterior density $f(\theta \mid x)$. Classical sampling properties of Bayesian inferences are therefore only of interest for purposes of comparison with classical procedures. Nevertheless, Bayesians regard x as random in the context of design. In deciding what data to observe, the Bayesian performs what is sometimes called *preposterior* analysis. This chapter continues with a brief discussion of sampling properties, then discusses design in general and the role of randomization in Bayesian design. Measures of the information in an experiment are also presented.

The chapter concludes with some theory of predictive inference, in which the objective is to make statements about future observations, rather than about parameters.

The Likelihood Principle

3.2 Writing Bayes' theorem as

$$f(\theta \mid x) = \frac{f(\theta) f(x \mid \theta)}{\int f(\theta) f(x \mid \theta) \, d\theta}, \tag{3.1}$$

it seems almost simplistic to remark that the data x only affect the posterior distribution through the likelihood $f(x \mid \theta)$. Yet this marks a profound difference between Bayesian and classical inference. Notice that $f(x \mid \theta)$ appears in both numerator and denominator of (3.1). Therefore if we modify $f(x \mid \theta)$ by multiplying it by an arbitrary constant, then that constant will cancel and leave the same posterior distribution. Even more generally, we can multiply $f(x \mid \theta)$ by an arbitrary function of x, and still obtain the same posterior distribution.

Consider two experiments, one yielding data x and the other yielding data y. If the two likelihoods $f(x \mid \theta)$ and $f(y \mid \theta)$ are identical up to multiplication by arbitrary functions of x or y, then they contain identical information about θ and lead to identical posterior distributions. The experiments might be very different in other respects, but those differences are irrelevant for inference about θ. This is the Likelihood Principle.

The treatment of the Likelihood Principle here is adequate to show its relevance to Bayesian statistics. However, the Likelihood Principle also represents a key difference between Bayesian and classical theory. As will be shown in **5.14**, classical inference does not follow the Likelihood Principle. Berger and Wolpert (1988) provide a much more careful and deep analysis than is given here. They develop a compelling argument for adopting the Likelihood Principle, and therefore for rejecting classical inference.

3.3 To illustrate this idea, let x be binomially distributed with parameters n and θ. That is, x is the number of successes in n Bernoulli trials with probability θ of success in each trial. If $x = r$ successes are observed, the likelihood is

$$f(x = r \mid \theta) = \binom{n}{r}\theta^r(1-\theta)^{n-r}. \tag{3.2}$$

Alternatively, let y be negative binomially distributed with parameters r and θ, i.e. y is the number of Bernoulli trials required, with probability θ of success in each, until the rth success is observed. The likelihood at $y = n$ is

$$f(y = n \mid \theta) = \binom{n-1}{r-1}\theta^r(1-\theta)^{n-r}. \tag{3.3}$$

Apart from the terms $\binom{n}{r}$ and $\binom{n-1}{r-1}$, which are functions of $x = r$ and $y = n$ but not of θ, the two likelihoods (3.2) and (3.3) are identical, and so will give the same inference. If, for instance, the prior distribution is beta, as in Example 1.4, the same beta posterior distribution will result from either experiment.

Some aspects of this example require further clarification. First, note that in general $f(x \mid \theta)$ and $f(y \mid \theta)$ are not identical for all x and y. We have selected $x = r$ and $y = n$ to make them the same. Two different experiments will generally provide different information. Yet if the actual results of those experiments are such that the likelihoods are proportional then, for these data, identical posterior distributions will result. When we are at the stage of deciding which experiment to perform, we do not know what data will result, and then the two experiments of taking a fixed number n of trials, or of continuing until we see a fixed number r of successes, are indeed different. Ideas of experimental design are introduced in **3.40**.

At this stage it may not seem very interesting if two experiments happen by mere chance to produce equivalent likelihoods, but the choice of $x = r$ and $y = n$ in (3.2) and (3.3) is not just an accident. For in both experiments we have then made the *same* observation, r successes out of n trials. The conclusion is that then it does not matter whether n was fixed and the number of successes random, or r was fixed and the number of trials random. Broadly speaking, the Likelihood Principle implies that it matters only what was observed, not what might have been observed but was not (like $r+1$ successes in n trials, or r successes in $n+1$ trials).

3.4 Consider an experiment yielding a sequence of data $\mathbf{x} = (x_1, x_2, \ldots, x_n)$. Then the obvious form for the likelihood is

$$f(\mathbf{x} \mid \theta) = f(x_1 \mid \theta) f(x_2 \mid x_1, \theta) \ldots f(x_n \mid x_1, x_2, \ldots, x_{n-1}, \theta). \tag{3.4}$$

However, the distinction between binomial and negative binomial experiments prompts the recognition that we observe more than $\mathbf{x}$. We also observe the decisions at each stage whether to continue sampling. Let E_i be the event that we take observation i, and its complement E_i^C is the event that we stop the experiment after observation $i-1$. Then we also observe $\mathbf{E} = (E_1, \ldots, E_n, E_{n+1}^C)$, and the full likelihood is

$$f(\mathbf{x}, \mathbf{E} \mid \theta) = P(E_1) f(x_1 \mid E_1, \theta) P(E_2 \mid x_1, E_1, \theta) \times \\ f(x_2 \mid x_1, E_1, E_2, \theta) \ldots f(x_n \mid x_1, x_2, \ldots, x_{n-1}, E_1, E_2, \ldots, E_n, \theta) \times \\ P(E_{n+1}^C \mid \mathbf{x}, E_1, E_2, \ldots, E_n, \theta) . \tag{3.5}$$

Now the fact that x_1 is observed entails E_1, so $f(x_1 \mid E_1, \theta) = f(x_1 \mid \theta)$. Similarly, the fact thar E_2 is observed entails E_1. So (3.5) will necessarily reduce to

$$f(\mathbf{x}, \mathbf{E} \mid \theta) = P(E_1) f(x_1 \mid \theta) P(E_2 \mid x_1, \theta) f(x_2 \mid x_1, \theta) \ldots \\ f(x_n \mid x_1, x_2, \ldots, x_{n-1}, \theta) P(E_{n+1}^C \mid \mathbf{x}, \theta) = f(\mathbf{x} \mid \theta) g(\mathbf{x}, \theta) , \tag{3.6}$$

where

$$g(\mathbf{x}, \theta) = P(E_1 \mid \theta) P(E_2 \mid x_1, \theta) P(E_3 \mid x_1, x_2, \theta) \ldots P(E_{n+1}^C \mid \mathbf{x}, \theta) . \tag{3.7}$$

Now the two likelihoods (3.4) and (3.6) will be equivalent according to the Likelihood Principle if $g(\mathbf{x}, \theta)$ is a function of $\mathbf{x}$ only. This will be the case if each term in (3.7) depends only on $\mathbf{x}$ and not on θ. That is, the probability of deciding to continue the experiment at each stage may depend on the observations available at that stage but not on θ. This is the case in most experiments.

Returning to the binomial and negative binomial experiments, consider first the binomial experiment. Then, whatever the data, we will decide to take the first, second, ..., nth observations and then stop. So $P(E_1) = P(E_2 \mid x_1, \theta) = \ldots = P(E_{n+1}^C \mid \mathbf{x}, \theta) = 1$ for all $\mathbf{x}$ and θ, so that $g(\mathbf{x}, \theta) = 1$. In the negative binomial experiment the decisions do depend on the data. E_k will occur if and only if $\sum_{i=1}^{k-1} x_i < r$. The decision still does not depend on θ, and for any $\mathbf{x}$ with $x_n = 1$ and $\sum_{i=1}^{n} x_i = r$ the probabilities in $g(\mathbf{x}, \theta)$ will again all equal 1.

In most planned experimentation, the decisions to continue experimenting are not obviously random, so that $g(\mathbf{x}, \theta) = 1$. However, the experiment might easily be stopped by some random, external event, such as an urgent telephone call. Indeed, if such possibilities are included it is hard to think of any real situation where randomness is totally absent. Even if it is planned to take exactly n observations (as in binomial sampling), there is always the chance of some unforeseen external catastrophe stopping it earlier. Fortunately, it is not necessary to worry about such things except in the rare circumstances that the stopping decisions in some way depend on θ.

An experiment in which $g(\mathbf{x}, \theta)$ does not depend on θ is said to be subject to noninformative stopping, in which case the simple likelihood (3.4) can be used. Otherwise, in the case of informative stopping, the full likelihood (3.6) is required. We shall always assume noninformative stopping in the remainder of this volume.

Example 3.1
In both the binomial and negative binomial experiments, the simple likelihood (3.4) is $\theta^r(1-\theta)^{n-r}$, since there will be r terms in (3.4) where x_i is a success with probability θ, and $n-r$ failures with probability $(1-\theta)$. However, the difference between this and (3.2) and (3.3) is not due to $g(\mathbf{x},\theta)$ which, as we have seen, is unity in both cases. The difference lies in defining the observation not as the sequence of successes and failures, but as the number of successes or the number of trials. (3.2) arises because there are $\binom{n}{r}$ sequences which have r successes in a fixed number n of trials. This distinction will be explored further with ideas of sufficiency in Example 3.5.

Example 3.2
An instrument produces readings $x_1, x_2, \ldots$ that are independently distributed as $N(\theta,1)$ given θ, except that if the instrument is called upon to produce a reading greater than 100 it will break down, with no reading produced. It is planned to take n observations, but the instrument breaks after $r<n$ observations. Then for any $i \leqslant n$, the probability of seeing observation i is

$$P(E_i \mid x_1, x_2, \ldots, x_{i-1}, \theta) = P(x_i \leqslant 100 \mid \theta) = \Phi(100-\theta),$$

where $\Phi(z)$ is the standard normal distribution function. If observation i is seen, it has the distribution of $N(\theta,1)$ conditioned on the fact that $x_i \leqslant 100$, i.e.

$$f(x_i \mid x_1, x_2, \ldots, x_{i-1}, \theta) = (2\pi)^{-1/2} \exp\{-(x_i-\theta)^2/2\}/\Phi(100-\theta)$$

(for $x_i \leqslant 100$, and zero for $x_i > 100$). The full likelihood (3.6) then becomes

$$f(\mathbf{x}, \mathbf{E} \mid \theta) = (2\pi)^{-r/2} \exp\{-\sum_{i=1}^{r}(x_i-\theta)^2/2\}\{1-\Phi(100-\theta)\},$$

whereas ignoring the possibility of instrument failure and using (3.4) would mean omitting the last term of this expression. In this example, if the instrument does not fail we can ignore the failure possibility, but if it does fail we have informative stopping.

Sequential use of Bayes' theorem
3.5 If the data can be divided into two parts in some way, $x=(x_1,x_2)$, then we can apply Bayes' theorem in two stages. The prior density $f(\theta)$ may first be updated to a posterior density $f(\theta \mid x_1)$ using data x_1. Thus

$$f(\theta \mid x_1) = f(\theta) f(x_1 \mid \theta)/f(x_1), \tag{3.8}$$

where $f(x_1 \mid \theta)$ is the likelihood for θ on data x_1 and $f(x_1)$ is the marginal density of x_1, given as usual by

$$f(x_1) = \int f(x_1, \theta)\,\mathrm{d}\theta = \int f(\theta) f(x_1 \mid \theta)\,\mathrm{d}\theta. \tag{3.9}$$

Next, we can update this to the final posterior density $f(\theta \mid x_1, x_2)$ using data x_2. Since x_1 has now been observed, the *prior* density for this application is $f(\theta \mid x_1)$. Remember that the terms 'prior' and 'posterior' must always be interpreted relative to some data.

$f(\theta \mid x_1)$ is the density of θ posterior to incorporating data x_1 but prior to x_2. Similarly, the likelihood in this application of Bayes' theorem is $f(x_2 \mid \theta, x_1)$. Then

$$f(\theta \mid x_1, x_2) = f(\theta \mid x_1) f(x_2 \mid \theta, x_1) / f(x_2 \mid x_1), \tag{3.10}$$

where

$$f(x_2 \mid x_1) = \int f(x_2, \theta \mid x_1)\, \mathrm{d}\theta = \int f(\theta \mid x_1) f(x_2 \mid \theta, x_1)\, \mathrm{d}\theta. \tag{3.11}$$

It is now simple to confirm that this two stage process is equivalent to updating directly from $f(\theta)$ to $f(\theta \mid x_1, x_2)$ by a single application of Bayes' theorem with the full data $x = (x_1, x_2)$. For (3.8) and (3.10) give

$$\begin{aligned} f(\theta \mid x_1, x_2) &= \{f(\theta) f(x_1 \mid \theta) f(x_2 \mid \theta, x_1)\} / \{f(x_1) f(x_2 \mid x_1)\} \\ &= f(\theta) f(x_1, x_2 \mid \theta) / f(x_1, x_2), \end{aligned} \tag{3.12}$$

and further confirmation of the consistency of these simple probability manipulations is provided by (3.9) and (3.11):

$$\begin{aligned} f(x_1, x_2) = f(x_1) f(x_2 \mid x_1) &= f(x_1) \int f(\theta \mid x_1) f(x_2 \mid \theta, x_1)\, \mathrm{d}\theta \\ &= \int f(\theta) f(x_1 \mid \theta) f(x_2 \mid \theta, x_1)\, \mathrm{d}\theta = \int f(\theta) f(x_1, x_2 \mid \theta)\, \mathrm{d}\theta, \end{aligned}$$

the usual formula for the denominator of Bayes' theorem, in the application (3.12).

3.6 Naturally, the data may be divided further, and the posterior density reached by a series of applications of Bayes' theorem. This may be appropriate when the data arrive sequentially in time. Then $f(\theta \mid x_1, x_2, \ldots, x_n)$ expresses all that is known about θ after observing data $x_1, x_2, \ldots, x_n$. When observation x_{n+1} becomes available, Bayes' theorem produces $f(\theta \mid x_1, x_2 \ldots, x_{n+1})$, expressing all that is now known about θ.

It is important to remember that at each stage the likelihood is conditional on all the data incorporated so far, as in the use of $f(x_2 \mid \theta, x_1)$ in (3.10). For data arriving sequentially in time, in particular, that branch of Statistics known as time series analysis is particularly concerned with describing the dependence of x_{n+1} on all the earlier data $x_1, x_2, \ldots, x_n$, in terms of unknown parameters θ. In other problems, there is no special ordering of the data, and in many models x_1 and x_2 will be independent given θ, so that $f(x_2 \mid \theta, x_1)$ reduces to $f(x_2 \mid \theta)$ in (3.10).

Example 3.3

Let x_1 and x_2 be independent given θ with distributions $N(\theta, v)$, and let the prior distribution for θ be $N(m, w)$. Then by Example 1.5 the posterior distribution after observing x_1 is $N(m_1, w_1)$, where $m_1 = (wx_1 + vm)/(w + v)$ and $w_1 = vw/(w + v)$. Since x_1 and x_2 are independent given θ, $f(x_2 \mid \theta, x_1) = f(x_2 \mid \theta)$, the $N(\theta, v)$ density, and we can apply the result of Example 1.5 again. The posterior density after observing $x = (x_1, x_2)$

is $N(m_2, w_2)$, where

$$m_2 = \frac{w_1 x_2 + v m_1}{w_1 + v} = \frac{v w x_2 + v(w x_1 + v m)}{v w + v(w + v)} = \frac{w(x_1 + x_2) + v m}{2w + v}$$

and

$$w_2 = \frac{v w_1}{w_1 + v} = \frac{v w}{2w + v},$$

Alternatively, we could have obtained the same result by a single application of Bayes' theorem with likelihood

$$f(x|\theta) = f(x_1|\theta) f(x_2|\theta) = (2\pi v)^{-1} \exp[-\{(x_1 - \theta)^2 + (x_2 - \theta)^2\}/(2v)].$$

Sufficiency

3.7 In general, data x are completely uninformative about θ if x is independent of θ. For then it immediately follows that $f(\theta \mid x) = f(\theta)$ for all x, and the posterior distribution after observing x is the same as the prior distribution. Whatever values the data might take do not change any beliefs or inferences about θ. Equivalently, θ and x are independent if and only if $f(x \mid \theta) = f(x)$ for all θ, so that the likelihood for θ on data x is independent of θ. Every value of θ has the same likelihood, and the data completely fail to discriminate between one value of θ and another.

3.8 The phenomenon of sufficiency, defined also in A**17.31**, arises when part of the data is uninformative. Divide the data by letting $x = (x_1, x_2)$ as in **3.5**, and let $f(x_2 \mid \theta, x_1) = f(x_2 \mid x_1)$. That is, given x_1, let the data x_2 be uninformative about θ. Then the second updating in **3.5** will not change the distribution of θ and $f(\theta \mid x_1, x_2) = f(\theta \mid x_1)$. So as long as x_1 is known, the value of x_2 is irrelevant to inference; we need not even observe its value. It is sufficient to observe x_1.

The division of x into x_1 and x_2 need not be in the sense that x comprises a number of distinct observations, of which some make up x_1 and the remainder x_2. x_1 and x_2 are simply two aspects of the data such that knowledge of both implies knowledge of x. We can regard x_1 as any function of x, and then x_2 is whatever else is needed to complete knowledge of x. x_2 might be x itself, for x_1 is automatically removed when we condition on it in $f(x_2 \mid \theta, x_1)$. Regarding x_1 as an arbitrary function of the data, it is often called a *sufficient statistic*.

Example 3.4

In Example 3.3 let $y = x_1 + x_2$ and we will show that y is sufficient. We can rewrite $x = (x_1, x_2)$ as $x = (y, z)$ where $z = x_1 - x_2$, since when y and z are both known we will know x. In order to prove that y is sufficient we must find $f(z \mid \theta, y)$. Now given θ, x_1 and x_2 are independent $N(\theta, w)$, and it is now straightforward to show that the distributions of y and z given θ are $N(2\theta, 2w)$ and $N(0, 2w)$ respectively. Furthermore, y and z are independent given θ. (See A**15.4**.) Therefore the distribution of z given θ and y is also $N(0, 2w)$, which does not involve θ. Therefore z is independent of θ given y and so y is a sufficient statistic.

3.9 In general, as Example 3.4 shows, it is not simple to demonstrate from the definition of **3.8** that a statistic is sufficient. Yet in Example 3.3, to which Example 3.4 refers, the posterior inference only depends on $y = x_1 + x_2$ and so it seems obvious that y is sufficient. We now develop a simpler test of sufficiency, based on this idea. Another derivation is given in A**17.32**.

The following statements are all equivalent, and so could each be used to define sufficiency of x_1.

(a) $f(x_2 \mid x_1, \theta)$ does not depend on θ.
(b) $f(\theta \mid x)$ does not depend on x_2.
(c) The likelihood of the full data x may be expressed as the product of two terms, one involving x_1 and θ but not x_2, and the other involving x_1 and x_2 but not θ.

The equivalence of (a) and (b) has been discussed in **3.8**. Formally, (a) is equivalent to $f(x_2|x_1,\theta) = f(x_2|x_1)$, which is equivalent to saying that x_2 and θ are independent given x_1, which is equivalent to $f(\theta \mid x) = f(\theta \mid x_1)$, which is equivalent to (b).

Now (a) clearly implies (c), since

$$f(x \mid \theta) = f(x_1 \mid \theta) f(x_2 \mid x_1, \theta) = f(x_1 \mid \theta) f(x_2 \mid x_1),$$

and the first term involves only x_1 and θ, whereas the second involves only x_1 and x_2. Since (b) implies (a), either (a) or (b) implies (c). Now (c) implies that $f(x \mid \theta)$ is equivalent under the Likelihood Principle to an expression involving only x_1 and θ. The second term will cancel in Bayes' theorem. Hence (b) will follow, which in turn implies (a). So (c) implies (a) and (b) and the proof is complete.

The classical proof of the equivalence of (a) and (c) in A**17.32** is more difficult because classical statistics does not allow θ to be treated as random, and so cannot proceed via (b). Yet (b) is the real inferential substance of sufficiency. (c) is useful only because it makes it easy to identify sufficient statistics.

Example 3.5

We saw in Example 3.3 that the binomial experiment produces likelihood

$$f(\mathbf{x} \mid \theta) = \theta^r (1-\theta)^{n-r}, \tag{3.13}$$

where $r = \sum_{i=1}^n x_i$ and where $x_i = 0$ for a failure and $x_i = 1$ for a success in the ith trial. This is a function only of r and θ. Therefore r is sufficient. Formally,

$$f(\mathbf{x} \mid \theta) = f(r \mid \theta) f(\mathbf{x} \mid r, \theta),$$

where $f(r \mid \theta)$ is the binomial likelihood (3.2) and $f(\mathbf{x} \mid r, \theta) = 1/\binom{n}{r}$ expresses the fact that given $\sum x_i = r$ all $\binom{n}{r}$ possible sequences $\mathbf{x}$ are equi-probable, independently of θ.

In negative binomial sampling we obtained the same likelihood (3.13), but now r is fixed and n is the (random) number of elements in $\mathbf{x}$. Now n is seen to be sufficient. We have $f(\mathbf{x} \mid \theta) = f(n \mid \theta) f(\mathbf{x} \mid n, \theta)$, where $f(n \mid \theta)$ is the negative binomial likelihood (3.3) and $f(\mathbf{x} \mid n, \theta)$ expresses the fact that given n all $\binom{n-1}{r-1}$ possible $\mathbf{x}$ are equi-probable.

Example 3.6
Let $x_1, x_2, \ldots, x_n$ be independently and identically disributed as $N(\theta, w)$ given θ. Then

$$f(x_1, x_2, \ldots, x_n \mid \theta) = \prod_{i=1}^{n} f(x_i \mid \theta) = \prod_{i=1}^{n} (2\pi w)^{-1/2} \exp\{-(x_i - \theta)^2/(2w)\}$$
$$= (2\pi w)^{-n/2} \exp\{-Q/(2w)\}$$

where $Q = \sum_{i=1}^{n}(x_i - \theta)^2 = n(\theta - \bar{x})^2 + \sum_{i=1}^{n}(x_i - \bar{x})^2$. Therefore $f(x_1, x_2, \ldots, x_n \mid \theta) = g_1(\bar{x}, \theta) g_2(x_1, x_2, \ldots, x_n)$, where $g_1(\bar{x}, \theta) = \exp\{-n(\theta - \bar{x})^2/(2w)\}$ and $g_2(x_1, x_2, \ldots, x_n) = (2\pi w)^{-n/2} \exp\{-\sum_{i=1}^{n}(x_1 - \bar{x})^2/(2w)\}$. Thus, $\bar{x}$ is sufficient.

Example 3.7
In Example 1.6, where $x_1, x_2, \ldots, x_n$ are independently and identically distributed as $N(\mu, \sigma^2)$, $(\bar{x}, s^2)$ is seen to be sufficient for $\theta = (\mu, \sigma^2)$.

3.10 Examples 3.4 and 3.6 demonstrate that a sufficient statistic is not unique. If x_1 is sufficient then $x_1' = g(x_1)$ is also sufficient if $g(x_1)$ is any one-to-one function of x_1. This is clear from either the definition of sufficiency in **3.8** or its characterization in **3.9**. Thus, in Example 3.4 we found that $y = x_1 + x_2$ is sufficient, but then so also is $\bar{x} = y/2$, agreeing with Example 3.6.

3.11 Further details about sufficiency with examples and exercises may be found in Chapter A17. Although sufficiency is an important concept in classical inference, in the Bayesian approach it is not strictly necessary to recognize where a sufficient statistic exists. Application of Bayes' theorem will automatically take account of sufficiency, so that if $x = (x_1, x_2)$ and x_1 is sufficient then posterior inference will automatically depend on x only through x_1. In classical inference if it were not recognized that x_1 is sufficient, then estimators might be proposed that depend on x_2, without realizing that they must be inefficient.

Ancillarity

3.12 As in **3.8** divide the data by $x = (x_1, x_2)$ but now let x_1 be uninformative. That is, let $f(x_1 \mid \theta) = f(x_1)$. Then $f(\theta \mid x_1) = f(\theta)$, and

$$f(\theta \mid x_1, x_2) = \frac{f(\theta) f(x_2 \mid x_1, \theta)}{\int f(\theta) f(x_2 \mid x_1, \theta) \, d\theta}.$$

The posterior distribution is still in general a function of both x_1 and x_2, so it is not immediately obvious how the fact that x_1 is uninformative is reflected in the posterior information about θ. In general, knowledge of x_1 affects the distribution of x_2 given θ, and so influences how x_2 is informative about θ. Information about θ derives entirely from x_2 through $f(x_2 \mid x_1, \theta)$, but x_1 can still give information in this indirect way.

Any function of the data that is independent of θ is said to be *ancillary*, or an ancillary statistic. Inference about θ will then derive only from the conditional distribution of the data given the ancillary statistic.

Example 3.8
A bag contains an unknown number θ of red balls. A die is thrown with result x_1 ($x_1 = 1, 2, \ldots, 6$), then x_1 white balls are added to the bag. The bag is shaken and a ball taken out; $x_2 = 1$ if the ball is red, otherwise $x_2 = 0$. Clearly, x_1 is ancillary and the posterior distribution proceeds from $P(x_2 = 1 \mid x_1, \theta) = \theta/(\theta + x_1)$. Inference will still make use of the known number x_1 of white balls in the bag, but will ignore the (ancillary) random mechanism by which x_1 was obtained. We would obtain the same inferences if x_1 had been fixed from the start.

Nuisance parameters
3.13 Consider $\theta = (\theta_1, \theta_2)$ and suppose that inference is required about θ_1 but not about θ_2. Then θ_2 is called a *nuisance parameter*. Inference always proceeds from the posterior distribution, so if we require inference only about θ_1 given data x we simply use its posterior distribution

$$f(\theta_1 \mid x) = \int f(\theta_1, \theta_2 \mid x)\, \mathrm{d}\theta_2 . \tag{3.14}$$

This is the obvious way to derive $f(\theta_1 \mid x)$: first derive the posterior distribution of the full parameter $\theta = (\theta_1, \theta_2)$, then integrate out θ_2 to obtain the marginal posterior distribution of θ_1. We have implicitly assumed this approach when discussing inference in Chapter 2.

In formal, decision-theoretic terms if a decision $d \in \mathscr{D}$ is to be made that is to be thought of as an inference about θ_1, then the loss function should not depend on θ_2. Thus, $L(d, \theta) = g(d, \theta_1)$, a function of d and θ_1 only. Then

$$E(L(d, \theta)) = \int\int g(d, \theta_1) f(\theta_1, \theta_2 \mid x)\, d\theta_1\, \mathrm{d}\theta_2 = \int g(d, \theta_1) f(\theta_1 \mid x) d\theta_1 .$$

Therefore, decision-making will also use only the marginal posterior distribution $f(\theta_1 \mid x)$.

3.14 If only $f(\theta_1 \mid x)$ is required then, at least in a formal sense, it can be obtained without introducing θ_2 at all. We need only use Bayes' theorem in the form

$$f(\theta_1 \mid x) = \frac{f(\theta_1) f(x \mid \theta_1)}{\int f(\theta_1) f(x \mid \theta_1)\, \mathrm{d}\theta_1} \tag{3.15}$$

However, the likelihood $f(x \mid \theta_1)$ is not readily specifiable without reference to θ_2. The essence of nuisance parameters is that, although their values are not of interest, they enter naturally in the formulation of a model for the data.

Consider the following example. An estimate is required of the resistivity μ of a new alloy. Measurements $x_1, x_2, \ldots, x_n$ are obtained. How should we specify $f(x_1, x_2, \ldots, x_n \mid \mu)$? In general it would not be realistic to assume that the x_is are independent given μ. For if we observe x_1 it is plausible that the distance of x_1 from μ (whose value is given in the conditional distribution) will give information about how far x_2 may be expected to be from μ. Unless we know the characteristics of the resistivity measurement process so well that only μ is unknown, then observing any x_is will help us learn about those characteristics, and so will provide information about the other x_is. In particular, if we are uncertain about the measurement error variance then observing x_1 will indeed (given

μ) provide information about the likely magnitude of $x_2 - \mu$. Therefore we introduce this variance as a second parameter σ^2. It is now more plausible that the x_is are independent given μ and σ_2, for instance if the shape of the error distribution may be assumed to be normal. In general, the specification of the likelihood is made via the model, in which parameters are introduced to represent all unknown aspects of the data-generating process.

Even though a subjective interpretation of probability allows distributions like $f(x \mid \theta_1)$ to be assessed without reference to θ_2, it is in practice much better to do so via the original model and likelihood $f(x \mid \theta_1, \theta_2)$.

Specifically,

$$f(x \mid \theta_1) = \int f(x, \theta_2 \mid \theta_1)\, d\theta_2 = \int f(x \mid \theta_1, \theta_2) f(\theta_2 \mid \theta_1)\, d\theta_2 . \tag{3.16}$$

acts as the likelihood in (3.15). It is obtained by taking the expectation of the likelihood $f(x \mid \theta)$ with respect to the conditional prior distribution $f(\theta_2 \mid \theta_1)$. Therefore, implementation of the Bayesian approach to nuisance parameters requires us to specify prior information about the nuisance parameters as well as about the parameters of interest, and to be able to do an integration, either (3.14) or (3.16).

Nonidentifiability

3.15 With $\theta = (\theta_1, \theta_2)$ suppose that the likelihood $f(x \mid \theta)$ does not depend on θ_1. That is, $f(x \mid \theta) = f(x \mid \theta_2)$, so that x and θ_1 are independent given θ_2. This is equivalent to $f(\theta_1 \mid x, \theta_2) = f(\theta_1 \mid \theta_2)$. The fact that the likelihood does not depend on θ_1 suggests that x provides no information about θ_1, but we see that the proper implication is that x provides no information about θ_1 *given* θ_2. For inference about θ_1, the marginal posterior

$$f(\theta_1 \mid x) = \frac{f(\theta_1) \int f(\theta_2 \mid \theta_1) f(x \mid \theta_2)\, d\theta_2}{\int f(\theta) f(x \mid \theta_2)\, d\theta}$$

does depend on x, but the information provided about θ_1 comes indirectly through θ_2. In doing so, it relies on prior information in the form of $f(\theta_2 \mid \theta_1)$. Because x provides no direct information about θ_1, θ_1 is said to be *nonidentifiable* from the data. The terminology will be better explained in **3.21** below.

3.16 We have seen both in ancillarity and nonidentifiability the phenomenon of indirect information. The passage of information through one or more indirect links can be nicely expressed with an *influence diagram*. See **6.41**.

Parameter constraints

3.17 Often there is prior information that constrains a parameter to a narrower range than the model alone would require. For instance if θ is the mean of a normal distribution we may know from the context of the problem that θ must be positive. Then we should use a prior distribution which is confined to $(0, \infty)$, unlike the normal prior distribution assumed in Example 1.5. A gamma prior distribution would be acceptable, for instance, but there is another approach that makes an interesting use of the decomposition of the data into $x = (x_1, x_2)$.

Let x_1 be the information that $\theta \in T$, representing a general constraint on θ. We can represent this formally by a likelihood

$$f(x_1 \mid \theta) = \begin{cases} 1 & \text{if} \quad \theta \in T \\ 0 & \text{if} \quad \theta \notin T \end{cases},$$

for then suppose that before learning of this constraint we had a prior density $f(\theta)$. After learning x_1 the posterior density is

$$f(\theta \mid x_1) = \begin{cases} f(\theta) / \int_T f(\theta)\, \mathrm{d}\theta & \text{if} \quad \theta \in T \\ 0 & \text{if} \quad \theta \notin T \end{cases},$$

which is simply the prior distribution conditioned on $\theta \in T$, or *truncated to* T. This is the real prior distribution, and any prior distribution on T may be thought of as an unconstrained prior distribution that has been truncated to T by 'observing' x_1.

Now let x_2 be the data, with likelihood $f(x_2 \mid \theta)$. Strictly we should use $f(x_2 \mid \theta, x_1)$, but we can assume that the constraint information x_1 is independent of the data x_2. Then we can obtain the posterior distribution $f(\theta \mid x_1, x_2)$ using the real (truncated) prior distribution $f(\theta \mid x_1)$ and the data likelihood $f(x_2 \mid \theta)$. This will naturally also be constrained to T. However, we can obtain the same result by introducing the information in the reverse order, taking x_2 first. We then obtain first the unconstrained posterior distribution

$$f(\theta \mid x_2) = \frac{f(\theta) f(x_2 \mid \theta)}{\int f(\theta) f(x_2 \mid \theta)\, \mathrm{d}\theta}.$$

Subsequently learning x_1 truncates this distribution to T, producing

$$f(\theta \mid x_1, x_2) = \begin{cases} f(\theta) f(x_2 \mid \theta) / \int_T f(\theta) f(x_2 \mid \theta)\, \mathrm{d}\theta & \text{if} \quad \theta \in T \\ 0 & \text{if} \quad \theta \notin T \end{cases}.$$

So a parameter constraint may be implemented simply by truncating the posterior distribution. It is necessary to remember, however, that the implied prior distribution is the result of truncating the stated prior.

Example 3.9

As in Example 1.5, let x_2 be distributed as $N(\theta, v)$ and let the prior distribution for θ be $N(m, w)$ except that we know that $\theta > 0$. The truncated prior distribution is therefore

$$f(\theta) = (2\pi w)^{-1/2} \exp\{-(\theta - m)^2/(2w)\} / \Phi(m/\sqrt{w})$$

for $\theta > 0$, and $f(\theta) = 0$ if $\theta \leqslant 0$, where $\Phi(t)$ is the standard normal distribution function. Using Example 1.5 we see immediately that the posterior distribution is $N(m_1, w_1)$ truncated to $(0, \infty)$, with m_1 and w_1 given in Example 1.5. Of course, m_1 and w_1 are not the mean and variance of this distribution. The mean is

$$\begin{aligned} E(\theta \mid x) &= \int_0^\infty \theta (2\pi w_1)^{-1/2} \exp\{-(\theta - m_1)^2/(2w_1)\} / \Phi(m_1/\sqrt{w_1})\, \mathrm{d}\theta \\ &= m_1 + w_1 (2\pi w_1)^{-1/2} \exp\{-m_1^2/(2w_1)\} / \Phi(m_1/\sqrt{w_1}). \end{aligned} \tag{3.17}$$

Similarly, the true prior mean is not m but (3.17) with the subscripts 1 removed. O'Hagan and Leonard (1976) present an alternative prior distribution for this example which allows for uncertainty about the parameter constraint.

Asymptotic normality

3.18 If θ is discrete then the event that $\theta = \theta_i$ can have zero posterior probability in one of two ways. Since $P(\theta = \theta_i \mid x) = P(\theta = \theta_i) f(x \mid \theta = \theta_i)/f(x)$, this will occur if either $f(x \mid \theta = \theta_i) = 0$ or $P(\theta = \theta_i) = 0$. In the first case, the data rule out the possibility that $\theta = \theta_i$, in the sense that if θ were really equal to θ_i then we could not have observed x as we did. The second case says that if we begin by believing that $\theta = \theta_i$ is impossible, then no data whatsoever can persuade us that $\theta = \theta_i$ has non-zero probability. This is rather a strong result, and means that we should not assign zero probability to any parameter values without being convinced, beyond the power of any evidence to dissuade us, that θ will not take those values. Although presented above for discrete θ, the argument works for continuous θ: if $f(\theta) = 0$ in some neighbourhood of $\theta = \theta_0$ then we are assigning zero probability to θ lying in that neighbourhood, and will continue to give that event zero probability after any data.

3.19 The admonition not to give zero probability to any event that is possible, however highly improbable, is sometimes called Cromwell's dictum. The attribution apparently is to a letter by Oliver Cromwell to the General Assembly of the Church of Scotland in 1650, in which he wrote 'I beseech you, in the bowels of Christ, think it possible that you may be mistaken'.

3.20 Let the true value of θ be θ_0. Provided we do not have zero prior probability that θ equals (or lies in a neighbourhood of) θ_0, if we acquire enough data then the posterior probability that θ equals (or lies in a neighbourhood of) θ_0 will tend to unity. This is a general property of Bayesian inference that is akin to the notion of consistency in classical inference (A**17.7**). Bayesian inference is inherently 'consistent'.

3.21 We first prove this result for the case of independent and identically distributed observations $x_1, x_2, x_3, \ldots$. The development is similar to the proof of consistency of maximum likelihood estimators in A**18.10**. Let the common distribution of the x_is be $g(x \mid \theta)$. After n observations, the posterior density is

$$\begin{aligned} f(\theta \mid x_1, x_2, \ldots, x_n) &\propto f(\theta) \prod_{i=1}^{n} g(x_i \mid \theta) = f(\theta) \exp\left\{ \sum_{i=1}^{n} \log g(x_i \mid \theta) \right\} \\ &= f(\theta) \exp\{ n L_n(\theta, \mathbf{x}) \}, \end{aligned} \tag{3.18}$$

say. For any fixed θ, $L_n(\theta, \mathbf{x})$ is the average of n independent and identically distributed random variables $\log g(x_i \mid \theta)$. By the Strong Law of Large Numbers (A**7.34**), if the true value of θ is θ_0, $L_n(\theta, \mathbf{x})$ converges in probability as $n \to \infty$ to its expectation

$$\bar{L}(\theta) = \int \{\log g(x \mid \theta)\} g(x \mid \theta_0) \, dx. \tag{3.19}$$

Furthermore, from the discussion of Kullback–Liebler divergence in **2.58**, (3.19) is maximized at $\theta = \theta_0$, the true value. Clearly, for any $\theta \neq \theta_0$, $\exp\{nL_n(\theta_0, \mathbf{x})\} / \exp\{nL_n(\theta, \mathbf{x})\}$ tends to infinity with probability one as $n \to \infty$. Therefore, provided $f(\theta_0) \neq 0$, the posterior f.f. (3.18) becomes infinitely higher at the true value θ_0 than anywhere else. So the posterior probability that θ equals, or lies in a neighbourhood of, θ_0 tends to unity.

One extra condition for this result should be mentioned, that (3.19) should be strictly greater at $\theta = \theta_0$ than at any other θ. This will not be true if there exists another value θ_1 such that for all x, $g(x \mid \theta_0) = g(x \mid \theta_1)$. Then we would observe identically distributed data when θ equals θ_0 or θ_1, so that the data cannot discriminate between these values, and the ratio of their posterior f.f.s (3.18) will always equal the prior ratio. This is the phenomenon of nonidentifiability. For if $\theta = (\phi, \psi)$ and ψ is nonidentifiable in the sense of **3.15**, then $g(x \mid \phi, \psi_0) = g(x \mid \phi, \psi_1)$ for all ψ_1 and x. Then the data cannot distinguish between values of ψ and the posterior distribution of ψ given ϕ will always equal its prior distribution.

3.22 For continuous θ we can go further and obtain the form of the posterior distribution as it approaches this degenerate limit. Assuming that $g(x \mid \theta)$ is continuous in θ, $\bar{L}(\theta)$ will be continuous in θ. Then for sufficiently large n, only values of θ in the immediate neighbourhood of θ_0 will be such that $\exp\{nL_n(\theta, \mathbf{x})\}/\exp\{nL_n(\theta_0, \mathbf{x})\}$ is not negligibly small. The posterior distribution becomes concentrated on a vanishingly small neighbourhood of θ_0. Assuming also continuity of the prior distribution $f(\theta)$ at θ_0, we can treat $f(\theta)$ as constant over this neighbourhood so that in the limit

$$f(\theta \mid x_1, x_2, \ldots, x_n) \propto \exp\{n\bar{L}(\theta)\}. \tag{3.20}$$

3.23 More specifically, if θ is scalar and $g(x \mid \theta)$ is three times differentiable in θ, then we can expand $\bar{L}(\theta)$ around θ_0 in a Taylor series:

$$\bar{L}(\theta) = \bar{L}(\theta_0) - (\theta - \theta_0)^2/(2v) + R, \tag{3.21}$$

where

$$v = -1/\bar{L}''(\theta_0), \tag{3.22}$$

$$\bar{L}''(\theta) = \int (\partial^2 \log g(x \mid \theta)/\partial\theta^2) g(x \mid \theta_0) \mathrm{d}x, \tag{3.23}$$

and R is the remainder term, $6R = (\theta - \theta_0)^3 \bar{L}'''(\theta^\star)$ for some $\theta^\star$ between θ and θ_0. We can absorb $\exp\{n\bar{L}(\theta_0)\}$ in the proportionality constant of (3.20). Then for $|\theta - \theta_0|$ of order $n^{-1/2}$, nR is of order $n^{-1/2}$ and may be ignored. Therefore the limiting form of the posterior distribution is a normal distribution with mean equal to the true value θ_0, and with variance v/n. Notice that the expression for v in (3.22) is the modal dispersion summary of the posterior distribution. Using (3.23) we also see it as the Fisher information I; see **A17.16**.

Generalizing to vector $\boldsymbol{\theta}$, the posterior distribution is asymptotically multivariate normal with mean equal to the true value $\boldsymbol{\theta}_0$ and with variance matrix $n^{-1}\mathbf{V}$, where $\mathbf{V}$ is minus the inverse of the matrix of second derivatives of $\bar{L}(\boldsymbol{\theta})$, evaluated at $\boldsymbol{\theta} = \boldsymbol{\theta}_0$. As before, this result assumes that there is positive prior probability in the neighbourhood of $\boldsymbol{\theta}_0$, and that $\bar{L}(\boldsymbol{\theta}_0)$ is strictly greater than $\bar{L}(\boldsymbol{\theta})$ at any $\boldsymbol{\theta} \neq \boldsymbol{\theta}_0$. In particular, $\mathbf{V}$ must be positive definite. Otherwise, we have non-identifiability and the prior information is not totally swamped by the data.

3.24 Having proved these results in the case of independent and identically distributed data, we now indicate how asymptotic normality can hold much more generally. Given an arbitrary sequence of data $x_1, x_2, x_3, \ldots$ we still have (3.18) if we define

$$L_n(\theta, \mathbf{x}) = \tfrac{1}{n} \sum_{i=1}^{n} \log f(x_i \mid \theta, x_1, x_2, \ldots, x_{i-1}). \tag{3.24}$$

Since the expectation of each component of the sum in (3.24) is maximized at $\theta = \theta_0$, the true value, the expectation of $L_n(\theta, \mathbf{x})$ is maximized at $\theta = \theta_0$. If some variant of the Strong Law of Large Numbers could be proved for this case, we would still have the posterior distribution concentrating asymptotically on θ_0. However, it is easy to see that such a law will not hold in complete generality. If the x_is are becoming less and less informative about θ, then later terms in the sum are degenerating to constants. Under such conditions we could not expect $L_n(\theta, \mathbf{x})$ to converge in probability to its mean.

For scalar θ, instead of expanding $\bar{L}(\theta)$ as in (3.21), consider the expansion of $L_n(\theta, \mathbf{x})$ around its maximum $\hat{\theta}$.

$$L_n(\theta, \mathbf{x}) = L_n(\hat{\theta}, \mathbf{x}) - (\theta - \hat{\theta})^2/(2v_n) + R,$$

where now $v_n = -1/L_n''(\hat{\theta}, \mathbf{x})$,

$$L_n''(\theta, \mathbf{x}) = \tfrac{1}{n} \sum_{i=1}^{n} \partial^2 \log f(x_i \mid \theta, x_1, x_2, \ldots, x_{i-1})/\partial \theta^2 \tag{3.25}$$

and the remainder term R involves third derivatives. Now we will have asymptotic normality if we can show that v_n converges in probability to some non-zero value. This condition is enough to rule out the possibility of the x_is becoming increasingly uninformative, for in that case the second derivatives in (3.25) would tend to zero.

For vector $\boldsymbol{\theta}$, subject to similar conditions, the posterior distribution will tend to normality with mean equal to its mode and variance matrix equal to its modal dispersion matrix.

3.25 Rigorous proofs of Bayesian consistency and asymptotic normality may be found in Johnson (1967, 1970), Walker (1969), Heyde and Johnstone (1979) and Poskitt (1987). Note, however, that the foregoing development applies only for finite-dimensional parameter spaces. Diaconis and Freedman (1986) give examples of inconsistency of Bayesian methods in 'non-parametric' problems. This includes models in which the unknown parameter is a function, as discussed in **10.22** and **10.43** to **10.48**.

3.26 In all these results notice that the prior distribution is essentially irrelevant. Provided that $f(\theta)$ is positive for all possible θ, and the data do not leave any aspect of θ nonidentifiable, then as we acquire more and more data the asymptotic posterior distribution is independent of the prior distribution. In particular, two individuals may have radically different prior opinions but, subject to the general conditions, after seeing sufficient data their posterior beliefs about θ converge. Prior information is eventually overwhelmed by the data. It is, of course, necessary that they agree about how the data are related to θ, as expressed in the likelihood, and that both give non-zero prior probability to (or in a neighbourhood of) θ_0.

Improper prior distributions

3.27 It is clear from the discussion of zero probabilities in **3.18** that the prior distribution can be made strong enough to overwhelm any data. If the prior distribution gives probability one to $\theta = \theta_i$, then the posterior distribution will do the same. It is usually more interesting to look at the opposite extreme, in which the prior distribution represents, in some sense, very weak prior information.

Consider for instance the problem in Example 1.5, where θ has the $N(m, w)$ prior distribution and a single observation x is obtained with distribution $N(\theta, v)$. Then the posterior distribution is $N(m_1, w_1)$ where $m_1 = (mv + xw)/(w + v)$ and $w_1 = wv/(w + v)$. The prior variance w represents the strength of the prior information, since the range of θ values with moderate prior density is $m \pm 2\sqrt{w}$. If w is large, a large range of values are plausible *a priori*, representing relatively weak prior information. Conversely, if w is small, prior information is strong, confining θ to a relatively narrow range. If v/w is large, prior information is much stronger than the data information. Then we have $m_1 \approx m$, $w_1 \approx w$, and beliefs about θ are almost unaffected by the data. In particular, if $w \to 0$ then both prior and posterior distribution assert with certainty that $\theta = m$, regardless of the data.

If $v/w \approx 0$, prior information is much weaker than the data. Then $m_1 \approx x$, $w_1 \approx v$. In particular, as $w \to \infty$ the posterior distribution tends to $N(x, v)$. Now we do not have a degenerate result, but one that is dominated by the data information. The mean x is the observation, and the variance v is the observation variance. However, the status of $N(x, v)$ as a genuine posterior distribution is open to debate. We have obtained it by a limiting process on the $N(m, w)$ prior distribution, but in the limit $N(m, \infty)$ is a non-existent distribution. As $w \to \infty$, $f(\theta) = (2\pi w)^{-1/2} \exp\{-(\theta - m)^2/(2w)\}$ tends to zero for all x, yet $f(\theta) = 0$ does not integrate to unity. In fact, the distribution $N(x, v)$ cannot be obtained as the posterior distribution for any genuine prior distribution, although we can get arbitrarily close to it with a normal prior having large variance. Using Bayes' theorem with $f(\theta) = 0$ for all θ, would result in $f(\theta \mid x) = 0$ for all θ, which is nonsense, and different from the claimed $N(x, v)$ distribution.

3.28 The problem is made more perplexing by the following argument. In computing $f(\theta \mid x)$ by

$$f(\theta \mid x) \propto f(\theta) f(x \mid \theta) \tag{3.26}$$

we need not use $f(\theta)$ and $f(x \mid \theta)$ exactly, but could replace them by any functions $f^\star(\theta)$ and $f^\star(x \mid \theta)$ to which they themselves are proportional. This was done with the likelihood in **3.2**, where we noted that two experiments yielding proportional likelihoods produce identical posterior distributions. In general, when applying Bayes' theorem it is convenient to use (3.26) and to drop any constants (i.e. factors not involving θ) from both $f(\theta)$ and $f(x \mid \theta)$. In the present example that gives

$$\begin{aligned} f(\theta \mid x) &\propto \exp\{-(\theta - m)^2/(2w)\} \exp\{-(x - \theta)^2/(2v)\} \\ &\propto \exp\{-(\theta - m_1)^2/(2w_1)\} \end{aligned} \tag{3.27}$$

and hence the posterior distribution is $N(m_1, w_1)$, inserting the necessary normalizing constant $(2\pi w_1)^{-1/2}$. Now if $w \to \infty$, the first term in (3.27) tends to 1 for all θ. We simply have $f(\theta \mid x) \propto \exp\{-(x-\theta)^2/(2v)\}$ and the posterior distribution is $N(x, v)$.

Here we appear to have obtained $N(x, v)$ by a direct application of Bayes' theorem; nonsense did not ensue from letting $w \to \infty$. However, we did so by asserting

$$f(\theta) \propto f^{\star}(\theta) = 1. \tag{3.28}$$

Bayes' theorem proceeds smoothly, this time oblivious of the fact that (3.28) is not strictly possible. There is no proper distribution $f(\theta)$ satisfying (3.28).

3.29 In general, the specification $f(\theta) \propto f^{\star}(\theta)$ is said to represent an *improper distribution* if $\int f^{\star}(\theta) d\theta = \infty$. For then there is no proportionality constant that will allow $f(\theta)$ to be a proper distribution, i.e. to integrate to one.

It is tempting to think that an improper prior distribution like (3.28) can be useful as a general expression of prior ignorance. Ignorance about the value of θ suggests regarding all possible values as equally likely, and this seems to be captured by (3.28). Yet it is also clear that an improper prior distribution cannot truly represent genuine prior information. In Chapter 4 we shall see that in discussing the different varieties of Bayesian philosophy one consideration is whether it is regarded as legitimate to use improper prior distributions, and if so under what conditions. We content ourselves in this chapter with displaying some of the pitfalls that arise when improper distributions are introduced. If their use is to be considered legitimate, these problems must be addressed. The discussion is generally confined for simplicity to scalar parameters, although improper distributions in higher dimensions are associated with still more difficulties; see for instance Dawid et al. (1973).

3.30 If θ is discrete, and takes only a finite number of possible values $\theta_1, \theta_2, \ldots, \theta_k$, there is no problem with (3.28). It simply implies the proper distribution $f(\theta_i) = P(\theta = \theta_i) = k^{-1}$ for $i = 1, 2, \ldots, k$. Even then, we cannot regard this as a recipe for describing ignorance. For, suppose that $k > 2$ and let $\phi = 1$ if $\theta = \theta_1$ and $\phi = 2$ if $\theta \neq \theta_1$. If I am ignorant of θ, then I am also ignorant of ϕ, and the recipe then gives two conflicting specifications, $P(\phi = 1) = 1/2$ and $P(\theta = \theta_1) = k^{-1}$.

In general, ignorance about θ surely implies ignorance about $\phi = g(\theta)$. We have seen that for discrete θ the recipe (3.28) breaks down if $g(\theta)$ is not a one-to-one transformation. For continuous θ it also breaks down for one-to-one nonlinear transformations. Suppose that $\theta \in [0, 1]$, so that (3.28) represents the proper prior distribution $f(\theta) = 1$, the uniform distribution on [0,1]. Now let $\phi = \theta^2$. The assertion that $f(\phi) = 1$ implies $f(\theta) = 2\theta$, θ and ϕ cannot both have uniform distributions on [0,1].

3.31 In the normal example, (3.28) is the limit of a series of prior distributions $N(m, w)$ as $w \to \infty$. The corresponding series of posterior distributions $N(m_1, w_1)$ tends to $N(x, v)$. Furthermore, if we formally apply Bayes' theorem with the limiting prior distribution (3.28) we obtain the limiting posterior distribution $N(x, v)$. However, limits need not be consistent in this sense.

In general, consider a series of prior disributions $f_1(\theta), f_2(\theta), \ldots$ and a corresponding series of posterior distributions $f_1(\theta \mid x), f_2(\theta \mid x) \ldots$ derived from any fixed likelihood

$f(x \mid \theta)$. If as $n \to \infty$, $f_n(\theta) \to f(\theta)$ and the posterior distribution corresponding to $f(\theta)$ is $f(\theta \mid x)$, then we do have $f_n(\theta \mid x) \to f(\theta \mid x)$ if $f(\theta)$ is a proper distribution. In that case we can also write $f_n(\theta) \propto f_n^\star(\theta)$ and suppose that $f_n^\star(\theta) \to f^\star(\theta)$. Then the posterior distribution derived from using $f_n^\star(\theta)$ as prior distribution is still $f(\theta \mid x)$, and using $f^\star(\theta)$ still gives the posterior $f(\theta \mid x)$ since $f^\star(\theta) \propto f(\theta)$. But if $f^\star(\theta)$ is improper the limit of the series of prior distributions $\{f_n(\theta)\}$, if it exists, is not a distribution. Then if $f(\theta \mid x)$ is the posterior distribution corresponding to $f^\star(\theta)$ we may find that $f_n(\theta \mid x) \not\to f(\theta \mid x)$.

Example 3.10

Continuing the problem of an observation x from the $N(\theta, v)$ distribution, let $f_n(\theta) \propto f_n^\star(\theta)$ where

$$f_n^\star(\theta) = \begin{cases} 0 & \text{if} \quad \theta < -n, \\ 1 & \text{if} \quad -n \leqslant \theta \leqslant n, \\ \exp(n^2) & \text{if} \quad n < \theta \leqslant n+1, \\ 0 & \text{if} \quad n+1 < \theta. \end{cases}$$

Then the limit of $f_n^\star(\theta)$ is the improper uniform distribution (3.28), but for any n, $f_n(\theta) \propto f_n^\star(\theta)$ is proper. However, the corresponding posterior distribution $f_n(\theta \mid x)$ asserts that the posterior probability that $-n < \theta \leqslant n$ tends to zero as $n \to \infty$. $f_n(\theta \mid x)$ does not tend to $N(x, v)$, which was obtained by applying Bayes' theorem to the improper limiting prior distribution.

3.32 The posterior distribution derived from an improper prior distribution may itself be improper.

Example 3.11

Let x be the number of successes in n independent trials with probability p of success in each trial, but suppose that both n and p are unknown. Thus $\theta = (n, p)$. Let n and p have independent uniform prior distributions. That is, $f(n, p) = f(n)f(p)$ with $f(p) = 1$ for $0 \leqslant p \leqslant 1$, the proper uniform distribution on [0,1], and $f(n) \propto 1$, the improper uniform distribution on the positive integers. Then

$$f(n, p \mid x) \propto \tbinom{n}{x} p^x (1-p)^{n-x} \tag{3.29}$$

for $n \geqslant x$. Integrating with respect to p, the marginal posterior distribution of n is

$$f(n \mid x) \propto \tbinom{n}{x} B(x+1, n-x+1) = (n+1)^{-1}$$

after expanding the binomial coefficient and beta function (and using the fact that x and n are integers). Now $\sum_{n=x}^{\infty} (n+1)^{-1} = \infty$, so this is an improper distribution.

The impropriety is not confined to the marginal disribution of n. Just as we could integrate (3.29) with respect to p, implying that the conditional distributions of p given n are proper, we can sum with respect to n. The conditional distributions of n given p are proper negative binomial distributions and we find $f(p \mid x) \propto p^{-1}$. This is also an improper distribution, its integral diverging at $p = 0$.

Although this example has been presented for a single observation x, the posterior impropriety persists even if a series of observations is available.

3.33 Let x have the distribution $N(\theta, v)$ as before. Now suppose that there is a prior probability p that $\theta = 0$. If $\theta \neq 0$ there is little prior feeling for how close to zero θ might be. We could try to represent this weak prior information by saying that if $\theta \neq 0$, θ has the $N(0, w)$ distribution, and let $w \to \infty$. The prior distribution is mixed, having a discrete probability at $\theta = 0$ and a probability density elsewhere, and we must therefore apply Bayes' theorem carefully. Consider the posterior probability that $\theta = 0$.

$$P(\theta = 0 \mid x) = P(\theta = 0) f(x \mid \theta = 0) / f(x),$$

where

$$f(x) = P(\theta = 0) f(x \mid \theta = 0) + P(\theta \neq 0) f(x \mid \theta \neq 0)$$

and

$$\begin{aligned} f(x \mid \theta \neq 0) &= \int f(\theta \mid \theta \neq 0) f(x \mid \theta) d\theta \\ &= \int (2\pi v)^{-1/2} (2\pi w)^{-1/2} \exp[-\{\theta^2/(2w)\} - \{(x-\theta)^2/(2v)\}] \, d\theta \\ &= \{2\pi(v+w)\}^{-1/2} \exp[-x^2/\{2(v+w)\}], \end{aligned}$$

which tends to zero as $w \to \infty$. Therefore $P(\theta = 0 \mid x) \to 1$ as $w \to \infty$, regardless of the data x or the prior probability p (provided $p \neq 0$). This example was apparently first given by Bartlett (1957). Similar problems arise whenever the prior distribution becomes improper over just one part of the parameter space, or if different improper distributions are defined on different parts of the space. For in the first case $f(\theta)$ is effectively zero where the improper distribution lies, and we cannot make $f(\theta) \propto 1$ because it is not zero elsewhere. In the second case $f(\theta) \to 0$ at different, arbitrary rates for different θ. Again we cannot just write $f(\theta) \propto 1$.

3.34 The paradox only arises at the limit when $w = \infty$. Otherwise $P(\theta = 0 \mid x) = (1 + y)^{-1}$, where

$$y = p^{-1}(1-p)(1 + w/v)^{-1/2} \exp[wx^2/\{2v(v+w)\}].$$

For fixed p, v and w, $P(\theta = 0 \mid x)$ is a decreasing function of $|x|$, which is entirely natural. Nor does $|x|$ have to be absurdly large to make $P(\theta = 0 \mid x)$ small, even for very large w. For instance, setting $p = 0.5$, $v = 1$ and $w = 10^6$, $P(\theta = 0 \mid x) = 0.001$ at $x = 5.25$. It is the attempt to avoid specifying how far θ might be from zero, by assuming the improper uniform prior distribution, that causes the 'paradox'.

There are a range of problems in which related phenomena arise, and these are discussed further in Chapter 7, beginning at **7.43**.

Conflicting information

3.35 Consider once again the simple normal example. x is distributed as $N(\theta, v)$ and θ as $N(m, w)$. The posterior distribution is $N(m_1, w_1)$ where $m_1 = (mv + xw)/(v + w)$ and $w_1 = vw/(v + w)$. The posterior mean is a compromise between the two sources of information – the prior information asserts that θ is near m, and the data that θ is near x.

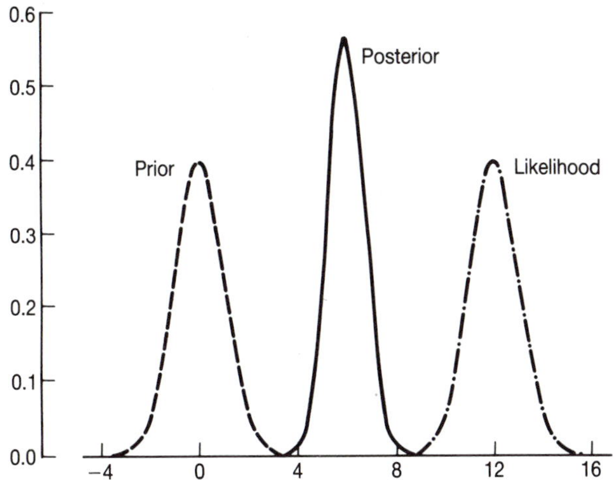

Fig. 3.1 Prior, likelihood and posterior in conflict

In fact, if $|m - x|$ is large, this compromise may be undesirable or unrealistic. For in that case the two information sources contradict each other. The prior information says that it is highly implausible that θ is further from m than a small multiple of $w^{1/2}$, yet the data says that it is also highly implausible that θ is further from x than a small multiple of $v^{1/2}$. If $|x - m|$ is large, one or both of the 'highly implausible' events has occurred, or one or both sources of information has been misspecified. The compromise posterior estimate m_1 effectively asserts that θ is so far from m and x as to be 'highly implausible' on both counts. Indeed, the entire posterior distribution conflicts with both prior distribution and data. Figure 3.1 shows this situation quite clearly, where prior, posterior and likelihood all support entirely different θ values.

Of course, if prior and likelihood are correct, then $N(m_1, w_1)$ is the right posterior distribution, but this is critically dependent on the shape of the normal distribution's tails. If we change either prior or likelihood to a distribution such as a t distribution with high degrees of freedom, which is very like the normal distribution except in the extreme tails, then the asymptotic behaviour of the posterior distribution as $|x - m| \to \infty$ is entirely different. Figure 3.2 shows such a case. See also Exercise 3.4.

3.36 For arbitrary prior $f(\theta)$ and likelihood of the location parameter form $f(x \mid \theta) = g(x - \theta)$, the asymptotic behaviour of the posterior distribution as $x \to \infty$ is studied by Dawid (1973), Hill (1974), O'Hagan (1979, 1981). In general terms, the information represented by the distribution with the thicker tails is ignored in the limit, and the posterior distribution converges to the distribution with the thinner tails. In Figure 3.2, for instance, the prior distribution is $N(0, 4)$ and has thinner tails than the t likelihood $f(x \mid \theta) \propto \{20 + (x - \theta)^2\}^{-11.5}$. We find that the posterior mean and variance both converge to their prior values: the data are ignored. Conversely, if we had let the prior distribution be t and the likelihood be $N(\theta, v)$ then the posterior distribution would converge to $N(x, v)$.

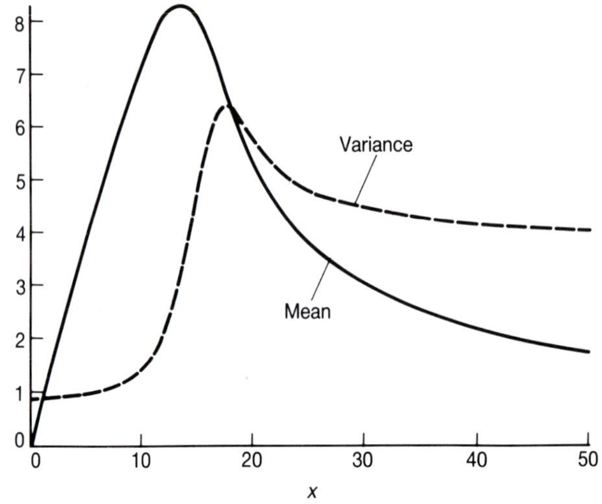

Fig. 3.2 Posterior mean and variance, normal prior and *t* likelihood

Remembering that this is the posterior distribution obtained from the improper uniform prior distribution, in this case it is the prior information that is ultimately rejected.

Only if both distributions have the same tail thickness is the conflict between prior and data not resolved by rejecting one information source. If both have thin tails like the normal distribution it is resolved by compromise. Otherwise, the posterior distribution becomes bimodal and continues to be divided between both sources of information as $x \to \infty$.

3.37 This example illustrates another general property of Bayesian inference, that tail shapes (of prior and likelihood) can strongly influence the posterior distribution when conflicts arise. We return to this topic in **7.22**.

Sampling properties

3.38 Classical inference is concerned with properties of inference rules under repeated sampling. An inference rule is a function, say t, which associates an inference $t(x)$ with every possible value x of the data. A classical inference rule must be propounded before the data are observed, at which point $t(x)$ is a random variable whose sampling distribution for any given θ may be deduced from that of the data, $f(x \mid \theta)$. The rule is said to be good if the properties of its sampling distribution $f(t(x) \mid \theta)$ are deemed to be desirable *for all* θ. Bayesian inference is concerned with features of the posterior distribution $f(\theta \mid x)$. There is no requirement that Bayesian inferences be determined by a rule of inference chosen before the data are observed, but in most problems we can define a rule of inference implicitly, so that for each x $t(x)$ is the Bayesian inference obtained from $f(\theta \mid x)$. A Bayesian inference rule is not constructed to have the repeated sampling properties which govern classical inference. Nor are those properties of interest in Bayesian inference, yet there are some contexts in which they have been studied by classical statisticians. First, they are a source of comparison between classical and Bayesian methods. Someone committed to the classical approach may take comfort from the fact that Bayesian inferences do not in general possess all the properties that classical theory regards as desirable. Of course, such a comparison is made on classical statistics' home ground, and does not question whether those properties are more or less desirable than those which Bayesians regard as important.

A different classical reaction is to use Bayesian methods as a source of ideas for classical inference. Having derived a Bayesian inference rule $t(x)$, its Bayesian origin is forgotten. It is then considered purely as a classical inference rule, and may be modified to achieve the derived properties. An example of this is the classical use of shrinkage estimators, discussed in **6.42**.

3.39 The bias of an estimator is an example of a sampling property which is important in classical inference. Consider $t(x)$ as a point estimator of θ. Its bias is then $b(\theta) = E(t(x) \mid \theta) - \theta$, and it is said to be unbiased if $b(\theta) = 0$ for all θ. A Bayesian estimator, derived either as a location summary of the posterior distribution or as a solution to a decision problem, will not typically be unbiased, and its bias will depend on the prior distribution. However, it is easy to show that the posterior mean has zero *expected* bias. For

$$E(b(\theta)) = E(t(x)) - E(\theta) = E\{t(x) - E(\theta \mid x)\},$$

which is zero identically if $t(x)$ is the posterior mean. Therefore, $E(\theta \mid x)$ is not biased consistently in one direction. Taking the expectation with respect to the prior distribution, the positive and negative biases cancel out.

A similar result can be shown for highest density intervals (or other posterior probability intervals); see Exercise 3.5.

Choice of experiments

3.40 Sampling properties arise much more naturally in Bayesian inference when we consider a choice between experiments. If, to aid inference about θ, we can choose between two or more experiments to perform, then this is clearly a decision problem. It is easy to formulate within the outline of decision theory in **2.26** to **2.43**. The current status is $S \mapsto \{S(c) : c \in \mathscr{C}\}$, where $c \in \mathscr{C}$ is the decision to carry out an experiment that will yield data x_c with likelihood $f(x_c \mid \theta)$. $S(c)$ is the status after taking this decision. Then after performing the experiment and observing x_c, the status is $S(c, x_c)$ and inference is then to be made about θ. This must now be cast as a statistical decision problem: $S(c, x_c) \mapsto \{S(c, x_c, d) : d \in \mathscr{D}\}$ where $\mathscr{D}$ is the set of possible inferences. Finally, let $S(c, x_c, d, \theta)$ be the status after choosing experiment c, observing x_c, making inference d and learning the true value of θ. Let $U(S(c, x_c, d, \theta)) = u(c, x_c, d, \theta)$ be the utility of being in this status. In Chapter 2 we expressed the utilities in a statistical decision problem in terms of a loss function $L(d, \theta)$, but must now allow utility to depend also on c and x_c. Typically, one experiment will cost more than another to perform, so it is clear that utility may depend on c. The fact that it may depend on x_c also is clear when we consider performing a negative binomial sampling experiment. Then the number of trials is random, depending on the sequence of observations, and if the cost of performing each trial is non zero the utility will depend on x_c.

With the general utility $u(c, x_c, d, \theta)$,

$$U(S(c, x_c, d)) = E\{u(c, x_c, d, \theta) \mid x_c\}, \tag{3.30}$$

the expectation being with respect to the posterior distribution $f(\theta \mid x_c)$. The optimal inference chooses d to maximize (3.30), giving

$$U(S(c, x_c)) = \sup_d E\{u(c, x_c, d, \theta) \mid x_c\}.$$

$$\therefore U(S(c)) = E[\sup_d E\{u(c, x_c, d, \theta) \mid x_c\}], \tag{3.31}$$

the outer expectation now being with respect to the distribution $f(x_c)$. The optimal choice of experiment is to maximize (3.31).

The distribution of $f(x_c)$ required here is its marginal, or *preposterior* distribution, given by $f(x_c) = \int f(x_c \mid \theta) f(\theta)\, d\theta$. Taking expectations with respect to the distribution of the data is a feature of classical inference, and violates the Likelihood Principle, whereas Bayesian inference conditions on the observed data. Design, however, takes place before the data are observed, as opposed to inference which occurs after observing the data. So it is both natural and necessary to perform these 'preposterior' calculations for the purposes of design.

3.41 It may be possible to separate $u(c, x_c, d, \theta)$ into additive components representing a cost of implementing experiment c, a cost of observing x_c and a loss for making inference d when the true parameter is θ:

$$u(c, x_c, d, \theta) = -L_1(c) - L_2(x_c) - L(d, \theta), \tag{3.32}$$

whereupon (3.30) and (3.31) become

$$U(S(c, x_c, d)) = -L_1(c) - L_2(x_c) - E\{L(d, \theta) \mid x_c\}, \tag{3.33}$$

$$U(S(c)) = -L_1(c) - E\{L_2(x_c)\} - E[\inf_d E\{L(d, \theta) \mid x_c\}]. \tag{3.34}$$

Then the inference is optimized as usual by minimizing posterior expected loss, the last term of (3.33). The utility of choosing experiment c is (3.34), and has three components. The first is the cost of implementing c and the second is the expected cost of observing the associated data x_c. Only the third term measures the value of the experiment in aiding inference about θ. Indeed, we can write (3.34) formally in terms of the value of the information x_c to the decision $\mathscr{D}$, as defined in **2.38**. In the definition (2.31), we take $u_0(X, d) = -E\{L(d, \theta) \mid x_c\}$, i.e. that part of (3.33) that relates to the inference problem. Then the value of information x_c to the inference problem $\mathscr{D}$ under the status $S(c)$ is

$$\begin{aligned} V(x_c, \mathscr{D}, S(c)) &= \inf_d E[E\{L(d, \theta) \mid x_c\}] - E[\inf_d E\{L(d, \theta) \mid x_c\}] \\ &= \inf_d E\{L(d, \theta)\} - E[\inf_d E\{L(d, \theta) \mid x_c\}]. \end{aligned} \tag{3.35}$$

The first term is the optimal expected loss if inference is required before performing any experiment. The expectation here is with respect to the prior distribution $f(\theta)$. The second term in (3.35) is the expected value of the optimal posterior expected loss, i.e. the preposterior expected loss.

Combining (3.34) and (3.35) we can write

$$U(S(c)) + A = V(x_c, \mathscr{D}, S(c)) - L_1(c) - E\{L_2(x_c)\}, \tag{3.36}$$

where $A = \inf_d E\{L(d, \theta)\}$. Since A does not depend on the experiment c, the optimal c will maximize (3.36).

The choice of experiment is a compromise between the value of an experiment to the inference or decision problem, and its cost (or expected cost) to perform.

Example 3.12

Independent observations $x_1, x_2, x_3, \ldots$ are available distributed as $N(\theta, v)$. The prior distribution is $N(m, w)$. It is required to estimate θ with quadratic loss $L(d, \theta) = (d - \theta)^2$. The experiment $n \in \{0, 1, 2, \ldots\}$ is to take n observations, so that the associated data are $\mathbf{x}_n = (x_1, x_2, \ldots, x_n)$. The posterior distribution given $\mathbf{x}_n$ is $N(m_n, w_n)$ with $m_n = (mv + n\bar{x}_n w)/(v + nw)$, $w_n = vw/(v + nw)$ and $\bar{x}_n = n^{-1} \sum_{i=1}^{n} x_i$. The optimal inference is the posterior mean m_n, and the posterior expected loss is then

$$\inf_d E\{L(d, \theta) \mid \mathbf{x}_n\} = E\{(m_n - \theta)^2 \mid \mathbf{x}_n\} = w_n.$$

Since this does not depend on $\mathbf{x}_n$, its expectation is also w_n. The optimal expected loss if no observations are taken is $w_0 = w$, and the value of observing x_n is thus $w - w_n = nw^2/(v + nw)$.

If now the cost of taking n observations is nc, the optimal sample size n maximizes $U(S(n)) = nw^2/(v + nw) - nc$. We find by examining $U(S(n+1)) - U(S(n))$ that the optimal n is the nearest integer to $\{(4v + c)/(4c)\}^{1/2} - (v/w)$.

3.42 If the required inference about θ can be formulated as a single, genuine decision problem with clearly defined loss function, and if this loss function can be extended satisfactorily to include the cost of experimentation, then choosing the experiment to maximize (3.31) is the optimal solution. However, experiments are often performed with the vaguer objective of 'learning about θ'. Posterior inference will not be a decision problem but will consist simply of summarizing the posterior distribution. In some sense, a good choice of experiment should help to make clear and informative inference about θ, but there is no single objective way of measuring how good an experiment is.

Measures of information

3.43 Even when inference is not expressed as a decision problem, we might adopt one of the formal loss structures in **2.45** to **2.58** to represent that inference. There is an obvious difficulty in this approach of extending the loss function to include the cost of experimentation. The latter may be well-defined, perhaps as a particular sum of money per observation. In contrast, when inference does not comprise a genuine decision problem the choice of a formal loss function is arbitrary, and in particular its scale is undefined. We cannot relate this abstract loss function to the concrete, financial cost of experiments.

Nevertheless, given any formulation of a loss function $L(d, \theta)$, calculating the value of the experiment that yields observation x_c, can be useful. We can think of $V(x_c, \mathscr{D}, S(c))$, equation (3.35), as a measure of the amount of information in the observation x_c. The choice of an experiment could then be made by informal comparison between the amounts of information yielded by the various experiments and their expected costs. In fact, if all experiments under consideration had the same costs, then the choice would reduce to maximizing information. A range of useful information measures result from adopting different formal loss functions.

3.44 Consider estimation of a scalar θ with quadratic loss $L(d, \theta) = (d - \theta)^2$ (as in Example 3.12). Then the optimal posterior estimate is the posterior mean, and the expected loss for this d is the posterior variance. We therefore have the *quadratic information* measure

$$I_Q(x) = \text{var}(\theta) - E\{\text{var}(\theta \mid x)\}. \tag{3.37}$$

Employing a standard identity, (3.37) may be rewritten as $I_Q(x) = \text{var}\{E(\theta \mid x)\}$. This expression relates nicely to the remark in **2.38**, that observing x only has zero value if knowing it does not change the optimal decision. The more the optimal posterior estimate $E(\theta \mid x)$ varies with x the more informative x is, according to this quadratic information measure. It also makes it clear that the preposterior expectation of the posterior variance must be less than (or equal to, if the data are so completely uninformative that the posterior mean is independent of x) the prior variance. In some simple models the posterior variance will always be less than the prior variance, as in the simple normal model with normal prior in Example 1.5, but this will not generally be true. There will typically be values of x for which $\text{var}(\theta \mid x) > \text{var}(\theta)$, but nevertheless the *expectation* is of a reduction in variance.

3.45 For vector parameter $\boldsymbol{\theta}$, we can use the more general quadratic loss $L(\mathbf{d}, \boldsymbol{\theta}) = (\mathbf{d} - \boldsymbol{\theta})'\mathbf{A}(\mathbf{d} - \boldsymbol{\theta})$, where $\mathbf{A}$ is a non-negative definite matrix. Then it was shown in **2.48** that the optimal estimate is $\mathbf{d} = E(\boldsymbol{\theta})$, with expected loss

$$E\{(\boldsymbol{\theta} - E(\boldsymbol{\theta}))'\mathbf{A}(\boldsymbol{\theta} - E(\boldsymbol{\theta}))\} = E\{\text{tr}\,\mathbf{A}(\boldsymbol{\theta} - E(\boldsymbol{\theta}))(\boldsymbol{\theta} - E(\boldsymbol{\theta}))'\} = \text{tr}\,\mathbf{A}\,\text{var}(\boldsymbol{\theta}),$$

where $\text{var}(\boldsymbol{\theta})$ is the covariance matrix of $\boldsymbol{\theta}$. Similarly, the optimal posterior estimate is the posterior mean, with expected loss $\text{tr}\,\mathbf{A}\,\text{var}(\boldsymbol{\theta} \mid x)$. We therefore have the general quadratic information measure

$$I_Q(x; \mathbf{A}) = \text{tr}\,A[\text{var}(\boldsymbol{\theta}) - E\{\text{var}(\boldsymbol{\theta} \mid x)\}]. \tag{3.38}$$

The choice of $\mathbf{A}$ should reflect the relative importance of accurately estimating the different elements of $\boldsymbol{\theta}$, or different linear combinations of those elements.

3.46 An alternative for vector $\boldsymbol{\theta}$ is offered by interval estimation with a fixed coverage probability p. In **2.51** it is shown that the optimal estimate is the highest density interval with coverage p, and the optimal expected loss is therefore the volume of this highest density interval. If $\boldsymbol{\theta}$ has a multivariate normal distribution, the volume of a highest density interval is proportional to $|\mathrm{var}(\boldsymbol{\theta})|^{1/2}$. This motivates

$$I_V(x) = |\mathrm{var}(\boldsymbol{\theta})|^{1/2} - E\{|\mathrm{var}(\boldsymbol{\theta} \mid x)|^{1/2}\} \tag{3.39}$$

as another measure of information, the *volume information* measure. An alternative approximation for non-normal distributions might be to replace $\mathrm{var}(\boldsymbol{\theta})$ by the modal dispersion matrix, since **2.16** and **2.17** show that this is asymptotically proportional to the volume of a highest density interval for small p.

3.47 The non-Bayesian theory of optimal experiments employs some closely related ideas. $I_V(x)$ and $I_Q(x; \mathbf{A})$ are respectively analogous to the classical D-optimality and A-optimality criteria; see Fedorov (1972). Bayesian designs chosen using quadratic measures in a variety of problems are given by Brooks (1972, 1976, 1980), Chaloner (1984, 1989), O'Hagan (1978), Owen (1970, 1975) and Pilz (1983). The volume measure, corresponding to classical D-optimality, is applied by Draper and Hunter (1967), Giovagnoli and Verdinelli (1985), Spezzaferri (1988), and Verdinelli (1983). Steinberg and Hunter (1984) and Verdinelli (1992) provide reviews. The classical approach cannot, of course, use $\mathrm{var}(\boldsymbol{\theta} \mid x)$, and instead is based on the sampling variance of an estimator $\hat{\boldsymbol{\theta}}$, i.e. $\mathrm{var}(\hat{\boldsymbol{\theta}} \mid \boldsymbol{\theta})$. In simple normal-theory problems with weak prior information, this may effectively coincide with $\mathrm{var}(\boldsymbol{\theta} \mid x)$.

3.48 In **2.54** we argued for the use of scoring rules as the most appropriate form of loss function when inference consists of summarizing the entire posterior distribution. In particular, the logarithmic scoring rule of **2.58** may be used for any θ, scalar or vector. The optimal expected loss is entropy. The prior entropy is $H(\theta) = -\int f(\theta) \log f(\theta)\, \mathrm{d}\theta$ and posterior entropy is $H(\theta \mid x) = -\int f(\theta \mid x) \log f(\theta \mid x)\, \mathrm{d}\theta$. Then the *entropy information* measure for x is

$$I_E(x) = H(\theta) - E\{H(\theta \mid x)\}, \tag{3.40}$$

first advocated in the Bayesian context by Lindley (1956) and further developed by Stone (1959). It is worth noting, however, that entropy of most standard distributions is not available in a simple closed-form expression.

Example 3.13

In Example 3.12 we found that the quadratic information about θ contained in n independent $N(\theta, v)$ observations, given a $N(m, w)$ prior distribution, was $I_Q(x) = w - w_n = nw^2/(v + nw)$. The entropy of the $N(m, w)$ distribution is also easily found to be $\{1 + \log(2\pi w)\}/2$. Therefore for the same experiment the entropy information is given by $2I_E(x) = \log w - \log w_n = \log(v + nw) - \log v$.

The two measures perform very differently. As $n \to \infty$, $I_Q(x) \to w$, but $I_E(x) \to \infty$. This is because the rate of increase of $I_Q(x)$ is of order n^{-2}, whereas that of $I_E(x)$ is of order n^{-1}. To see how this might lead to very different choices of experiment, consider Example 3.12, where quadratic information was used and a cost of cn was assumed for n

observations. For small c, the optimal n was found to be of order $c^{-1/2}$. Therefore halving c will (asymptotically) result in a 40% increase in optimal sample size. If we use entropy information instead, then halving the cost of an observation will (again asymptotically) double the optimal sample size.

Example 3.14
In this example the cost of each experiment is the same, and so can be ignored. We have a fixed number n, of independent observations available. Each observation may be made either with distribution $N(\theta_1, 1)$ or $N(\theta_2, 1)$. Inference is required about $\boldsymbol{\theta} = (\theta_1, \theta_2)$, and in particular about $\phi = \theta_1 - \theta_2$. Prior distributions for θ_1 and θ_2 are independent, $N(0, w_1)$ and $N(0, w_2)$ respectively. If we take m observations from $N(\theta_1, 1)$ and $n - m$ from $N(\theta_2, 1)$, the posterior distributions of θ_1 and θ_2 will be independent with variances $\text{var}(\theta_1 \mid x) = w_1/(1 + mw_1)$ and $\text{var}(\theta_2 \mid x) = w_2/\{1 + (n - m)w_2\}$.

If we concentrate on inference about ϕ, then both quadratic and entropy measures suggest choosing m to minimize $\text{var}(\phi \mid x) = \text{var}(\theta_1 \mid x) + \text{var}(\theta_2 \mid x)$. Treated as a continuous function of m, $\text{var}(\phi \mid x)$ is minimized at $m_0 = (n/2) + (w_1 - w_2)/(2w_1 w_2)$. The optimal m is therefore the next integer above or below m_0. In particular, if n is even and $w_1 = w_2$ the optimal m is $n/2$.

If, instead of concentrating on ϕ, we consider inference about $\boldsymbol{\theta}$, both the volume and entropy measures lead to minimizing $|\text{var}(\boldsymbol{\theta} \mid x)| = \text{var}(\theta_1 \mid x)\text{var}(\theta_2 \mid x)$. This function is also minimized at $m = m_0$, so giving essentially the same optimal choice of experiment.

Randomization

3.49 Suppose that we wish to know which of two treatments is best for a particular disease. We have n patients available, and can choose which treatment to apply to each patient. Suppose that if a patient is given treatment i his or her response is distributed as $N(\theta_i, 1)$, where θ_i is the true mean response under treatment i. For deciding which treatment is best we require inference about $\phi = \theta_1 - \theta_2$ and, assuming normal prior distributions, this is precisely the problem that is considered in Example 3.14. The solution there says how many patients to allocate to treatment 1, but not which patients.

In fact, the posterior distribution of $\boldsymbol{\theta} = (\theta_1, \theta_2)$ depends only on how many patients receive each treatment. Once we have decided to give m patients treatment 1, then all of the $\binom{n}{m}$ allocations of patients to treatments are equally good for the purposes of posterior inference. The allocation can be made at random or arbitrarily. It might, for instance, be convenient to assign the first m patients to treatment 1 and then switch to treatment 2.

3.50 This conclusion runs counter to the conventional wisdom in the conduct of clinical trials, and to the whole of non-Bayesian experimental design theory. Both clinicians and classical statisticians stress the importance of making the allocation random. Randomization is an essential part of classical experimental design, and yet seems to play no part in our Bayesian theory. Indeed, it is easy to show that a randomized choice of design can never be better than making a non-randomized choice. More generally, a randomized decision cannot beat the best non-randomized decision.

Let d_0 and d_1 be two possible decisions, and consider d_p which chooses decision d_0 with probability p and d_1 with probability $1 - p$ $(0 < p < 1)$. Then by extending the argument, (2.21).

$$U(S(d_p)) = p\,U(S(d_0)) + (1 - p)U(S(d_1))$$

where $S(d_i)$ is the status after making decision d_i and $S(d_p)$ is the status after choosing to make the randomized decision d_p. If d_0 and d_1 are not equally good decisions, $U(S(dp))$ lies between $U(S(d_0))$ and $U(S(d_1))$, and d_p is inferior to the better of d_0 and d_1. If d_0 and d_1 are equally good then $U(S(d_p)) = U(S(d_0)) = U(S(d_1))$, and d_p is also just as good as d_0 or d_1. A randomized decision can never be preferred to the optimal non-randomized decision. We therefore never need to consider randomizing when making decisions, and the same applies to choosing an experiment.

3.51 This apparent conflict between theory and established practice has evoked a variety of responses. See for example Rubin (1978), Basu (1980), Kadane and Seidenfeld (1990).

3.52 To introduce the role of randomization in Bayesian design of experiments we return to the example of a clinical trial to compare two treatments. To assume that if treatment 1 were given to patient i, his or her response would have distribution $N(\theta_1, 1)$ for any i is to suppose that patients are indistinguishable. In practice, a doctor will very often expect certain patients to respond better than certain others. It might therefore be more realistic to say that the response of patient i to treatment j has the distribution $N(\theta_j + c_i, 1)$, where c_i represents the doctor's judgement of how much better patient i will respond than the average patient. This modification changes nothing, because we simply define the corrected response for patient i as his or her measured response minus c_i. The model in terms of the corrected responses is as before. We can still make an arbitrary choice of one of the $\binom{n}{m}$ patient-treatment assignments, and there is still no need to randomize.

Now although a doctor may expect one patient to respond better than another, it is difficult to quantify this judgement in the numbers $c_1, c_2, \ldots, c_n$. In practical clinical trials those symptoms and characteristics that are thought likely to affect response are recognized by grouping the patients, such that the patients within a group have similar symptoms and characteristics. Then patients within a group are regarded as indistinguishable, and within a group we still have conflict between theory and practice: if patients are indistinguishable then in theory they are equivalent and there is still no justification for randomization, yet the accepted practice is to assign patients to treatments at random within each group.

The practical reason for randomizing amongst theoretically equivalent patients lies in the fact that they are similar but not completely indistinguishable. Even though it may be difficult to identify relevant differences in a group of similar patients, there is a risk in allowing the assignment of patients to treatments to be a matter of arbitrary choice. In exercizing that choice it is quite possible for unconscious biases to enter. If the patients on treatment 1 perform better than those on treatment 2, we could not confidently attribute this to treatment 1 being superior. They may simply be the better responding patients, and might have responded even better had they been given treatment 2. It is to guard against the systematic intrusion of unconscious biases that randomization is needed, regardless of whether classical or Bayesian inference is to be used.

A related problem is Simpson's paradox; see Simpson (1951), Blyth (1971), Lindley and Novick (1981). This describes a phenomenon which can for instance imply that Treatment A is better on average for male patients and also for female patients, but when averaged over both sexes Treatment B is better. A sample with an atypical balance of men and women could then lead to entirely the wrong inference. Randomization will guard against a sample being selected with such undesirable properties.

3.53 Randomization should be used whenever there are variations on the optimal experiment that are theoretically equivalent. Another example is the choice of sample in a sample survey. Having grouped, or 'stratified', the members of the population, the optimal design will specify how many indiviuals should be sampled from each group, but not which ones. The individuals in a group are being regarded as equivalent, where in fact there are inevitably differences that may cause unconscious bias if the sample is chosen judgementally. An interviewer, for instance, if allowed to choose interviewees is likely to avoid those whose dress, manner or actions indicate that they may become aggressive or abusive.

3.54 It must be stressed that randomization is not a panacea. First, it should not be used as an excuse for not thinking. Patients or individuals in a population should not be modelled as equivalent when they have identifiable characteristics that may be expected to influence their response in particular ways. Treating them as equivalent and randomizing will throw away that information.

Second, it is perfectly possible for a randomly chosen assignment to be the same as a judgementally chosen one, or to have even more obvious risk of bias. For instance, even if there is no reason to suppose that male patients will respond in some experiments better (or worse) on average than female patients, we would not want to assign all the males to treatment 1 and the females to treatment 2. The fact that such an assignment came about through randomizing would not make it any better. When randomization has been used, we should check that its outcome does not appear to have any imbalances that could suggest any bias in the experiment.

Sequential decisions and experiments

3.55 An experiment may be planned in several stages, such that after each stage a decision is taken whether to stop the experiment then or to continue with the next stage. Let $\mathbf{y}_i$ denote the data obtained at stage i, let $\mathbf{x}_i = (\mathbf{y}_1, \mathbf{y}_2, \ldots, \mathbf{y}_i)$ represent all the data available after stage i, and let $S(\mathbf{x}_i)$ be the corresponding status. Then $S(\mathbf{x}_i) \mapsto \{S_s(\mathbf{x}_i), S_c(\mathbf{x}_i)\}$, where $S_s(\mathbf{x}_i)$ is the status after deciding to stop the experiment and $S_c(\mathbf{x}_i)$ is the status after deciding to continue to the next stage. In that case, further data $\mathbf{y}_{i+1}$ will be obtained, leading to status $S(\mathbf{x}_{i+1})$. Now the utility associated with stopping the experiment, $U(S_s(\mathbf{x}_i))$, may be obtained from the associated statistical decision problem. An 'inference' $d \in \mathscr{D}$

will be taken, such that if the true parameter value is θ, a utility $u(\mathbf{x}_i, d, \theta)$ will be obtained. Then

$$U(S_s(\mathbf{x}_i)) = \sup_{d \in \mathscr{D}} \int u(\mathbf{x}_i, d, \theta) f(\theta \mid \mathbf{x}_i)\, \mathrm{d}\theta. \tag{3.41}$$

If the decision is to continue, the utility

$$U(S_c(\mathbf{x}_i)) = \int U(S(\mathbf{x}_{i+1})) f(\mathbf{y}_{i+1} \mid \mathbf{x}_i)\, \mathrm{d}\mathbf{y}_{i+1} \tag{3.42}$$

is obtained by taking an expectation over the preposterior distribution of the stage $i+1$ data $\mathbf{y}_{i+1}$, given $\mathbf{x}_i$.

Then the decision will be to stop if (3.41) exceeds (3.42). Notice that (3.41) gives an explicit computation for the utility of stopping, in terms of the utility $u(\mathbf{x}_i, d, \theta)$ which is determined by the statistical decision problem. In particular, $u(\mathbf{x}_i, d, \theta)$ may be expressed similarly to (3.32), with components representing the cost of performing i stages of experimentation, the cost of obtaining the resulting data, and a loss function depending only on d and θ. In contrast, (3.42) expresses $U(S_c(\mathbf{x}_i))$ only implicitly in terms of the utility obtaining after $i+1$ stages of the experiment. This forms the basis of a recurrence relation

$$U(S(\mathbf{x}_i)) = \max\{U(S_s(\mathbf{x}_i)), \textstyle\int U(S(\mathbf{x}_{i+1})) f(\mathbf{y}_{i+1} \mid \mathbf{x}_i)\, \mathrm{d}\mathbf{y}_{i+1}\} \tag{3.43}$$

in which the utility after i stages is expressed in terms of the utility after $i+1$ stages. (3.43) is called the *dynamic programming* equation for this sequential experimental design problem.

3.56 A sequential experiment is specified by a *policy*, which says, for each stage i and each possible $\mathbf{x}_i$, whether to continue the experiment or to stop, *and* if the experiment is to stop which 'inference' $d \in \mathscr{D}$ should be made. The dynamic programming method allows an optimal policy to be determined recursively. If the optimal policy has been determined for stages $i+1, i+2, \ldots$, and the consequent utility $U(S(\mathbf{x}_{i+1}))$ has been determined for all $\mathbf{x}_{i+1}$, then (3.43) gives the utility $U(S(\mathbf{x}_i))$ for all $\mathbf{x}_i$ at stage i. It also implicitly determines for each $\mathbf{x}_i$ whether to stop or to continue. For each $\mathbf{x}_i$ where the decision is to stop, (3.41) determines the optimal 'inference'. Therefore given an optimal policy and utility for stages $i+1, i+2, \ldots$ an optimal policy and utility for stages $i, i+1, i+2, \ldots$ can be derived.

If the maximum possible number of stages is n, then the experiment must stop after n stages and $U(S(\mathbf{x}_n)) = U(S_s(\mathbf{x}_n))$ can be determined from (3.41). The dynamic programming approach then works backwards through stages $n-1, n-2, \ldots, 1$ to determine a complete optimal policy. This is the theoretical solution, but in practice the calculations escalate rapidly in complexity as each successive stage is introduced. Solutions are very difficult to obtain in practice except for the simplest problems.

3.57 The problem can be made more complex if a choice of stage $i+1$ experiments is allowed if the decision is to continue after stage i. These are examples of a much more general class of sequential decision problems. The dynamic programming method has been applied extensively to the optimization and control of stochastic processes; see Bellman

(1957), Whittle (1983) and Gittins (1989). It was first presented in a Bayesian context by Lindley (1960). Examples of specific applications in Bayesian sequential design are given by Freeman (1970) and Glazebrook (1978), who consider allocation of treatments in sequential experiments (see also **3.49**).

Predictive inference

3.58 It is often of interest to make inferences about some future observation y, after observing data x. Although this problem is given the distinctive name of *predictive* inference, it is really no different from any other inference problem, and its solution is the same. Such inference must be based on the posterior distribution of y, $f(y \mid x)$. Now all of the parameters are nuisance parameters, and $f(y \mid x)$ will usually be derived by integrating out θ from their joint posterior density. Thus,

$$f(y \mid x) = \int f(y, \theta \mid x)\,\mathrm{d}\theta = \int f(y \mid \theta, x) f(\theta \mid x)\,\mathrm{d}\theta. \tag{3.44}$$

The integral on the right of (3.44) is the product of the full posterior distribution $f(\theta \mid x)$ with the likelihood for the future observation, $f(y \mid \theta, x)$. In many problems this likelihood will reduce to $f(y \mid \theta)$ because x and y will be independent given θ.

If the required inference about y is expressed through a decision problem with loss function $L(d, y)$, then the posterior expected loss is

$$\begin{aligned} E(L(d,y) \mid x) &= \int L(d,y) f(y \mid x)\,\mathrm{d}y \\ &= \int \left[\int L(d,y) f(y \mid \theta, x) f(\theta \mid x)\,\mathrm{d}\theta \right] \mathrm{d}y. \end{aligned} \tag{3.45}$$

Now, subject to appropriate regularity conditions, we can reverse the order of integration in (3.45). If $f(y \mid \theta, x) = f(y \mid \theta)$ also holds, then

$$E(L(d,y) \mid x) = \int L^{\star}(d,\theta) f(\theta \mid x)\,\mathrm{d}\theta = E(L^{\star}(d,\theta) \mid x), \tag{3.46}$$

where

$$L^{\star}(d,\theta) = \int L(d,y) f(y \mid \theta)\,\mathrm{d}y = E(L(d,y) \mid \theta) \tag{3.47}$$

acts as a loss function for an equivalent inference about θ. Therefore, in this case we can reexpress the predictive inference problem as a parametric inference problem with loss function $L^{\star}(d,\theta)$. If x and y are not independent given θ, then $L^{\star}(d,\theta)$ will depend also on x.

Example 3.15

Suppose that a future observation y has mean θ and variance v (given θ and data x). It is desired to estimate y with quadratic loss $L(d,y) = (d-y)^2$. Then

$$L^{\star}(d,\theta) = E((d-y)^2 \mid \theta) = v + (d-\theta)^2,$$

and the problem is identical to estimating θ with quadratic loss.

3.59 The predictive framework is the most natural way to formulate a real decision problem. If the loss function is to represent a loss that the decision-maker can actually incur at some point in the future, it must depend only on quantities whose values are known at that time. That is, it depends only on future observations. In contrast, the parameters in most statistical models are unobservable.

If we take the view that all inference is directly or indirectly motivated by decision problems, then it can be argued that all inference should be predictive, since inference about unobservable parameters has no direct relevance to decisions. This viewpoint is advocated by Geisser (1971, 1980, 1985) and Aitchison and Dunsmore (1975). An extreme version of the predictivist approach is to regard parameters as neither meaningful nor necessary. Then the predictive distribution for a future y is obtained by

$$f(y \mid x) = f(x, y)/f(x), \tag{3.48}$$

without reference to a parameter θ. The difficulty with this method lies in defining $f(x, y)$. If x is to provide information about y, they cannot be independent, yet reliable subjective evaluation of joint distributions of non-independent variables is extremely difficult. As we have seen in **3.14**, introducing parameters is a natural way of representing the joint distribution of the data. In the more pragmatic predictivist formulation, all unobservable parameters are nuisance parameters, but they are still employed in constructing the basic model. The predictive distribution is obtained by (3.44) rather than (3.48).

Strong justification for introducing parameters in this way, even if they are not strictly meaningful or of interest in themselves, is provided by De Finetti's theorem: see **4.39**.

3.60 The alternative view of inference is that it is concerned with learning about the process generating the data. The unknown aspects of this process are described in the most expressive and parsimonious way by the parameters, which therefore become the most natural subjects for inference. Even if inference of this kind is ultimately motivated by decisions that depend on future observations, those decision problems cannot be realistically formulated in advance, and could depend on any combination of future observations. In the absence of a specific decision formulation, the most constructive way to use the available data is to learn about the parameters.

EXERCISES

3.1 In a Poisson process of unknown rate λ, x events are observed in time t. Show that the same inferences would be reached regardless of whether t was fixed (so that x had a Poisson distribution) or x was fixed (so that t had a gamma distribution). With t fixed, x/t is the classical minimum variance unbiased estimator of λ; prove that this is not the case for x fixed.

3.2 Observations x_1 and x_2 are independently distributed given (θ_1, θ_2) with distributions $N(\theta_i, 1)$, $i = 1, 2$. The prior distribution is the (improper) uniform distribution

$$f(\theta_1, \theta_2) \propto 1,$$

but constrained by $\theta_1 > \theta_2$. Show that the posterior distribution is a truncated bivariate normal distribution and that

$$f(\theta_1 \mid x_1, x_2) \propto \phi(\theta_1 - x_1)\Phi(\theta_1 - x_2),$$

where ϕ and Φ are the standard normal f.f. and d.f. respectively. Show also that

$$E(\theta_1 \mid x_1, x_2) = x_1 + \phi(d)/\{\sqrt{2}\Phi(d)\},$$

where $d = (x_1 - x_2)/\sqrt{2}$.

3.3 Let x be the number of successes in n independent trials with probability θ of success in each trial. The improper prior distribution

$$f(\theta) \propto \theta^{-1}(1-\theta)^{-1}$$

is assumed. Prove that this corresponds to assuming a uniform prior distribution for $\log \theta/(1-\theta)$. Show that this prior distribution results in $E(\theta \mid x) = x/n$, so agreeing with the usual classical estimator in this case, but that the posterior distribution may be improper.

3.4 Let x be distributed as $N(\theta, 1)$ given θ, which has the double exponential prior density

$$f(\theta) = \exp(-|\theta|)/2.$$

Prove that as $x \to \pm\infty$ the posterior distribution tends to $N(x + \text{sign}(x), 1)$.

3.5 Let $R(x)$ be such that $P(\theta \in R(x) \mid x) = \gamma$ for all x. That is, for each possible x, the posterior probability that θ lies in $R(x)$ is γ. If $R(x)$ is regarded as a random set, define

$$\beta(\theta) = P(\theta \in R(x) \mid \theta)$$

to be its classical confidence coefficient as a function of θ. Prove that $E(\beta(\theta)) = \gamma$.

3.6 A parameter θ has the gamma prior distribution with density

$$f(\theta) = \{\Gamma(b)\}^{-1} a^b \theta^{b-1} c^{-a\theta},$$

for $\theta > 0$, and with $a > 0$, $b > 0$. It is proposed to conduct an experiment in which an observation x is Poisson distributed with mean $n\theta$. After the experiment, θ will be estimated by d, with quadratic loss $(d-\theta)^2$, but the experiment will incur a cost which is represented by nc on the same utility scale. Show that the preposterior expected loss is $nc + b\{a(a+n)\}^{-1}$, and thereby that the optimal n is the nearest integer to $(\frac{b}{ac} + \frac{1}{4})^{1/2} - a$.

3.7 Generalizing Example 3.14, each of n observations may be made from any of k distributions $N(\theta_i, 1)$, $i = 1, 2, \ldots, k$. Independent prior distributions $N(0, w_i)$ are expressed for the θ_is. Show that for large n the optimal allocation is to take a proportion

$$P_i = k^{-1} - n^{-1}(h_i - \bar{h})$$

from the ith distribution, where $h_i = w_i^{-1}$ and $\bar{h} = k^{-1}\sum_{i=1}^{n} h_i$, when the loss function is either the trace or the determinant of the posterior variance matrix of the θ_is.

3.8 Let $x_1, x_2, x_3, \ldots$ be identically and independently distributed given θ as $N(\theta, 1)$. The prior distribution for θ is $N(m, v)$. Show that the prior predictive distribution for $\mathbf{x}_n = (x_1, x_2, \ldots, x_n)$ is $N(m\mathbf{1}, \mathbf{I} + v\mathbf{J})$, where $\mathbf{1}$ is the $n \times 1$ vector of ones, $\mathbf{I}$ is the $n \times n$ identity matrix and $\mathbf{J} = \mathbf{1}\mathbf{1}'$ is the $n \times n$ matrix of ones. Show also that the posterior predictive distribution of x_{n+1} given $\mathbf{x}_n$ is $N(m_n, 1 + v_n)$, where $m_n = (m + nv\bar{x})/(1 + nv)$ and $v_n = v/(1 + nv)$. Confirm that this distribution may be derived from the first as $f(x_{n+1} \mid \mathbf{x}_n) = f(\mathbf{x}_{n+1})/f(\mathbf{x}_n)$.

3.9 Let $x_1, x_2, x_3, \ldots$ be identically and independently distributed given θ and data y with $E(x_i \mid \theta) = \theta$ and $E(|x_i|^d \mid \theta) < \infty$ for some $d > 2$. Prove that the predictive distribution $f(\bar{x} \mid y)$ of $\bar{x} = n^{-1} \sum_{i=1}^{n} x_i$ tends to $f(\theta \mid y)$, with $\bar{x}$ substituted for θ, as $n \to \infty$.

CHAPTER 4

SUBJECTIVE PROBABILITY

Probability as a degree of belief

4.1 Acceptance of the Bayesian method as the natural and proper approach to statistical inference has become almost synonymous with the adoption of a subjective interpretation of probability. Most of the modern literature on Bayesian statistics is written from this subjective perspective. In this chapter, we explore the theory of subjective probability, and the various practical issues arising when we try to assign numerical values to probabilities. Chapter 5 considers the Bayesian method from the point of view of other interpretations of probability.

4.2 Subjective probability, also known as personal probability, concerns the judgements of a given person, conveniently called You, about uncertain events or propositions. Thus if A is an event, $P(A)$ measures the strength of Your degree of belief that A will occur. If $P(A) = 1$, You are certain that A will occur, and if $P(A) = 0$ You are certain that it will not occur. As $P(A)$ increases from 0 to 1, it describes an increasing degree of belief in the occurrence of A. This is an entirely different interpretation from frequency probability, in which $P(A)$ is the limiting frequency of occurrence of A in an infinite sequence of hypothetical repetitions of A. A can only have a frequency probability if it is, at least in principle, repeatable in this way, whereas subjective probability can apply to any event at all.

Frequency probability is closely connected with classical inference, because classical inference requires the data to be repeatable. An unbiased estimator, for instance, is defined to have expected value equal to the parameter being estimated. The statement is conditional on the parameter taking a fixed, but unknown, value, while the data are imagined as repeatable. Although experimental data can often be thought of as repeatable, and therefore having frequency probability distributions, parameters are generally unique and unrepeatable.

Example 4.1

$X_1, X_2, \ldots, X_n$ are experimental measurements of the resistivity of a particular alloy, whose true resistivity is θ. Here, the data are clearly repeatable, and have well-defined frequency probabilities, such as the probability that a measurement exceeds the true value θ. In contrast, this alloy is unique and has a unique resistivity. It is hard to conceive of a frequency probability that θ is less than some specific value.

4.3 In classical statistics a parameter is fixed (but unknown) and is not a random variable. All probability statements are therefore derived from the probability distribution of the data given the parameters, i.e. $f(x \mid \theta)$. From a subjective view of probability, this is a very distorted approach. The value of a parameter is unknown, and it therefore has a subjective probability distribution. Once the data are observed, they cease to be unknown, and so cease to be regarded as random variables. Bayesian inference therefore proceeds from the posterior distribution, i.e. the distribution of the (random) parameters conditional on (the fixed observed values of) the data.

Of course, before the data are observed, they are unknown and therefore random. The prior distribution $f(\theta)$ and the likelihood $f(x \mid \theta)$ are both subjective distributions that express Your beliefs about θ and x prior to observing the data. Once the data have been observed, only θ is unknown, and Your beliefs about θ are now expressed by $f(\theta \mid x)$.

Conditional probability

4.4 All subjective probabilities are conditional. $P(A)$ expresses Your degree of belief about A based on the totality of Your current information. If You then observe the occurrence of another event B, Your probability for A becomes $P(A \mid B)$, because B has been added to Your information. A more explicit notation would recognize the initial information as, say, H and write $P(A \mid H)$ and $P(A \mid B, H)$ for these probabilities. However, it is convenient to simplify the notation by suppressing H. In any particular context, the background information H may be taken to comprise all that You know prior to learning the values of any of the data, i.e. H is the prior information.

The likelihood $f(x \mid \theta)$, for instance, represents Your beliefs about the data x if, in addition to Your prior information, You learnt the value of the parameter θ. This is a hypothetical state of knowledge, because You do not know θ, nor will You learn its value before observing these data. In developing a theory of subjective probability, conditional probabilities need to be considered carefully. In frequency probability theory it is usual to *define* $P(A \mid B)$ as $P(A, B)/P(B)$. A subjective theory must justify $P(A \mid B) = P(A, B)/P(B)$ as expressing Your belief about A after observing B, not just as a mathematical definition.

A second point is to consider the status of a hypothetical conditional probability such as $f(x \mid \theta)$. If You are required to assert what You would believe about A if You were to observe B, is it the same as the belief You subsequently have about A when You have observed B? This is rather a fine philosophical point, whose implications are discussed in detail by Goldstein (1985).

4.5 The subjective notion of conditional probability induces a very simple concept of independence. Two random quantities x and y are independent if learning the value of one does not change Your beliefs about the other. This is the literal interpretation of the statement that $f(x \mid y) = f(x)$. The discussion of sufficiency in **3.7**, and related ideas of ancillarity and nonidentifiability, made use of this concept of independence. Since in classical inference, all probabilities are conditional on the parameters, the often made classical assertion that observations are independent must be understood as saying that they are independent *given* the parameters. Indeed, the discussion of nuisance parameters in **3.14** makes it clear that data will typically not be independent if we do not condition on all the parameters.

Measuring probabilities

4.6 How is a subjective probability $P(A)$ to be determined? The simplest way is a direct measurement technique, analogous to weighing an object by placing it on one side of a traditional balance, with some known weight on the other side. Suppose, for instance, that A is the event of snow falling in London on Christmas Day next year. To measure my own personal probability for A, I could begin by comparing it with a single coin toss. I regard A as less likely than obtaining Heads on a single toss, so $P(A) < 1/2$. In fact, I regard A as less likely than obtaining Heads on three successive coin tosses, but more likely than four successive Heads, so $0.0625 < P(A) < 0.125$. I now compare A with the event of getting five or six Heads in six tosses, and decide that A is less likely than this, so that $P(A) < 0.109\,375$. Continuing in this way, I might determine $P(A)$ to any desired accuracy, at least in principle.

A formal derivation of subjective probability based on this approach would need to assume that any two events can be compared to say which You regard as the more probable, and also that there exists a set of standard events with given probabilities against which any other event can be compared to determine its probability. In order for probabilities to be derivable that follow the usual laws of probability, it is necessary to make additional assumptions about the consistency of comparisons.

In the following development the assumptions are formulated as given by De Groot (1970). Numerous other formulations are possible, and a large and complex literature has resulted. To pursue the mathematical and philosophical details, see Savage (1972), Fine, (1973), Howson and Urbach (1989), Walley (1991) and references therein. However, insofar as the different formulations imply differences in the practice of Bayesian inference, those differences may be discussed simply in terms of De Groot's formulation.

An axiomatic formulation

4.7 Let $A \leqslant B$ denote the judgement that A is not more probable than B, given a common underlying information base H. We follow the notation of A**7.5**. That is, an *event* A is a set containing one or more elementary outcomes from the set Ω of all elementary outcomes. We could equally easily think of A as a more general *proposition*. Whereas an event is said to *occur* if the true elementary outcome lies in A, a proposition is simply a statement which is either true or false. To each of the set theoretic operations between events corresponds a propositional operation. Thus, the intersection $A \cap B$ corresponds to the logical conjunction 'A and B', the union $A \cup B$ correponds to the logical disjunction 'A or B', the complement A^c corresponds to the logical negation 'not A', and the superset relation $A \supset B$ correponds to the logical expression 'B implies A'. The empty set ϕ correponds to a proposition that is certainly false, and the universal set Ω to a proposition that is certainly true.

The first assumption is that all events may be compared:

(A1) For any A, B, either $A \leqslant B$ or $B \leqslant A$, or both.

We write $A < B$ in the case that $A \leqslant B$ but it is not also judged that $B \leqslant A$. The next two assumptions are to ensure that the comparisons are made in a logically consistent way, and simply reflect some obvious properties of any notion of 'not more probable than':

(A2) If A_1, A_2, B_1, B_2 are four events such that $A_1 \cap A_2 = B_1 \cap B_2 = \phi$, $A_1 \leqslant B_1$ and $A_2 \leqslant B_2$, then $A_1 \cup A_2 \leqslant B_1 \cup B_2$. If, in addition, either $A_1 < B_1$ or $A_2 < B_2$ then $A_1 \cup A_2 < B_1 \cup B_2$.
(A3) For any A, $\phi \leqslant A$. Furthermore, $\phi < \Omega$.

The fourth assumption is rather stronger.

(A4) If $A_1 \supset A_2 \supset \ldots$ is a decreasing sequence of events with limit $\bigcap_{i=1}^{\infty} A_i$, and B is some fixed event such that $B \leqslant A_i$ for all $i = 1, 2, \ldots$, then $B \leqslant \bigcap_{i=1}^{\infty} A_i$.

The final assumption is essentially providing the set of standard propositions with fixed probabilities.

(A5) There exists a random variable $\mathcal{U}$ taking values in [0,1] such that if A_1 and A_2 are the events that $\mathcal{U}$ falls in given subintervals of [0,1] with lengths ℓ_1 and ℓ_2 respectively, then $A_1 \leqslant A_2$ if and only if $\ell_1 \leqslant \ell_2$.

The relationship $\ell_1 \leqslant \ell_2$ in (A5) is, of course, just a normal inequality between numbers.

4.8 De Groot (1970) proves that assumptions (A1) to (A5) imply that for every event A there is a unique probability $P(A)$ (or more strictly $P(A \mid H)$, recognizing the underlying information H) such that the usual probability axioms are satisfied:

(P1) $P(A) \geqslant 0$ and $P(\Omega) = 1$,
(P2) If $A \cap B = \phi$, then $P(A \cup B) = P(A) + P(B)$,
(P3) If $A_1, A_2, \ldots$ is an infinite sequence of events such that $A_i \cap A_j = 0$ for all $i \neq j$, then $P(\bigcup_{i=1}^{\infty} A_i) = \sum_{i=1}^{\infty} P(A_i)$.

Although (P2) implies that $P(\bigcup_{i=1}^{n} A_i) = \sum_{i=1}^{n} P(A_i)$ for any finite disjoint sequence, (P3) asserts this for infinite sequences.

Not only do such probabilities exist, but they are easily constructed. Using the random variable $\mathcal{U}$ of (A5), consider the set of events $B(u)$, that $\mathcal{U}$ lies between 0 and u, for $0 \leqslant u \leqslant 1$. $B(0)$ is ϕ, the impossible event, and $B(1)$ is Ω, the certain event. For A, there is a unique p such that $A \leqslant B(u)$ if and only if $u \geqslant p$. Then define $P(A) = p$. This is precisely the process of comparing A with the set of standard events. In this way, $P(A)$ can be established. Since it is based on personal judgements it is a subjective probability.

4.9 The implications of Your accepting the assumptions or *axioms* (A1) to (A5) are easily seen. You are prepared to compare any two events to say which You judge to be more probable, and can do so in a consistent way. Furthermore, since You have the uniform random variable $\mathcal{U}$ as a yardstick, enough comparisons can be made to determine Your probability for any event uniquely and exactly. Those probabilities obey the usual laws of probability and therefore all the theorems of probability theory apply to them.

As was remarked in **4.4**, we must now go a little further and justify $P(A \cap B)/P(B)$ as $P(A \mid B)$, the probability You would give to A if, in addition to background information, You were to observe B. First note that (A1) to (A5) will be assumed to hold on any information base, and will therefore apply to $P(A \mid B)$. Therefore (P1) to (P3) will also apply to these conditional probabilities. Also, we clearly have $P(A_1 \mid B) \geqslant P(A_2 \mid B)$ if and only if $P(A_1 \cap B) \geqslant P(A_2 \cap B)$. Putting these considerations together will prove that $P(A \mid B) = P(A \cap B)/P(B)$. Therefore Bayes' theorem applies to subjective probabilities. Your judgements about a parameter θ prior to observing data x may be expressed as a prior distribution for θ, and this distribution must be proper. Your judgements about the data, conditional on any given value of θ can be expressed as a distribution $f(x \mid \theta)$ and the likelihood function is thereby defined. Your posterior distribution of θ could now be derived by Bayes' theorem, but this step is strictly unnecessary; Your beliefs about θ after observing x also directly determine $f(\theta \mid x)$, and this must agree with the result of applying Bayes' theorem.

Practical considerations will nevertheless limit to some extent what a real subjective Bayesian can do as opposed to this ideal. As we have seen, any simple probability can be measured directly by reference to the standard uniform random variable $\mathscr{U}$. To determine a distribution for a continuous random variable, such as a prior or posterior distribution for a continuous parameter, requires infinitely many such judgements. In practice only a finite number of judgements can ever be made, so that complete specification of such distributions is a practical impossibility. The implications of this fact are taken up in **4.31**.

Qualitative probability

4.10 We now consider what happens if one of the five axioms is rejected, beginning with the last, (A5). Note that it is important to distinguish between criticism of whether probabilities and comparisons should ideally behave as in the axioms, and criticism of whether a real person can actually make comparisons that satisfy them. Assumptions (A1) to (A4) ensure that whatever comparisons are made between events are consistent. If they hold, then it is possible to assign probabilities to the events that agree with those comparisons. But unless the set of comparisons is sufficiently rich, the assignment of probabilities is not unique. (A5) provides the necessary structure to obtain unique, quantitative probability values. There are alternative axioms that would serve the same purpose, such as postulate P6 of Savage (1972), but without them the set of preferences between events given by (A1) to (A4) imply only a *qualitative* probability structure.

Almost all of the theory of probability disappears if we are restricted to qualitative probabilities. Although some authors have criticized the fifth assumption, it is generally seen as relatively mild, and few propose doing without it completely.

In fact, most people can readily bring to mind a random variable like $\mathscr{U}$ to act as a fixed yardstick against which to compare other events. One might, for instance, imagine a pointer spinning freely and coming to rest pointing in a direction $u/360$ degrees clockwise from some fixed point. If (A5) is thereby achievable in practice, it is hard to criticize it theoretically.

Finite additivity

4.11 Assumption (A4) directly produces the property (P3) of probabilities, which is known as the countable additivity axiom. Countable additivity is a convenient mathematical property, without which some familiar parts of probability theory concerned with continuous random variables, or discrete random variables with infinitely many possible values, take on a strange aspect. In particular if probability is only *finitely additive*, then what we have called improper distributions become acceptable. Some implications of finite additivity are presented by Stone (1979) and Hill and Lane (1984).

4.12 Consider a random variable x whose possible values are the positive integers $1, 2, 3, \ldots$. The assertion that x has a uniform distribution, $f(x) \propto 1$, is unacceptable if probability is countably additive. For $f(x) \propto 1$ implies that there exists a number p such that $P(x = k) = p$ for all $k = 1, 2, 3, \ldots$. But if $p > 0$ then if a is an integer greater than p^{-1} we have $P(x \leqslant a) = ap > 1$, which is not possible (violating either (P1) or (P2)). We cannot have $p > 0$ whether we accept countable additivity or not, nor can we have $p < 0$ because it breaks axiom (P1). Setting $p = 0$ violates (P1) also if we have countable additivity because then $P(\Omega) = \sum_{k=1}^{\infty} P(X = k) = 0$. Yet $p = 0$ is allowable if probability is only finitely additive. We could have $P(x = k) = 0$ for all k, so that the probability of x taking a value in any finite set is zero, but have non-zero probabilities for x falling in infinite sets. We could, for instance say $P(x \text{ is even}) = 0.5$.

The idea of a uniform random variable on the positive integers, with the above properties, is attractive. It could serve as a formulation of ignorance about such a quantity, as discussed in **3.29**. However, there are a number of difficulties associated with finite additivity. If probability is countably additive, the distribution of x is completely specified by stating $P(x = k)$ for all k, i.e. the probability of any proposition concerning x is derivable from these basic probabilities. That is clearly not the case if probability is finitely additive, for then probabilities of infinite sets are not determined. Many more probability statements are needed before the distribution of x is completely determined. Some are readily obtained from the notion that x should have a uniform distribution, such as $P(x \text{ divides by } m) = m^{-1}$ for any finite m, but this is not enough. There are many events whose probability cannot be deduced from these statements, so that to determine such a probability we must add more probability statements. However many such statements we include, there remain many probabilities still undefined. Nevertheless, Kadane and O'Hagan (1993) show that some progress can be made with defining a uniform distribution in this way.

4.13 Further complications arise if we wish to use this kind of uniform distribution as a prior distribution in Bayes' theorem. The denominator may not be written as $f(x) = \int f(\theta) f(x \mid \theta)\, d\theta$ under finite additivity, nor would the posterior density be enough to specify the posterior distribution. Finite additivity is not just a licence to use improper distributions as in **3.28**.

De Finetti (1974) found no axiomatic argument in favour of countable additivity. Indeed, there seems no direct reason to suppose that comparisons between events ought to satisfy (A4). Yet we have seen that the properties of finitely additive distributions are complex and may be counter-intuitive. This, then, is an indirect justification for (A4); probabilities ought to be countably additive because the alternative is unacceptable.

4.14 Strictly, we could retain countable additivity and still allow improper distributions if we drop the assertion in (P1) that $P(\Omega) = 1$. Nothing in the assumptions (A1) to (A5) requires this, although without it those assumptions do not imply a unique probability assignment. For if we simply double all the probabilities they will still be consistent with (A1) to (A5) but we would now have $P(\Omega) = 2$. Any probability assignment with finite $P(\Omega)$ could therefore be reduced to an equivalent assignment with $P(\Omega) = 1$ and nothing is gained. However, Hartigan (1983) suggests allowing infinite $P(\Omega)$ with the result that we can now have $P(x = k) = p > 0$ for all $k = 1, 2, 3, \ldots$. In such a distribution, any event with finite probability is practically impossible. The probability values allow at least qualitative comparison of those extremely improbable events, but allow no distinctions between more probable events, all of which are given infinite probabilities. There are various other complications associated with 'improper' distributions in Hartigan's theory.

Prior ignorance

4.15 Improper prior distributions are sometimes advocated as representations of prior ignorance. The notion of prior ignorance is itself appealing for several reasons. In particular, when the use of genuine prior information may be controversial (perhaps because of accusations from non-Bayesians that subjective probability is unscientific or merely personal bias), a distribution representing ignorance might be an acceptable compromise. Prior ignorance is a basic feature of the objectivist view of probability, which is considered in **5.32** to **5.49**. Distributions representing ignorance developed there are typically improper, but from a subjectivist viewpoint we have seen that improper distributions are difficult to justify. If we accept countable additivity improper distributions do not exist, so it is not possible to have subjective beliefs that conform to this notion of prior ignorance. Improper distributions can exist if we only require finite additivity, but now they correspond to possibly unintuitive expressions of belief.

A purely subjectivist view is that complete ignorance cannot exist. Poincaré (1905) wrote

> "If we were not ignorant there would be no probability, there could only be certainty. But our ignorance cannot be absolute for then there would be no longer any probability at all. Thus the problems of probability may be classed according to the greater or less depth of this ignorance."

If there is a random quantity x about which we wish to express prior probabilities, then the context in which this wish arises gives meaning to x which, alone, is enough to say that we cannot be totally ignorant of it. However, if in Poincaré's terms Your depth of ignorance is great, then there is a sense in which Your beliefs *approximate* to some ideal (if unattainable) state of total ignorance. The use of improper prior distributions as expressions of *weak* prior information is discussed in **4.35**.

4.16 Assumption (A3) is quite innocuous. $\phi \leqslant A$ simply says that You cannot judge anything less probable than the empty set, an event which logically cannot occur. To add $\phi < \Omega$, rather than $\phi \leqslant \Omega$, only prevents the degenerate case of judging all events to be equivalent.

Comparability of events

4.17 Assumption (A2) is a very desirable property for a relationship '$\leqslant$' that is intended to mean 'is not more probable than'. However, it has strong consequences. It can be shown, for instance, to imply transitivity. That is, if $A \leqslant B$ and $B \leqslant C$ then $A \leqslant C$.

Another implication of (A2) is seen as follows. Let $A_2 = \phi$ and B_2 be some extremely improbable, but not impossible, event. Then $A_2 < B_2$, and now $A_1 \leqslant B_1$ implies $A_1 < B_1 \cup B_2$, since $A_1 \cup A_2 = A_1$. Therefore if A_1 is judged to be no more probable than B_1 it will be judged strictly less probable than $B_1 \cup B_2$, an event that differs from B_1 only through the inclusion of the extremely improbable B_2. Therefore (A2) implies that arbitrarily fine comparisons can be made between events. (The existence of an arbitrarily improbable B_2 also is provided by (A5).)

Such fine comparisons are needed to give a precise numerical probability to an event. For You to assign a probability of 0.85 to event A, You must be prepared to say that A is strictly more probable than the event that the uniform random variable $\mathcal{U}$ lies in [0, 0.849 999 9] and strictly less probable than $\mathcal{U}$ lying in [0, 0.850 000 1] (and even finer judgements than these). However desirable such precision may be, it is all too rarely achievable in practice.

4.18 One response to this is to say that instead of the probability of A having a unique value $P(A)$, it can only be represented by a *lower probability* $\underline{P}(A)$ and an *upper probability* $\overline{P}(A)$. A variety of axiomatic formulations leading to such a conclusion are examined thoroughly by Walley (1991). See also Dempster (1968, 1985), Shafer (1976, 1981) and Wasserman (1990). The attempt to build a theory of probability in which, for instance, the upper and lower probabilities of A and B are somehow combined to produce upper and lower proabilities for the event "A or B", leads to some difficult mathematics and rather counter-intuitive results. One of these is the phenomenon of *dilation* in which, for every one of a set of mutually exclusive and exhaustive events $B_1, B_2, \ldots, B_k$, the difference between the upper and lower conditional probabilities $\overline{P}(A \mid B_i)$ and $\underline{P}(A \mid B_i)$ is greater than between $\overline{P}(A)$ and $\underline{P}(A)$. The following example is given by Walley.

Example 4.2

Let H_1 and H_2 be the events of obtaining 'Heads' in the first and second tosses of a coin. You judge the coin to be fair, but believe that the tosses may not be independent. You are unwilling to assert anything beyond a belief that the probabilities of H_1 and H_2 are both one half. Then the following conclusions are intuitively reasonable, and justified formally by Walley (1991).

$$\underline{P}(H_2) = \overline{P}(H_2) = 1/2,$$

$$\underline{P}(H_2 \mid H_1) = 0, \qquad \overline{P}(H_2 \mid H_1) = 1,$$

$$\underline{P}(H_2 \mid H_1^c) = 0, \qquad \overline{P}(H_2 \mid H_1^c) = 1.$$

That is, having observed the result of the first toss, no matter whether it is a Head or Tail, You are unwilling now to assert anything about the probability of H_2. Observing the first toss, no matter what its outcome, reduces Your level of knowledge about H_2.

4.19 An obvious objection to upper and lower probabilities is that, if it is not practical to evaluate $P(A)$ exactly, it is no more practical to assign exact values to $\underline{P}(A)$and $\overline{P}(A)$. We take the view in this book that assumption ($A2$) is acceptable in the sense that comparisons between events really should act in accordance with ($A2$). An ideal person might be able to make the necessary fine and accurate probability judgements, and Your 'true' probability $P(A)$ is the unique value which accurately represents Your beliefs. In practice You generally cannot evaluate $P(A)$ accurately. We explore the practical aspects of personal probability later in this chapter beginning at **4.26**.

4.20 Assumption (A1) says that all events may be compared. It is certainly arguable that this is too strong, that some events are just not comparable. In that case, the theory would only allow the derivation of precise, consistent probabilities for those events which are comparable with events defined on the uniform random variable $\mathscr{U}$ required by (A5). There would be other events that are not comparable with these. Probabilities might be assigned to incomparable groups of events, but would not necessarily be consistent with each other.

4.21 In support of (A1), we could suggest that any two events A and B may be compared by offering You a choice between receiving a reward if and only if A occurs, or receiving the same reward if and only if B occurs. If You choose the first option then presumably You regard A as being at least as probable as B. That this is not necessarily true is easily established by an example. Let A be the event of global nuclear war next year and B the event that Fermat's last theorem is proved next year. Many mathematicians would regard A as more probable than B, but still would choose to have a reward if B occurs rather than A. This is obviously because the same reward is worth much less to someone in the event of nuclear war than otherwise.

This example may cause us to restrict comparisons to events that are *ethically neutral*, that is events which do not in themselves alter Your utilities appreciably. Events concerning the random variable $\mathscr{U}$ are ethically neutral, so probabilities could be derived for ethically neutral events but not for others. The restriction would be important in problems of statistical decision theory in a business context, for it would rule out comparisons of events such as a change in bank interest rates. Indeed, it is precisely those events which change a person's or a company's financial status whose probabilities are needed in decision theory.

4.22 Therefore, forcing events A and B to be compared by a choice of bets fails when either event is not ethically neutral. Nevertheless, there is little practical objection to (A1). The simple judgement that A is more probable than B does not seem to be affected by utilities, and despite the theoretical objections of some philosophers there do not seem in practice to be categories of events that are incomparable. It may be more difficult to make clear comparisons between very dissimilar events, but this is associated with the difficulty of making arbirarily fine comparisons in general, i.e. the failure in practice of (A2).

Other formulations of subjective probability

4.23 We have discussed in detail one axiomatic derivation of subjective probability. Two other formulations should be mentioned. De Finetti (1974) defines probability through the quadratic scoring rule, Example 2.16. Your probability $P(A)$ is defined as follows. You may choose a value x and, having chosen, will suffer a penalty, either of $(1-x)^2$ if A turns out to be true or of x^2 if A is false. The chosen value x is *defined* to be $P(A)$. There is an implicit axiom analogous to (A1), that $P(A)$ can always be determined for any A. De Finetti adds the axiom of coherence, that for any events $A_1, A_2, \ldots, A_n$ the total penalty implied by the chosen values $P(A_1), P(A_2), \ldots, P(A_n)$ cannot be uniformly reduced by some alternative set of choices. From this he deduces the probability properties (P1) and (P2). For instance, suppose that B is the negation of A, $B = A^c$. The total penalty is $(1-P(A))^2 + P(B)^2$ if A is true (so B is false) and $P(A)^2 + (1-P(B))^2$ if B is true. Unless $P(A) + P(B) = 1$, this penalty can be reduced whether A is true or false by replacing $P(A)$ by $x = (P(A) + 1 - P(B))/2$ and $P(B)$ by $1 - x$. Therefore, $P(A^c) = 1 - P(A)$.

De Finetti's approach is *operational*, in the sense that it defines probability as implicit in Your actions, and those actions have real consequences, in the form of penalties. The approach based on the axioms (A1) to (A5) is not operational, because a judgement that A is more probable than B is not an action chosen in awareness of the consequences of making this judgement rather than that B is more probable than A. The difficulty of an operational approach is that probabilities become bound up with utilities of the consequences. If, for instance, De Finetti's penalties are paid in money, the choice of x will be affected by Your utility for money. This will generally lead to Your chosen x being closer to 0.5 the larger the units of money used, because $x = 0.5$ minimizes the maximum potential penalty. To avoid this risk-aversion phenomenon, penalties must be paid in utility units, which means knowing Your utility function. The definition of utility in **2.27** was in terms of probability, which is not satisfactory when we are trying to define probability.

4.24 The approach of Ferguson (1967) attempts to solve this problem by merging the definitions of probability and utility into one development. He first defines utilities, through assumptions that imply a definition as in **2.27**, but using predefined, or 'standard' probabilities. As we have seen, this is analogous to postulating the existence of the uniform variable $\mathscr{U}$ in (A5). Now let a set of decisions $\mathscr{D}$ be such that for any $d \in \mathscr{D}$ and any possible value x of a random variable X, the utility of d after observing $X = x$ is defined in this way. Ferguson postulates axioms concerning preferences among the decisions in $\mathscr{D}$ and among randomized choices of those decisions (with 'standard' probabilities). He then derives unique (subjective) probabilities $P(X = x \mid H)$ and utilities for each $d \in \mathscr{D}$, such that the latter are their 'expected utilities' given by extending the argument to include X.

The implication is that a person who acts according to Ferguson's system must act *as if* he or she had specific subjective probabilities for any event A or random variable X, and was choosing decisions by maximizing (expected) utility. The development is a powerful argument in favour of the use of subjective probability, but does not offer a construction of probabilities that is convenient to apply in practice.

4.25 In summary, there are a variety of theoretical formulations of subjective probability. There is some debate at a philosophical level over whether probabilities can be defined quantitatively or only qualitatively, whether the domain of probability is all events (or propositions) or is confined to a subset, such as the set of ethically neutral events, and whether probability should be countably additive or only finitely additive. The latter issue, in particular, offers the theoretical possibility of admitting 'improper' distributions, although this seems to be gained at the expense of having the more complex and often counter-intuitive properties of finite additivity.

In practical Bayesian Statistics, subjective probabilities are invariably accepted to be quantitative, unbounded in domain and countably additive. The practical difficulties lie in making the fine judgements necessary to obtain a precise quantitative measurement of any probability, and in specifying a subjective probability distribution for a random variable taking a large, or even infinite, number of possible values.

Limitations to direct measurement

4.26 Let A be the event that more than half of all patients treated with a certain drug will recover. A doctor judges that A is about as likely as getting two heads when tossing two coins, and so the doctor's personal probability is given as $P(A) = 0.25$. This is clearly not a precise measurement, but the doctor may be unwilling to try to measure $P(A)$ more accurately. To do so would require comparing A with other events having probabilities close to 0.25, but the doctor may feel unable to make such fine comparisons in this case. Any practical usage of Bayesian methods must acknowledge that probabilities cannot be asserted with perfect accuracy.

4.27 It should be stressed that exactly analogous difficulties apply to classical frequency probability. A frequency probability is defined as the limiting frequency of occurrence of the event in an infinite series of independent repetitions. Since an infinite sequence of repetitions can never be carried out, no frequency probability can ever be known either.

It is interesting to explore, however, why in both theories it is common to assert that the probability of throwing any particular score with a single die is one-sixth. To assert this as a frequency probability means believing that if the die were tossed an infinite number of times then the relative frequencies of the different scores would all converge to one-sixth. To assert it as a subjective probability requires no such act of faith. It is only necessary to have an equal degree of belief in each of the possible scores occurring on this single toss. Indeed, one might expect that the die will not be perfectly fair, in the sense that some scores may occur more frequently than the others, but lack any knowledge of *which* scores. In this case, the symmetry of beliefs justifies very accurate measurement of the subjective probabilities, for if all scores have the same probability on this single toss, each must have a probability of one-sixth. To justify the same as frequency probabilities requires the much stronger belief that the die is perfectly evenly balanced and a perfect cube. We return to the distinction between these two kinds of symmetry of beliefs in **4.39**.

Elaboration

4.28 A key technique for making accurate judgements of probabilities arises from the recognition that some probabilities are easier to judge accurately by direct measurement than others. Consider ten tosses of a coin. It is easy to assign the probability 0.5 to the event that any single toss results in Heads, and to be confident of this judgement. A coin may not be perfectly symmetric, but however hard and carefully I contemplated it, it is highly unlikely that I would settle on a probability greater than 0.51 or less than 0.49. Now let A be the event of four heads in ten tosses. I might judge $P(A) = 0.2$, but could accept any alternative value in the range 0.15 to 0.25. However, it is not sensible to try to make a direct evaluation of $P(A)$ in this way, when I can calculate it as $\binom{10}{4}p^4(1-p)^6$, where p is the probability of Heads on a single toss. As p varies from 0.49 to 0.51, this calculation varies only from 0.1966 to 0.2130. $P(A)$ cannot be measured as accurately as p by a direct judgement, but the more accurate measurement of p can be used to build a more accurate measurement of $P(A)$.

4.29 Wherever a theorem of probability asserts that a certain *target probability* can be expressed in terms of one or more other *component probabilities*, then the target probability may be measured by inserting judgements or measurements of the component probabilities into the theorem. This process is known as *elaboration*. Not only is this a widely used technique but, at least from the subjective viewpoint, it is the primary role of probability theory. If, in the above example, I could have measured $P(A)$ accurately by a direct judgement, then I have no need of the theorem that tells me it equals $\binom{10}{4}p^4(1-p)^6$. If all events could be given arbitrarily accurate probabilities simply by considering each alone, there would be no need for any probability theory. The fact is, however, that it is generally wise to split any complicated problem into smaller components wherever possible.

4.30 Bayes' theorem is an elaboration. Given a substantial quantity of information, comprising both prior information and data, it is difficult to make direct posterior judgements about the parameters. Instead, we first use Bayes' theorem to represent the posterior distribution in terms of the prior distribution and likelihood. The advantage of Bayes' theorem is that it separates the prior information about the parameters from the data. Specifying either the prior distribution or the likelihood is typically a simpler task than mentally synthesizing the two sources of information directly into a posterior distribution, and so they may typically be specified more accurately and reliably.

Similarly, the likelihood is invariably only specified after further elaboration. This elaboration is called the statistical model, and its purpose is to enable the full joint distribution of the data given the parameters to be specified in terms of a series of simpler judgements. For instance, a model which includes a statement that the observations are identically and independently distributed given θ implies that $f(x_1, x_2, \ldots, x_n \mid \theta) = \prod_{i=1}^{n} g(x_i \mid \theta)$, where g is their common distribution.

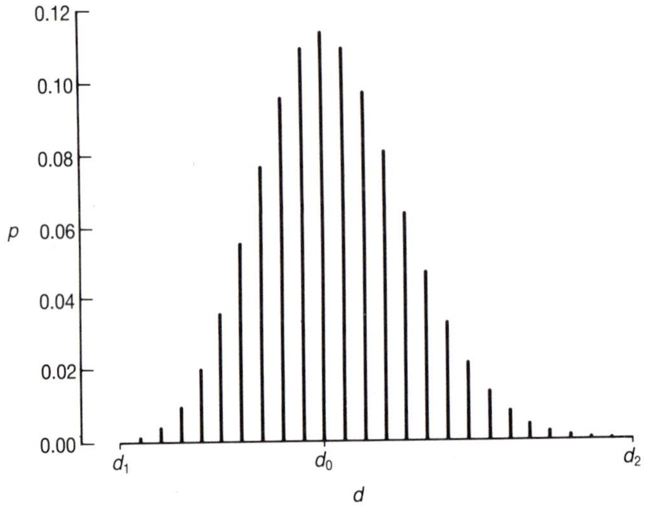

Fig. 4.1 Fitting probabilities to shape

Specifying distributions

4.31 The number of loaves that a baker will sell on a given day is a random variable, say D, and the baker's uncertainty about D is described by his probability distribution $f(d) = P(D = d)$. These probabilities will be hard for the baker to assess because there are a large number of possible values of d, and because that also means that each probability will be small. To assess the distribution of a continuous random variable X by separately considering each probability $P(X \leqslant x)$ is not just difficult but impossible, since there are now an infinite number of probabilities.

If it is difficult or impossible to specify a distribution by specifying all of its probabilities, some other approach is needed. The baker may begin by making an estimate, a guess at D. Specifically, he may state that the most probable value of D, the mode of the distribution, is d_0. So, $P(D = d)$ will rise steadily as d increases to d_0, and then will fall away again. The rates of increase and decrease can be roughly determined by the baker stating two more values, d_1 and d_2, such that for all $d \leqslant d_1$ and $d \geqslant d_2$, $P(D = d)$ is neglible. This might be made more explicit by saying that for these values, $P(D = d) \leqslant 0.001P(D = d_0)$. It is reasonable now to suppose that the baker's distribution will look like Figure 4.1. Having specified d_0, d_1, d_2 and this unimodal shape, the individual probability values are effectively very accurately specified. Given any arbitrary distribution $P^*(D = d)$ that fits this specification, the baker's actual $P(D = d)$ cannot be far from $P^*(D = d)$ without violating one of his stated judgements. If, therefore, such a $P^*(D = d)$ is found, the baker may set $P(D = d) = P^*(D = d)$, and this will be a very accurate specification.

4.32 The various standard families of probability distributions — binomial, Poisson, hypergeometric, Pareto, normal, gamma, beta, t, chi-squares, F, Weibull, et cetera — exist to allow You to fit a distribution to whatever specifications You wish to make about it.

This is the role of standard distributions, just as elaboration is the true role of probability theory.

In general, a subjective probability distribution is specified using summaries. In Chapter 2 we considered in some depth how summaries may be used as inference, describing a posterior distribution in terms that are readily understood as specific posterior beliefs. Conversely, beliefs about a random variable are best elicited as summaries. The process of specifying a subjective probability distribution is summarization in reverse. This point is developed further in **6.9**.

Example 4.3
X is binomially distributed with parameters n and θ. In specifying a prior distribution for θ, first note that θ is a probability, and so is a continuous random variable on the interval [0,1]. A convenient family of distributions is therefore the beta family. Since this family itself has two parameters, a prior distribution from the beta family may be specified by giving just two summaries. These might, for instance, be mean and variance, or mode and modal dispersion.

Sensitivity analysis
4.33 The impossibility of making arbitrarily precise and accurate judgements of probability implies that whenever a probability value is assigned, say $P(A) = 0.25$, we must recognize that there is nothing sacred about this particular value. Values sufficiently close to it, such as $P(A) = 0.22$ or $P(A) = 0.27$, would be more or less equally acceptable. Any calculation that is then performed using this probability could also be performed using plausible alternative values. If the result of the calculation is affected only superficially by this change, then we will say that it is *robust* to mis-specification of $P(A)$. Otherwise it is *sensitive* to $P(A)$, in which case You should try to improve the measurement of $P(A)$.

In the Bayesian method, we must similarly acknowledge that both the prior distribution and likelihood may be subject to specification or measurement errors, and perform a full sensitivity analysis. This means determining the effect on posterior inferences of altering the summaries used to specify the prior and likelihood, and the choice of distributions to fit those summaries, within reasonable limits.

Example 4.4
Suppose that the prior distribution of θ is asserted to be $N(m, v)$, and the distribution of x given θ is $N(\theta, w)$. Consider a sensitivity analysis of the posterior mean. Using the above specification, $E(\theta \mid x) = (mw + xv)/(w + v)$. To examine the effect of mis-specification of the prior mean, we can change m by an amount d to $m + d$, whereupon $E(\theta \mid x)$ changes by the amount $dw/(w + v)$. If the prior variance v is sufficiently large relative to the observation variance w, then $E(\theta \mid x)$ will be robust to mis-specifying m. Similarly, changing w by an amount d changes $E(\theta \mid x)$ by $d(m - x)v/\{(w + v)(w + v + d)\}$. If $|m - x|$ is sufficiently small, or w is sufficiently large, $E(\theta \mid x)$ will be robust to mis-specifying w. In general, if prior information is relatively weak (in this case, if v is large), posterior inference will be robust to mis-specification of the prior. Conversely, if the data are weak, inferences will be robust to mis-specification of the likelihood.

Part of the specification is that the prior distribution and likelihood are normal. We have seen in **3.35** that posterior inference can be highly sensitive to certain changes in the shape of the prior density or likelihood if $|x - m|$ is sufficiently large.

4.34 The degree to which different summaries of prior and likelihood should be varied in a sensitivity analysis depends on how accurate the initial specification is, which can itself only be assessed subjectively.

Example 4.5

Suppose that a new drug is proposed for treating a certain disease, and θ is the proportion of patients who achieve a given degree of improvement with the new drug. A physician asserts his prior distribution to be

$$f(\theta) = B(7,3)^{-1}\theta^6(1-\theta)^2, \qquad (0 \leqslant \theta \leqslant 1). \tag{4.1}$$

The physician may be very confident of some of the prior judgements used to specify this distribution, but less sure of others. He might, for instance, be confident that a unimodal distribution is a true reflection of his prior beliefs about θ. Values of θ around a single mode he judges to be most probable, and other values should be less probable the further they are from the mode. In that case, there is no need to include bimodal prior distributions in a sensitivity analysis. The modal value, which he has judged in (4.1) to be at $\theta = 0.75$, is perhaps less well determined. Nevertheless, it may be based on experience with existing treatments, or with drugs of this type, to the extent that the physician is confident that his modal value should lie between 0.7 and 0.8. In addition to stating the mode to be 0.75, the physician might have been quite confident that $\theta \geqslant 0.4$, and formalized this belief by $P(\theta \leqslant 0.4) = 0.025$. This is a less precise assertion, and he would probably be willing to allow $P(\theta \leqslant 0.4)$ to be as high at 0.075. Posterior inference could well be sensitive to such changes.

Weak prior information

4.35 It was remarked in Example 4.4 that if prior information is weak, relative to the information in the data, then inferences will generally be robust to mis-specification of the prior distribution. This conclusion is supported by the argument in **3.22** and **3.23**, that as the quantity of data increases the asymptotic posterior distribution is independent of the prior. It follows that the prior distribution may be specified very crudely, provided that prior information is genuinely very weak in comparison with the information in the data. This is a context in which improper prior distributions may be acceptable. The argument is not that prior information is completely absent and improper distributions represent prior ignorance, but that prior information is too weak to need specifying carefully, and that improper distributions can be convenient approximations to proper distributions representing weak information.

Improper distributions cannot be used in this way if any of the problems considered in **3.30** to **3.33** arise, because this implies that the hoped-for robustness does not hold. If the posterior distribution is improper (**3.32**) or the degenerate posterior of **3.33** results, then the improper prior distribution is unacceptable; any proper prior distribution would produce completely different posterior inference. In **3.30** it is pointed out that an improper prior distribution purporting to represent weak information about θ may not appear to represent weak information about $\phi = g(\theta)$ in the same way. For most such transformations the distinction will be unimportant, because with strong data posterior inference will be robust as to whether we choose to express weak prior information about θ or about ϕ.

In the case of an extreme function like $g(\theta) = \theta^{100}$ robustness may not hold. A uniform prior for $\phi = \theta^{100}$, $f(\phi) \propto 1$, translates into $f(\theta) \propto \theta^{99}$, which is very unlike a uniform prior for θ. The posterior distribution would quite possibly be very sensitive to such a change. However, weak prior information about θ that could reasonably be approximated by a uniform distribution simply does not translate into information about θ^{100} that could also be approximated by a uniform distribution. If prior information is genuinely weak relative to the data, the posterior distribution should be robust to any *reasonable* choice of prior distribution.

4.36 A variety of approaches to the formal study of sensitivity and robustness are considered in Chapter 7. Given an asserted prior distribution $f_0(\theta)$ and likelihood $f_0(x \mid \theta)$, the ranges of possible posterior distributions and inferences are considered as the actual prior distribution $f(\theta)$ ranges over some neighbourhood Γ of $f_0(\theta)$ and as $f(x \mid \theta)$ ranges over a neighbourhood Λ of $f_0(x \mid \theta)$. There are undoubted similarities between this approach and the theory of upper and lower probabilities mentioned in **4.18**, and in particular Seidenfeld and Wasserman (1991) exhibit dilation effects, but there are important differences. First, it is as impractical to identify exactly the set Γ of possible prior distributions as it is to evaluate $f(\theta)$ exactly. Second, if inferences do not appear to be sufficiently robust there is the possibility of narrowing Γ or Λ by thinking more carefully about Your probability specifications.

Resolving sensitivity

4.37 If all posterior inferences of interest are found to be sufficiently robust to all reasonable variations in both prior distribution and likelihood, then the accuracy of specification is adequate for the purposes of making those inferences. If unacceptable sensitivities are found, then reliable inferences cannot be drawn unless the accuracy of specification can be improved.

Suppose that a sensitivity to the prior distribution is found: the stated prior is $f_0(\theta)$ but if this is varied to $f_1(\theta)$ then posterior inferences change in some significant way. The obvious way for You to resolve the sensitivity is to think further about Your prior beliefs to determine which of $f_0(\theta)$ or $f_1(\theta)$ is in fact inconsistent with those beliefs. This may be possible in practice if Your initial assessment $f_0(\theta)$ was not the result of the most careful consideration possible. Faced with the need to think further, You may decide that $f_0(\theta)$ is in fact a sufficiently accurate representation of Your prior information that varying it as far as $f_1(\theta)$ is not reasonable. Or You may now feel that $f_1(\theta)$ is the better representation and $f_0(\theta)$ is not after all acceptable.

4.38 Special difficulties arise in going back to reconsider the prior distribution after the data have been seen and an analysis performed to obtain posterior inferences. The prior distribution represents beliefs about θ before the data are obtained. Once they have been obtained it may no longer be possible in practice to make reliable judgements about prior beliefs.

There are both philosophical and practical difficulties associated with this problem. In defining subjective probability, we have taken Your current base of information to be implicit. It is arguable that subjective probability cannot be defined conditional on a hypothetical state of information. If that were the case, the likelihood would not exist, because it is based on the hypothetical knowledge of the value of θ. After observing the data, the prior distribution is now based on a hypothetical state of information. The two situations are not entirely comparable, however. In one case we hypothesize the addition of knowledge (about θ), while in the other we hypothesize a loss of knowledge (about the data). Whatever the philosophical arguments, it is certainly easier in practice to contemplate what You would believe if some new information were to be available than to forget some piece of information that You have already acquired.

If reliable reassessment of prior information is not possible, the sensitivity can only be resolved by direct consideration of the posterior. If the inferences produced by $f_0(\theta)$ and $f_1(\theta)$ are noticeably different, then You may decide that one or other does not accord with Your beliefs about θ now that You have observed the data.

Exchangeable events

4.39 Symmetry is a powerful aid to specifying probabilities. A judgement that a die is equally likely to fall with any of its six faces uppermost leads immediately to the assignment of a probability of one-sixth to each event. This argument does not actually require You to believe that the die is unbiased in the sense that each face will occur an equal number of times in an infinite series of throws of the die, only that on this throw You see no reason to believe that one particular face is more likely to occur than any other. Consider a die that You can sense is unequally balanced, but You cannot judge which faces are thereby more likely to occur. You would still judge the possible faces to be equally probable because You have no reason to assign a higher probability to one than another. It may seem that somehow probability alone is not enough to describe adequately Your uncertainty in these situations. In one case Your probability of one-sixth is a result of a strong belief that the six faces will occur equally often because of the symmetry of the die, whereas in the other the same probability appears to arise from a *lack* of information, leading to the much weaker belief that You have no reason to give a higher probability to one face than another. This argument suggests that something more than probability alone is needed to distinguish these two kinds of belief.

The resolution of this argument does not lie, however, in looking at only a single throw of the die. On a single throw, You genuinely would give a probability of one-sixth to throwing a six in each case, and there really is no difference in Your beliefs about this one throw. We need to look at a series of throws. Let A_i be the event of throwing a six on the ith occasion that the die is thrown. Whether You believe the die to be unbiased or biased in an unknown way, You would assign a probability of one-sixth to each of the

A_is, because each A_i in isolation is the result of a single throw. But consider $P(A_2 \mid A_1)$. If You believe strongly in an unbiased die then You will still be inclined to give a probability of one-sixth to A_2 even after observing that A_1 is true, but if You suspect a bias then the observation of a six on the first throw will increase Your probability of a six on the second, because You now have some slight evidence to suggest that the bias favours sixes. It is in the joint probability distribution of a series of throws that the two cases are distinguishable, and we still need nothing beyond probability to describe Your knowledge or beliefs.

Despite the differences between these two kinds of beliefs about the die, there is a stronger symmetry that they share. Consider the probability $P(A_1 \cap A_2)$ of getting a six on the first two throws. This value would differ between the two cases, but in either case You would assign the same value to the probability $P(A_i \cap A_j)$ of getting sixes on any two specified throws. More generally, Your probability for throwing six on each of k specified throws depends on k and on Your beliefs about the die, but not on *which* throws are specified.

4.40 Events $A_1, A_2, \ldots, A_n$ are said to be exchangeable if

$$P\left(\bigcap_{j=1}^{k} A_{i_j}\right) = P\left(\bigcap_{j=1}^{k} A_j\right) \tag{4.2}$$

for all $k = 1, 2, \ldots, n$ and for all $1 \leqslant i_1 < i_2 < \ldots < i_k \leqslant n$. The space of all possible outcomes of the n events has 2^n elements, and in general it would take $2^n - 1$ probabilities to specify the probabilities of all possible outcomes. In the case of exchangeability, however, the symmetry implied by (4.2) reduces this number dramatically. All that is needed is the n probabilities $q_1 = P(A_1)$, $q_2 = P(A_1 \cap A_2), \ldots, q_n = P(\bigcap_{i=1}^{n} A_i)$ on the right hand side of (4.2). For by the Inclusion-Exclusion Law (Exercise A7.1), the probability of any outcome in which r specified A_is occur and the other $n - r$ do not occur is $\sum_{k=0}^{n-r} (-1)^k \binom{n-r}{k} q_{r+k}$. Therefore probabilities of all possible outcomes can be expressed in terms of the q_ks.

4.41 The symmetry can be seen even more clearly in this result, since the probability of any outcome depends only on how many events occur, not on which events occur. This means that probabilities do not depend on the sequence in which the A_is are arranged. If they were permuted in any way, probabilities would be unchanged. This is the origin of the term 'exchangeable'.

4.42 Let the random variable r be the number of A_is that occur, so that r takes values $0, 1, 2, \ldots, n$ with probabilities p_r. In terms of the q_ks

$$p_r = \binom{n}{r} \sum_{k=0}^{n-r} (-1)^k \binom{n-r}{k} q_{r+k} \,.$$

Alternatively it is simple to express the q_ks in terms of the p_rs. For consider the probabilities of the A_is conditional on r. Then all outcomes in which r A_is occur and

$n-r$ do not are equi-probable, with probabilities $\binom{n}{r}^{-1}$. Therefore the probability that any specific k A_is occur given r is $\binom{n-k}{r-k}\binom{n}{r}^{-1}$ for $r \geqslant k$. Therefore

$$q_k = \sum_{r=k}^{n} \binom{n-k}{r-k}\binom{n}{r}^{-1} p_r. \qquad (4.3)$$

The p_rs characterize the probabilities of A_is as effectively as the q_ks, since again only n probabilities are required to determine the probability of any event concerning the A_is. In general, if B is any such event, $P(B) = \sum_{r=0}^{n} P(B \mid r)p_r$ and $P(B \mid r)$ is the probability of event B if it is known that exactly r A_is will occur, and that all outcomes having exactly r occurrences are equi-probable.

Example 4.6

A pack of 52 cards is shuffled thoroughly. Let A_i be the event that the ith card in the pack is a Spade. Then $P(A_i) = 1/4$ for any i, also $P(A_i \cap A_j) = (1/4) \times (12/51)$ for any $i \neq j$, and so on. The A_is are exchangeable with $n = 52$. Clearly, $p_{13} = 1$ and $p_r = 0$ for any other r. The model of a shuffled pack of n cards with r of one type and $n-r$ of the other precisely fits the general exchangeable sequence conditional on r.

Example 4.7

Another special case of exchangeability is independence. Let $A_1, A_2, \ldots, A_n$ be independent with $P(A_i) = p$ for all i. Then $q_k = p^k$ and p_r is the binomial probability $\binom{n}{r}p^r(1-p)^{n-r}$.

De Finetti's theorem for exchangeable events

4.43 An infinite set of events $A_1, A_2, A_3, \ldots$ are exchangeable if $A_1, A_2, \ldots, A_n$ are exchangeable for every $n = 1, 2, 3, \ldots$. We now consider the limiting form of the characterization of **4.42**. For almost any exchangeable sequence, as $n \to \infty$ the probability that r is finite will tend to zero, so instead let $t_n = r_n/n$, where r_n is the number of occurrences in $A_1, A_2, \ldots, A_n$, and let $t = \lim_{n\to\infty} t_n$. It can be proved (De Finetti (1937)) that the limit exists with probability one for any exchangeable sequence, so t is well defined. In place of the discrete probability distribution $p_0, p_1, \ldots, p_n$, the probabilities of the A_is are in the limit characterized by the distribution of t which takes values in the interval [0,1].

Given t, the proportion of A_is that occur in the sequence is t, with all such sequences equi-probable. This is seen to imply that the A_is are conditionally independent with $P(A_i \mid t) = t$. For in particular, the probability of k specific A_is occuring given t is t^k, the limit of the finite exchangeability expression $\binom{n-k}{r-k}\binom{n}{r}^{-1}$ as $n \to \infty$ with $r/n \to t$. (4.3) becomes $q_k = E(t^k)$, the expectation being with respect to the distribution of t.

The characterization of an infinite set of exchangeable events as independent with common probability t given that the limiting frequency is t (together with the fact that this limit exists with probability one) is known as *De Finetti's theorem.*

Example 4.8
A zoologist studying a genetic mutation believes that the proportion of individuals who will show the physical sign of the mutation will be one quarter or three quarters. Then if A_i is the event that individual i exhibits the mutation, the A_is may be regarded as an infinite exchangeable sequence and $t = 0.25$ or 0.75. If the zoologist gives $P(t = 0.25) = p$ then

$$P(A_i) = q_1 = E(t) = (3 - 2p)/4,$$

$$P(A_i \mid A_j) = q_2/q_1 = (9 - 8p)/(12 - 8p).$$

In particular, if $p = 1/2$, $P(A_i) = 1/2$ and $P(A_i \mid A_j) = 5/8$.

Example 4.9
A_i is the event that the ith component coming off a production line is faulty. Your beliefs about the proportion t of faulty items in an infinite sequence of components is expressed in the distribution $f(t) = 0.01(1-t)^{99}$. Your probability for an individual component being faulty is therefore $P(A_i) = E(t) = 1/101$.

4.44 Exchangeable events are essentially what the classical statistician calls independent events with unknown probability of success on each trial. The limiting frequency t of successes is the classical frequency probability of a success on any one trial. If t were known, then $P(A_i) = t$ is the frequency probability. From the subjective point of view, t is not a probability but a limiting frequency. Nevertheless, the two approaches agree exactly in that if t is known, De Finetti's theorem states that $P(A_i \mid t) = t$ and the A_is are also independent given t.

If, therefore, a frequency probability were known to You for any event then it would agree with Your subjective probability for that event. In practice, frequency probabilities are invariably unknown because an infinite sequence of trials has not been observed. Then the frequentist can only regard trials of that event as independent with unknown probability of success. To the subjectivist, those trials are exchangeable. That is, they are conditionally independent given an unknown limiting frequency t of successes. Subjectivity then enters through a personal distribution for t.

4.45 In classical statistics, the 'unknown probability' of success in a trial would become a parameter, and it is reasonable now to denote the unknown limiting frequency by the symbol θ. De Finetti's theorem can be seen as justifying the imposition of the classical *model* of independent trials conditional on the unknown parameter θ. The obvious difference is that the classical approach has no counterpart to the subjective *prior* distribution for θ.

If You now observe n events from an infinite exchangeable sequence, and there are r successes, Your distribution for θ changes. The likelihood is $f(r \mid \theta) = \binom{n}{r}\theta^r(1-\theta)^{n-r}$, using De Finetti's theorem. Combining this with Your prior distribution for θ through Bayes' theorem yields Your posterior distribution.

Example 4.10

Following Example 4.8, if n individuals are observed and r exhibit the mutation, the zoologist's posterior probability for $t = 0.25$ is

$$P(t = 0.25 \mid r) = \frac{p\binom{n}{r}(0.25)^r(0.75)^{n-r}}{p\binom{n}{r}(0.25)^r(0.75)^{n-r} + (1-p)\binom{n}{r}(0.75)^r(0.25)^{n-r}}$$
$$= 3^{n-r}p/\{3^{n-r}p + 3^r(1-p)\}.$$

For instance, if $p = 0.5$, $n = 5$ and $r = 4$, the posterior probability that $t = 0.25$ is $1/27 = 0.037$, and for any unobserved individual j, $P(A_j \mid r) = (0.25 \times 0.037) + (0.75 \times 0.963) = 0.73$. After only five observations, the zoologist's belief is now very strongly in favour of the proportion exhibiting the mutation in the population being three quarters.

Example 4.11

Let x be the number of successes in n trials with limiting frequency θ of success. The prior distribution is the beta distribution $f(\theta) = \{B(p,q)\}^{-1}\theta^{p-1}(1-\theta)^{q-1}$. Then it is easily seen that the posterior distribution is also beta with parameters $p + x$ and $q + n - x$ in place of p and q. That is,

$$f(\theta \mid x) = \{B(p + x, q + n - x)\}^{-1}\theta^{p+x-1}(1-\theta)^{q+n-x-1}.$$

4.46 The power of De Finetti's theorem arises from the fact that it only requires a series of exchangeable events. The circumstances in which events may be judged exchangeable are essentially those which a frequentist would regard as trials of a certain generic event. From this simple assumption the conventional parametric model arises as a natural characterization of exchangeability.

The predictive approach to inference in which only statements about observable quantities are regarded as meaningful was discussed in **3.59**. The limiting frequency θ, like parameters of models generally, is unobservable. Its value can only be known after making an infinite number of observations, which is not possible in practice. The exchangeable events themselves are observable, and the strict predictive view of inference would demand that inference after observing some n of the events should consist of statements of posterior probabilities for unobserved events. However, De Finetti's theorem says that the posterior probability of an event A_i is the posterior expectation of θ, which may be obtained in the conventional Bayesian way using the binomial likelihood and a prior distribution for θ. Furthermore, prior moments of θ are meaningful in the predictive view because $E(\theta^k) = q_k$, which is the probability of k events all occurring, an observable event in itself. The prior distribution will be specified in practice by specifying a few summaries, such as moments, and then fitting a convenient standard distribution. For example, mean and variance would be specified and the beta distribution having these moments selected as the prior distribution. This procedure can be seen as a simple and effective way of specifying the q_ks, satisfying the requirements of the predictive approach.

For those who see no difficulty in assigning a distribution to θ, which although strictly unobservable is still clearly defined, there is no need to think about $f(\theta)$ in this oblique way.

Exchangeable random variables

4.47 Random variables $X_1, X_2, X_3, \ldots$, are said to be exchangeable if for all $k = 1, 2, 3, \ldots$, for all y_1, y_2, $y_3, \ldots$ and all $1 \leqslant i_1 < i_2 < \ldots < i_k$,

$$P(X_1 \leqslant y_1, X_2 \leqslant y_2, \ldots, X_k \leqslant y_k) = P(X_{i_1} \leqslant y_1, X_{i_2} \leqslant y_2, \ldots, X_{i_k} \leqslant y_k). \quad (4.4)$$

In words, (4.4) says that the joint distribution of the first k X_is is the same as for any other k X_is. In particular, all X_is have the same marginal distribution (the case $k = 1$). It is quite straightforward to generalize the theory of exchangeable events to exchangeable random variables.

4.48 It is immediately clear that independently and identically distributed (iid) random variables are exchangeable. Furthermore, if conditional on $Z = z$, the random variables X_1, X_2, $X_3, \ldots$ are iid with common distribution function G_z, and Z has a distribution with f.f. $f(z)$, then since

$$\begin{aligned} P(X_{i_1} \leqslant y_1, X_{i_2} \leqslant y_2, \ldots, X_{i_k} \leqslant y_k) &= \int P(X_{i_1} \leqslant y_1, X_{i_2} \leqslant y_2, \ldots, X_{i_k} \leqslant y_k \mid z) f(z)\,\mathrm{d}z \\ &= \int \prod_{j=1}^{k} G_z(y_j) f(z)\,\mathrm{d}z \end{aligned} \quad (4.5)$$

and this does not depend on the choice of i_1, i_2, $i_3, \ldots$, the X_is are exchangeable. If we can now generalize De Finetti's theorem to the case of exchangeable random variables then the converse of this result will hold. That is, if X_1, X_2, $X_3, \ldots$ are exchangeable they can be represented as conditionally iid.

4.49 We can argue informally as follows. Define the events $A_i(y)$ that $X_i \leqslant y$, for $i = 1, 2, 3, \ldots$ and $-\infty < y < \infty$. Then for any fixed y, $A_1(y)$, $A_2(y)$, $A_3(y), \ldots$ is an infinite sequence of exchangeable events if X_1, X_2, $X_3, \ldots$ is an infinte sequence of exchangeable random variables. Therefore by De Finetti's theorem the proportion of successes in the first n events in this sequence tends to a limit $F(y)$ as $n \to \infty$, and conditional on $F(y)$ the $A_i(y)$s are independent with $P(A_i(y) \mid F(y) = t) = t$. Then conditional on the *function* F, the random variables X_1, X_2, $X_3, \ldots$ become iid with distribution function F. The joint distribution of the X_is is therefore characterized by this conditional independence together with a unique distribution on the unknown distribution function F. Rigorous treatment of such results for quite general random quantities is given by Hewitt and Savage (1955).

This result is rather more complicated than (4.5) suggests, because to express the X_is as conditionally iid we must in general condition not on a scalar or vector random variable Z, in finite dimensions, but on the random distribution function F in the infinite dimensional space of all possible distribution functions. If Z is a random vector in (4.5) this corresponds to the distribution on F giving probability one to the event that F lies in the family of distributions $\{G_z\}$ indexed by z. *Within* that family, the distribution on F can be reduced to a distribution $f(z)$ on Z. More discussion of exchangeable random variables in general is given in **6.37**.

4.50 We have seen that De Finetti's theorem for exchangeable events relates classical frequency probabilities to subjective probabilities, and explains in subjectivist terms the classical model of trials, in which events are independent with common but unknown (frequency) probability. Similarly, extending De Finetti's theorem to exchangeable random variables allows us to examine another of the fundamental classical statistical models, in which observations $X_1, X_2, \ldots, X_n$ are said to be a *random sample* from some population. This is expressed in classical terms by saying that the X_is are iid with common distribution F, the population distribution. From a subjectivist viewpoint we need only judge the X_is to be exchangeable (and part of a potentially infinite sequence of observations of this type, the 'population') to conclude that they can be expressed as iid with unknown common distribution F. The situation in which a classical statistician might model observations as a random sample is precisely that in which a Bayesian might judge them to be exchangeable, i.e. there is a natural symmetry of beliefs about them.

However, assuming exchangeability alone means that the parameter in this model is the entire function F, the limiting distribution formed by the infinite sequence $X_1, X_2, X_3, \ldots$. It is not a simple matter to formulate prior information about a random distribution function, and further assumptions are typically made. Thus, the conventional model that $X_1, X_2, \ldots, X_n$ are independent and distributed normally with unknown mean μ and variance σ^2 says much more than that the X_is are exchangeable. It assumes also that the limiting distribution in the infinite sequence of which $X_1, X_2, \ldots, X_n$ are a part will certainly take a normal form, but with unknown location and scale. Implicitly, the prior distribution for F places zero prior probability everywhere except on the tiny subspace of normal distributions. By means of this strong prior assumption, the problem of specifying a prior distribution over the infinite-dimensional space of distribution functions is reduced to that of specifying a prior distribution on the two-dimensional space of the reduced parameter $\theta = (\mu, \sigma^2)$.

4.51 The classical model that the X_is are independent and identically distributed is generally accompanied by an assumption that F is restricted to some standard parametric family, so that the prior distribution reduces to a distribution over the parameters of that family. It is clearly appropriate, then, to examine the sensitivity of posterior inferences to changes in the prior distribution of F. F might be restricted to an alternative family of distributions, or the original family might be widened. In the second case, for instance, F might originally be restricted to the class of normal distributions with given variance, and this might be widened to include all normal distributions. Then the original reduced parameter μ is enlarged to (μ, σ^2). The sensitivity analysis might then explore whether inferences about μ are affected by this change.

If F is restricted to an entirely different family of distributions, it is meaningless to consider sensitivity of inferences about the reduced parameters, because they are different. Instead it is appropriate to examine the sensitivity of predictive inference.

Classical *nonparametric inference* is briefly mentioned in A**21.3**, and corresponds to making no assumptions about F. The Bayesian analogue then requires a prior distribution over the space of all distributions, and this approach is considered almost as briefly in **10.20** and **10.24**.

EXERCISES

4.1 Suppose that Your probabilities are only finitely additive, and that in specifying Your distribution for a random variable x taking positive integer values You state $P(x=n)=p_n$, $n=1,2\ldots$ such that $\sum_{n=1}^{\infty} p_n = q$. Prove that if $q=1$ then Your stated probabilities imply a unique probability $P(A)$ for every event A, but that otherwise they only place bounds $\underline{P}(A) \leqslant P(A) \leqslant \overline{P}(A)$ such that $\overline{P}(A) - \underline{P}(A) = 1-q$ for events A with infinitely many elements.

4.2 Let Your probability assertion $P(A)=p$ be interpreted as saying that You would regard a bet on A at odds $(1-p)/p$ as fair. That is, You would be willing to make a bet in which You gain $(1-p)s$ (in some money units) if A occurs and lose ps if it does not occur, for all sufficiently small s. Furthermore You would be willing instead to take the other side of this bet, so that You will lose $(1-p)s$ if A occurs and gain ps if it does not.

Suppose that You make probability assertions $P(A_i)=p_i$ for a set of mutually exclusive and exhaustive events $A_1, A_2, \ldots, A_n$. Prove that unless $\sum_{i=1}^{n} p_i = 1$, You would be willing to take a combination of bets such that You will lose no matter which A_i occurs.

4.3 Following Exercise 4.1, prove that if $q<1$ dilation is possible. That is, for two sets A and B, wider upper and lower bounds apply to both $P(A \mid B)$ and $P(A \mid B^c)$ than to $P(A)$.

4.4 A bag initially contains p red balls and q white balls. A ball is taken from the bag, with all balls equally likely to be drawn. The selected ball is returned to the bag and another ball of the same colour is also then added to the bag. This process is repeated with a series of balls being drawn from the bag, and each time the selected ball is returned to the bag plus another of the same colour. Let A_i be the event that the ith ball to be drawn is red. Therefore, conditional on the first n draws yielding r red balls and $n-r$ white balls, the probability of A_{n+1} becomes $(p+r)/(p+q+n)$. Prove that $A_1, A_2, A_3, \ldots$ form an infinite sequence of exchangeable events, characterized by the beta prior distribution as in Example 4.11.

4.5 Let $A_1, A_2, \ldots, A_n$ be exchangeable events and consider $P(A_2 \mid A_1)$. Prove that in general it is possible for $P(A_2 \mid A_1) < P(A_2)$, but if these events form part of an infinite exchangeable sequence then $P(A_2 \mid A_1) \geqslant P(A_2)$.

4.6 Consider an infinite sequence $x_1, x_2, x_3, \ldots$ of exchangeable random variables, with $E(x_i^2) < \infty$. Prove that $\operatorname{cov}(x_i, x_j) \geqslant 0$.

4.7 Show that if $x_1, x_2, \ldots, x_n$ have the exchangeable joint density function

$$f(x_1, x_2, \ldots, x_n) = n!\ \left(1 + \textstyle\sum_{i=1}^{n} x_i\right)^{-(n+1)}$$

they can be represented as conditionally independent and exponentially distributed.

CHAPTER 5

NON-SUBJECTIVE THEORIES

5.1 To someone who accepts the subjective interpretation of probability, the Bayesian method is by far the most natural approach to statistical inference. Parameters are unknown, and are therefore random variables, and the most natural way to say what is known about a random variable is to summarize its distribution. In this chapter, we examine other interpretations of probability and other approaches to statistical inference. Of particular interest is the extent to which they accommodate some or all aspects of the Bayesian method.

We begin by considering classical inference, which is associated with the frequentist interpretation of probability. Classical decision theory, in which parameters do not have distributions, is based on the notion of admissibility, and there is a close connection between admissible decision rules and Bayes rules. Arguments in favour of the Likelihood Principle, discussed from the Bayesian point of view in **3.2** to **3.4**, were presented from a more classical perspective in A**31.21**. However, most classical methods violate the Likelihood Principle. Some disturbing consequences of such violations cannot be avoided without adopting the Bayesian method. We also briefly consider so called *empirical Bayes* methods and fiducial inference.

The logical theory of probability is of most interest to Bayesian statistics because of its emphasis on representing prior ignorance. A variety of methods have been proposed for generating prior distributions to represent ignorance, and these are reviewed at the end of this chapter.

Admissibility

5.2 Consider a statistical decision problem with a loss function $L(d, \theta)$, for d in the set $\mathscr{D}$ of possible decisions, and for parameter θ. Observation x is made with likelihood $f(x \mid \theta)$. The Bayes decision rule, see **3.27**, would minimize the posterior expected loss $E\{L(d, \theta) \mid x\}$. In the classical theory, parameters do not have frequentist probabilities. Therefore a decision must be chosen using only $L(d, \theta)$ and $f(x \mid \theta)$.

A function $s(x)$ assigning a decision $s(x) \in \mathscr{D}$ for every possible observation x, is called a *(decision) rule*. The *risk* of rule $s(x)$ is $R_s(\theta) = E\{L(s(x), \theta) \mid \theta\}$. In contrast to the posterior expected loss, where expectation of $L(d, \theta)$ is taken with respect to the posterior distribution $f(\theta \mid x)$, the risk is the expectation with respect to the likelihood $f(x \mid \theta)$. It is therefore a function of the unknown parameter θ. A rule $s(x)$ is said to be *inadmissible* if there exists another rule $t(x)$ such that $R_s(\theta) \geqslant R_t(\theta)$ for all θ, with strict inequality for some θ. A rule which is not inadmissible is admissible. Only admissible rules are acceptable for decision making.

5.3 Using loss functions of the forms presented in **2.46**, **2.50** and **2.52**, the classical inference problems of point estimation, interval estimation and hypothesis testing can be expressed as statistical decision problems. In point estimation $\mathscr{D}$ is the set of possible values for θ and $L(d, \theta)$ will be zero at $d = \theta$ and increase as d moves further from θ. A rule $s(x)$ is called an estimator of θ. For instance, the quadratic loss function is $L(d, \theta) = (d - \theta)^2$. In place of (2.38), the classical risk calculation is

$$E\{L(s(x), \theta) \mid \theta\} = E\{s(x)^2 \mid \theta\} - 2\theta E\{s(x) \mid \theta\} + \theta^2 = \text{var}\{s(x) \mid \theta\} + \{b_s(\theta)\}^2, \qquad (5.1)$$

where $b_s(\theta) = E\{s(x) \mid \theta\} - \theta$ is the bias of the estimator $s(x)$. (5.1) is the mean-square-error of $s(x)$. An estimator will be inadmissible if its mean-square-error is uniformly higher than (or at least as high as) that of another estimator. Classical estimation theory based on mean-square-error is considered in A**17.30**.

5.4 Inadmissible estimators are unacceptable, but this criterion does not itself greatly restrict the class of estimators. For instance, the degenerate estimator $s(x) = a$ (for all x) has risk $R_s(\theta) = (a - \theta)^2$ which is zero at $a = \theta$. No other estimator can achieve such a low mean-square-error at this point, so this estimator is admissible. In order to remove estimators like this, which classical statisticians regard as undesirable, the set of possible estimators (or strategies) is usually first restricted according to some other, more or less arbitrary, criterion. The most common criterion to apply is unbiasedness. Attention is restricted to estimators for which $b_s(\theta) = 0$ for all θ. Then mean-square-error reduces to variance. A large part of Chapter A17, from A**17.7** to A**17.25**, is concerned with achieving unbiasedness and with finding the estimator with minimum variance in the class of unbiased estimators. Yet insistence on unbiasedness typically removes a great variety of potentially useful estimators. Exercise A17.12 shows, for instance, that if X is binomially distributed with parameters n and θ then the *unique* unbiased estimator of θ is x/n (making Example A17.9 rather superfluous). An unbiased estimator need not be sensible, and Exercise A17.26 even shows an instance where the only unbiased estimator is absurd. Example A17.14, Exercise A17.16 and Example 5.2 show that the minimum variance unbiased estimator can have uniformly higher mean-square-error than other reasonable estimators, and therefore is inadmissible under squared error loss. Example 5.1 presents a simple unbiased estimator that is inadmissible under any loss.. See also Exercise 5.2. Furthermore, the unbiasedness criterion violates the Likelihood Principle; see **5.14**.

Example 5.1

Let x be distributed as $N(\sqrt{\theta}, 1)$, given parameter $\theta > 0$. The natural unbiased estimator of θ is $s(x) = x^2 - 1$, but for $|x| < 1$, $s(x)$ is negative. Therefore for any loss function appropriate for estimating θ the (biased) estimator $t(x) = x^2 - 1$ if $|x| \geqslant 1$ and $t(x) = 0$ otherwise, will have smaller loss than $s(x)$ when $|x| < 1$. It therefore has uniformly lower risk and $s(x)$ is inadmissible under any such loss function.

5.5 Interval estimation has d a set of possible values of θ, and $\mathscr{D}$ is the set of all such sets. With the loss function (2.43), the risk of a rule $s(x)$ is

$$R_s(\theta) = P(\theta \notin s(x) \mid \theta) + E\{w(s(x)) \mid \theta\}, \qquad (5.2)$$

where $w(d)$ is a loss associated with the 'width' of the set d. Then the first term in (5.2) penalizes a rule if the probability that $s(x)$ contains θ is small, whereas the second term penalizes it if its expected size, or 'width', is large. Classical interval estimation theory has concentrated on restricting attention to *confidence intervals*. $s(x)$ is a confidence interval with confidence coefficient γ if $P(\theta \in s(x) \mid \theta) = \gamma$ for all θ. Then the risk (5.2) reduces to the 'width' term. Chapter A20 is devoted almost exclusively to confidence intervals, and although attention mostly focuses on the problems of constructing them some theory of 'shortest' confidence intervals is given in A**20.10**.

5.6 For a hypothesis test, $\mathscr{D} = \{d_o, d_1\}$, where d_0 means in some way to favour the hypothesis H_0 that $\theta \in \Omega_0$, whereas decision d_1 favours the alternative hypothesis that $\theta \in \Omega_1 = \Omega_o^c$. A rule $s(x)$ is called a test, and is equivalently defined by a critical region $S = \{x : s(x) = d_1\}$. With the natural loss function (2.47), the risk is

$$R_s(\theta) = \begin{cases} a_0 P(x \in S \mid \theta) & \text{if} \quad \theta \in \Omega_0, \\ a_1 P(x \notin S \mid \theta) & \text{if} \quad \theta \in \Omega_1. \end{cases}$$

In classical terminology, $P(x \in S \mid \theta)$ is a probability of the first kind of error if $\theta \in \Omega_0$, whereas $P(x \notin S \mid \theta)$ is a probability of the second kind of error if $\theta \in \Omega_1$. Again, classical theory conventionally reduces attention to a subset of available strategies. In this case, a test has size α if $P(x \in S \mid \theta) = \alpha$ for all $\theta \in \Omega_0$, and only tests of a given size are considered. Then risk reduces to probabilities of second kind of error, which are to be minimized. The relevant classical theory is given in Chapters A21 to A23, beginning with the simplest case in A**21.8**. In the case of two simple hypotheses the effect of the Neyman–Pearson Lemma, A**21.10**, is to characterize all admissible tests.

Admissible rules and Bayes rules

5.7 The rule defined by letting $s(x)$ be the decision that minimizes the posterior expected loss, derived from a prior distribution $f(\theta)$, is called the Bayes rule for that prior. If $f(\theta)$ is improper the posterior distribution may still be proper for all x, in which case the resulting rule $s(x)$ is called a generalized Bayes rule. It is now of interest to enquire whether classical statisticians and Bayesians will use the same, or essentially the same, set of rules in any problem. That is, what is the relationship between the set of admissible rules and the set of Bayes rules, possibly augmented by the generalized Bayes rules?

In general terms, the agreement between the two approaches is close but not perfect.

5.8 The strongest part of the relationship is that, subject only to mild conditions, Bayes rules are always admissible. Notice that for any rule $s(x)$,

$$\int R_s(\theta) f(\theta)\, d\theta = \int \left\{ \int L(s(x), \theta) f(x \mid \theta)\, dx \right\} f(\theta)\, d\theta \tag{5.3}$$

$$= \int \left\{ \int L(s(x), \theta) f(\theta \mid x)\, d\theta \right\} f(x)\, dx \tag{5.4}$$

$$= \int E\{L(s(x), \theta) \mid x\} f(x)\, dx\,, \tag{5.5}$$

assuming that the reversal of the order of integration in (5.4) is valid, which essentially means that the expected risk (5.3) exists. (Exercise 5.3 shows that a Bayes rule may be inadmissible if expected risk is infinite.) Now if $s(x)$ is the Bayes rule for prior $f(\theta)$ it minimizes the posterior expected loss $E\{L(s(x),\theta)\,|\,x\}$ for all x. Therefore it minimizes (5.5), and there cannot exist another rule $t(x)$ with $R_t(\theta) \leqslant R_s(\theta)$ for all θ and strict inequality for some θ. So the Bayes rule is admissible.

There is an assumption in this proof that such a rule $t(x)$ would yield $\int R_t(\theta)f(\theta)\,d\theta < \int R_s(\theta)f(\theta)\,d\theta$, but this is not strictly true unless $\int_T f(\theta)d\theta > 0$, where $T = \{\theta : R_t(\theta) < R_s(\theta)\}$. Although the Bayes rule $s(x)$ may not be admissible if this condition does not hold, the dominating rule $t(x)$ has strictly smaller risk only on a set T of zero prior probability, which the Bayesian would not regard as a problem. An example of formally sufficient conditions for the Bayes rule to be admissible are that $f(\theta)$ is strictly positive for all θ and that the risk function is continuous for any rule.

5.9 The converse, that every admissible rule is a Bayes rule for some prior distribution, is generally untrue. Generalized Bayes rules may also be admissible. A simple example is found in the case of a single observation x distributed as $N(\theta, 1)$ given the parameter θ, with quadratic loss $L(d,\theta) = (d-\theta)^2$. It may be shown that $t(x) = x$ is admissible. With quadratic loss the Bayes rule is $s(x) = E(\theta\,|\,x)$; see **2.46**. In this problem there is no proper prior distribution such that $E(\theta\,|\,x) = x$ for all x, but in **3.27** we found that this posterior mean arises if $f(\theta) \propto 1$, the improper, uniform distribution.

The following also gives a simple example of inadmissible generalized Bayes rules.

Example 5.2

Let $x_1, x_2, \ldots, x_n$ be independently distributed as $N(0,\theta)$ given the parameter $\theta > 0$, and assume quadratic loss. The prior distributions $f(\theta) \propto \theta^r$ are improper for all r. The posterior distribution is given by

$$f(\theta\,|\,\mathbf{x}) \propto \theta^r \theta^{-n/2} \exp -q/(2\theta), \tag{5.6}$$

where $q = \sum_{i=1}^n x_i^2$. Now (5.6) is proper for all $\mathbf{x}$ if $r < (n-2)/2$, and the posterior expected loss exists if $r < (n-6)/2$. Then the generalized Bayes rule $E(\theta\,|\,\mathbf{x}) = q/(n-2r-4)$ is obtained.

Now consider the class of estimators $t(x) = aq$. The risk of $t(x)$ is $R_t(\theta) = E\{(aq-\theta)^2\,|\,\mathbf{x}\}$, which is easily found to be $\theta^2(n^2a^2 + 2na^2 - 2na + 1)$. This is minimized at $a = (n+2)^{-1}$, therefore $t(x)$ is inadmissible for every other value of a. In particular, the minimum variance unbiased estimator, $a = n^{-1}$, and the generalized Bayes rules above for $r \neq -3$ are all inadmissible.

5.10 In some problems, there even exist admissible rules that are neither Bayes nor generalized Bayes. Such cases are unusual and the rules concerned are typically rather strange. Nevertheless, it is mathematically difficult to establish regularity conditions under which the set of admissible rules coincides with the set of Bayes rules, or with the Bayes rules augmented by some or all of the generalized Bayes rules. A thorough examination of the relationship between Bayes rules and admissible rules is given by Berger (1985).

The James–Stein Estimator

5.11 In many simple problems, standard classical estimators are found to be generalized Bayes. An example is the problem of estimating k normal means. Suppose that $x_1, x_2, \ldots, x_k$ are independent given $\boldsymbol{\theta} = (\theta_1, \theta_2, \ldots, \theta_k)$ with x_i having the $N(\theta_i, 1)$ distribution. We wish to estimate $\boldsymbol{\theta}$ with the loss function $L(\mathbf{d}, \boldsymbol{\theta}) = \sum_{i=1}^{k}(d_i - \theta_i)^2$. The standard classical estimator is $\mathbf{t}(\mathbf{x})$, where $t_i(\mathbf{x}) = x_i$ is the estimator of θ_i. This estimator is Maximum Likelihood and unbiased. It also has smallest risk, uniformly in $\boldsymbol{\theta}$, of all unbiased estimators. It is easily seen to be the generalized Bayes estimator under the uniform prior $f(\boldsymbol{\theta}) \propto 1$. Yet it was shown by Stein (1956) to be inadmissible for $k > 2$. An estimator with uniformly smaller risk derived by James and Stein (1961) is given in A**31.56**.

Discussion

5.12 The fact that, under only mild conditions, Bayes rules are admissible offers classical statisticians a way of finding large numbers of admissible rules. This is convenient because in general it is rather difficult to prove that any given rule is admissible. However, the statistician must ultimately select just one rule from the large class of admissible rules to use in a given practical problem. Classical statistics conventionally attempts to resolve the problem by applying other criteria, such as unbiasedness. Yet it is possible to make the choice by reference to the risk function alone. If the choice lies between two rules, $s(x)$ and $t(x)$ say, then if neither dominates the other there will be some values of θ for which $R_s(\theta) < R_t(\theta)$ and others for which $R_t(\theta) < R_s(\theta)$. If $t(x)$ is preferred to $s(x)$ then implicitly it is more important to have a rule that performs well for θ in $T = \{\theta | R_t(\theta) < R_s(\theta)\}$ than for θ in the complement of T. If we extend this idea and express the relative importance of risk at different values of θ in a weight function $w(\theta)$, then we could define the best rule to be that which minimizes $\int R(\theta)w(\theta)\mathrm{d}\theta$. It is now clear from (5.5) that the chosen rule would be the Bayes rule for prior distribution $f(\theta) \propto w(\theta)$. Interpreting the prior distribution as a weight function in this way provides one route for classical statisticians to adopt the Bayesian method.

The fact that by no means all generalized Bayes rules are admissible is another argument in favour of caution in using improper prior distributions. It also casts some doubt on classical estimation theory, since many of the standard classical estimators are only generalized Bayes and as the James–Stein result shows, may not be admissible. In that particular example the usual estimator is only inadmissible under a certain class of loss function, and A**31.57** suggests that caution must be used over the choice of loss function, but the inadmissibility in Example 5.1 holds for any reasonable loss function. Also, it has to be said that the emphasis on variances of unbiased estimators in classical theory implies that quadratic loss is the accepted norm. The multivariate form is also adopted in the classical Gauss–Markov theory, see A**19.6**.

Sufficiency, ancillarity and likelihood

5.13 The concepts of sufficiency and ancillarity have been discussed in **3.7** to **3.12**, and the Likelihood Principle in **3.2** to **3.4**. Sufficiency was considered from the classical viewpoint in A**17.31** and plays a major role in the theory of classical inference. Ancillarity was considered briefly in A**23.37**. However, these issues were also brought together in A**31.12** to A**31.25** in a discussion of the result of Birnbaum (1962), which stated that if classical statisticians were to adopt certain abstractions of their own teaching on sufficiency and ancillarity then they should also adopt the Likelihood Principle (LP). This was an important conclusion because, as we shall see, a great deal of classical inference theory violates the LP. Birnbaum's theorem initiated a substantial controversy, in which the consequences of various alternative principles of sufficiency and ancillarity (the latter often recast as a 'conditionality' principle) were explored. To the references given in Chapter A31 may be added the more Bayesian perspectives of Dawid (1980). We shall not go over the arguments again here, but simply remark that within the classical inference framework may be found strong, but not conclusive, support for the LP. A full and wide-ranging analysis of the LP is given by Berger and Wolpert (1988).

Classical violations of the LP

5.14 The LP asserts that if two experiments yield likelihood functions that are proportional to one another, then You should make identical inferences from these two experiments. A simple example is given in **3.3**, where a sequence of independent trials with probability θ of success in each is considered. In one experiment You observe the number of successes in n trials, and the result x is therefore binomially distributed with parameters n and θ. In the other experiment, y is the number of trials before the rth success is observed, which has a negative binomial distribution. If in the first experiment You observe $x = r$, or in the second You observe $y = n$, You obtain proportional likelihoods, and so should make identical inferences. Bayesian inference always accords with the LP, and therefore does so in this case. The LP is also reasonable in this case because in both experiments the same observation has been made, r successes in n trials; only the stopping rule is different. Yet, as is shown in A**31.6** and Example A31.5, the classical minimum variance unbiased estimators of θ are different in the two experiments.

This is not surprising because the criterion of unbiasedness itself violates the LP. The bias of an estimator $t(x)$ is

$$b(x) = E\{t(x) \mid \theta\} - \theta \tag{5.7}$$

and the expectation in (5.7) uses values of $f(x \mid \theta)$ for all x. The LP requires inference to be based only on the value of $f(x \mid \theta)$ for the observed x. So the recommendation to use the minimum variance unbiased estimator r/n in the binomial experiment when we observe $x = r$ is based not just on the likelihood $f(x = r \mid \theta) \propto \theta^r(1-\theta)^{n-r}$ but on the distribution $f(x \mid \theta)$ at values of x that were not observed. The reason why this differs from the minimum variance unbiased estimator $(r-1)/(n-1)$ in the negative binomial experiment is because the two experiments differ over what might have been observed.

5.15 Taking any expectation with respect to the distribution of x, $f(x \mid \theta)$, is in contravention of the LP. The essence of classical inference is to choose an inference rule on the basis of its performance in repeated sampling, which always entails an expectation of this kind. Bias is just one example of such an expectation; others are the variance of an estimator, the probabilities of first and second kinds of errors in hypothesis testing, and in a decision theory setting the risk function. Classical inference inevitably depends on what might have been observed but was not. The LP insists that inference use only the probability of the actual observation.

Regardless of any arguments in favour of the LP, the principle of judging an inference rule on the basis of its repeated sampling behaviour seems equally natural to classical statisticians. Faced with the conflict between the two, most seem happy to abandon the LP.

Conditioning

5.16 The result of abandoning the LP in favour of repeated sampling arguments is that classical inference rules only have desirable properties 'on average' over the sample space of possible observations. It may be clear that their properties are actually undesirable in some parts of that space, as the following example shows.

Example 5.3

The estimator $s(x) = x^2 - 1$, of the square root of a normal mean in Example 5.1, is negative for $|x| < 1$. $s(x)$ may be unbiased when averaged over the whole sample space, but it is clearly biased for $|x| < 1$. Since,

$$E(s(x) \mid \theta) = P(|x| < 1 \mid \theta)E(s(x) \mid \theta,\ |x| < 1) + \{1 - P(|x| < 1 \mid \theta)\}E(s(x) \mid \theta,\ |x| \geqslant 1),$$

the fact that $s(x)$ is biased downwards when $|x| < 1$ implies that it is biased upwards when $|x| \geqslant 1$. Therefore, even when it gives a sensible estimate, the claim of unbiasedness is rather misleading.

5.17 In general consider a set A in the sample space and define the conditional sampling distribution $f_A(x \mid \theta) = f(x \mid \theta, x \in A) = f(x \mid \theta)/P(x \in A \mid \theta)$ for $x \in A$ (and $f_A(x \mid \theta) = 0$ for $x \notin A$). The properties of an inference rule in sampling from $f_A(x \mid \theta)$ will typically be different from those it has in sampling from $f(x \mid \theta)$. Yet when we perform a specific experiment and observe $x \in A$, it seems that the behaviour of the rule in repeated sampling constrained to A is more relevant than averaging over the entire sample space.

Buehler and Fedderson (1963) proved the following result. If x_1 and x_2 are independently distributed as $N(\mu, \sigma^2)$ then the interval $I(x_1, x_2) = (\min\{x_1, x_2\}, \max\{x_1, x_2\})$ is a classical 50% confidence interval for μ. But the probability that $I(x_1, x_2)$ contains μ given that $(x_1, x_2) \in A$, where A is the set of points for which $3|x_1 - x_2| \geqslant 2|x_1 + x_2|$, is at least 0.5181 for all μ and σ^2. Therefore if we also observe that $x \in A$ we should have confidence of 51.81%, not 50%, that $\mu \in I(x_1, x_2)$. Such a conditioning set is called a positively biased relevant subset, in the context of classical confidence intervals. A negatively biased relevant subset is one for which the conditional confidence coefficient is always less than or equal to a value which is strictly less than the unconditional confidence coefficient. Robinson (1975) presents examples with relevant subsets having extreme positive and negative bias. Another example is given in Exercise 5.4. Robinson (1979) proves that Bayesian intervals do not have analogous conditioning problems (subject to very mild conditions).

Of course, given an observed x there is not a unique A for us to use to condition on $x \in A$. If the above argument holds, the *most* relevant subset is $A = \{x\}$, but the conditional distribution of x given $x \in A$ is now degenerate and the idea of an inference rule and repeated sampling become meaningless. Conditioning on the specific observed x does not cause classical inference to conform to the LP or to approach the Bayesian method; instead it collapses.

5.18 The usual classical response is that one must average over the full sample space of all possible observations. There are, however, many situations in which it is not entirely clear what the sample space is. Most practical measurements are bounded, and in particular measuring instruments typically can only produce readings in a limited range. Yet such bounds are rarely acknowledged, as the great amount of theory based on normal distributions shows. If bounds are explicitly recognized, as in Example A31.2, the properties of inference rules should be calculated afresh.

Sequential inference

5.19 Another cause of difficulty over the sample space is the stopping rule. Experiments are often performed without a well-defined stopping rule. Data may be accumulated until some more or less unpredictable event intervenes, such as the project being terminated, or a report on progress is requested. Classical inference depends upon the stopping rule, in contravention of the LP, and so is not generally possible when the stopping rule is not explicitly stated.

Another notable context in which the stopping rule affects classical methods is sequential inference. Sequential experimental design from a Bayesian perspective is developed in **3.55** to **3.57**. There we consider at various stages during an experiment deciding whether to continue the experiment by obtaining more data, or to stop and make an inference or decision based on the data available up to that point. If the decision is to stop at a point where n observations have been made, then the inference is based upon the posterior distribution of the unknown parameters, based on the n observations, and is exactly the same as would have been obtained if a nonsequential experiment had been conducted, with the intention from the outset having been to take exactly n observations.

However, classical inference based upon the same data would be different if the nonsequential experiment were performed. If a hypothesis test is required, for instance, the sequential experiment results in a lower degree of significance from the same data, because the probability of the first kind of error is inflated by the chance of rejecting the null hypothesis when it is true at some other stage of the sequential experiment. The difference between classical and Bayesian inference in this context can be quite striking. To a Bayesian it seems absurd that classical inference when the experiment has stopped after n observations depends not only on whether a decision was taken at some earlier stage not to stop the experiment then, but also on whether the decision at this stage might have been to continue and defer inference to a later stage.

5.20 Bayesian methods are unaffected by these problems over the sample space. Whether we truncate to $x \in A$ or whatever the stopping rule (provided it is noninformative:

see **3.4**) the likelihood is proportional to $f(x \mid \theta)$ for the specific observed x. The classical approach cannot avoid difficulties with the sample space as long as it rejects the Likelihood Principle in favour of repeated sampling arguments.

Finite population sampling

5.21 A similar phenomenon arises in inference from a finite population. Let the N individuals of the population be characterized by some measurement of interest taking value ξ_i on individual $i, i = 1, 2, \ldots, N$. We wish to estimate $\theta = \sum_{i=1}^{N} \xi_i$, the population total (or equivalently the population mean θ/N). For instance, the population might be the voters in some referendum and $\xi_i = 1$ if voter i will vote 'Yes' and $\xi_i = 0$ otherwise. Then θ is the number of 'Yes' votes that will be cast.

The data typically comprise observations of the measurement of interest in a sample of n individuals chosen by some random mechanism. Consider the case $n = 1$, where the sampled individual is individual i with probability $p_i, i = 1, 2, \ldots, N$, $\sum_{i=1}^{N} p_i = 1$. Let x be the resulting observation. The estimator $t(x)$ will have expectation $\sum_{i=1}^{N} p_i t(\xi_i)$ which equals θ identically if and only if $t(\xi_i) = \xi_i / p_i$. This is the unique unbiased estimator, yet it is not obviously sensible. Suppose that individual 1 is selected, so that the estimate of θ is ξ_1 / p_1. Now suppose that an observation were selected with different probabilities $q_1, q_2, \ldots, q_N$, then if individual 1 is selected in this experiment the estimate is ξ_1 / q_1. The same observation, ξ_1, has been made in the two experiments but different estimates are made.

5.22 This example is more than just another instance of classical inference violating the LP. It is, however, similar to the example of binomial versus negative binomial sampling, which was explored in **3.4** by recognizing that we also observe the outcome of the stopping rule. In the case of sampling from a finite population, we observe also the outcome of the sample selection. In the example we observe $s = 1$, where the random variable s is the number of the sampled individual. In one case $P(s = 1) = p_1$ and in the other $P(s = 1) = q_1$. Like a noninformative stopping rule, s is ancillary. Inference should therefore proceed conditionally on s. The special nature of sampling from a finite population is that when we condition on the ancillary statistic, the outcome of the sample selection, there are no probabilities left. The likelihood becomes, for a sample of one individual,

$$P(X = x \mid \xi_1, \xi_2, \ldots, \xi_N, s = i) = \begin{cases} 1 & \text{if } x = \xi_i, \\ 0 & \text{otherwise}, \end{cases} \tag{5.8}$$

and is degenerate in the same way as **5.17**. Classical inference collapses, but the Bayesian method proceeds without difficulty, as is shown in **10.25**.

5.23 The degeneracy of the likelihood (5.8) (and for samples of any size) was pointed out by Godambe (1966). In the subsequent debate, other frameworks and other justifications for the classical framework were proposed. Royall (1968) argued that information as to which units were selected should often be irrelevant, and so could be ignored. Hartley and Rao (1968) and Royall (1970a) considered classes of estimators that did not depend on the sampled units.

Basu (1969) and Ericson (1969a) concluded that only a Bayesian approach was usable in finite population sampling. A different classical approach is based on the notion of a *superpopulation*, from which the actual population is supposed to have been drawn, and was proposed by Royall (1970b). To see how this idea works, and how it can reproduce results obtained with the more standard classical sample survey theory, we examine the simplest case.

First, suppose that a simple random sample of size n is drawn without replacement from a finite population of N members. Let the population mean be $\mu = N^{-1}\sum_{i=1}^{N}\xi_i$, and let the population variance be $\sigma^2 = (N-1)^{-1}\sum_{i=1}^{N}(\xi_i - \mu)^2$. Then it is a standard result that the best linear unbiased estimator of μ is the sample mean $\bar{x}$, with variance $(N-n)\sigma^2/(Nn)$. The factor $(N-n)/N$ arises because the sampling is without replacement. See for instance Barnett (1991). This theory, as we have seen, is dependent on the sampling probabilities.

Now suppose instead that the members of the population are themselves a sample from some infinite superpopulation with mean μ_0 and variance σ_0^2. We still wish to estimate the population mean μ using a sample of n observations. The sample mean is an unbiased estimator of μ_0 with variance σ_0^2/n. To estimate μ we need to take the observed total $n\bar{x}$, add an estimate of the total of the $N-n$ unsampled members, and then divide by N. The total of unsampled members is itself random with mean $(N-n)\mu_0$ and variance $(N-n)\sigma_0^2$. We can estimate it in the same classical way as predictive inference is made about future observations from a linear model; see A**28.11**. Thus, this total is estimated by $(N-n)\bar{x}$, so that we still use $\bar{x}$ as an unbiased estimator of μ. Its variance is N^{-2} times the sum of the sampling variance of $(N-n)\bar{x}$, which is $(N-n)^2\sigma_0^2/n$, and the predictive variance $(N-n)\sigma_0^2$ of the total of unsampled members. The resulting variance of $\bar{x}$ is therefore $(N-n)\sigma_0^2/(Nn)$. The superpopulation theory yields the same estimator as the standard result, with the same variance except that the population variance σ^2 is replaced by the superpopulation variance σ_0^2. However, the expectation of σ^2 is σ_0^2, and the corresponding sample quantity $(n-1)^{-1}\sum_{i=1}^{n}(x_i - \bar{x})^2$ is an unbiased estimator of either. The primary difference between the two theories is that in the superpopulation approach the method of selecting the sample is not relevant. The dependence on the (ancillary) sampling mechanism is removed.

5.24 The superpopulation approach is very like the Bayesian method presented in **10.26**, since the assumed way in which the population has been drawn from the superpopulation yields the equivalent of a prior distribution for the ξ_is except that some aspects (parameters) of the superpopulation are still assumed to be unknown. However, this highlights the problem with it as a classical theory. In order for the assumed probabilities, with which the population is drawn from the superpopulation, to be acceptable as frequency probabilities the population itself must in some sense be just one trial of a potentially infinitely repeatable event. This is rarely acceptable. To adopt a superpopulation approach to sample survey theory typically entails abandoning the classical statistician's strict reliance on frequency probability. Then the main classical objection to Bayesian inference also disappears. The superpopulation model reflects only the way that the statistician perceives the structure of the population, using implicitly subjective probabilities. The classical statistician has avoided dependence on the sampling scheme, and the problem of degeneracy when conditioning on the sample, but only by accepting this key tenet of Bayesian statistics!

Further details of this debate can be found in Cassel et al. (1977).

Empirical Bayes

5.25 Similar considerations arise in the methodology termed 'empirical Bayes' by Robbins (1955). Suppose that a prior distribution is hypothesized for the parameters θ, but is only specified to be in a certain class of prior distributions. We can represent this by a prior distribution $f(\theta \mid \phi)$, where the *hyperparameters* ϕ index the family of priors. We can then construct a 'likelihood'

$$f(x \mid \phi) = \int f(x \mid \theta) f(\theta \mid \phi)\, \mathrm{d}\theta \tag{5.9}$$

relating the data to the hyperparameters. In the basic parametric empirical Bayes approach, ϕ is estimated from (5.9) by classical methods, such as an unbiased estimator, yielding an estimate $\hat{\phi}$. The prior is then taken to be $f(\theta \mid \hat{\phi})$ and inference about θ is by the appropriate Bayes rule for this prior.

Example 5.4

Suppose that observations x_i, $i = 1, 2, \ldots, n$, are independently distributed as $N(\theta_i,\ 1)$ given $\boldsymbol{\theta} = (\theta_1, \theta_2, \ldots, \theta_n)$. Now let the prior distribution for $\boldsymbol{\theta}$ be such that θ_i is distributed as $N(\mu,\ 1)$ and the θ_is are independent given the unknown parameter μ. Then from (5.9), the x_is are independent $N(\mu,\ 2)$ given μ, and so their mean $\bar{x}$ is the optimal classical estimator of μ. We now take $N(\bar{x},\ 1)$ as the prior distribution of each θ_i independently. The posterior distribution would then be such that θ_i is distributed as $N((x_i + \bar{x})/2,\ 1/2)$ and the θ_is are independent. If the decision problem were to estimate the θ_is with quadratic loss, the empirical Bayes estimator would be $\hat{\theta}_i = (x_i + \bar{x})/2$. This is the James–Stein problem and, unlike the usual classical estimator, the empirical Bayes estimator can be shown to be admissible. This estimator shrinks the classical estimates towards the sample mean, a property which is considered more generally in **6.42**.

5.26 The superpopulation approach of **5.23** can be expressed in a similar way, with $\theta = (\xi_1, \xi_2, \ldots, \xi_N)$ and $\phi = (\mu_0, \sigma_0^2)$. The superpopulation model provides $f(\theta \mid \phi)$, while $f(x \mid \theta)$ is the degenerate likelihood (5.8). There are nevertheless some differences. First, the approach used in **5.23** to finding the variance of the estimator $\bar{x}$ relies on all the information about the unobserved sample members coming direct from the hyperparameters: the inference is purely predictive here. In empirical Bayes as described in **5.25** no account is taken of the uncertainty in the estimate $\hat{\phi}$; the prior $f(\theta \mid \hat{\phi})$ is treated as *the* prior. While Morris (1983) has some suggestions, allowing for uncertainty in $\hat{\phi}$ remains a problem with the empirical Bayes method.

5.27 It may seem that the same objection applies to empirical Bayes as was raised for superpopulations in **5.24**, that to adopt $f(\theta \mid \phi)$ as an at least partially specified prior distribution means abandoning frequentist probability, and hence one might as well become fully Bayesian. After all, the usual classical objection to putting a distribution on parameters is that parameters are not random variables in the frequentist sense – they are not repeatable. Yet the status of parameters is not so clear cut. Example 5.4 is a typical empirical Bayes scenario, in which the θ_is can be thought of as a sample from some other distribution, $f(\theta \mid \phi)$. The x_is might, for instance, be observed yields of several varieties of some food crop. The θ_is are true underlying yield capabilities of the varieties, and to the extent that these varieties are merely a sample from the great collection of possible varieties that breeders might create, $f(\theta \mid \phi)$ could be a distribution based on frequency probabilities. Empirical Bayes was originally proposed in this context, in which case it is a purely classical statistical technique. The use of apparently Bayesian methods to estimate θ disguises the non-Bayesian treatment of the fundamental parameters ϕ. Empirical Bayes is not Bayesian because it does not admit a distribution for ϕ.

5.28 Further details of the empirical Bayes approach may be found in Berger (1986) and Deeley and Lindley (1981) and in references in A**31.51**. See also **6.35**, where the prior distribution is modelled in two stages beginning with $f(\theta \mid \phi)$ and continuing with $f(\phi)$. The second stage completes the full Bayesian treatment of models such as in Example 5.4. Another related idea is the Type II Maximum Likelihood (ML-II) method of Berger and Berliner (1986). They call (5.9) a Type II likelihood and set $\hat{\phi}$ to the maximum of (5.9). They then perform a Bayesian analysis with $f(\theta \mid \hat{\phi})$ as the prior. The differences from empirical Bayes are that $f(\theta \mid \phi)$ need not have any frequentist interpretation, and that the resulting posterior distribution is used for any appropriate posterior inference or summarization, without the need for a formal decision structure. ML-II shares with empirical Bayes the possibly substantial drawback that no account is generally taken of uncertainty in $\hat{\phi}$.

Fiducial inference

5.29 The fiducial approach has been described in A**31.26** to A**31.30**. The objective is to obtain a *fiducial distribution* for the parameters based on the data, without bringing in a prior disribution. The fiducial distribution is nevertheless intended to be used for inference in the same way as a posterior distribution. It is supposed to represent the information about θ given by the data alone, whereas the posterior distribution represents both data and prior information. Unfortunately, the mechanism for producing the fiducial distribution is ad hoc and has no natural status in probability theory. The numerous difficulties with the theory, outlined in Chapter A31, have meant that the fiducial approach has attracted only sporadic interest. The Bayesian method remains the only unambiguous, consistent way of attaching a probability distribution to parameters.

Pivots

5.30 The concept of a pivot has sometimes been used in arguments in favour of fiducial inference. A pivot is a function $P(x, \theta)$ of both data and parameter(s), whose distribution given θ is independent of θ. This property is deduced from the sampling distribution $f(x \mid \theta)$. Consider, for example, a single observation x, distributed as $N(\theta, 1)$ given θ. Then

$P(x, \theta) = x - \theta$ is distributed as $N(0, 1)$ independent of θ. The fiducial argument is then that this justifies a *fiducial* dsitribution for θ of $N(x, 1)$ based on the observed x.

To study this argument it is helpful to distinguish the random variables X and Θ from their possible realizations x and θ. Now the model asserts that X is distributed as $N(\theta, 1)$ given $\Theta = \theta$, for all θ. Therefore $X - \Theta$ is distributed as $N(0, 1)$ given $\Theta = \theta$, for all θ. Since this does not depend on θ, the random variable $X - \Theta$ is also distributed as $N(0, 1)$, as is $\Theta - X$. Now this is consistent with, but does not imply, the statement that $\Theta - X$ is distributed as $N(0, 1)$ given $X = x$, for all x. That would be equivalent to the distribution of Θ given $X = x$ being $N(x, 1)$ for all x, but as we have seen, this is not justified by the pivotal property alone. Indeed, there are many possibilities for the distribution of Θ given $X = x$, which is the posterior distribution, and this particular posterior distribution only arises from giving Θ the improper, uniform prior. Example 5.5 shows another possible solution. The fiducial distribution is intended to be a probability distribution for the parameter based only on the data x. The step of going from the pivotal property to the asserted fiducial distribution has, as we have said in **5.29**, no natural status in probability theory.

Example 5.5
Let the prior distribution of Θ be also $N(0, 1)$. Then we find the posterior distribution to be $N(x/2,\ 1/2)$. Therefore in this case the distribution of $\Theta - X$ given $X = x$ is $N(-x/2,\ 1/2)$, rather than $N(0, 1)$. This is still consistent with the unconditional distribution of $\Theta - X$ being $N(0, 1)$. For, the marginal distribution of X is $N(0, 2)$. Then

$$E(\Theta - X) = E\{E(\Theta - X \mid X)\} = E(-X/2) = 0,$$

$$\begin{aligned} \operatorname{var}(\Theta - X) &= E\{\operatorname{var}(\Theta - X \mid X)\} + \operatorname{var}\{E(\Theta - X \mid X)\} \\ &= E(1/2) + \operatorname{var}(-X/2) = 1/2 + 2(1/2)^2 = 1. \end{aligned}$$

5.31 Barnard (1980, 1985) proposes a mode of inference which he calls pivotal inference, the result of which is a statement of the distribution of the pivot $P(x, \theta)$ together with the observed value of x. He argues that the fiducial step of taking the pivotal distribution conditional on x is then only one interpretation which may be given to the pivotal inference. Pivots are used in classical inference, in a way which is perfectly proper in that theory, to construct confidence intervals. For instance if $X - \theta$ is distributed as $N(0, 1)$ for all θ then $P(|x - \theta| \leqslant 1.96 \mid \theta) = 0.95$ for all θ, leading to the 95% confidence interval $(x - 1.96, x + 1.96)$ for θ.

Objective probability

5.32 As mentioned in **1.16** and also in A**31.38**. it is quite possible to follow the Bayesian method in almost every respect without holding either a frequentist or a subjective view of probability. The *objective* notion of probability, also called 'logical' or 'necessary' probability is that $P(A \mid H)$ represents a degree of belief in the event A based on information H, but that this is not a subjective matter of an individual's personal degree of belief. Instead it is a unique, objective measure of the degree to which A is logically entailed by the evidence H. Objective probability is obviously also very diferent from frequentist probability, and like subjective probability does not require A to be repeatable. Objective probability can therefore apply to parameters in statistical models, Bayes' theorem is used to construct a posterior distribution, and inferences are drawn from it in the ways we have described as 'the Bayesian method'. Of course, the posterior distribution and inferences are now interpreted as the unique, objective probabilities and inferences for θ that are logically implied by the data and prior information. Conversely, the prior distribution must also be the unique distribution implied by prior information.

5.33 It is in the prior probability distribution that objective theorists have difficulty. From the subjective point of view, any probability can be specified by direct, subjective introspection, and Bayes' theorem is only a device to improve the accuracy of that process. The objective view must be that prior distributions can be derived uniquely by some formula. Bayes' theorem now becomes an essential part of that formula. For if any substantive prior information is available, we can regard the prior distribution as instead a posterior distribution, posterior to the acquisition of that prior information. It should then be possible to strip away that information with Bayes' theorem (used in reverse to deduce prior from posterior) and so arrive at a state of no information. The objective approach therefore begins with the apparently simpler task of finding logically consistent and realistic representations of complete prior ignorance about θ.

5.34 It is important to note that if this approach is to be practical then it must always be possible to construct a likelihood function for any information (as a unique, objective probability distribution $f(x \mid \theta)$). Much of what a subjectivist would regard as prior information is not readily treated in this way. It may be difficult to set down the prior information itself explicitly, yet a subjectivist could regard it as legitimate for the inference ultimately to be influenced by all kinds of prior information. The objectivist view would be that such prior opinions are too insubstantial to be treated as prior information in an 'objective' scientific theory. Therefore, any prior information which is not as explicit as the data, and does not admit a suitably 'objective' likelihood, is ignored and the prior distribution is specified using the rules for prior ignorance.

5.35 The task of finding 'logically consistent and realistic' representations of prior ignorance, as referred to in **5.33**, meets with the difficulties described in Chapter 3. In particular, we saw in **3.30** that the uniform distribution would not represent ignorance in this way. Ignorance about θ implies ignorance about $\phi = g(\theta)$ and this implies the following *invariance* condition. If the ignorance prior distribution for θ is $f_0(\theta)$, the ignorance prior for ϕ should be $f_0(g^{-1}(\phi))\,|dg^{-1}(\phi)/d\phi|$.

Jeffreys (1967, section 3.10) proposed a solution as follows. Let $I(\theta)$ be the Fisher information.

$$I(\theta) = -E\{\partial^2 \log f(x \mid \theta)/\partial\theta^2\} \tag{5.10}$$

In the case of a vector parameter, $I(\theta)$ is the matrix formed as minus the expectation of the matrix of second order partial derivatives of $\log f(x \mid \theta)$. The Jeffreys prior distribution is then

$$f_0(\theta) \propto |I(\theta)|^{1/2} \tag{5.11}$$

Example 5.6
If $x_1, x_2, \ldots, x_n$ are normally distributed with mean θ and known variance v, then

$$f(x \mid \theta) \propto \exp\{-n(\bar{x}-\theta)^2/(2v)\}$$

and $\partial^2 \log f(x \mid \theta)/\partial\theta^2 = -n/v$. Therefore, $f_0(\theta) \propto (n/v)^{1/2} \propto 1$, and the uniform distribution is the Jeffreys prior.

Example 5.7
If $x_1, x_2, \ldots, x_n$ are normally distributed with known mean m and variance θ, then

$$f(x \mid \theta) \propto \theta^{-n/2} \exp\{-s/(2\theta)\},$$

where $s = \sum_i (x_i - m)^2$. Then $\partial^2 \log f(x \mid \theta)/\partial\theta^2 = \{n/(2\theta^2)\} - \{s/\theta^3\}$, and since $E(s \mid \theta) = n\theta$, the Jeffreys prior is $f_0(\theta) \propto \theta^{-1}$.

Example 5.8
If $x_1, x_2, \ldots, x_n$ are distributed as $N(\mu, \sigma^2)$ with $\theta = (\mu, \sigma^2)$ then

$$f(x \mid \theta) \propto \sigma^{-n} \exp[-n\{s + (\bar{x}-\mu)^2\}/(2\sigma^2)\}],$$

where $s = n^{-1} \sum_i (x_i - \bar{x})^2$. Then

$$I(\theta) = E\begin{pmatrix} n/\sigma^2 & n(\bar{x}-\mu)/\sigma^4 \\ n(\bar{x}-\mu)/\sigma^4 & -\{n/(2\sigma^4)\} + n\{s + (\bar{x}-\mu)^2\}/\sigma^6 \end{pmatrix} = \begin{pmatrix} n/\sigma^2 & 0 \\ 0 & n/(2\sigma^4) \end{pmatrix}$$

and the Jeffreys prior is $f_0(\mu, \sigma^2) \propto \sigma^{-3}$.

5.36 Other justifications have been proposed for the Jeffreys prior. In particular, Box and Tiao (1973, section 1.3) argue that it assigns a uniform prior distribution to a function $\phi = g(\theta)$ in terms of which the likelihood is approximately 'data translated'. That is, $f(x \mid \phi)$ is approximately of the form $h(x - \phi)$ so that ϕ behaves like a location parameter. The use of the uniform prior distribution for a location parameter is supported by several authors, and is a particular case of the Jeffreys prior. Akaike (1978) arrives at Jeffreys prior as what he calls an 'impartial' distribution.

5.37 A number of objections can be made to the Jeffreys prior, the most important of which is that it depends on the form of the data. The prior distribution should only represent the prior information, and not be influenced by what data are to be collected. For instance, **A31.39** points out that there are different Jeffreys priors for the binomial and negative binomial experiments of **3.3**, so that posterior inference using the Jeffreys prior in each case will violate the Likelihood Principle.

5.38 Other objections concern the form taken by the Jeffreys prior in certain cases. Jeffreys himself found Examples 5.6 and 5.8 inconsistent, preferring in the case of Example 5.8 to use $f_0(\mu, \sigma^2) \propto \sigma^{-2}$, the product of the priors implied by Examples 5.6 and 5.7 (Jeffreys, 1967, page 182). He argued that in this case ignorance about μ and σ^2 should be represented by independent ignorance priors for the two parameters separately. It may be acceptable from the objectivist viewpoint to impose prior constraints like this as a form of prior information, but it is not clear under what circumstances independence should be imposed. It cannot be done consistently as a general rule, for instance if we transform from $\theta = (\theta_1, \theta_2)$ to $\phi = (\theta_1\theta_2, \theta_1/\theta_2)$. Villegas (1977a, 1981) discusses $f_0(\mu, \sigma^2) \propto \sigma^{-2}$ and $f_0(\mu, \sigma^2) \propto \sigma^{-3}$ as cases of the right- and left-invariant Haar measures, preferring the latter.

5.39 Another noteworthy case is the James–Stein problem. For if x_i are independently distributed as $N(\theta_i, 1)$, $i = 1, 2, \ldots, n$, the Jeffreys prior is $f(\boldsymbol{\theta}) \propto 1$, uniform over R^n. Then the posterior mean of each θ_i is x_i, but this estimator is inadmissible in the classical sense under quadratic loss, see **5.11**.

Maximum entropy

5.40 A different approach to the formal representation of ignorance is based on entropy. We derived entropy in **2.58** as the minimum of the expected loss of a logarithmic scoring rule, the unique impartial, symmetric, proper scoring rule. The entropy $H(f) = -\int_{-\infty}^{\infty} f(\theta) \log f(\theta) d\theta$ of the density $f(\theta)$ can be thought of as a measure of how uninformative $f(\theta)$ is about θ. For if we try to convey our information about θ as a general form of inference in the scoring rule framework, $H(f)$ is the lowest obtainable expected loss. If $H(f)$ is high, then the best decision is still poor. Now to represent prior ignorance we could use the prior density $f(\theta)$ which maximizes the entropy.

Example 5.9

Suppose that θ is discrete, with possible values $\theta_1, \theta_2, \ldots, \theta_k$. The prior distribution with maximum entropy will then maximize $\sum_{i=1}^{k} p_i \log p_i$, where $p_i = P(\theta = \theta_i) = f(\theta_i)$, subject to the condition that $\sum_{i=1}^{k} p_i = 1$. Defining $F = \sum p_i \log p_i + \lambda \sum p_i$, where λ is a Lagrange multiplier, $\partial F/\partial p_i = \log p_i + 1 + \lambda$. Equating this to zero yields the solution $p_i = k^{-1}$, $i = 1, 2, \ldots, k$. That is, the maximum entropy prior is the uniform distribution.

5.41 Entropy is a very familiar concept in physics as a measure of the amount of disorder and unpredictability in a system. Jeffreys was primarily a physicist, and the maximum entropy approach has been advocated forcefully by another physicist, Jaynes (1968, 1981, 1983). The primary criticism of this approach is that it is not invariant under change of parametrization, the problem which the Jeffreys prior was designed to avoid. In general, unrestricted maximization of entropy leads to a uniform prior distribution, which was shown to be sensitive to parametrization in **3.30**. However, the maximum entropy approach yields interesting results if we allow constraints to be applied.

Example 5.10

Suppose that θ is discrete as in Example 5.9, but that we add the further constraints that $E\{g_t(\theta)\} = r_t$, for $t = 1, 2, \ldots, s$. Then maximizing $F = \sum_i p_i \log p_i + \lambda \sum_i p_i + \sum_t \mu_t \sum_i p_i g_{ti}$ with $g_{ti} = g_t(\theta_i)$ and Lagrange multipliers λ, $\mu_1, \mu_2, \ldots, \mu_s$, we find $f(\theta) \propto \exp\{-\sum_{t=1}^{s} \mu_t g_t(\theta)\}$. The proportionality constant and the values of the multipliers μ_t are then found by applying the constraints. For instance if $s = 1$ and we apply the single constraint that $E(\theta) = m$ we have $f(\theta) \propto \exp(-\mu\theta)$ and μ is given by $\sum_{i=1}^{k} \theta_i \exp(-\mu\theta_i) = m \sum_{i=1}^{k} \exp(-\mu\theta_i)$.

5.42 Although the mathematics is more difficult for continuous θ, the same result can be shown to hold as in Example 5.11. Indeed it is a straightforward excercise in the calculus of variations to show that entropy is maximized subject to $E\{g_t(\theta)\} = r_t, t = 1, 2, \ldots, s$, by

$$f(\theta) \propto \exp\{\textstyle\sum_{t=1}^{s} c_t g_t(\theta)\}, \tag{5.12}$$

with the constants c_t determined by the constraints.

Example 5.11
For $\theta \geqslant 0$, the prior distribution with maximum entropy subject to $E(\theta) = m$ is given by (5.12) as $f(\theta) \propto e^{c\theta}$, which gives the exponential distribution with mean m, $f(\theta) = m^{-1}\exp(-\theta/m)$.

Example 5.12
The maximum entropy prior distribution for scalar θ subject to having specified mean and variance is the normal distribution. If all prior first and second order moments are specified for a vector θ, the maximum entropy distribution is again easily seen to be the multivariate normal distribution with those moments.

Example 5.13
For scalar $\theta \in [a, b]$ suppose that we specify the prior distribution function $P(\theta \leqslant c_t) = r_t$ at s points $a < c_1 < c_2 < \ldots < c_s < b$, $0 < r_1 < r_2 < \ldots < r_s < 1$. Then we have $g_t(\theta) = 1$ for $\theta \leqslant c_t$ and zero otherwise. Applying (5.12) gives the result that $f(\theta)$ is piecewise uniform on the subintervals $(a, c_1), (c_1, c_2), \ldots, (c_s, b)$.

5.43 It is not clear how $E\{g_t(\theta)\} = r_t$ could be an objective part of prior information, and it is therefore unclear what role (5.12) could play in a fully objectivist theory of inference. It is, however, of interest within more usual Bayesian theory based on subjective probability; see **6.21**

Reference priors
5.44 Bernardo (1979a) develops the entropy approach further using the entropy information measure developed in **3.26**. The expected amount of information provided by observing x is given by

$$H\{f(\theta)\} - E[H\{f(\theta \mid x)\}], \tag{5.13}$$

where the expectation is over the preposterior distribution $f(x)$ of x. If the experiment yielding x were to be repeated, giving a new observation independent of x given θ and with the same distribution, the posterior distribution would be expected to show a further reduction in entropy, representing the expected information in the second observation. If this were repeated indefinitely we would eventually learn θ exactly and so remove all the entropy in the original prior distribution.

5.45 Let $f_k(\theta)$ be the prior distribution maximizing the expected information about θ to be provided by a hypothetical k independent replications of the experiment yielding data $(x_1, x_2, \ldots, x_k)$. Then the *reference* posterior distribution after the actual experiment has produced the single observation x is $f_0(\theta \mid x) = \lim_{k\to\infty} f_k(\theta \mid x)$, where $f_k(\theta \mid x) \propto f_k(\theta) f(x \mid \theta)$. Finally, the reference prior distribution $f_0(\theta)$ is defined as satisfying $f_0(\theta \mid x) \propto f_0(\theta) f(x \mid \theta)$. This is not necessarily $\lim_{k\to\infty} f_k(\theta)$, because of the possibility of improper distributions arising. $f_0(\theta \mid x)$ will in general be a proper limit of proper posterior distributions, and $f_0(\theta)$ is then defined relative to this well-behaved limit.

5.46 In the case of discrete θ taking a finite number of possible values, this process reduces simply to maximizing prior entropy, and so gives the uniform distribution. However, Bernardo (1979a) shows that this is not the case for continuous θ. Using the limiting posterior distribution, which **3.23** shows depends on the Fisher information matrix (5.10), it is shown that under appropriate regularity conditions the reference prior distribution is the Jeffreys prior (5.11). This derivation of the Jeffreys prior from an entropy formulation nicely unifies the two approaches to specifying prior ignorance. Notice that although entropy is affected by reparametrization, the expected entropy information (5.13) is not. This fact makes it possible for Jeffreys' invariant prior to be achieved.

Further development of reference priors is given by Berger and Bernardo (1992), and related ideas by Clarke and Wasserman (1992). The relationship between entropy and the Jeffreys prior is further explored by Clarke and Barron (1990).

Implications

5.47 Although the objective view of probability leads to the use of Bayes' theorem, with all the characteristics of the Bayesian method for posterior inference, the way in which the prior distribution is derived implies some important differences.

(a) If the Jeffreys prior is used, the posterior distribution will depend on the form of the experiment in a way that violates the Likelihood Principle.

(b) If the prior distribution is defined by maximizing entropy, then posterior inference will not be coherent in respect of reparametrization. For instance, uniform priors for a positive parameter θ and for $\phi = \theta^2$ produce the result that $P(\theta \leqslant a \mid x) \neq P(\phi \leqslant a^2 \mid x)$. This approach requires a judgment (which it is hard to see as objective) of a natural parametrization.

(c) If subjective prior information exists then it may not be reflected in the posterior distribution.

5.48 The claims of the Jeffreys prior or maximum entropy prior to provide objective representations of prior ignorance must fail because of points (a) and (b) above. The ideal of inference based on objective probability has yet to be given a satisfactory practical form.

5.49 Nevertheless, formulations which may be interpreted as objective prior distributions representing prior ignorance can be of value within a subjective Bayesian framework.

As discussed in **4.35** and in **3.29**, such distributions (even when improper) may be used to represent *weak* prior information. Reference prior distributions were proposed by Bernardo (1979a) partly for this purpose and partly to act as a reference point in the following sense. When You have genuine prior information expressed in a proper prior distribution $f(\theta)$, Bernardo argues that the strength of that prior information might be assessed by comparing the inferences You thereby obtain with those You would have obtained by using a reference prior. This indicates another possible use of 'ignorance priors' in a subjective Bayesian analysis.

EXERCISES

5.1 Let x be the number of successes in n independent trials with probability θ of success in each trial, but censored so that only positive x is observed. That is

$$P(X = x \mid \theta) = \binom{n}{x}\theta^x(1-\theta)^{n-x}/\{1-(1-\theta)^n\}$$

for $x = 1, 2, \ldots, n$. Prove that no unbiased estimator of θ exists.

5.2 A single observation is distributed given θ as $N(\sqrt{\theta}, \theta)$. Show that the estimator of the form $t(x) = ax^2$ with minimum mean square error is $t(x) = x^2/5$, and hence that both the maximum likelihood estimator and the unbiased estimator $x^2/2$ are inadmissible with respect to squared error loss. Show further by means of the minimum variance bound (A**17.15**) that every unbiased estimator is inadmissible.

5.3 Let x be distributed as $N(\theta, 1)$ given θ, and let the prior distribution for θ be $N(0, 1)$. Consider estimating θ by d with loss function

$$L(d, \theta) = (d-\theta)^2 \exp(3\theta^2/4)$$

Prove that the Bayes estimator is then $d(x) = 2x$, but that this has uniformly higher risk than $t(x) = x$, and is therefore inadmissible. Confirm that in this problem the expected risk is infinite. (Berger (1985), page 254)

5.4 Let x_1 and x_2 be independently distributed as $N(\theta_i, 1), i = 1, 2$, given (θ_1, θ_2). Show that

$$R(x_1, x_2) = \{(x_1, x_2) : (x_1 - \rho x_2)^2 \leqslant c_\gamma^2(1+\rho^2)\}$$

is a classical $100\gamma\%$ confidence interval for $\rho = \theta_1/\theta_2$, where c_γ is the symmetric $100\gamma\%$ standard normal point, i.e. $\Phi(c_\gamma) = (1+\gamma)/2$. Prove, however, that if $x_1^2 + x_2^2 < c_\gamma^2$ then $R(x_1, x_2) = (-\infty, \infty)$, and hence that this defines a positively biased relevant subset.

5.5 Let x be the number of successes in n independent trials with probability $c\theta$ of success in each trial, where $c \in [0, 1]$ is a known constant. Show that the Jeffreys prior distribution is

$$f(\theta) \propto \theta^{-1/2}(1-c\theta)^{-1/2},$$

and so depends on c.

5.6 In the same context as Exercise 3.1, prove that different Jeffreys prior distributions apply when x is fixed or when t is fixed.

5.7 Consider specifying a prior distribution for a parameter $\theta \in (-\infty, \infty)$ by maximum entropy subject to the constraint that $E\{\log(a+\theta^2)\} = t$. Prove that the maximum entropy distribution is $f(\theta) \propto (a+\theta^2)^{-b}$, where $b > 1/2$ satisfies

$$\begin{aligned} t &= \{\textstyle\int_{-\infty}^{\infty}(a+\theta^2)^{-b}\log(a+\theta^2)\,\mathrm{d}\theta\}/\{\textstyle\int_{-\infty}^{\infty}(a+\theta^2)^{-b}\,\mathrm{d}\theta\} \\ &= \log a + \psi(b) - \psi(b-1/2), \end{aligned}$$

where ψ is the digamma function $\psi(x) = \mathrm{d}\log\Gamma(x)/\mathrm{d}x$.

CHAPTER 6

SUBJECTIVE PRIOR DISTRIBUTIONS

6.1 The most obvious practical difference between classical and Bayesian inference lies in the Bayesian's use of prior information, and careful specification of the prior distribution is of great importance. This chapter is concerned with techniques for prior specification. We begin by reviewing ideas for specifying individual probabilities and the basic method of specifying a prior distribution by stating some appropriate properties or summaries. The concepts of conjugacy and conditional conjugancy are introduced at this point, to provide convenient families of prior distributions. The range of useful and usable prior distributions for a given problem is further increased by consideration of hierarchical prior modelling and the use of mixture distributions. The chapter ends with discussion of the Bayes Linear Estimator, which represents a compromise of the full Bayesian method, allowing much more limited specification of prior information and likelihood.

Specifying probabilities

6.2 Previous chapters have touched on the problems of specifying individual prior probabilities. The inevitable imprecision of subjective probability judgements is described in **4.26** and the technique of elaboration to improve such judgments is introduced in **4.28**. The decision-theoretic device of scoring rules is considered in **2.54**. The connection between probability and frequency, explored in **4.39** to **4.46**, is also valuable, since one natural way of specifying a probability is to think about what proportion of events 'of this type' one would expect to occur. De Finetti's theorem says that the probability of a single event equals the expectation of the limiting frequency of occurrence of events with which it is exchangeable.

6.3 In order to improve techniques of subjective probability specification, psychologists have studied how people make probability judgements in practice. Edwards and Phillips (1964), Tversky (1974), Tversky and Kahneman (1974), Lichtenstein *et al.* (1982), Alpert and Raiffa (1982) and Garthwaite (1989) described several common errors of judgement. It is clear that in general people do not naturally have good habits of subjective probability assessment. Statisticians and others who regularly need to make careful, formal specifications of probabilities should be explicitly trained to do so, and to avoid known pitfalls.

More often, of course, a statistical analysis of real data demands not the prior probability judgments of the statistician but of the scientist, experimenter or decision-maker who has collected the data, and who has an interest in the inferences. Then the statistician must also develop skills of eliciting prior probabilities from others. Methods for eliciting probability distributions, notably for the parameters of linear models, are proposed and discussed by Kadane (1980), Winkler (1980), Kadane *et al.* (1980), Garthwaite and Dickey (1988, 1992) and Garthwaite (1992).

Calibration

6.4 A person's probability judgements are said to be *calibrated* if they agree with acutal frequencies of occurrence. That is, if we consider the sequence of all events $A_{p,1}$, $A_{p,2}\ldots$ to which you assign the probability p, then you are calibrated if the limiting frequency of occurrence of events in this sequence is p, and if this is true for all p in [0,1]. A person who is not calibrated may reasonably be supposed to be a poor judge of probability. To see the strength of this argument, suppose that You judge the events $A_{p,1}$, $A_{p,2}\ldots$ to be exchangeable. You have given each of these a probability p, but in a *long sequence* of such events You observe that a proportion $q \neq p$ of them occur. Then Your posterior distribution for the true limiting frequency, say θ_p, will be concentrated on the observed frequency q. Your posterior probability that θ_p equals p will be small, regardless of Your prior beliefs, and You will therefore believe that You are not calibrated. Furthermore, Your probability now for any future event in this sequence is Your posterior expectation of θ_p, which will also be close to q. So you cannot continue to assign probability p to events in this sequence after You have enough evidence to convince You that these judgements are not calibrated. Instead, You must recalibrate Yourself by assigning probabilities in future which agree with the observed frequency.

6.5 In reality, the events $A_{p,i}$ will not be genuinely exchangeable. Although $E(A_{p,i}) = p$ for all i, it will not be true that $E(A_{p,i}$ and $A_{p,j})$ is the same for all i and $j \neq i$. Within the sequence there may be one or more subsequences of related events, where the probability p has been assigned for similar reasons, so that events in a subsequence may be exchangeable. But You would not necessarily assign the same probability to two related events both occurring as to two unrelated events.

Nevertheless, suppose that You observe a long sequence of such events that is typical of the sequence as a whole, so that for instance the average value of $P(A_{p,i}$ and $A_{p,j})$ in the observed sequence is the same as in the sequence as a whole. Then it is generally true that if the proportion of events that occur in the observed sequence is q, it can still be shown that You should now believe that the proportion that will occur in the remainder of the sequence as a whole will be close to q. Therefore it is not possible for You to continue to assert a probability p for all future events in the sequence.

6.6 Calibration is no guarantee of good probability judgements. The fact that a proportion p of the events to which you assign probability p occur does not imply that p is an accurate measure of Your degree of belief in those events. Consider the sequence of *all* events for which You are called upon to specify Your probability. Suppose that You decide to give a probability of 0.5 to the first such event that You feel certain will occur, and to the first event arising *after* that one that You feel certain will *not* occur, and then to the first subsequent event that You judge is certain to occur, and so on. Such a rule would not change Your calibration, but would be poor probability specification.

De Groot and Fienberg (1983) introduce the idea of *refinement* of probability assessors. Assessor R is more refined than assessor S if Ss probability specifications can be represented

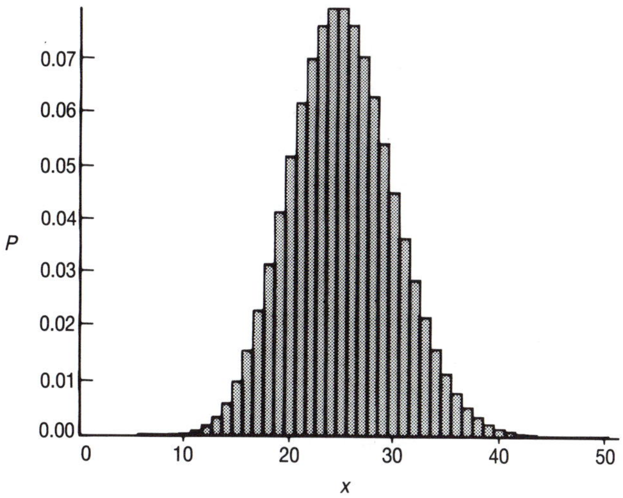

Fig. 6.1 Poisson distribution with mean 25

as being random perturbations of Rs. As we have seen, both R and S can be calibrated, but Rs probabilities represent better judgment than Ss.

Specifying distributions by summaries

6.7 We now consider specification of prior distributions. For a discrete θ taking only a few possible values, it is possible to specify its distribution by individually specifying its probabilities $P(\theta = \theta_i)$. Otherwise, this procedure is unnecessarily inaccurate, even for discrete θ. For continuous θ it is obviously impossible. Consider instead the effect of specifying a few summaries of Your prior distribution for θ. For a scalar θ these might include mean, mode, standard deviation. If we add to these Your assessment that the distribution is unimodal, $f(\theta)$ is already rather precisely defined. Any two distributions satisfying these conditions would generally look somewhat similar. Individual probabilities would not vary greatly. More important, we would not expect posterior inferences to be sensitive to the precise choice of prior from among the class satisfying these conditions.

Example 6.1

Figure 6.1 shows the Poisson distribution with mean 25. It has mode at 24 and 25 and standard deviation 5. It is hard to construct another distribution with these values and the same general, unimodal shape as the Poisson distribution but with radically different probabilities. Within those constraints any plausible specification of a precise distribution would be expected to produce very similar posterior inferences to those induced by the Poisson prior.

6.8 A simple and effective technique for specifying a prior distribution, therefore, is to specify an appropriate selection of summaries, and then to adopt *any* convenient distribution that conforms to these conditions. It is, of course, always possible to find models and data for which posterior inference is sensitive to even very small changes in the prior distribution, so this choice should always be subject to sensitivity analysis. Nevertheless. provided enough well-chosen summaries are specified then, for most forms of data, posterior inference should be insensitive to the arbitrary final choice of distribution. This general approach, and the ideas of sensitivity analysis, are also discussed in **4.31–4.34**. Sensitivity and the related concept of robustness are fully developed in Chapter 7.

6.9 Subjective specification of a distribution in this way is essentially the converse of summarization. Summarization seeks to express the principal features of a distribution in readily understood terms. The specification process takes information expressed in those same readily understood terms and tries to find an appropriate distribution. In either direction, summaries are the natural way of conveying information about θ. For instance, to be told that the posterior mean of θ is 1.5 conveys a useful point estimate of θ. Conversely, one may make a prior estimate of θ by specifying the prior mean $E(\theta)$. It is, of course, important to understand the sense in which the mean represents an estimate, or location summary, of a distribution. To be told that the mode is 1.5 conveys slightly different information, so it is not enough just to assert a prior estimate when specifying the prior distribution. Is it to be the prior mean, mode or some other location summary?

Similar considerations arise in eliciting a prior distribution from somebody else. A request for a prior estimate of θ should explain what sort of estimate is needed — the most probable value (mode), a value such that θ is equally likely to lie above or below (median), or an expected value in the sense of a long-run average (mean) — and also needs to explain how these things are different. In the same way that the posterior mean is easy to misinterpret as a location summary, because of its dependence on tail weight, it can be hard to assess a prior mean reliably. Unfortunately, it is usually much easier to find a distribution with a given mean (among other conditions) than with a given median.

Overfitting and feed-back

6.10 The question of how many summaries to use can be answered, albeit imprecisely. The objective should be to use all those summaries that express Your most clearly held and easily expressed beliefs about θ. It is almost always possible to assert a range of probable values for θ, in terms of a location summary and a dispersion summary. It is also generally reasonable to suppose that $f(\theta)$ is unimodal. A bimodal distribution would only arise in response to a firmly held belief that θ is likely to fall in one of two ranges but less likely to be in between.

Having formulated every important feature of Your prior information in this way, then what is left to specify represents opinions that are less easy to formulate and to specify reliably. A distribution $f(\theta)$ could be chosen now to fit the stated summaries. It is, however, valuable to take the process further in order to verify that this distribution indeed represents Your prior information. *Overfitting* consists of specifying some further summaries and checking that the chosen distribution also fits these closely enough. Since these extra summaries will typically not represent such firmly held views, close agreement may not be required. But a substantial difference would warn that an alternative distribution should be chosen.

Feedback consists of computing further summaries of the chosen distribution and checking that they conform to Your actual prior information. Again, close agreement is not generally expected, but any surprising implication of the chosen distribution would indicate a lack of fit of the distribution to the prior information.

Thus the role of summaries, of communicating information in both directions, can be used quite flexibly in the prior specification process.

Families of distributions closed under sampling

6.11 We now address the question of how to select a distribution satisfying the specified prior summaries. Given that at this point it is acceptable to use any distribution that agrees with those summaries, the choice is arbitrary. It is usually made on grounds of convenience, in one of two senses. It may just be the first distribution that springs to mind to fit the summaries, or a distribution of a general form for which it is easy to fit them. For instance, if prior mean and variance of a scalar θ are given, then an obvious choice is the normal distribution with the given mean and variance. Or if θ is necessarily positive, we could instead fit a gamma distribution having those moments.

6.12 Although prior distributions are very often chosen in this way, particularly in complex problems, there is a more important sense in which a choice of prior can be convenient. Suppose that $x_1, x_2, \dots, x_n$ are independent given θ and distributed as $N(\theta, v)$. Then we have seen, for instance in Examples 1.5 and 3.3, that if the prior distribution of θ is normal the posterior is also normal. A normal posterior distribution is convenient in the sense that its summaries are well known, so this posterior distribution is easily understood. If we had used some other form of prior distribution, as for instance a t distribution, then the task of posterior summarization would have been much more difficult. In general, if a prior distribution could be chosen such that, whatever actual values of the data might later be observed, the posterior will be easy to summarize, then this would be a very convenient choice.

6.13 Suppose that data x are to be observed with distribution $f(x \mid \theta)$. A family $\mathscr{F}$ of prior distributions for θ is said to be *closed under sampling* from $f(x \mid \theta)$ if for every prior distribution $f(\theta) \in \mathscr{F}$ the posterior distribution $f(\theta \mid x) \propto f(\theta)f(x \mid \theta)$ is also in $\mathscr{F}$. If $\mathscr{F}$ is closed under sampling from $f(x \mid \theta)$ it follows immediately that if we obtain any number of sets of data $x_1, x_2, \dots, x_n$, all independently distributed as $f(x \mid \theta)$ given θ, then the posterior distribution will remain in $\mathscr{F}$. (Consider updating the posterior sequentially as described in **3.6**.)

Example 6.2

Example 1.5 shows that the normal family is closed under sampling of a single observation from $f(x \mid \theta) \propto \exp\{-(x-\theta)^2/(2v)\}$. The fact that it remains closed under sampling for data $x_1, x_2, \dots, x_n$ independently distributed as $N(\theta, v)$ is then immediate. In fact, Example

1.5 shows that the posterior variance is always less than the prior variance. Therefore, the subfamily of all normal distributions having variance less than or equal to any fixed σ_0^2 is also closed under sampling in this problem.

Example 6.3
If x is binomially or negative binomially distributed with $f(x \mid \theta) \propto \theta^x(1-\theta)^{n-x}$ then the family of beta distributions is closed under sampling; see Example 1.4.

Example 6.4
Let x be uniformly distributed given θ on the range $(0, \theta)$. Thus, $f(x \mid \theta) = \theta^{-1}$ for $0 < x < \theta$. Let $g(\theta; a, r) = ra^r\theta^{-r-1}$ for $\theta > a$ (and $g(\theta; a, r) = 0$ otherwise). Then the family $\mathscr{F} = \{g(\theta; a, r) : a \geqslant 0, r > 0\}$ is easily seen to be closed under sampling. If the prior is $f(\theta) = g(\theta; a, r)$ then the posterior is $f(\theta \mid x) = g(\theta; a^\star, r+1)$, where $a^\star = \max(a, x)$.

6.14 If the distributions in $\mathscr{F}$ are easy to summarize, and $\mathscr{F}$ is closed under sampling in the problem at hand, then whatever prior distribution we choose in $\mathscr{F}$ the posterior will be easy to summarize, for any observed data. We may also have the other form of convenience described in **6.11**. For if these distributions are easy to summarize the reverse process, of finding a distribution in $\mathscr{F}$ to match specified prior summaries, will often be straightforward. However, this will depend on how rich a diversity of distributions is contained in $\mathscr{F}$. For instance, if You specify that Your prior distribution for θ is skew or bimodal then no member of the normal family will fit that summary.

The ideal is to find a family $\mathscr{F}$ that is closed under sampling, easy to summarize, and rich enough to contain distributions matching almost any prior beliefs. Unfortunately, the last two objectives tend to conflict. The set of all distributions is closed under sampling and rich enough to match any prior beliefs, but its members are not uniformly easy to summarize!

6.15 Also, given any family $\mathscr{F}$ that is closed under sampling, it is easy to make that family larger. Given $f(\theta) \in \mathscr{F}$ define $f^\star(\theta) = g(\theta)f(\theta) / \int g(\theta)f(\theta)\,\mathrm{d}\theta$, where $g(\theta)$ is any given, positive, bounded function. Then the family $\mathscr{F}^\star$ of $f^\star(\theta)$ generated by $f(\theta) \in \mathscr{F}$ is also closed under sampling, and so is $\mathscr{F} \cup \mathscr{F}^\star$. However, for arbitrary $g(\theta)$ it is likely that $f^\star(\theta) \propto g(\theta)f(\theta)$ will not be simple to summarize, and in particular that the normalizing constant $\int g(\theta)f(\theta)\,\mathrm{d}\theta$ cannot be obtained analytically.

Conjugate families

6.16 Suppose that $f(x \mid \theta)$ is a member of the k-parameter exponential family, see **A17.39**, with density

$$f(x \mid \theta) = \exp\{\textstyle\sum_{j=1}^k A_j(\theta)B_j(x) + C(x) + D(\theta)\}. \tag{6.1}$$

Then a family $\mathscr{F}$ that is closed under sampling is easily identified with members

$$f(\theta; a_1, a_2, \ldots, a_k, d) = \exp\{\textstyle\sum_{j=1}^k a_j A_j(\theta) + d\,D(\theta) + c(a_1, a_2, \ldots, a_k, d)\}, \tag{6.2}$$

where

$$c(a_1, a_2, \ldots, a_k, d) = -\log \int \exp\{\textstyle\sum_{j=1}^k a_j A_j(\theta) + d\, D(\theta)\}\, d\theta$$

is the normalizing constant. The posterior arising from the prior (6.2) is then $f(\theta; a_1 + B_1(x), a_2 + B_2(x), \ldots, a_k + B_k(x), d + 1)$. We can recognize (6.2) as a $(k+1)$-parameter exponential family for θ. The family is defined formally as containing all those distributions (6.2) indexed by values $a_1, a_2, \ldots, a_k, d$ leading to proper density functions, i.e. those for which $c(a_1, a_2, \ldots, a_k, d)$ exists as defined.

6.17 A more general $(k+1)$-parameter exponential family than (6.2) can be written as

$$f^\star(\theta; a_1, a_2, \ldots, a_k, d, e_1, e_2, \ldots, e_s) = \exp\{\textstyle\sum_{j=1}^k a_j A_j(\theta) + d\, D(\theta) + \sum_{r=1}^s e_r E_r(\theta) + c^\star\} \quad (6.3)$$

with

$$\begin{aligned} c^\star &= c^\star(a_1, a_2, \ldots, a_k, d, e_1, e_2, \ldots, e_s) \\ &= -\log \int \exp\{\textstyle\sum_{j=1}^k a_j A_j(\theta) + d\, D(\theta) + \sum_{r=1}^s e_r E_r(\theta)\}\, d\theta. \end{aligned}$$

The family $\mathscr{F}^\star$ with members $f^\star$ is also an exponential family and is also closed under sampling from (6.1), and can be seen to be obtained from $\mathscr{F}$ by the device of **6.15**, with $g(\theta) = \exp\{\sum_{r=1}^s e_r E_r(\theta)\}$. A family $\mathscr{F}^\star$ with members (6.3) is called *conjugate* to (6.1), and the special case of $\mathscr{F}$ with members (6.2) is called the *natural conjugate* family to (6.1).

Example 6.5

Example 6.3 shows clearly a natural conjugate family. Observations are taken from the Bernoulli distribution $f(x \mid \theta) = \theta^x (1-\theta)^{1-x}$, $x = 0, 1$, which is a case of (6.1) with $k = 1$, $A_1(\theta) = \log\{\theta/(1-\theta)\}$, $B_1(x) = x$, $C(x) = 0$ and $D(\theta) = \log(1-\theta)$. The natural conjugate form (6.2) is

$$f(\theta; a_1, d) = \exp\{a_1 A_1(\theta) + d\, D(\theta) + c(a_1, d)\} = \theta^{a_1}(1-\theta)^{d-a_1} \exp\{c(a_1, d)\},$$

which is the family of beta distributions normalized by $c(a_1, d) = -\log B(a_1 + 1, d - a_1 + 1)$. The original exponential family has one parameter so the natural conjugate family has two.

Example 6.6

Suppose that observations $x_1, x_2, \ldots, x_n$ are independently distributed as $N(\mu, \sigma^2)$. This is an exponential family with $k = 2$ parameters, $A_1(\theta) = \sigma^{-2}$, $B_1(x) = -x^2/2$, $A_2(\theta) = \mu/\sigma^2$, $B_2(x) = x$, $C(x) = 0$, $D(\theta) = -(\log(2\pi\sigma^2) + \mu^2/\sigma^2)/2$. The natural conjugate family is therefore

$$\begin{aligned} f(\theta; a_1, a_2, d) &= \exp\{a_1 A_1(\theta) + a_2 A_2(\theta) + d\, D(\theta) + c(a_1, a_2, d)\} \\ &\propto (\sigma^2)^{-d/2} \exp\{-d(\mu - m)^2/(2\sigma^2) - b\sigma^{-2}\}, \end{aligned} \quad (6.4)$$

where $m = a_2/d$ and $b = -a_1 - a_2^2/(2d)$. Although this is a three-parameter family, it is equally convenient to work with a four-parameter family in which the two occurences of d in (6.4) are allowed to differ. We are therefore led from the natural conjugate (6.4) to the more general conjugate family given by (6.3) with $s = 1$, $E_1(\theta) = \log(\sigma^2)$ and $e_1 = -t + d/2$,

$$f(\theta; m, d, b, t) \propto (\sigma^2)^{-t} \exp\{-d(\mu - m)^2/(2\sigma^2) - b\sigma^{-2}\}. \tag{6.5}$$

6.18 The examples we have given of families closed under sampling have all themselves been expressed in terms of small numbers of parameters, the members of the family being indexed by the different possible values of these parameters. To distinguish these parameters indexing the family of prior distributions from the parameters θ about which we wish to make inference, we call them *hyperparameters*. Thus in Example 6.6 we consider inference about $\boldsymbol{\theta} = (\mu, \sigma^2)$, so that μ and σ^2 are *the* parameters, while the quantities m, d, b and t appearing in (6.5) are hyperparameters. In general, the conjugate family (6.3) has $k + 1 + s$ hyperparameters.

6.19 It is clearly desirable to have a prior family which is indexed by only a few hyperparameters. Otherwise specification of the prior distribution becomes a laborious task. In general, to identify a particular prior distribution within a family indexed by n hyperparameters requires n prior summaries to be given and then a (possibly very complex) system of n equations to be solved. Also, posterior summarization will tend to become correspondingly difficult.

Suppose, therefore, that we wish the family $\mathscr{F}$ to be indexed by only a finite number of parameters. In order for $\mathscr{F}$ to be closed under sampling from $f(x \mid \theta)$, it is clear that the likelihood $\prod_{i=1}^{n} f(x_i \mid \theta)$ must involve only a finite number of different functions of $\mathbf{x} = (x_1, x_2, \ldots, x_n)$ for arbitrarily large n. That is, this likelihood must possess a finite number of sufficient statistics, which is shown in A**17.39** to imply (given regularity conditions) that $f(x \mid \theta)$ is a member of the general exponential family (6.1). Therefore, the conjugate prior family (6.3) for sampling from the exponential family (6.1) is (subject to those regularity conditions) the only instance of a family which is closed under sampling and also finitely parametrizable.

Example 6.4 shows an exceptional case where the usual regularity conditions are broken by the range of possible x values being dependent on θ. Sufficiency in such cases is dealt with in A**17.40** and A**17.41**. Except in such cases, if $f(x \mid \theta)$ is not an exponential family distribution any family which is closed under sampling from $f(x \mid \theta)$ must be so large and complex that it cannot be indexed by a finite number of hyperparameters. This makes such families typically impractical to work with. In practice, then, the notion of a family that is closed under sampling is useful only in the case of sampling from exponential families.

6.20 Greater flexibility in accommodating prior beliefs can be achieved by increasing the number of hyperparameters in a conjugate family. Consider the family (6.5) in Example 6.6. It has the property that the conditional distribution of μ given σ^2 is normal with mean m and variance σ^2/d. Therefore all members of this family incorporate a particular form of dependence between the two parameters μ and σ^2, in which the value of σ^2 affects conditional uncertainty about μ. It is not possible within this family to represent prior information about μ and σ^2 that does not conform to this dependence. It is, for instance, not possible to accommodate prior independence between μ and σ^2. Consider now the six-parameter family

$$f(\theta; m_1, m_2, v, b, d, t) \propto (\sigma^2)^{-t} \exp[-(\mu - m_1)^2/(2v) - \{b + d(\mu - m_2)^2\}/(2\sigma^2)], \qquad (6.6)$$

which is easily shown to follow the general conjugate form (6.3). If $d = 0$ we now have independence between μ and σ^2. Having moved from the natural conjugate family (6.4) with three hyperparameters to this more general conjugate family with six hyperparameters, summarization is inevitably more difficult. Within the subfamily given by $d = 0$, of course, we have independent distributions for μ and σ^2 that are easy to summarize. Therefore it is simple to specify a member of this family to represent prior beliefs in the case of prior independence (which motivated our use of (6.6)). However, the posterior distribution will not be a member of this subfamily (i.e. the subfamily is not closed under sampling), and posterior summarization will not be so straightforward. We shall return to this issue in **6.33**.

Choice of prior by maximum entropy

6.21 Another way to select a prior distribution from among the class of distributions satisfying some given prior summaries is to use maximum entropy. According to the development in **2.58**, this will represent in some sense a distribution with as little prior information as possible, apart from the information supplied by the given prior summaries. We are then faced with a constrained optimization problem, to choose $f(\theta)$ to maximize the prior entropy (2.54) subject to specified constraints. In the case where those constraints are all in the form of expectations,

$$\int A_j(\theta) f(\theta)\, d\theta = c_j, \qquad j = 1, 2, \ldots, k \qquad (6.7)$$

it is printed in **5.41** that the solution has the exponential family form

$$f(\theta) = \exp\{\textstyle\sum_{j=1}^{k} a_j A_j(\theta) + c\},$$

where the a_js are chosen to satisfy the constraints (6.7) and c to normalize the density. For example, for a scalar θ taking values in $(-\infty, \infty)$ if the prior mean and variance are specified the maximum entropy prior is the normal distribution with the given mean and variance (Example 5.12).

Specification by posterior properties

6.22 In addition to using prior summaries, the prior distribution may be specified by stating some required summaries of the posterior distribution. Before developing this idea we should consider whether in general it is acceptable to base the choice of prior distribution on what You would like to see in the posterior distribution. To do so might appear to be 'cheating'. Data are often collected in the hope of being able to support a particular conclusion. They should of course also be able to provide evidence against the desired conclusion, but if You are allowed to choose the prior distribution to achieve a particular posterior inference, then You could always report that inference regardless of the data. To report that inference without explaining how it was obtained would indeed be dishonest. But in a Bayesian analysis both likelihood and prior distribution should not only be set out but also justified. For no scientist can expect to convince his colleagues that his conclusion is correct unless he can persuade them to accept the premises upon which it is based.

Clearly, to choose the prior distribution to achieve a specific posterior inference regardless of the data is unacceptable. In particular it means that the prior distribution would depend on the data, which it cannot do. In considering use of properties of posterior distributions to specify prior distributions, we must be careful to avoid such 'cheating'.

6.23 It is quite acceptable, however, to ask what prior distribution would be necessary to achieve a particular posterior inference from the observed data. As a simple example, consider the comparison of two simple hypotheses. The parameter θ takes two possible values, θ_0 and θ_1. Then it is easy to derive from Bayes' theorem that

$$\frac{P(\theta=\theta_1 \mid x)}{P(\theta=\theta_0 \mid x)} = \frac{P(\theta=\theta_1)}{P(\theta=\theta_0)} \frac{f(x \mid \theta=\theta_1)}{f(x \mid \theta=\theta_0)}. \tag{6.8}$$

On the left hand side is the *posterior odds* in favour of $\theta = \theta_1$ (or against $\theta = \theta_0$). (6.1) expresses the posterior odds as the product of the prior odds and the *likelihood ratio* $f(x \mid \theta = \theta_1)/f(x \mid \theta = \theta_0)$. Classical hypothesis testing is based on the likelihood ratio without, of course, referring to prior probabilities; see **A21.10**. In Bayesian terms, this likelihood ratio is a special case of the *Bayes factor*, discussed in Chapter 7, beginning at **7.34**.

Now if You observe data x such that the likelihood ratio has the value 100 then You can remark that the posterior odds will exceed 1 unless the prior odds are less than 0.01. So in order to have a posterior belief that $\theta = \theta_0$ is more probable than $\theta = \theta_1$ You must have a prior belief given by $P(\theta = \theta_1) < 0.01P(\theta = \theta_0)$, or $P(\theta = \theta_1) < 0.0099$. In showing what an extreme prior belief against $\theta = \theta_1$ is needed in order to favour $\theta = \theta_0$ after observing x, we demonstrate how strongly the data support $\theta = \theta_1$. Such an argument can be useful, particularly in sensitivity analysis.

6.24 In the above argument we are looking at the prior distribution which would be needed to achieve a particular posterior inference from the observed data, not as a way of determining Your actual prior distribution, but as a description of the strength of the data. We now turn to ways of using posterior distributions in order to help You determine Your actual prior distribution. However, we do not use the actual observed data for this, since we have seen that to do so could lead to a dependence of the prior distribution on the data. Instead, we consider the posterior distribution which would arise from hypothetical data, or the form of some posterior inference as a function of the data.

Characterization by posterior moments

6.25 Suppose that we observe x distributed as $N(\theta, 1)$. We have seen in Example 1.5 that if the prior distribution is normal then the posterior mean increases linearly with x. This seems to be a reasonable way for the posterior distribution to behave. Clearly, since $E(x \mid \theta) = \theta$, larger x somehow suggests that θ is larger, and it is not implausible that this relationship should be linear. We have also seen that in practice You would specify Your prior distribution by specifying some prior summaries, such as mean and variance, and then choosing the precise distribution to fit those summaries in a more or less arbitrary way. Putting these arguments together, we can say that if You choose to complete the prior specification by adopting a normal prior distribution (as opposed to any other distribution conforming to the stated prior summaries), then the posterior mean will be a linear function of x. If this is acceptable, then the choice of a normal distribution is also acceptable. If, however, You felt that the posterior mean should not increase linearly with x then You should find an alternative prior distribution which leads to a posterior mean having the desired form as a function of x.

Therefore the qualitative behaviour of the posterior distribution as a function of x can be an aid to specifying the form of the prior distribution. Continuing the above example we can ask several questions. Are there other forms of prior distribution for which $E(\theta \mid x)$ is linear in x, or does linearity characterize the normal prior? Given another desired form for $E(\theta \mid x)$, e.g. cubic, what prior distributions would produce that form? What functional forms are possible for $E(\theta \mid x)$?

6.26 We first note some simple results about exponential families and conjugate families. Consider the distribution (6.1). Since $\int f(x \mid \theta)\,\mathrm{d}x = 1$,

$$\exp\{-D(\theta)\} = \int \exp\{\textstyle\sum_{j=1}^{k} A_j(\theta) B_j(x) + C(x)\}\,\mathrm{d}x. \tag{6.9}$$

Therefore, $D(\theta)$ is a function only of the quantities $A_1(\theta), A_2(\theta), \ldots, A_k(\theta)$. The whole family (6.1) can be thought of as indexed by the possible values of these k quantities, which are called the *natural parameters* of this exponential family. Writing $\phi_j = A_j(\theta)$ and $\boldsymbol{\phi} = (\phi_1, \phi_2, \ldots, \phi_k)'$, we can rewrite (6.1) in terms of its natural parameters as

$$f(x \mid \boldsymbol{\phi}) = \exp\{\textstyle\sum_{j=1}^{k} \phi_j B_j(x) + C(x) - M(\boldsymbol{\phi})\}, \tag{6.10}$$

where $M(\boldsymbol{\phi})$ is now $-D(\theta)$ expressed in terms of $\boldsymbol{\phi}$. The corresponding natural conjugate family is

$$f(\boldsymbol{\phi}; \mathbf{a}, d) = \exp\{\mathbf{a}'\boldsymbol{\phi} - d\,M(\boldsymbol{\phi}) + c^\star(\mathbf{a}, d)\}, \tag{6.11}$$

where $\mathbf{a} = (a_1, a_2, \ldots, a_k)'$ and d are the hyperparameters.

6.27 Now the moments of $B_j(x)$ can be obtained from its moment generating function

$$E[\exp\{t\,B_j(x)\} \mid \boldsymbol{\phi}] = \exp\{M(\phi_1, \ldots, \phi_{j-1},\ \phi_j + t,\ \phi_{j+1}, \ldots, \phi_k) - M(\boldsymbol{\phi})\}$$

using (6.10) and (6.9). In particular, the expectation of $B_j(x)$ given $\boldsymbol{\phi}$ is found by differentiating the above expression with respect to t and setting $t = 0$. This gives $E\{B_j(x) \mid \boldsymbol{\phi}\} = M^j(\boldsymbol{\phi}) = \partial M(\boldsymbol{\phi})/\partial\phi_j$.

In the case of the normal observations considered in **6.25** we find $k = 1$, $B_1(x) = x$, $\phi_1 = \theta = M^1(\phi)$. There we found that the posterior mean of θ was linear in x when a natural conjugate prior is used. We now generalize this to show that posterior means of the quantities $M^j(\boldsymbol{\phi})$ are linear functions of the sufficient statistics $B_j(x)$.

Consider the prior expectation of $M^j(\boldsymbol{\phi})$ under the natural conjugate (6.11).

$$E\{M^j(\boldsymbol{\phi})\} = \int M^j(\boldsymbol{\phi}) f(\boldsymbol{\phi}; \mathbf{a}, d)\, \mathrm{d}\boldsymbol{\phi}$$

$$= -d^{-1} \int \{\partial f(\boldsymbol{\phi}; \mathbf{a}, d)/\partial \phi_j\}\, \mathrm{d}\boldsymbol{\phi} + d^{-1} a_j \int f(\boldsymbol{\phi}; \mathbf{a}, d)\, \mathrm{d}\boldsymbol{\phi}. \qquad (6.12)$$

The second term in (6.12) is a_j/d and we assume for the moment that the first term is zero, so that $E\{M^j(\boldsymbol{\phi})\} = a_j/d$.

Then since the posterior density combines (6.10) and (6.11) to produce a member of the family (6.11) in which a_j becomes $a_j + B_j(x)$ and d becomes $d+1$ (the property of the conjugate family being closed under sampling), the posterior expectation is

$$E\{M^j(\boldsymbol{\phi}) \mid x\} = \{a_j + B_j(x)\}/(d+1), \qquad (6.13)$$

a linear function of $B_j(x)$.

If $x_1, x_2, \ldots, x_n$ are a sample from the distribution (6.10) and $\bar{B}_j = n^{-1} \sum_{i=1}^{n} B_j(x_i)$, repeated application of (6.13) gives

$$E\{M^j(\boldsymbol{\phi}) \mid \mathbf{x}\} = \{a_j + n\bar{B}_j\}/(d+n),$$

which is linear function of the sufficient statistic $\bar{B}_j$, and explains why this weighted average form appears frequently in Bayesian analysis.

6.28 The above result is presented rigorously in Diaconis and Ylvisaker (1985), who give conditions for the first term in (6.12) vanishing. Diaconis and Ylvisaker (1979, 1985) also prove the converse, that in a wide variety of cases if $E\{M^j(\boldsymbol{\phi}) \mid x\}$ is a linear function of $B_j(x)$ as in (6.13), for all j and with the same slope for each j, then the prior distribution is a member of the natural conjugate family. This allows us to answer one of the questions in **6.25**; the posterior mean will be linear for the normal mean problem if and only if the prior distribution is normal. It is not possible to say in general, however, what forms of posterior mean are obtainable with non-conjugate priors.

Example 6.7

Let $f(x \mid \theta)$ be the $N(\theta, 1)$ density and the prior density be given by $f(\theta) \propto \theta e^{-\theta^2/2}$. Then we easily find that

$$E(\theta \mid x) = E^\star(\theta^2 \mid x)/E^\star(\theta \mid x),$$

where $E^\star(g(\theta) \mid x)$ is the posterior expectation of $g(\theta)$ that would arise from the conjugate prior $f^\star(\theta) \propto e^{-\theta^2/2}$. Therefore $E(\theta \mid x) = (x^2+2)/(2x)$.

6.29 Various other results characterizing prior distributions in terms of posterior moments may be found in Goldstein (1977), Goel and De Groot (1980) and Ralescu and Ralescu (1981).

Imaginary observations

6.30 Another use of posterior distributions to specify the prior distribution is to imagine what Your posterior beliefs would be if You were to observe various hypothetical observations. Summaries of a posterior distribution based on imaginary observations can be used to aid in specifying the prior distribution, but cannot be expected to be helpful in all circumstances. For if You could reliably specify arbitrary posterior summaries based on any desired imaginary observations You would not need to consider Your prior distribution at all. You would simply summarize directly Your posterior distribution based on the actual data. Nevertheless, suitably chosen summaries of posterior distributions based on selected imaginary observations can often usefully be specified in practice. See also **7.55**.

Example 6.8

Let $x_1, x_2, \ldots,$ be distributed independently as $N(\theta, w)$ given parameter θ. Let the prior mean be $E(\theta) = m$. Now if $\bar{x}_n$ is the mean of the first n observations consider the posterior mean $E(\theta \mid \mathbf{x})$ after the imaginary observation $\bar{x}_n = m + 2k$ for some given k. You might judge that the posterior mean should be $m + k$, so that it lies midway between prior mean m and sample mean $\bar{x}_n$, if the sample mean were based on a specific number $n = n_0$ of observations. For this number of observations, the sample information suggesting that θ is close to $\bar{x}_n$ is equally balanced in Your judgement by Your prior information suggesting that θ is closer to m. Then if You are also willing to accept a normal prior distribution, it must be $N(m, w/n_0)$ to fit these specifications.

Multivariate distributions

6.31 In principle the methods discussed so far in this chapter are applicable to specifying multivariate prior distributions, i.e. distributions for more than one parameter $(\theta, \phi, \ldots)$ or a vector parameter $\boldsymbol{\theta}$. However, the examples we have given are predominantly for a scalar parameter. The problem of representing prior beliefs adequately in a multivariate prior distribution is much more complex than the scalar case. Consider just the case of two parameters θ and ϕ. To specify the joint prior distribution $f(\theta, \phi)$ implicitly entails specifying a conditional distribution $f(\theta \mid \phi)$ for each possible value of ϕ, as well as the marginal distribution $f(\phi)$.

We have seen, conversely, in Chapter 2 that to summarize a multivariate distribution adequately requires a large number and variety of summaries. If we are to specify a multivariate prior distribution by first specifying a number of summary values and then finding a convenient distribution to fit those summaries then two things are clear. We will need to make a much larger number of prior summary judgements than for a scalar parameter, and we will need a much larger variety of standard multivariate distributions to be available to fit this greater number of prior summaries. Unfortunately, both these facts cause real difficulties in practice. It is not surprising that practical subjective specification of a relatively large number of aspects of prior belief is difficult, particularly when some of the quantities involved, like summaries of association, are themselves quite complex. It is doubly unfortunate, then, that the range of standard multivariate distributional forms available is very limited. Of the four volumes of the major reference work by Johnson and

Kotz (1969, 1970a, 1970b, 1972), the first three deal with standard univariate distributions and only the last is concerned with multivariate distributions.

6.32 This problem is well illustrated by the forms of natural conjugate distributions. **6.27** shows that the hyperparameters $a_1, a_2, \ldots, a_k$ of the general natural conjugate family (6.2) simply serve as prior means of the natural parameters $A_1(\theta), A_2(\theta), \ldots, A_k(\theta)$. Only one more hyperparameter, d, is available and it will clearly not be possible to represent a suitably wide range of prior summary specifications within any natural conjugate family. To summarize a k-dimensional distribution even superficially requires at least k location summaries (such as means) and k univariate dispersion summaries (such as variances), and really should include a $k \times k$ matrix dispersion summary. The natural conjugate family, with only $k+1$ hyperparameters, cannot possibly fit even the most superficial prior summarization if $k > 1$. This fact is shown clearly in Example 6.5, which concerns the very basic model of a sample from $N(\mu, \sigma^2)$. The natural conjugate family for (μ, σ^2) has only three hyperparameters, and imposes a dependence structure between μ and σ^2 that is generally unrealistic. In particular note that the variance of μ given σ^2 is a multiple of σ^2.

Conditional conjugacy

6.33 It is clear that for practical purposes we must work with much larger families of distributions than the natural conjugate families. This will inevitably mean greatly increased complexity, both in prior specification and in posterior summarization. Natural conjugates are in general particularly well behaved and easy to summarize, the results of **6.27** being examples of this. Larger families will typically be analytically intractable to some extent. One way of enlarging a natural conjugate family whilst retaining some of its tractability is to employ the idea of conditional conjugacy.

Let $\boldsymbol{\theta}_1$ be composed of a subset of the elements of the parameter vector $\boldsymbol{\theta}$, and consider distributions of the form

$$f(\boldsymbol{\theta}) \propto \exp\{\textstyle\sum_{j=1}^{k} a_j(\boldsymbol{\theta}_1) A_j(\boldsymbol{\theta}) + d(\boldsymbol{\theta}_1) D(\boldsymbol{\theta})\} \qquad (6.14)$$

This is the natural conjugate form (6.2) except that we now allow $a_1, a_2, \ldots, a_k$ and d to be functions of $\boldsymbol{\theta}_1$ instead of constants. The term 'conditional conjugate' arises because if $\boldsymbol{\theta}_1$ were known, so that we would regard (6.14) as a prior distribution for the other elements of $\boldsymbol{\theta}$ only, then it would be a member of the natural conjugate family. Under the condition of known $\boldsymbol{\theta}_1$, the $a_1(\boldsymbol{\theta}_1), a_2(\boldsymbol{\theta}_1) \ldots, a_k(\boldsymbol{\theta}_1)$ and $d(\boldsymbol{\theta}_1)$ revert to constants. (Typically, we will also be able to remove some of the $A_j(\boldsymbol{\theta})$s if they are functions only of $\boldsymbol{\theta}_1$.) Therefore the conditional distributions of (6.14) obtained by conditioning on $\boldsymbol{\theta}_1$ are natural conjugates and will generally be tractable and easy to summarize.

We can define a family of distributions (6.14) by letting the $a_j(\boldsymbol{\theta}_1)$s and $d(\boldsymbol{\theta}_1)$ be members of suitable sets of functions of $\boldsymbol{\theta}_1$. The family will then be closed under sampling provided each such family is closed under the addition of constants. That is, if the set of functions $a_j(\boldsymbol{\theta}_1)$ comprising the family is $\mathscr{S}_j$, then $\mathscr{S}_j$ must have the property that if $a_j(\boldsymbol{\theta}_1) \in \mathscr{S}_j$ then $a_j(\boldsymbol{\theta}_1) + b \in \mathscr{S}_j$ also. The natural conjugate family arises when each S_j is just a set of constants (and similarly the set of $d(\boldsymbol{\theta}_1)$ is a set of constants).

Example 6.9
The family (6.6) in **6.20** exhibits a double conditional conjugacy. As a function of μ for fixed σ^2 it is a normal distribution, and so a member of the natural conjugate family for the mean μ when σ^2 is known. As a function of σ^2 for fixed μ it is again a member of the natural conjugate family for the variance σ^2 when μ is known. This means that both conditional distributions are tractable and easy to summarize. In particular, we can easily integrate out either μ or σ^2 to obtain marginal distributions. However, those marginal distributions,

$$f(\mu) \propto \{b + d(\mu - m_2)^2\}^{-(t+1/2)} \exp\{-(\mu - m_1)^2/(2v)\}$$

and

$$f(\sigma^2) \propto (\sigma^2)^{-t}(v^{-1} + d\sigma^{-2})^{-1/2} \exp\{-b/(2\sigma^2) - (v + d^{-1}\sigma^2)^{-1}(m_1 - m_2)^2/2\},$$

are complex and very difficult to summarize analytically. So the conditional conjugate family (6.6) retains some but not all of the analytical convenience of the natural conjugate (6.5).

Prior modelling

6.34 The extra complexity of multivariate specification is primarily caused by the need to specify the form of association between the parameters (and the great variety of forms that association can take). The problem disappears in the special case of independence. If we can assert that ϕ and θ are independent *a priori* then $f(\theta, \phi) = f(\theta)f(\phi)$ and we need only specify the two univariate priors $f(\theta)$ and $f(\phi)$. Now although it will very often be unreasonable to assert that parameters are independent *a priori*, it may be reasonable to assert a form of conditional independence. The idea is in fact identical to the technique invariably used in constructing statistical models.

In order to write down the likelihood $f(\mathbf{x} \mid \boldsymbol{\theta})$ for a vector $\mathbf{x} = (x_1, x_2, \ldots, x_n)$ of observations we need to specify a multivariate distribution, which is potentially a very difficult task. Yet we often suppose that the x_is are independent and identically distributed (iid) as $g(x \mid \boldsymbol{\theta})$, so the task reduces to specifying a single univariate distribution. The x_is are not actually independent but only conditionally independent *given* the unknown parameters $\boldsymbol{\theta}$. Indeed, the parameters $\boldsymbol{\theta}$ are generally introduced specifically to achieve this conditional independence, as described in **3.14**. More complex statistical models may not express the x_is as themselves independent and identically distributed given $\boldsymbol{\theta}$, but almost invariably are expressed in terms of random errors or disturbances which are then assumed to be iid given $\boldsymbol{\theta}$ (as in the linear model, **9.2**).

6.35 The same ideas of statistical modelling can be applied to prior distributions, to give techniques of prior modelling. The prior distribution of $\boldsymbol{\theta}$ is first expressed conditional on some *unknown* hyperparameters $\boldsymbol{\phi}$ as $f(\boldsymbol{\theta} \mid \boldsymbol{\phi})$. The prior specification must then be completed by a prior distribution for $\boldsymbol{\phi}$, $f(\boldsymbol{\phi})$. Then the prior for $\boldsymbol{\theta}$ is implicitly

$$f(\boldsymbol{\theta}) = \int f(\boldsymbol{\theta} \mid \boldsymbol{\phi}) f(\boldsymbol{\phi})\, \mathrm{d}\boldsymbol{\phi}. \tag{6.15}$$

If, for instance, the prior model is that $\theta_1, \theta_2, \ldots, \theta_k$ are iid given $\boldsymbol{\phi}$, the first step is $f(\boldsymbol{\theta} \mid \boldsymbol{\phi}) = \prod_{i=1}^k g(\theta_i \mid \boldsymbol{\phi})$. The use of the term 'hyperparameter' here is consistent with its use in the context of a family of prior distributions. The conditional distributions $f(\boldsymbol{\theta} \mid \boldsymbol{\phi})$ do comprise a family of distributions indexed by the possible values of the hyperparameters $\boldsymbol{\phi}$. However, in **6.17** the hyperparameters were to be fixed to select the particular member of the family that has specified prior summaries, and so it was *the* prior distribution $f(\boldsymbol{\theta})$. Now we let the hyperparameters be unknown and subject to what might be called a 'hyperprior distribution' $f(\boldsymbol{\phi})$. It is this that must now be specified in order to define the 'marginal' prior distribution (6.15). It will, of course, typically be chosen from some other family of distributions (perhaps indexed by 'hyper-hyper-parameters'!) to fit some specified prior summaries.

Example 6.10

Parameters $\theta_1, \theta_2, \ldots, \theta_k$ are stated to be independently distributed as $N(\phi, v)$ conditional on an unknown ϕ. Therefore

$$f(\boldsymbol{\theta} \mid \phi) = (2\pi v)^{-k/2} \exp\{-\textstyle\sum_{i=1}^k (\theta_i - \phi)^2/(2v)\}. \tag{6.16}$$

The prior distribution for ϕ is then specified as $N(m, t)$. Combining this with (6.16) through (6.15) gives the marginal prior distribution

$$\begin{aligned} f(\boldsymbol{\theta}) &= \int (2\pi v)^{-k/2}(2\pi t)^{-1/2} \exp\{-\textstyle\sum_{i=1}^k (\theta_i - \phi)^2/(2v) - (\phi - m)^2/(2t)\}\, d\phi \\ &= (2\pi v)^{-k/2}\{v/(v + tk)\}^{1/2} \\ &\quad \times \exp[-\textstyle\sum_{i=1}^k (\theta_i - \theta_.)^2/(2v) - k(\theta_. - m)^2/\{2(v + tk)\}] \end{aligned} \tag{6.17}$$

where $\theta_. = k^{-1} \sum_{i=1}^k \theta_i$. The θ_is are not unconditionally independent in (6.17). Instead this can be expressed as a multivariate normal prior distribution in which $E(\theta_i) = m$, $\text{var}(\theta_i) = v + t$, $\text{cov}(\theta_i, \theta_j) = t$; see **A15.1**.

Exchangeable parameters

6.36 Suppose that Your prior knowledge about parameters $\theta_1, \theta_2, \ldots, \theta_k$ is such that You have the same beliefs about any θ_i as about any other, and the same beliefs about any group of θ_is as about any other group of the same size. This symmetry property of prior knowledge of the θ_is agrees precisely with the definition of exchangeability in **4.47**. We therefore assume that the θ_is are exchangeable random variables.

Example 6.11

The parameters $\theta_1, \theta_2, \ldots, \theta_k$ in Example 6.10 are exchangeable. This follows from the argument of **4.48** and more directly by observing the symmetry in (6.17). In general if random variables have a multivariate normal distribution they are exchangeable if and only if their means are equal, all variances are equal and all covariances are equal.

6.37 From the generalization of De Finetti's theorem outlined in **4.49** we can conclude that if the θ_is are exchangeable (and if they can be considered as part of an infinite population of parameters of this type) then they can be represented as conditionally iid with unknown common distribution function F, together with a unique prior distribution on F.

Example 6.12
The multivariate normal joint distribution (6.17) is exchangeable and is uniquely characterized by the derivation given in Example 6.10, in which the θ_is are iid with distribution $N(\phi, v)$ conditional on ϕ and ϕ has the distribution $N(m, t)$. In terms of the general characterization, F is confined with probability one to the family of normal distributions with variance v.

Example 6.13
If the distribution of F is degenerate, assigning probability one to the single distribution G, then the exchangeable θ_is become (unconditionally) iid with common distribution G.

Example 6.14
In example 6.10, $t \geqslant 0$ and therefore θ_i and θ_j have a non-negative correlation. (The case $t = 0$ is covered by Example 6.13.) This is a special case of a general result that if the θ_is can be considered as part of an infinite sequence of exchangeable parameters, so that the generalization of De Finetti's theorem applies, then the correlation between any pair must be positive, see Exercise 4.6. However, $\theta_1, \theta_2, \ldots, \theta_n$ could have an exchangeable joint distribution in which their common covariance is negative. In that case there is no equivalent representation of the θ_is as conditionally iid. If the θ_is have such a joint distribution they cannot form part of an infinite exchangeable sequence. The explanation is clear when we note that their covariance matrix must be non-negative definite, from which we can deduce that their correlation $r \geqslant -1/(n-1)$; see Exercise A15.3. Therefore a finite number n of θ_is can be exchangeable with negative covariances but an infinite number cannot. Conversely, if n θ_is have a common negative correlation r they cannot be considered as part of a larger set of m exchangeable θ_is unless $m \leqslant (r-1)/r$, and can never be part of an infinite exchangeable sequence.

6.38 The role of exchangeability in the sense of exchangeable observations is discussed in **4.50**, where it is seen as justifying conventional statistical models. A judgement that $\theta_1, \theta_2, \ldots, \theta_k$ are exchangeable similarly justifies modelling them as iid with unknown distribution F. Statistical models usually go on to assume that F lies in some finitely parametrized set, leading to a conventional *parametric* model in which those parameters θ become the subject of *parametric inference*. Similarly, in prior modelling we will generally go on to restrict F to a family of distributions indexed by a hyperparameter ϕ. To complete the prior specification we must then provide a prior distribution for ϕ.

Hierarchical models

6.39 The statistical model and prior model together form an ordered structure in which the distribution of the data is written conditionally on parameters θ as $f(x \mid \theta)$, the distribution of θ is written conditionally on hyperparameters ϕ as $f(\theta \mid \phi)$, and is completed by the distribution of $\phi, f(\phi)$. We could go further and write the distribution of ϕ conditionally on some more ('hyper-hyper') parameters ψ as $f(\phi \mid \psi)$, and this process could continue as far as is needed. Such models are called *hierarchical* models, because of the way in which the distribution of parameters in each level of the hierarchy depends on the parameters in the next level.

In fact it is usual to say that the distribution of parameters at any level of the hierarchy depends on parameters at the next lower level *and*, conditional on those parameters, is independent of parameters at all levels below that. For instance, if we model the distribution of θ in terms of $f(\theta \mid \phi)$ and $f(\phi)$, the likelihood $f(x \mid \theta)$ is formally the distribution of x given θ *and* ϕ. Writing it as $f(x \mid \theta)$ incorporates a judgement that if we know θ then knowing ϕ would not add any information about x. This is reasonable because ϕ has been introduced merely as a way of formulating $f(\theta)$ as in (6.15). The reason for making this interpretation of $f(x \mid \theta)$ is that otherwise the distributions $f(x \mid \theta)$, $f(\theta \mid \phi)$ and $f(\phi)$ together do not completely specify the joint distribution of x, θ and ϕ. The extra assumption allows us to write

$$f(x, \theta, \phi) = f(x \mid \theta) f(\theta \mid \phi) f(\phi). \tag{6.18}$$

We shall always assume that a hierarchical model does specify the full joint distribution of all quantities in this way.

6.40 The hyperparameters are often of interest in their own right. Just as the parameters characterize the structure of the underlying data generating process, the hyperparameters in turn characterize the structure of the parameters. In a three-level hierarchy with the form (6.18), inference about θ and ϕ is simply obtained through their joint posterior distribution

$$\begin{aligned} f(\theta, \phi \mid x) &= f(\theta,\ \phi,\ x)/f(x) \\ &= \frac{f(x \mid \theta) f(\theta \mid \phi) f(\phi)}{\int\int f(x \mid \theta) f(\theta \mid \phi) f(\phi)\, \mathrm{d}\theta\, \mathrm{d}\phi} \\ &\propto f(x \mid \theta) f(\theta \mid \phi) f(\phi), \end{aligned} \tag{6.19}$$

and inference about ϕ is given by its marginal posterior

$$\begin{aligned} f(\phi \mid x) &= \int f(\theta, \phi \mid x)\, \mathrm{d}\theta \propto f(\phi) \int f(x \mid \theta) f(\theta \mid \phi)\, \mathrm{d}\theta \\ &= f(\phi) f(x \mid \phi). \end{aligned}$$

6.41 Notice, however, that ϕ is unidentifiable in the sense of **3.15**. The likelihood involves only θ, so prior belief about ϕ is influenced by the data x only indirectly through θ. As a formal expression of this, notice that (6.18) says that $f(x \mid \theta, \phi) = f(x \mid \theta)$ and hence that x and ϕ are independent given θ. Therefore $f(\phi \mid \theta, x) = f(\phi \mid \theta)$, i.e. the prior and posterior distributions of ϕ given θ are the same.

Shrinkage

6.42 The influence of the prior distribution in general is to 'pull' the likelihood towards the prior, and in particular, a posterior estimate can often be seen in terms of a classical estimate being 'pulled' towards the corresponding prior estimate. When several parameters have the same prior mean, their various classical estimates may all be pulled towards the common prior mean with the result that the Bayesian estimates are less spread than their classical counterparts. This general phenomenon is known as shrinkage.

Example 6.15

Let $x_1, x_2, \ldots, x_n$ be distributed as $N(\theta_i, v)$, $i = 1, 2, \ldots, n$, given $\boldsymbol{\theta} = (\theta_1, \theta_2, \ldots, \theta_n)$. Suppose that the θ_is have independent and identical $N(m, w)$ prior distributions. Then the (x_i, θ_i) pairs are independent, and therefore the θ_is have independent $N(m_i^\star, w^\star)$ posterior distributions, as in Example 1.5, where $m_i^\star = (wx_i + vm)/(w + v)$ and $w^\star = vw/(w + v)$. The classical estimate of θ_i is x_i, and $m_i^\star$ represents a shift from x_i towards m. The range of the posterior means is $w/(w + v)$ times the range of the classical estimates. The shrinkage factor $w/(w + v)$ is greater the stronger the prior information is relative to the data, as represented by their respective precisions w^{-1} and v^{-1}.

Example 6.16

Following Example 6.15 assume instead a hierarchical prior distribution, such that the θ_is are independently distributed as $N(\phi, w)$ given the hyperparameter ϕ. Let the prior distribution for ϕ be the improper uniform distribution $f(\phi) \propto 1$. This is the limiting case $t \to \infty$ of Example 6.10. As in Example 5.4, we then find that the posterior mean of θ_i becomes $(wx_i + v\bar{x})/(w + v)$. The shrinkage factor is the same as in Example 6.15, but instead of shrinking the estimates towards the fixed point m (which need not even be within the range of the x_is) the posterior mean now shrinks towards the sample mean $\bar{x}$.

6.43 Although shrinkage may seem strange and unintuitive to a classical statistician, it is easy to justify and explain in less strictly Bayesian terms. Referring to the above example, let the range of the θ_is, i.e. the difference between the largest and the smallest of the x_is, be d. Let the range of the x_is be y. Now it is straightforward to show that the expectation of y given $\boldsymbol{\theta}$ is greater than d. Therefore in strictly classical terms to estimate each θ_i by x_i overestimates their range, and this alone will justify shrinking the estimates. Intuitively, if x_1 is the largest of the x_is, we should expect that it is so large partly because θ_1 is large, but also partly by chance through the observation error $x_1 - \theta_1$ also being positive. Similarly one would expect that the smallest x_i will be so small partly because its θ_i is small and partly because of chance. Shrinkage estimators and their properties have therefore been studied also in non-Bayesian ways; see James and Stein (1961), Efron and Morris (1973), Copas (1983, 1988) and Stigler (1990).

Shrinkage arises naturally in many Bayesian analyses, and is not only a property of estimates such as the posterior mean. It is generally associated with a pulling together of the posterior distributions, and will affect many other posterior summaries. Wherever prior information suggests that two parameters are expected *a priori* to have similar values, or in some more general way to be 'close together', this will tend to cause shrinkage of Bayesian inferences relative to the classical inferences that do not use that prior information. Shrinkage is particularly common in hierarchical models, where relationships or correlations are typically induced between the elements of θ through the hyperparameter ϕ. Recognizing that shrinkage is to be expected in a given problem can thereby be an aid to specifying an appropriate hierarchical prior distribution, and constitutes use of posterior properties as in **6.22** or **6.30**.

Mixture priors

6.44 Suppose that in the three-level hierarchical model ϕ is a discrete variable taking values $\phi_1, \phi_2, \ldots, \phi_m$. Writing its prior distribution as $P(\phi = \phi_i) = p_i$ and the conditional distributions of θ as $f(\theta \mid \phi = \phi_i) = f_i(\theta)$, for $i = 1, 2, \ldots, m$, the unconditional prior distribution (6.15) is

$$f(\theta) = \sum_{i=1}^{m} p_i f_i(\theta). \tag{6.20}$$

This is called a mixture of the distributions $\{f_i(\theta)\}$ with weights $\{p_i\}$. The more general form (6.15) can also be thought of as a mixture prior, but as a continuous mixture of densities $f(\theta \mid \phi)$ with weight function $f(\phi)$.

The posterior distribution is then also a mixture.

$$\begin{aligned} f(\theta \mid x) &= f(\theta) f(x \mid \theta) / \textstyle\int f(\theta) f(x \mid \theta)\, d\theta \\ &= \sum_{i=1}^{m} p_i^{\star} f_i(\theta \mid x), \end{aligned} \tag{6.21}$$

where

$$f_i(\theta \mid x) = f_i(\theta) f(x \mid \theta) / \textstyle\int f_i(\theta) f(x \mid \theta)\, d\theta \tag{6.22}$$

is the posterior distribution that would obtain if $f_i(\theta)$ were the prior distribution, and

$$p_i^{\star} = \frac{p_i \int f_i(\theta) f(x \mid \theta)\, d\theta}{\sum_{i=1}^{m} p_i \int f_i(\theta) f(x \mid \theta)\, d\theta}. \tag{6.23}$$

is the updated weight. By comparison with the more general formula $f(\theta \mid x) = \int f(\theta \mid x, \phi)$ $f(\phi \mid x)\, d\phi$ we can identify $p_i^{\star} = P(\phi = \phi_i \mid x)$, the posterior distribution of the hyperparameter.

6.45 A mixture of a small number of standard distributions can be a simple and convenient way of obtaining a prior distribution to fit a set of stated prior summaries that cannot be satisfied by any of those distributions alone.

Example 6.17

For a parameter θ taking values in $[0, 1]$, the beta family of distributions is convenient, but these are all unimodal, and are positively or negatively skewed according as the mean is less than or greater than 0.5. Many other shapes can be obtained with a mixture of just two or three beta distributions. Figure 6.2 shows a bimodal mixture of beta distributions

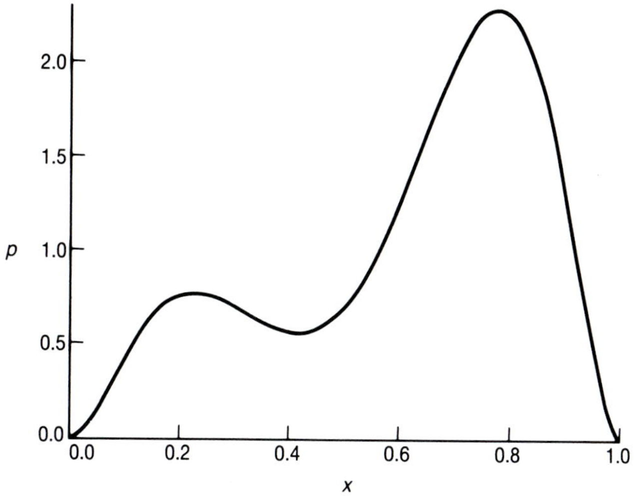

Fig. 6.2 Density of 0.25 *Be* (3, 8) + 0.75 *Be* (8, 3)

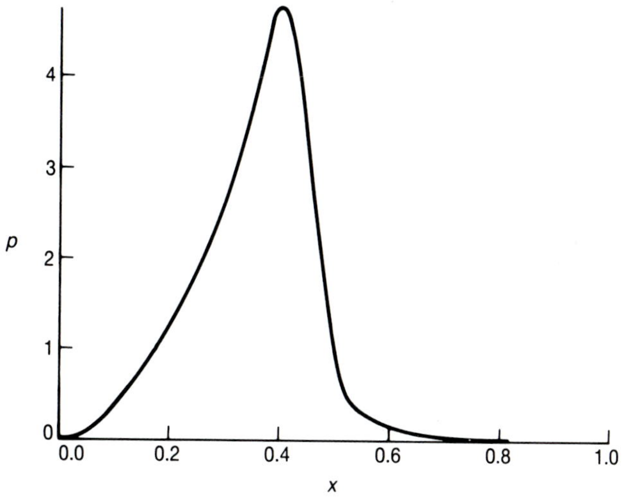

Fig. 6.3 Density of 0.33 *Be* (4, 10) + 0.33 *Be* (15, 28) + 0.33 *Be* (50, 70)

with parameters (3,8) and (8,3) with weights 0.25 and 0.75. It represents a belief that θ is most likely to be between 0.6 and 0.9 but that it could also lie between 0.1 and 0.3. Figure 6.3 shows an equally weighted mixture of beta distributions with parameters (4,10), (15,28) and (50,70). It has a mean of 0.35 but is strongly negatively skewed.

6.46 If each $f_i(\theta)$ in the mixture is a member of a family that is closed under sampling from $f(x \mid \theta)$, then the much larger family of mixtures (6.20) is also closed under sampling.

Suppose that x follows the exponential family distribution

$$f(x \mid \theta) = \exp\{\textstyle\sum_{j=1}^{k} A_j(\theta)B_j(x) + C(x) + D(\theta)\}$$

given θ, and that the prior distribution is a mixture of natural conjugates,

$$f(\theta) = \sum_{i=1}^{m} p_i \exp\{\textstyle\sum_{j=1}^{k} a_{ij}A_j(\theta) + d_i D(\theta) + c(\mathbf{a}_i, d_i)\}.$$

Then the posterior distribution is the mixture (6.21) with components

$$f_i(\theta \mid x) = \exp\{\textstyle\sum_{j=1}^{k} a_{ij}^{\star}A_j(\theta) + d_i^{\star} D(\theta) + c(\mathbf{a}_i^{\star}, d_i^{\star})\}$$

and $a_{ij}^{\star} = a_{ij} + B_j(x)$, $d_i^{\star} = d_i + 1$. The weights (6.23) are proportional to $\exp\{c(\mathbf{a}_i, d_i) - c(\mathbf{a}_i^{\star}, d_i^{\star})\}$.

Example 6.18

Let x be binomially distributed given θ with $f(x \mid \theta) \propto \theta^x(1-\theta)^{n-x}$. This is the one-parameter exponential family with $A(\theta) = \log\{\theta/(1-\theta)\}$, $B(x) = x$ and $D(\theta) = n\log(1-\theta)$. The natural conjugate family is the beta family $f(\theta) \propto \theta^{r-1}(1-\theta)^{s-1} \propto \exp\{aA(\theta)+dD(\theta)\}$, where $a = r-1$ and $d = (s+r-2)/n$, and the normalizing constant is $\exp\{c(a,d)\} = B(r,s)^{-1}$. If the prior distribution is a mixture of beta distributions with weights p_i and parameters (r_i, s_i), then letting $a_i = r_i - 1$ and $d_i = (s_i + r_i - 2)/n$, the posterior is also a mixture of beta distributions with $a_i^{\star} = a_i + B(x) = x + r_i - 1$ and $d_i^{\star} = d_i + 1 = 1 + (s_i + r_i - 2)/n$, corresponding to $r_i^{\star} = x + r_i$ and $s_i^{\star} = n - x + s_i$. The weights are

$$p_i^{\star} \propto B(x + r_i, n - x + s_i)/B(r_i, s_i).$$

Example 6.19

Let x be distributed as $N(0, \theta)$ given θ, which is a one-parameter exponential family with $A(\theta) = -\theta^{-1}$, $B(x) = x^2/2$ and $D(\theta) = -\log(2\pi\theta)/2$. The natural conjugate family is

$$f(\theta) = (2\pi\theta)^{-d/2}\exp\{-a/\theta + c(a,d)\} \tag{6.24}$$

and the normalizing constant is found to be

$$\exp\{c(a,d)\} = (2\pi)^{d/2}a^{(d-2)/2}/\Gamma\{(d-2)/2\}.$$

Therefore if the prior is a mixture of distributions (6.24) with parameters a_i and d_i and weights p_i, the posterior weights are

$$\begin{aligned} p_i^{\star} &\propto \exp\{c(a_i + x^2/2, d_i + 1) - c(a_i, d_i)\} \\ &\propto \frac{(a_i + x^2/2)^{(d_i-1)/2}\Gamma\{(d_i-2)/2\}}{a_i^{(d_i-2)/2}\Gamma\{(d_i-1)/2\}}. \end{aligned}$$

6.47 Mixtures of natural conjugate priors offer a very diverse family of distributions that is capable of representing much more varied prior beliefs than a single natural conjugate. Indeed, although we have pointed out in **6.32** that natural conjugate families in more than one dimension are extremely limited, Diaconis and Ylvisaker (1988) and Dalal and

Hall (1983) prove that it is possible to approximate *any* distribution arbitrarily accurately with a finite mixture of natural conjugate distributions. Of course, to approximate accurately some forms of distribution would need a mixture of very many natural conjugates, but in practice mixtures with small numbers of distributions are extremely diverse and can represent a very wide variety of other distributions. Example 6.17 illustrates the approach.

Bayes linear estimation

6.48 A completely different approach to specifying prior distributions is adopted in *Bayes linear* methods. To the problem that an infinite number of subjective judgements are required to specify a prior distribution for a continuous parameter, the conventional Bayesian response described in **6.8** is to make a small number of actual prior summary statements and then to complete the specification by an arbitrary choice of distribution to fit those summaries. The Bayes linear response is to use only the stated prior summaries. The prior distribution $f(\theta)$ remains unspecified, and consequently we can no longer use Bayes' theorem. Can any methods of posterior inference be derived that depend only on a finite number of prior summaries, and not on the detailed form of $f(\theta)$?

The answer in general is negative. It is clear, for instance, that the posterior mean is not simply a function of certain prior summaries. Since $E(\theta \mid x) = \int \theta f(\theta \mid x)\,\mathrm{d}\theta$, the posterior mean is affected by any change in $f(\theta \mid x)$, which can be caused by a corresponding change in $f(\theta)$. Whatever prior summaries You specify, it will be possible to vary $f(\theta)$ such that those summaries are unchanged yet $E(\theta \mid x)$ is affected. The same argument applies to almost any posterior summary. Any form of posterior inference that depends only on a limited number of prior summaries must represent only an approximation to a full Bayesian analysis.

6.49 The Bayes linear methods rely on the following argument. As was shown in **2.46**, the mean minimizes expected quadratic loss, therefore the posterior mean $d(x) = E(\theta \mid x)$ minimizes

$$D(x) = E[\{d(x) - \theta\}^2 \mid x]$$

for any given x. Therefore the posterior mean as a function of x minimizes

$$D = E\{D(x)\} = E[\{d(x) - \theta\}^2]. \tag{6.25}$$

We have used here the general result that $E(y) = E\{E(y \mid z)\}$. Now suppose that $d(x)$ is constrained to be a linear function

$$d(x) = a + \mathbf{b}'\mathbf{z} \tag{6.26}$$

of a vector $\mathbf{z} = \mathbf{z}(x)$ of specified functions of x. Substituting (6.26) into (6.25) and expanding,

$$\begin{aligned} D &= E(a^2 + 2a\mathbf{b}'\mathbf{z} + \mathbf{b}'\mathbf{z}\mathbf{z}'\mathbf{b} - 2a\theta - 2\mathbf{b}'\mathbf{z}\theta + \theta^2) \\ &= a^2 + 2a\mathbf{b}'E(\mathbf{z}) + \mathbf{b}'E(\mathbf{z}\mathbf{z}')\mathbf{b} - 2aE(\theta) - 2\mathbf{b}'E(\mathbf{z}\theta) + E(\theta^2) \\ &= \{a + \mathbf{b}'E(\mathbf{z}) - E(\theta)\}^2 + \mathbf{b}'\mathrm{var}\,(\mathbf{z})\mathbf{b} - 2\mathbf{b}'\mathrm{cov}\,(\mathbf{z}, \theta) + \mathrm{var}\,(\theta), \end{aligned}$$

where $\text{var}(\mathbf{z}) = E(\mathbf{z}\mathbf{z}') - E(\mathbf{z})E(\mathbf{z})'$ is the variance matrix of the vector $\mathbf{z}$, and $\text{cov}(\mathbf{z}, \theta) = E(\mathbf{z}\theta) - E(\mathbf{z})E(\theta)$ is the vector of covariances between θ and elements of $\mathbf{z}$. Then

$$D = \{a + \mathbf{b}'E(\mathbf{z}) - E(\theta)\}^2 + (\mathbf{b} - \mathbf{b}^\star)'\text{var}(\mathbf{z})(\mathbf{b} - \mathbf{b}^\star) + D^\star, \tag{6.27}$$

where

$$\mathbf{b}^\star = \text{var}(\mathbf{z})^{-1}\text{cov}(\mathbf{z}, \theta),$$
$$D^\star = \text{var}(\theta) - \text{cov}(\mathbf{z}, \theta)'\text{var}(\mathbf{z})^{-1}\text{cov}(\mathbf{z}, \theta). \tag{6.28}$$

The first two terms in (6.27) are seen to be non-negative. Therefore the unconditional expected quadratic loss (6.25) is minimized in the class of linear functions (6.26) by setting $\mathbf{b} = \mathbf{b}^\star$ and $a = -\mathbf{b}^{\star\prime}E(\mathbf{z}) + E(\theta)$, i.e. by

$$d(x) = E(\theta) + \text{cov}(\mathbf{z}, \theta)'\text{var}(\mathbf{z})^{-1}\{\mathbf{z} - E(\mathbf{z})\}. \tag{6.29}$$

6.50 (6.29) is known as the *Bayes linear estimator* of θ based on data $\mathbf{z}$. The posterior mean is a location summary and is often thought of as a point estimate of θ, and the Bayes linear estimator is the closest approximation to $E(\theta \mid x)$ in the class of linear estimators (6.26). We are using 'closest' here in the sense of expected squared difference, because following (2.38) we can write

$$D(x) = \{d(x) - E(\theta \mid x)\}^2 + \text{var}(\theta \mid x)$$
$$\therefore D = E[\{d(x) - E(\theta \mid x)\}^2] + E\{\text{var}(\theta \mid x)\}, \tag{6.30}$$

so that minimizing D means minimizing the expectation of $\{d(x) - E(\theta \mid x)\}^2$. So the Bayes linear estimator can be thought of as a linear approximation to the posterior mean.

The key property of the Bayes linear estimator is that it depends only on $E(\theta)$, $E(\mathbf{z})$, $\text{cov}(\mathbf{z}, \theta)$ and $\text{var}(\mathbf{z})$. Not only do we not need a full specification of the prior distribution $f(\theta)$, we do not need a full specification of the likelihood $f(x \mid \theta)$ either. All that is required is first and second order moments of the joint distribution of θ and those functions $\mathbf{z}(x)$ of the data that we have chosen to use in the Bayes linear estimator.

6.51 Clearly, if the posterior mean is exactly linear in $\mathbf{z}(x)$ then it will equal the Bayes linear estimator for all x. **6.28** describes how a judgement of linearity of the posterior mean can be used to help determine the prior distribution. In the case of sampling from an exponential family distribution, exact linearity can be used to imply that the prior must be a member of the natural conjugate family. Given that enough prior summaries are stated to identify a specific member of the natural conjugate family, then an assertion of linearity of the posterior mean removes the need for any further prior specification — $f(\theta)$ is identified uniquely. Bayes linear estimation is a similar idea. If we are prepared to assert approximate posterior linearity, and to use the Bayes linear estimator as a surrogate posterior location summary, then only a very limited specification of both prior and likelihood is needed. The argument now applies much more generally than **6.28**, which concerns linearity of the posterior mean in the particular functions $B_j(x)$ when x follows the exponential family distribution (6.10).

6.52 The quantity $D^\star$ in (6.28) can serve in a similar way as a surrogate dispersion summary. It is the expected squared error (6.25) for the Bayes linear estimator (6.29), and so represents a kind of variance. However, it is important to recognize two limitations to giving $D^\star$ this interpretation. First, the posterior variance $\text{var}(\theta \mid x)$ is specific to the observed x. It measures the posterior expected squared distance of θ from its mean for this x. In contrast, $D^\star$ is an average expected squared distance of θ from the Bayes linear estimator, averaged over all x. It is therefore a less precise summary of dispersion. Also, (6.30) shows that $D^\star \geqslant E\{\text{var}(\theta \mid x)\}$, with equality if and only if $E(\theta \mid x)$ is exactly linear in $\mathbf{z}$. $D^\star$ is typically larger than the posterior variance because the Bayes linear estimator is typically further from θ than $E(\theta \mid x)$.

Example 6.20
Suppose that $x_1, x_2, \ldots, x_n$ are judged to be exchangeable given θ with $E(x_i \mid \theta) = \theta$, $\text{var}(x_i \mid \theta) = v$ and $\text{cov}(x_i, x_j \mid \theta) = c$. If $E(\theta) = m$ and $\text{var}(\theta) = w$, then $E(x_i) = E\{E(x_i \mid \theta)\} = m$, $\text{var}(x_i) = E\{\text{var}(x_i \mid \theta)\} + \text{var}\{E(x_i \mid \theta)\} = v + w$, $\text{cov}(x_i, x_j) = E\{\text{cov}(x_i, x_j \mid \theta)\} + \text{cov}\{E(x_i \mid \theta), E(x_j \mid \theta)\} = c + w$ and $\text{cov}(x_i, \theta) = E\{\theta E(x_i \mid \theta)\} - E(x_i)E(\theta) = E(\theta^2) - E(x_i)E(\theta) = w$. The Bayes linear estimator of θ based on $\mathbf{z}(\mathbf{x}) = \mathbf{x}$ is therefore

$$d(\mathbf{x}) = m + w\mathbf{1}'\mathbf{V}^{-1}(\mathbf{x} - m\mathbf{1}),$$

where $\mathbf{1}$ is vector of n ones, $\mathbf{V} = (v - c)\mathbf{I} + (c + w)\mathbf{1}\mathbf{1}'$ and $\mathbf{I}$ is the $n \times n$ identity matrix. Now

$$\mathbf{V}^{-1} = (v - c)^{-1}[\mathbf{I} - (c + w)\{(v - c) + n(c + w)\}^{-1}\mathbf{1}\mathbf{1}'],$$

so that

$$d(\mathbf{x}) = (1 - a)m + a\bar{x}$$

where $\bar{x} = n^{-1}\mathbf{1}'\mathbf{x}$ is the mean of the x_is and $a = nw\{(v - c) + n(c + w)\}^{-1}$.

Therefore the Bayes linear estimator based on $\mathbf{x}$ is linear in $\bar{x}$, and this must equal the Bayes linear estimator based on $\bar{x}$ alone. This is easily verified by deriving $\text{var}(\bar{x}) = n^{-2}\{n(v + w) + n(n - 1)(c + w)\} = n^{-1}\{v + (n - 1)c\} + w$ and $\text{cov}(\theta, \bar{x}) = w$.

6.53 Bayes linear estimation relies on unconditional expectations, taken with respect to the joint distribution of θ and x. For instance, (6.25) expresses the criterion D as a preposterior expectation of the posterior expected squared error $D(x)$. The resulting estimator (6.29) thereby depends not only on the actually observed $\mathbf{z}$ but also, through $E(\mathbf{z}), \text{cov}(\mathbf{z}, \theta)$ and $\text{var}(\mathbf{z})$, on what might have been observed. Therefore Bayes linear estimation violates the Likelihood Principle. For instance, Exercise 6.8 derives the Bayes linear estimator for a binomial parameter θ, but a different estimator would be obtained from the same data in the negative binomial case (where the number of successes, x, is fixed and the number of trials, n, is random).

6.54 Bayes linear estimators have been rediscovered independently several times, in different contexts. See Whittle (1958), Stone (1963), Hartigan (1969), Goldstein (1975a), Smouse (1984). The fact that if the posterior mean is linear it must have the form (6.29) was proved for scalar $\mathbf{z}$ by Ericson (1969b). The linear Bayes approach has been developed both technically and philosophically in a series of papers by Goldstein (1975b, 1979, 1981, 1985, 1986, 1990).

Goldstein (1981) presents the Bayes linear estimator as a projection of θ into the space spanned by $\mathbf{z}$, using covariance as the metric. He argues in Goldstein (1988a) that this is the natural mode of posterior 'inference' when only first and second order moments are accepted. Through studying the projection operator, and in particular its eigenvalues, Goldstein (1988b, 1991) presents a range of technical tools to facilitate practical use of linear Bayes methods in complex problems. Recent case studies are Farrow and Goldstein (1993) and O'Hagan and Wells (1993).

EXERCISES

6.1 Suppose that two families $\mathscr{F}$ and $\mathscr{G}$ of prior distribution are both closed under sampling from $f(x \mid \theta)$. Prove that both the union of $\mathscr{F}$ and $\mathscr{G}$ and the intersection of $\mathscr{F}$ and $\mathscr{G}$ are also closed under sampling from $f(x \mid \theta)$.

6.2 Consider sampling from the gamma density

$$f(x \mid \theta, \phi) = \theta^{\phi} x^{\phi-1} e^{-\theta x} / \Gamma(\phi)$$

where $\theta > 0$ and $\phi > 0$. Show that in the natural conjugate family for (θ, ϕ) the marginal densities of ϕ have the form

$$f(\phi) \propto c^{\phi} \Gamma(1 + d\phi) / \{\Gamma(\phi)\}^d.$$

6.3 Let $f(x \mid \theta)$ be a member of the two-parameter exponential family

$$f(x \mid \theta) = \exp\{A_1(\theta)B_1(x) + A_2(\theta)B_2(x) + C(x) + D(\theta)\},$$

then the family of prior distributions

$$f(\theta) \propto \exp\{a_1 A_1(\theta) + a_2 A_2(\theta) + bA_1(\theta)A_2(\theta) + dD(\theta)\} \tag{6.31}$$

is conjugate to $f(x \mid \theta)$ as a special case of (6.3). Show that it is also a conditional conjugate family both for $A_1(\theta)$ given $A_2(\theta)$, and for $A_2(\theta)$ given $A_1(\theta)$. Obtain the conditional conjugacy properties of (6.31) if the product $A_1(\theta)A_2(\theta)$ is replaced by the ratio $A_1(\theta)/A_2(\theta)$, and relate this distribution to (6.6).

6.4 Let x have the $N(\theta, w)$ distribution given θ. Prove that $E(\theta g(\theta) \mid x)/E(g(\theta) \mid x)$ will be linear in x if and only if the prior distribution has the form $f(\theta) \propto n(\theta)/g(\theta)$, where $n(\theta)$ is a normal density function.

6.5 For the one-parameter exponential family $f(x \mid \theta) = \exp\{A(\theta)B(x) + C(x) + D(\theta)\}$ prove (subject to appropriate regularity conditions) that $\mathrm{d}E\{A(\theta) \mid x\}/\mathrm{d}B(x) = \mathrm{var}\{A(\theta) \mid x\}$, and hence that $E\{A(\theta) \mid x\}$ is an increasing function of $B(x)$ for any prior distribution. (O'Hagan, 1979)

6.6 Observations $x_1, x_2, \ldots, x_n$ are independently distributed given parameters $\theta_1, \theta_2, \ldots, \theta_n$ according to the Poisson distributions $f(x_i \mid \boldsymbol{\theta}) = \theta_i^{x_i} e^{-\theta_i}/x_i!$. The prior distribution for $\boldsymbol{\theta}$ is constructed hierarchically. First the θ_is are assumed to be identically and independently distributed given a hyperparameter ϕ according to the exponential distribution $f(\theta_i \mid \phi) = \phi e^{-\phi\theta_i}$, $\theta_i \geqslant 0$ and then ϕ is given the improper uniform prior $f(\phi) \propto 1$, $\phi \geqslant 0$. Provided that $n\bar{x} > 1$, prove that the posterior distribution of $z = (1 + \phi)^{-1}$ has the beta form.

$$f(z \mid \mathbf{x}) \propto z^{n\bar{x}-2}(1 - z)^n.$$

Thereby show that the posterior means of the θ_is are shrunk by a factor $(n\bar{x} - 1)/(n\bar{x} + n)$ relative to the usual classical procedure which estimates each θ_i by x_i.

What happens if $n\bar{x} \leqslant 1$?

6.7 Consider a mixture prior distribution $f(\phi, \psi) = \sum_{i=1}^m p_i f_i(\phi, \psi)$ for parameters (ϕ, ψ), where $p_i \geqslant 0$ and $\sum_{i=1}^m p_i = 1$. If $E_i(\phi)$, $\mathrm{var}_i(\phi)$ and $\mathrm{cov}_i(\phi, \psi)$ denote the corresponding expectation, variance and covariance of $f_i(\phi, \psi)$, it is easy to see that if $E_i(\phi) \geqslant c$ for every $i = 1, 2, \ldots, m$ then $E(\phi) \geqslant c$. Prove that if $\mathrm{var}_i(\phi) \geqslant c$ for every $i = 1, 2, \ldots, m$ then $\mathrm{var}(\phi) \geqslant c$.

Prove, however, that if $\mathrm{cov}_i(\phi, \psi) \geqslant c$ it is not necessarily true that $\mathrm{cov}(\phi, \psi) \geqslant c$.

6.8 Let x be the number of successes in n trials with probability θ of success in each trial. Let $\mathbf{z}$ in (6.26) be the scalar x, so that we are interested in estimating θ by a linear function of x. Show then that the Bayes linear estimator is

$$\hat{\theta}(x) = \alpha\frac{x}{n} + (1-\alpha)E(\theta)$$

where $\alpha = \text{var}(\theta)/\{\text{var}(\theta) + n^{-1}E(\theta(1-\theta))\}$.

6.9 Let $\boldsymbol{\theta}$ be a vector parameter and we wish to estimate $\boldsymbol{\theta}$ by a linear estimator of the form. $\mathbf{d}(x) = \mathbf{a} + \mathbf{B}\mathbf{z}$, where $\mathbf{z} = \mathbf{z}(x)$ is a specified vector of functions of the data x. Let

$$\mathbf{D}(\mathbf{a}, \mathbf{B}) = E[\{\mathbf{d}(x) - \boldsymbol{\theta}\}'\{\mathbf{d}(x) - \boldsymbol{\theta}\}]. \tag{6.32}$$

Show that if

$$\mathbf{B}^\star = \text{var}(\mathbf{z})^{-1}\text{cov}(\mathbf{z}, \boldsymbol{\theta}),$$

$$\mathbf{a}^\star = E(\boldsymbol{\theta}) - \mathbf{B}^\star E(\mathbf{z})$$

then

$$\mathbf{D}(\mathbf{a}^\star, \mathbf{B}^\star) = \text{var}(\boldsymbol{\theta}) - \text{cov}(\mathbf{z}, \boldsymbol{\theta})'\text{var}(\mathbf{z})^{-1}\text{cov}(\mathbf{z}, \boldsymbol{\theta}) \tag{6.33}$$

and $\mathbf{D}(\mathbf{a}, \mathbf{B}) - \mathbf{D}(\mathbf{a}^\star, \mathbf{B}^\star)$ is non-negative definite for all $\mathbf{a}$ and $\mathbf{B}$.

(The resulting generalized Bayes linear estimator

$$\mathbf{d}(x) = E(\boldsymbol{\theta}) + \text{cov}(\mathbf{z}, \boldsymbol{\theta})'\text{var}(\mathbf{z})^{-1}\{\mathbf{z} - E(\mathbf{z})\} \tag{6.34}$$

is therefore optimal in terms of a variety of different definitions of 'minimizing' the expected squared difference matrix (6.32).)

CHAPTER 7

ROBUSTNESS AND MODEL COMPARISON

7.1 A major question in any application of Bayesian methods is the extent to which the inferences are sensitive to possible mis-specification of the prior distribution or the likelihood. And if different specifications lead to different inferences, can we determine which is the 'correct' or 'better' specification? This chapter considers a number of ways of looking at such questions.

Robustness and sensitivity

7.2 As discussed in **4.33**, any real application of Bayesian methods must acknowledge that both prior distribution and likelihood have only been specified as more or less convenient approximations to whatever the investigator's true beliefs might be. If the inferences from the Bayesian analysis are to be trusted, it is important to determine that they are not sensitive to such variations of prior and likelihood as might also be consistent with the investigator's stated beliefs. Formally, although You may have specified a particular prior distribution $f_0(\theta)$ and a likelihood $f_0(x \mid \theta)$, we suppose that Your true prior distribution is a member of some set Γ of distributions containing $f_0(\theta)$, and that the likelihood lies in a set Λ of functions containing $f_0(x \mid \theta)$. For any given $f(\theta) \in \Gamma$ and $f(x \mid \theta) \in \Lambda$ we can derive the implied posterior distribution $f(\theta \mid x)$ and the set of all such posteriors is $\Gamma^\star$. The particular posterior $f_0(\theta \mid x)$, arising from the specified $f_0(\theta)$ and $f_0(x \mid \theta)$, is obviously a member of $\Gamma^\star$. For any required inference, we can calculate the inference which would be obtained for each $f(\theta \mid x) \in \Gamma^\star$ and so obtain a set of possible inferences. Again, this set contains the inference actually derived from $f_0(\theta \mid x)$. If all inferences in the set are sufficiently close to the inference from $f_0(\theta \mid x)$, then we will say that that inference is *robust*. Otherwise, there is sensitivity which You should attempt to resolve as described in **4.37**.

In order to carry out such an analysis You must specify Γ, Λ, the inference required, and the accuracy required in that inference. It is important to recognize that it is just as impossible to specify Γ or Λ precisely as to specify the prior or likelihood precisely. If, for instance, You make the prior assertion that $P(\theta < 0) = 0.5$, we have to allow that Your true prior probability that $\theta < 0$ may not be exactly 0.5, because You cannot make arbitrarily fine judgements of probability reliably. As part of specifying Γ You should consider how far from 0.5 the true probability might be. But it is unrealistic then to place bounds such as $0.47 \leqslant P(\theta < 0) \leqslant 0.53$ and treat the 0.47 and 0.53 as exact, for the same reason – You cannot make arbitrarily fine judgements of probability. A formal sensitivity analysis based on a specific Γ and Λ cannot conclusively demonstrate robustness or sensitivity because of this uncertainty about the boundaries of Γ and Λ. To suppose that the stated Γ is therefore just one member of some wider class of sets (in the same way as we first supposed that $f_0(\theta)$ was just one member of the set Γ) leads to unprofitable infinite regress. Instead we should recognize that any formal sensitivity analysis is only indicative of actual robustness or sensitivity.

7.3 The formulation of robustness in **7.2** is termed *inference robustness* by Box and Tiao (1964), who contrast it with *criterion robustness*. Criterion robustness, which is often employed in non-Bayesian statistics, considers an inference criterion (such as an estimator) appropriate to the specified model $f_0(x \mid \theta)$ and examines how sensitive the distribution of that criterion is to changes in the model. Thus, if the specified model is that $x_1, x_2, \ldots, x_n$ are independently and identically distributed as $N(\mu, \sigma^2)$, an appropriate classical criterion for inference about μ is $\sqrt{n}(\bar{x} - \mu)/s$. Criterion robustness would then examine how the sampling distribution of this criterion varies from the t_{n-1} distribution as the model is perturbed. Box and Tiao argue that this is an inadequate examination of robustness because as the model varies the inference criterion should also change. Inference robustness examines whether the optimal inference is sensitive to changes in the model, and is the natural way to examine robustness in the Bayesian framework.

Classes of prior distributions

7.4 Although it is natural to consider mis-specification of both the prior distribution and likelihood through classes Γ and Λ, in practice it is not always easy to think about them separately. If we consider a possibly different model for the data x, this will often involve not just a change of likelihood $f(x \mid \theta)$ but even a change in the parameter θ.

Example 7.1

Suppose that observations $x_1, x_2, \ldots, x_n$ are the electricity consumption of an office building on n different days. It is desired to relate electricity consumption to temperature, and the average temperatures $t_1, t_2, \ldots, t_n$ on those n days are also recorded. A simple statistical model would be a linear regression

$$x_i = \alpha + \beta t_i + e_i, \qquad i = 1, 2, \ldots, n, \tag{7.1}$$

where α and β are unknown parameters, the e_is are independent and identically distributed errors with each e_i having the $N(0, \sigma^2)$ distribution, and σ^2 is a third unknown parameter. The likelihood function is

$$f(\mathbf{x} \mid \alpha, \beta, \sigma^2) = \prod_{i=1}^{n} (2\pi\sigma^2)^{-1/2} \exp\{-(x_i - \alpha - \beta t_i)^2/(2\sigma^2)\}.$$

A prior distribution $f(\alpha, \beta, \sigma^2)$ would represent prior beliefs about the parameters, such as a belief that β should be negative since electricity consumption should be less on warmer days.

Now we consider two alternative models. On very hot days, electricity consumption might actually be higher because of the use of fans or air-conditioning, so we might alternatively imagine a quadratic regression model

$$x_i = \alpha + \beta t_i + \gamma t_i^2 + e_i,$$

where the e_is are $N(0, \sigma^2)$ as before but now there are four parameters – α, β, γ and σ^2. Not only is the likelihood function different,

$$f(\mathbf{x} \mid \alpha, \beta, \gamma, \sigma^2) = \prod_{i=1}^{n} (2\pi\sigma^2)^{-1/2} \exp\{-(x_i - \alpha - \beta t_i - \gamma t_i^2)^2/(2\sigma^2)\},$$

but the parameter space changes. A prior distribution for the first model cannot serve as a prior distribution for the second. However, we can think of less radical changes of model, such as to accept the original representation (7.1) but to suppose that the distribution of the errors is not normal. If, for instance, we assumed a Cauchy error distribution the likelihood would become

$$f(\mathbf{x} \mid \alpha, \beta, \sigma^2) = \prod_{i=1}^{n} \sigma\pi^{-1}\{\sigma^2 + (x_i - \alpha - \beta t_i)^2\}^{-1}.$$

The parameters are the same, and therefore a prior distribution for the first model could serve as a prior distribution for this model. But it is still not necessarily reasonable to suppose that the same set Γ of prior distributions is appropriate in both cases, because the meanings of the parameters change when the model changes. In particular, σ^2 is the variance of the errors in the first model, and prior information about σ^2 would reflect that interpretation, whereas if errors are Cauchy distributed they do not have a finite variance. σ^2 now represents a scale parameter in a different sense, and prior beliefs about it must be expected to be different.

7.5 Because of this difficulty in general of separating the parameters from the model, we begin this chapter by supposing the likelihood $f(x \mid \theta)$ to be fixed. We therefore consider sensitivity of posterior inferences to variation of the prior distribution over some class Γ only. Berger (1990) reviews work in this area.

The specified prior distribution $f_0(\theta)$ is in Γ, and we can think of Γ as some suitable neighbourhood of $f_0(\theta)$. We now give a number of examples of specifications of Γ. To simplify the specifications we use the notation that P_0 and E_0 denote probabilities and expectations derived from the specified prior distribution $f_0(\theta)$, while P and E denote probabilities and expectations from any other prior distribution $f(\theta) \in \Gamma$.

Example 7.2

The total variation class defines $f(\theta)$ to be in Γ if the probability it gives to θ lying in any given set A is within ϵ of the probability given to that event by $f_0(\theta)$. That is, for all possible sets A,

$$|P(\theta \in A) - P_0(\theta \in A)| \leqslant \epsilon. \tag{7.2}$$

The total variation class simply supposes that the probabilities implied by $f_0(\theta)$ could be mis-specified to within plus or minus ϵ. Obviously, the size of Γ is governed by the the size of ϵ. This class of prior distributions is considered by Wasserman and Kadane (1990, 1992).

Example 7.3
The density ratio class is defined by the condition that for all sets A and B having non-zero probability under $f_0(\theta)$,

$$\frac{P(\theta \in A)}{P(\theta \in B)} \leqslant k \frac{P_0(\theta \in A)}{P_0(\theta \in B)}. \tag{7.3}$$

In particular, by choosing A or B to be the set of all possible θ values, we have

$$|\log P(\theta \in A) - \log P_0(\theta \in A)| \leqslant \log k, \tag{7.4}$$

which expresses the similarity of this class to the total variation class of Example 7.2. However, (7.3) is stronger than (7.4) and allows the density ratio class to be expressed in an equivalent form: $f(\theta) \in \Gamma$ if there exists a constant c such that

$$f_0(\theta) \leqslant cf(\theta) \leqslant kf_0(\theta) \tag{7.5}$$

for all θ. This formulation can then be further generalized by replacing the constant k by two functions $l(\theta)$ and $u(\theta)$ such that for all θ, $l(\theta) \leqslant f_0(\theta) \leqslant u(\theta)$, and defining $f(\theta) \in \Gamma$ if there exists a constant c such that $l(\theta) \leqslant cf(\theta) \leqslant u(\theta)$. (7.5) then corresponds to $l(\theta) = f_0(\theta)$ and $u(\theta) = kf_0(\theta)$. Density ratio classes were first investigated by De Robertis and Hartigan (1981), and subsequently by Wasserman (1992b).

Example 7.4
The density bounded class is closely related to the density ratio class. Again we have a lower and upper function $l(\theta)$ and $u(\theta)$ such that $l(\theta) \leqslant f_0(\theta) \leqslant u(\theta)$ for all θ. Then $f(\theta) \in \Gamma$ if $l(\theta) \leqslant f(\theta) \leqslant u(\theta)$ for all θ. The difference between this and the density ratio class is that the density ratio class allows $f(\theta)$ to be scaled by an arbitrary constant c to fit between $l(\theta)$ and $u(\theta)$, while for the density bounded class c must be 1. The density bounded class is somewhat easier to visualize, since densities $f(\theta)$ in Γ are explicitly bounded by $l(\theta)$ and $u(\theta)$. This class is considered by Lavine (1991).

Epsilon contamination

7.6 Another approach to defining Γ is to let

$$f(\theta) = (1 - \epsilon)f_0(\theta) + \epsilon q(\theta), \tag{7.6}$$

where $q(\theta)$ is a 'contaminating' prior density in some class Q. Therefore, for given Q and ϵ the ϵ-contamination class is

$$\Gamma = \{(1 - \epsilon)f_0(\theta) + \epsilon q(\theta) : q(\theta) \in Q\}. \tag{7.7}$$

Q might be the set of all probability distributions or some smaller set. As long as $f_0(\theta) \in Q$ then $f_0(\theta) \in \Gamma$. The ϵ-contamination class has been extensively studied using different contamination classes Q; see, for instance, Berger and Berliner (1986), Sivaganesan and Berger (1989). Notice that ϵ-contamination distributions (7.6) satisfy the total variation condition (7.2), so that for any Q the family (7.7) is a subset of a total variation class. If the densities $q(\theta) \in Q$ are bounded for all θ then it is also a subset of a density bounded class.

7.7 Given a likelihood $f(x \mid \theta)$, the posterior density from prior (7.6) is easily obtained. Writing

$$m_0(x) = \int f_0(\theta) f(x \mid \theta) \, d\theta, \tag{7.8}$$

$$m_q(x) = \int q(\theta) f(x \mid \theta) \, d\theta,$$

as the marginal distributions of x derived from the prior distributions $f_0(\theta)$ and $q(\theta)$, then

$$\int f(\theta) f(x \mid \theta) \, d\theta = (1-\epsilon) m_0(x) + \epsilon m_q(x).$$

Therefore the posterior density is

$$\begin{aligned} f(\theta \mid x) &= \frac{(1-\epsilon) f_0(\theta) f(x \mid \theta) + \epsilon q(\theta) f(x \mid \theta)}{(1-\epsilon) m_0(x) + \epsilon m_q(x)} \\ &= (1-\epsilon^\star) f_0(\theta \mid x) + \epsilon^\star q(\theta \mid x), \end{aligned} \tag{7.9}$$

where $f_0(\theta \mid x)$ and $q(\theta \mid x)$ are posterior distributions derived from the prior distributions $f_0(\theta)$ and $q(\theta)$, i.e.

$$f_0(\theta \mid x) = f_0(\theta) f(x \mid \theta) / m_0(x),$$

$$q(\theta \mid x) = q(\theta) f(x \mid \theta) / m_q(x),$$

and

$$\epsilon^\star = \frac{\epsilon m_q(x)}{(1-\epsilon) m_0(x) + \epsilon m_q(x)}. \tag{7.10}$$

Although (7.9) seems to express the posterior class $\Gamma^\star$ as an ϵ-contamination class, this is not the case because $\epsilon^\star$ is not fixed: (7.10) shows that it depends on $q(\theta)$.

Classes with fixed summaries

7.8 An approach to defining Γ which is more in keeping with our discussion in **6.7** of specifying prior distributions, is to let Γ comprise all distributions having specified values for certain summaries. For instance, Berliner and Goel (1986) and Ruggeri (1990) let Γ be the class of all distributions having certain specified quantiles. Berger and O'Hagan (1986) consider the same class with the added condition of unimodality. Goutis (1993) discusses the class of distributions with given moments.

7.9 ϵ-contamination classes with specified summaries can also be defined. For instance, if Q is the class of distributions with certain quantiles constrained to equal those specified for $f_0(\theta)$, then all $f(\theta)$ in Γ will have those specified quantiles, and Γ is an ϵ-contamination sub-class of Q; see Moreno and Cano (1991), Sivaganesan (1991).

Sensitivity of integration-based summaries

7.10 In general, we wish to find the extreme values of a posterior summary of interest, as $f(\theta)$ ranges over Γ. Let the summary of interest be the posterior expectation of a function $h(\theta)$. For instance $h(\theta) = \theta$ represents interest in the posterior mean, and if $h(\theta)$ is the indicator function of a set A then interest lies in the posterior probability that $\theta \in A$. Therefore consider

$$m = \sup_{\Gamma} E\{h(\theta) \mid x\} = \sup_{\Gamma} \frac{\int h(\theta) f(\theta) f(x \mid \theta)\, \mathrm{d}\theta}{\int f(\theta) f(x \mid \theta)\, \mathrm{d}\theta}. \qquad (7.11)$$

Sensitivity of this summary, or inference, to the specification of the prior distribution will be described by both the supremum m and the infimum of $E\{h(\theta) \mid x\}$, but the latter may be found by analogous methods. We therefore concentrate on (7.11).

7.11 For most Γ and h, the solution of (7.11) is difficult, particularly because as $f(\theta)$ varies both the numerator and denominator change. However, the problem may be linearized as follows. For all $f(\theta) \in \Gamma$,

$$\int h(\theta) f(\theta) f(x \mid \theta)\, \mathrm{d}\theta \leqslant m \int f(\theta) f(x \mid \theta)\, \mathrm{d}\theta$$

$$\therefore \int h^{\star}(\theta, m) f(\theta) f(x \mid \theta)\, \mathrm{d}\theta \leqslant 0, \qquad (7.12)$$

where $h^{\star}(\theta, m) = h(\theta) - m$. Now let

$$t(m) = \sup_{\Gamma} \int h^{\star}(\theta, m) f(\theta) f(x \mid \theta)\, \mathrm{d}\theta. \qquad (7.13)$$

Then m is a solution of the equation $t(m) = 0$.

7.12 The optimization in (7.13) is generally much simpler than (7.11). It may be made simpler still if Γ is a convex set. For then any $f(\theta)$ in Γ may be written as a weighted average of the bounding vertices of Γ. Formally, suppose that

$$\Gamma = \{\textstyle\int f_{\delta}(\theta)\, \mathrm{d}F(\delta) : F \in \Delta\}, \qquad (7.14)$$

where $f_{\delta}(\theta)$ are the vertices of Γ, and Δ is the set of all possible distributions for δ. Then

$$\int h^{\star}(\theta, m) f(\theta) f(x \mid \theta)\, \mathrm{d}\theta = \int \left\{ \int h^{\star}(\theta, m) f_{\delta}(\theta) f(x \mid \theta)\, \mathrm{d}\theta \right\} \mathrm{d}F(\delta), \qquad (7.15)$$

which is the expectation of the expression in braces with respect to the distribution $F(\delta)$. It is bounded above by the supremum of that expression. Also, since F can range over all possible distributions, including degenerate distributions giving unit mass to an individual δ, that bound is also the supremum of (7.15). Thus

$$t(m) = \sup_{\delta} \int h^{\star}(\theta, m) f_{\delta}(\theta) f(x \mid \theta)\, \mathrm{d}\theta. \qquad (7.16)$$

Furthermore, since the original problem (7.11) is solved by the solution of (7.13) for some m,

$$m = \sup_{\delta} \frac{\int h(\theta) f_{\delta}(\theta) f(x \mid \theta)\, \mathrm{d}\theta}{\int f_{\delta}(\theta) f(x \mid \theta)\, \mathrm{d}\theta}. \tag{7.17}$$

Example 7.5

Consider the extreme case where Γ is the set of all possible prior distributions. Its vertices are the degenerate distributions, and so from (7.17) it follows that $m = \sup_{\theta} h(\theta)$, regardless of $f(x \mid \theta)$. The effect of allowing all possible prior distributions is to allow the degenerate prior distributions that assert a specific value for θ with probability one. Then all posterior distributions are possible and no upper bound can be placed on $E\{h(\theta) \mid x\}$ other than the logical bound of the supremum of $h(\theta)$.

7.13 If Γ is an ϵ-contamination class (7.7), then using (7.9) we have

$$E\{h(\theta) \mid x\} = (1 - \epsilon^{\star}) E_0\{h(\theta) \mid x\} + \epsilon^{\star} E_q\{h(\theta) \mid x\}, \tag{7.18}$$

where $E_0(\cdot \mid x)$ denotes expectation with respect to the posterior distribution $f_0(\theta \mid x)$ and $E_q(\cdot \mid x)$ denotes expectation with respect to $q(\theta \mid x)$. (7.18) expresses $E\{h(\theta) \mid x\}$ as a contaminated form of the inference $E_0\{h(\theta) \mid x\}$ implied by the stated prior. However, the contamination fraction $\epsilon^{\star}$ depends on $q(\theta)$ in a rather complex way, and although (7.18) may be a useful form for direct solution of (7.11), when applying the linearized approach (7.3) another expression is more helpful.

$$E\{h(\theta) \mid x\} = \frac{a + \int h(\theta) q(\theta) f(x \mid \theta)\, \mathrm{d}\theta}{b + \int q(\theta) f(x \mid \theta)\, \mathrm{d}\theta}, \tag{7.19}$$

where $a = \epsilon^{-1}(1 - \epsilon) \int h(\theta) f_0(\theta) f(x \mid \theta)\, \mathrm{d}\theta$ and $b = \epsilon^{-1}(1 - \epsilon) \int f_0(\theta) f(x \mid \theta)\, \mathrm{d}\theta$. Now a and b are constants, and (7.19) may be optimized over $f(\theta) \in \Gamma$ by noting that instead of (7.12) we have

$$\int h^{\star}(\theta, m) q(\theta) f(x \mid \theta)\, \mathrm{d}\theta \leqslant bm - a,$$

Therefore if

$$t_1(m) = \sup_{Q} \int h^{\star}(\theta, m) q(\theta) f(x \mid \theta)\, \mathrm{d}\theta$$

the supremum of $E\{h(\theta) \mid x\}$ over Γ is a solution of $t_1(m) = bm - a$.

Example 7.6

Let Γ be given by the ϵ-contamination class (7.7), where Q is the set of all distributions. Then

$$\sup_{\Gamma} E\{h(\theta) \mid x) = \sup_{\theta} \frac{a + h(\theta) f(x \mid \theta)}{b + f(x \mid \theta)}. \tag{7.20}$$

It may be possible to solve this explicitly, otherwise it can be linearized.

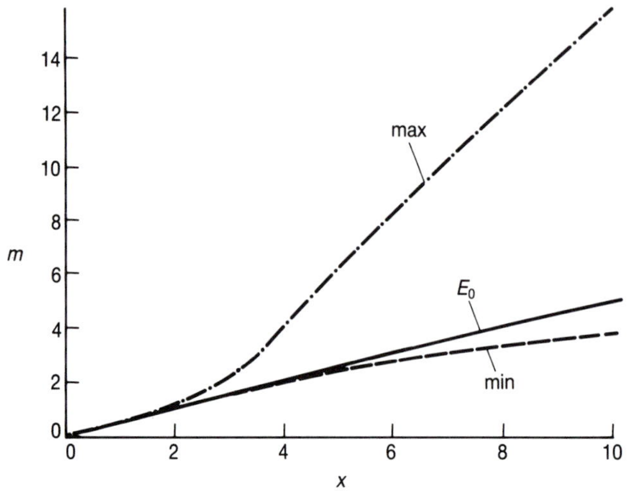

Fig. 7.1 Bounds on posterior mean, Example 7.7

Example 7.7

Following Example 7.6, let $h(\theta) = \theta$, $f(x \mid \theta) = \exp\{-(\theta - x)^2/2\}$, $f_0(\theta) = \exp(-\theta^2/2)$ and $\epsilon = 0.1$. (7.20) cannot be solved explicitly, but a simple numerical search for maxima and minima produces Figure 7.1, where these are plotted against x. The stated prior yields $E_0(\theta \mid x) = x/2$, and for x not too far from zero the posterior bounds lie tightly around this value, showing robustness of $E(\theta \mid x)$ to the class of prior distributions. For larger x, however, the bounds separate rapidly.

The same behaviour is found for any ϵ, although the smaller ϵ is, the larger x needs to be before sensitivity of $E(\theta \mid x)$ becomes substantial. This finding is closely related to the discussion of conflicting information in **3.35**. Figure 7.1 resembles the converse of Figure 3.2. Thus, if we replace the normal stated prior distribution $f_0(\theta)$ by a t distribution then as $x \to \infty$, $E(\theta \mid x) - x \to 0$. By allowing the contamination of $f_0(\theta)$ instead we find this behaviour is contained within the bounds on $E(\theta \mid x)$. It is contamination of the extreme tails of $f_0(\theta)$ which is responsible for the sensitivity for large x. The contamination is allowing the tails of $f(\theta)$ to be much heavier than the normal $f_0(\theta)$, and so to behave like a t distribution.

7.14 General work on probabilities which can only be asserted to lie between upper and lower bounds is to be found in Walley (1991). Lavine (1991a) and Lavine et al. (1991) consider robustness in multiparameter problems. An approach to robustness with reference to both the prior and likelihood is given by Lavine (1991b).

Local sensitivity

7.15 The approach of considering the range of posterior inferences which arise from a class of prior distributions may be called *global* sensitivity analysis. An alternative is to consider the effect on posterior inferences of infinitessimal changes in the prior distribution, which we can call *local* sensitivity analysis. One simple way to formulate local sensitivity is to let the prior distribution be

$$f(\theta) = (1-\epsilon)f_0(\theta) + \epsilon q(\theta), \tag{7.21}$$

where $f_0(\theta)$ is the stated prior distribution and $q(\theta)$ describes the particular way in which we propose to perturb $f_0(\theta)$. There is obvious similarity between (7.21) and the definition of the ϵ-contamination class Γ in **7.6**, but instead of considering the class Γ of $f(\theta)$ generated by letting $q(\theta)$ range over a set Q, we let $q(\theta)$ be fixed and consider posterior inference as $\epsilon \to 0$. However, the formal equivalence of (7.21) and (7.6) means that (7.9) and (7.10) again describe the posterior distribution arising from (7.21).

For a given inference problem, let the inference from $f_0(\theta \mid x)$ be d_0. This is the inference deriving from the stated prior distribution $f_0(\theta)$. If the corresponding inference from $f(\theta \mid x)$ is $d(\epsilon)$, suppose that

$$d(\epsilon) = (1-\epsilon^\star)d_0 + \epsilon^\star d_q, \tag{7.22}$$

where d_q is the inference resulting from the prior distribution $q(\theta)$. (7.22) will hold for many kinds of inferences, and in particular for integration-based summaries via (7.18). Then using (7.10) we can define

$$\begin{aligned} D(q) &= \lim_{\epsilon\to 0} \epsilon^{-1}\{d(\epsilon) - d_0\} = (d_q - d_0)\lim_{\epsilon\to 0}\epsilon^{-1}\epsilon^\star \\ &= (d_q - d_0)m_q(x)/m_0(x). \end{aligned} \tag{7.23}$$

We can regard $|D(q)|$ as a measure of local sensitivity of the posterior inference to mis-specification of the prior density in the direction defined by $q(\theta)$.

7.16 Rather than concentrating on $D(q)$ for a specific $q(\theta)$, we may ask what is the maximum sensitivity $\sup_q |D(q)|$ for all possible distributions $q(\theta)$. Or, making a further connection with ϵ-contamination, we can ask for

$$D(Q) = \sup_{q\in Q} |D(q)|. \tag{7.24}$$

Now since it is easy to establish that for any mixture of distributions $q_1(\theta)$ and $q_2(\theta)$,

$$D(aq_1 + (1-a)q_2) = aD(q_1) + (1-a)D(q_2),$$

we have

$$D(q) = \int D^\star(\theta)q(\theta)\,\mathrm{d}\theta,$$

where $D^\star(\theta)$ is the value of $D(q)$ for a degenerate distribution placing probability one at the single point θ. Therefore the unconstrained maximum sensitivity is

$$\sup_q |D(q)| = \sup_\theta |D^\star(\theta)|. \tag{7.25}$$

Similarly (7.24) is solved for convex Q by maximizing over the set of vertices of Q. The sensitivity measure $D(Q)$ is introduced and discussed in more detail by Sivaganesan (1993).

Example 7.8
If the inference required is the posterior mean, then if $q(\theta)$ is the degenerate distribution giving probability one to a specific θ the posterior $q(\theta \mid x)$ will also give probability one to that θ, and $d_q = \theta$. We also have $m_q(x) = f(x \mid \theta)$. Therefore (7.25) reduces to

$$\sup_q |D(q)| = \{m_0(x)\}^{-1} \sup_\theta |\theta - d_0| f(x \mid \theta). \tag{7.26}$$

Example 7.9
Following Example 7.8 let x be distributed as $N(\theta, 1)$, i.e. $f(x \mid \theta) = (2\pi)^{-1/2} \exp\{-(x - \theta)^2/2\}$. Then $|\theta - d_0| f(x \mid \theta)$ is maximized at

$$\theta = y + (1 + z^2)^{1/2} \text{sign}(z), \tag{7.27}$$

where $y = (x + d_0)/2$ and $z = (x - d_0)/2$. If we further suppose that the stated prior distribution is $N(0, 1)$ so that $d_0 = x/2$ and $m_0(x) = (4\pi)^{-1/2} \exp(-x^2/4)$, (7.26) is simple to calculate for any x. We find that the maximum sensitivity grows very rapidly with x, being 0.86 at $x = 0$, 5.14 at $x = 2$ but 5.2×10^{11} at $x = 10$. If x is large, the posterior mean is extremely sensitive to the slightest inaccuracy in specification of $f(\theta)$ in the neighbourhood of $\theta = x$ (to which (7.27) reduces for large x). This conclusion is very similar to that of Example 7.7.

7.17 There are many other ways in which local sensitivity can be formulated, essentially as some kind of derivative of a posterior inference or expected utility with respect to the prior distribution. See Diaconis and Freedman (1986), Ruggeri and Wasserman (1993) and Sivaganesan (1993). An important question is whether such derivatives necessarily exist, or whether the effect on the posterior inference might tend to a non-zero value as the perturbation of the prior tends to zero. The following example, based on Example 3 of Kadane and Chuang (1978), suggests what might be possible.

Example 7.10
Suppose that the posterior density from the stated prior is $f_0(\theta \mid x)$, but that by some kind of ϵ perturbation of the stated prior the posterior becomes $(1-\epsilon) f_0(\theta \mid x) + \epsilon q(\epsilon^{-1})$ where $q(\epsilon^{-1})$ is the degenerate distribution giving probability one to $\theta = \epsilon^{-1}$. Then as $\epsilon \to 0$ the posterior density tends to $f_0(\theta \mid x)$, but the posterior mean tends to $d_0 + 1$ where d_0 is the mean of $f_0(\theta \mid x)$. So it is possible at least in principle for an infinitessimal change in the prior distribution to produce a non-infinitessimal change in the inference. Kadane and Chuang (1978) call this *unstable* inference. However, we must first determine whether this kind of perturbation of the posterior distribution could ever be achieved by realistic perturbation of the prior. It can certainly be achieved by a prior distribution of the form

$$f(\theta) = (1 - a(\epsilon)) f_0(\theta) + a(\epsilon) q(\epsilon^{-1})$$

in which there is a prior probability $a(\epsilon)$ attached to the point $\theta = \epsilon^{-1}$ by the degenerate distribution $q(\epsilon^{-1})$, but the prior probability needs to increase rapidly towards 1 as ϵ tends to zero in order for the corresponding posterior probability to be ϵ. Then it is no longer true that $f(\theta) \to f_0(\theta)$ as $\epsilon \to 0$.

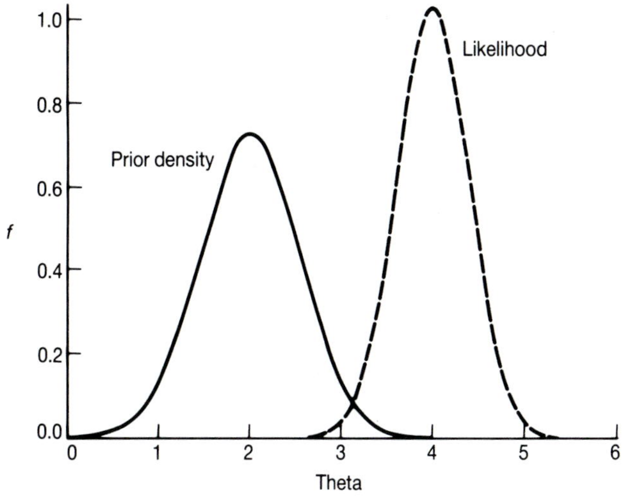

Fig. 7.2 Conflict between likelihood and prior

7.18 The continuity of posterior inferences as functions of the prior distribution is in fact proved under quite general conditions by Cuevas and Sanz (1988) and Salinetti (1992).

Surprise

7.19 Equation (7.23) has in its denominator the predictive density $m_0(x)$ of the data x under the prior distribution $f_0(\theta)$, equation (7.8). We must generally expect posterior inference to be sensitive to perturbation of the prior when $m_0(x)$ is small, for we can write the posterior distribution arising from $f_0(\theta)$ as

$$f_0(\theta \mid x) = f_0(\theta) f(x \mid \theta)/m_0(x).$$

Dividing by a small $m_0(x)$ will greatly magnify the effect of small changes in the numerator caused by perturbing either $f_0(\theta)$ or $f(x|\theta)$. (7.23) formalizes this intuitive argument, and Example 7.9 confirms that extreme sensitivity arises when the data x are such that $m_0(x)$ is very small.

Since $m_0(x)$ is the predictive distribution of x, small $m_0(x)$ denotes data whose prior probability is low. The fact that such an improbable value of x has been observed can therefore be considered surprising. Whenever surprising data are observed, sensitivity of posterior inferences must be suspected.

7.20 Equation (7.8) shows that this situation arises when the product $f(x \mid \theta) f_0(\theta)$ is small for all θ. This in turn denotes a conflict between the likelihood and the prior, since every θ for which the likelihood is high has low prior density, and whenever the prior density is high the likelihood of θ is small. The situation is illustrated by Figure 7.2. When prior and likelihood conflict, the form of the posterior density can change quite dramatically in response to any change to the very small values in the tails of either $f(x \mid \theta)$ or $f(\theta)$.

7.21 When we speak of $m_0(x)$ being small we must recognize that there is no absolute measure of smallness. If the data comprise a very large number of observations then the probability of obtaining any particular set of data x will be small for every x. But then $m_q(x)$ will also be small. Equation (7.23) shows that sensitivity can be expected when $m_0(x)$ is *relatively* small. Following **7.16** and Example 7.8, the maximum value of $m_q(x)$ over all possible distributions $q(\theta)$ occurs when $q(\theta)$ is the degenerate distribution giving probability one to the value of θ maximizing $f(x \mid \theta)$. Therefore a possible index of surprise is $\{m_0(x)\}^{-1} \sup_\theta f(x \mid \theta)$. This measures how surprising the observed x is according to the prior $f_0(\theta)$, relative to all other possible prior distributions.

Alternatively, we could measure how surprising the observation x is compared to other possible observations, according to the prior $f_0(\theta)$, by $\{m_0(x)\}^{-1} \sup_x m_0(x)$. This suggestion contrasts with Box's (1980) proposal of using the tail-area probability $\int_Y m_0(y)dy$ where Y is the set $\{y : m_0(y) \leqslant m_0(x)\}$. This is therefore the probability of observing data with marginal density less than or equal to the value $m_0(x)$ for the data we actually observed. However, both of these measures violate the Likelihood Principle by referring to data other than the actually observed x.

Limiting sensitivity

7.22 The discussion in **7.20** suggests that when sensitivity arises because of conflict the scale of the sensitivity and the posterior inferences can depend heavily on the shapes of the tails of the prior density and likelihood. In particular, the high sensitivity in Example 7.9 relates to the very thin tails of normal distributions. There may be less sensitivity to perturbation of the prior in general if the prior density $f_0(\theta)$ is given heavier tails.

Example 7.11

If in Example 7.9 we let the stated prior distribution be the Cauchy f.f. $f_0(\theta) = \pi^{-1}(1+\theta^2)^{-1}$ we can no longer write down d_0 and $m_0(x)$ explicitly, but must compute them numerically for individual x. We find that the maximum sensitivity (7.26) is 1.16 at $x = 0$, 5.85 at $x = 2$ and 482 at $x = 10$. It therefore grows much less rapidly with x than in the case of the normal prior density of Example 7.9. Figure 7.3 plots the logarithm of the maximum sensitivity against x for both prior distributions.

7.23 Example 7.11 suggests a general strategy for achieving more robust posterior inference, i.e. always ensure that Your stated prior distribution has heavy tails. That will reduce the surprise level when data are observed that conflict with Your stated prior beliefs, and so reduce the sensitivity of posterior inferences when such data arise. This idea raises a number of points for consideration.

In specifying Your stated distribution You will have first given values to a number of prior summaries and then completed the specification with a convenient distribution $f_0(\theta)$ to fit those values. If Your stated summaries consist of a prior mean and variance

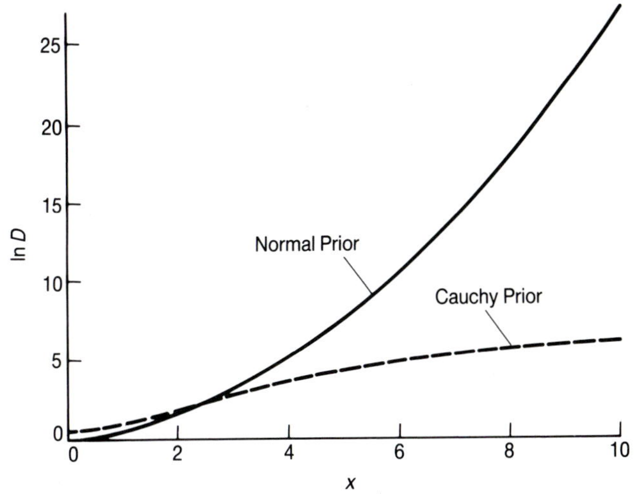

Fig. 7.3 Log maximum sensitivity, Examples 7.9 and 7.11

and a statement of unimodality, then a heavy-tailed t distribution can be chosen to fit those summaries just as well as a thin-tailed normal distribution. In this circumstance it is attractive to choose a heavy-tailed distribution to reduce sensitivity in the event of unexpected data. If, on the other hand, Your prior specification included a firm view that Your prior beliefs about θ should have tails like a normal distribution then it would not be legitimate to over-ride this choice. Sensitivity is not an issue in this case because You would not be willing to consider prior distributions whose tails varied from the normal form of Your stated $f_0(\theta)$. The kinds of perturbations of $f_0(\theta)$ that You would be willing to consider (as described by a set Γ, a set Q of epsilon contaminations or a set Q of directions for local sensitivity study) would not then cause large changes in posterior inference in the way we have been considering.

7.24 One way in which You could state a preference for a normal prior distribution in Example 7.9 is to believe that the posterior mean should be linear in x, even for large x. We noted in **6.28** that this linearity only holds if the prior distribution is normal. Other forms of prior distribution can produce very different behaviour of the posterior mean as a function of x. This is particularly true if the prior distribution has non-normal tails, as the following examples show.

Example 7.12
Let x have the $N(\theta, 1)$ distribution given θ, and let θ have a uniform prior distribution over $[a, b]$. Then the posterior density is

$$\begin{aligned} f(\theta \mid x) &\propto f(x \mid \theta) f(\theta) = (b-a)^{-1}(2\pi)^{-1/2} \exp\{-(x-\theta)^2/2\} \\ &\propto \exp\{-(x-\theta)^2/2\} \end{aligned}$$

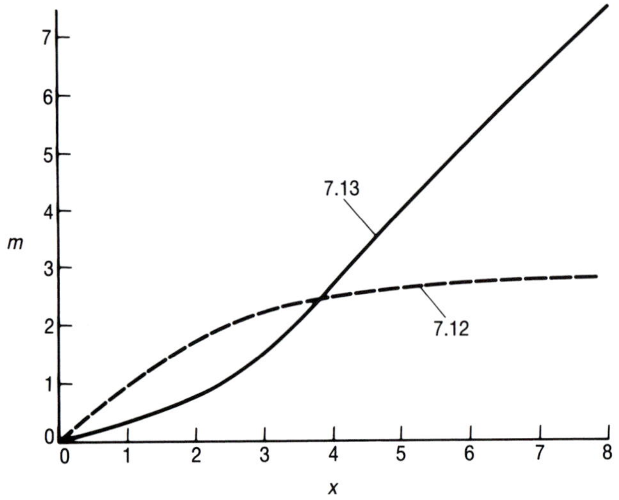

Fig. 7.4 Posterior means for Examples 7.12 and 7.13

for $a \leqslant \theta \leqslant b$, and $f(\theta \mid x) = 0$ otherwise. So the posterior distribution is $N(x, 1)$ truncated to the interval $[a, b]$. In fact

$$f(\theta \mid x) = \{\Phi(b-x) - \Phi(a-x)\}^{-1}(2\pi)^{-1/2}\exp\{-(\theta - x)^2/2\}$$

for $a \leqslant \theta \leqslant b$, where Φ denotes the standard normal d.f. as usual. Clearly, $E(\theta \mid x)$ must also lie in $[a, b]$. The exact form of $E(\theta \mid x)$ is found by direct integration.

$$\begin{aligned} E(\theta - x \mid x) &= \{\Phi(b-x) - \Phi(a-x)\}^{-1}(2\pi)^{-1/2}\int_a^b (\theta - x)\exp\{-(\theta-x)^2/2\}\,\mathrm{d}\theta \\ &= \{\Phi(b-x) - \Phi(a-x)\}^{-1}(2\pi)^{-1/2}[-\exp\{-(\theta-x)^2/2\}]_a^b \\ &= \{\Phi(b-x) - \Phi(a-x)\}^{-1}\{\phi(a-x) - \phi(b-x)\}, \end{aligned}$$

where ϕ denotes the standard normal f.f. Therefore

$$\begin{aligned} E(\theta \mid x) &= x + E(\theta - x \mid x) \\ &= x - \{\Phi(b-x) - \Phi(a-x)\}^{-1}\{\phi(b-x) - \phi(a-x)\}. \end{aligned}$$

This function is plotted in Figure 7.4 for the case $a = -3, b = 3$ and is clearly not linear. $E(\theta \mid x) \to b$ as $x \to \infty$, and in fact the limit of the posterior distribution as $x \to \infty$ is degenerate with probability one at $\theta = b$.

Example 7.13

Let x have the $N(\theta, 1)$ distribution given θ, and let θ have a t prior distribution of the form

$$f(\theta) \propto (1 + \theta^2)^{-c}. \tag{7.28}$$

Then

$$f(\theta \mid x) \propto (1 + \theta^2)^{-c}\exp\{-(x - \theta)^2/2\}. \tag{7.29}$$

In examining this posterior density, first note that

$$\mathrm{d}\log f(\theta \mid x)/\mathrm{d}\theta = -2\theta c(1+\theta^2)^{-1} + x - \theta.$$

This is clearly negative for all $\theta > x$, and since $2\theta(1+\theta^2)^{-1} = 1-(1-\theta)^2(1+\theta^2)^{-1} \leqslant 1$ for all θ, it is positive for all $\theta < x-c$. Therefore the posterior mode (or modes) must lie between $x-c$ and x. Now the tails of the prior distribution (7.28) are very flat, such that for any constants a and b, $\{1+(x+a)^2\}^{-1}/\{1+(x+b)^2\}^{-1} \to 1$ as $x \to \infty$. Therefore the posterior density (7.29) has the property that as $x \to \infty$ the ratio $f(\theta = x+a \mid x)/f(\theta = x+b \mid x)$ tends to the ratio of likelihoods at those two points, i.e. $\exp\{(b^2-a^2)/2\}$. Therefore the posterior density over any range $x-A \leqslant \theta \leqslant x+B$, for arbitrarily large A and B, becomes proportional to the likelihood $\exp\{-(x-\theta)^2/2\}$. Since we began by showing that the posterior density falls away uniformly outside that range, it follows that as $x \to \infty$ the posterior distribution tends to $N(x, 1)$. We can show similarly that $E(\theta \mid x)$ tends to x as x tends to infinity, but we certainly do not have the strictly linear form $E(\theta \mid x) = x$ for all x which would result from the improper uniform prior distribution.

The actual form of $E(\theta \mid x)$, obtained by numerical calculations, is plotted for the case $c = 2$ in Figure 7.4 and is clearly non-linear.

7.25 These examples show just two different forms of asymptotic behaviour for the posterior distribution and posterior mean as a normal observation x tends to infinity. Characterizing prior distributions giving rise to different posterior responses to extreme observations is an aid to prior specification in the spirit of **6.28**.

Suppose that the likelihood has the location parameter form $f(x \mid \theta) = g(x-\theta)$. Then as x increases a conflict arises between the likelihood, which is asserting that θ is most likely to be near x, and the prior distribution which says that θ is believed to be near the prior mean. If both are normal distributions we have seen that the posterior mean is a weighted average of x and the prior mean, and so is a linear function of x. For every x this represents a compromise between the prior and likelihood information. In general, the conflict will be resolved by compromise if the prior and likelihood have tails of equal thickness. If the prior has thicker tails, as in the case of a normal likelihood and Cauchy prior, then as x increases the posterior distribution may become essentially the likelihood (normalized to integrate to one). The effect is the same as if a uniform prior had been used, and corresponds to rejecting the prior information. The posterior mean tends to x. This is the case of Example 7.13. Conversely, if the prior has thinner tails than the likelihood, the conflict may be resolved by rejecting the data. The posterior distribution converges to the prior distribution and the posterior mean to the prior mean. Proofs of these results under appropriate definitions of tail thickness are given by Dawid (1973), Hill (1974) and O'Hagan (1979, 1990) and Le and O'Hagan (1993). See also De Finetti (1961) and Goldstein (1982) for related theory.

This discussion has concentrated on asymptotic behaviour of the posterior d.f. or f.f., but the above references show that asymptotic behaviour of moments need not be the same. Progression towards the limit is considered by O'Hagan (1981), who shows that in a particular class of problems the posterior variance reaches a peak before falling back to its limiting value. The peak occurs when the posterior is still trying to accommodate both conflicting sources of information, and the subsequent reduction of variance corresponds to the conflict being resolved. If the conflict is never resolved, because prior and likelihood have equally heavy tails, the posterior variance can be unbounded (see Fan and Berger (1992)), in striking contrast to the compromise which results when both are normal.

Outliers

7.26 A major preoccupation of non-Bayesian approaches to robustness is to develop inference procedures that are robust to the occurrence of outliers in the data. (The concern is with inference robustness; see **7.3**). Outliers are 'rogue observations' which lie, in some sense, far from the bulk of the data. Classically robust procedures aim to reduce the influence which extreme observations have on inference, often going to the extent of *rejecting* outliers. The rejected observations are removed from the data before calculating the required inference. O'Hagan (1979, 1990) proves that if the data are assumed to be distributed with heavy tails (as opposed to the thin-tailed normal distribution) then the influence of outliers in Bayesian inference will be reduced automatically. The extreme observations are 'rejected' because they are much less informative about θ than the non-outlying data.

Heavy-tailed distributions, and t distributions particularly, have been advocated in a variety of application areas, principally to obtain more robust inference. Application in linear models is considered by West (1984, 1985), Angers and Berger (1991) and Fan and Berger (1992). Meinhold and Singpurwalla (1989) apply them to Kalman filtering and dynamic linear models (see **10.37**). Application in econometric models is considered by Geweke (1992b), who argues that heavy-tailed data arise naturally in some econometric problems. He and Carlin and Polson (1991) apply Gibbs sampling methods (see **8.48**) to analyse the resulting posterior distributions, which are generally intractable. Note, however, useful analytical results and approximations in Pericchi and Smith (1992) and Angers and Berger (1991).

7.27 A conceptually quite different approach to outliers is to represent them as deriving from an alternative distribution to that generating the non-outlying data. Analysis with a variety of contamination distributions is presented by Box and Tiao (1968), Abraham and Box (1978), Guttman et al. (1978), Sharples (1990), Verdinelli and Wasserman (1991) and Peña and Tiao (1992). Contamination distributions generally are considered by Goldstein (1982). An important ingredient of such approaches is that they give posterior probabilities of particular observations being outliers, and are often proposed with that as their principal aim. Other methods of outlier detection are found in Pettit and Smith (1985), Geisser (1987,1989), Chaloner and Brant (1988), Guttman and Peña (1988), Pettit (1988,1990), Girón et al. (1991) and Bayarri and Berger (1992).

7.28 Adopting heavy-tailed prior distributions as a routine practice we achieve more robustness in the face of surprising data primarily by forcing a particular resolution of prior/likelihood conflicts. It means that when such a conflict arises You will be prepared to believe the data and relinquish the originally stated prior distribution. In practice, that is often a very reasonable attitude to adopt. Most investigators would be unwilling to hold to their original prior beliefs when data are obtained that simply do not accord with those beliefs. But we should be aware that there are alternatives. You can assert that the prior information is sufficiently reliable to justify compromise with the data. You can even believe the prior distribution in preference to the likelihood.

The latter position is defensible if You suspect that the data are prone to containing outliers. A single observation x could then be an outlier if it conflicts with prior information, and the appropriate response would then be to ignore x. If, on the other hand, several observations $x_1, x_2, \ldots, x_n$ agree well with each other but conflict with the prior, You will be inclined to believe the data (on the grounds that they are most unlikely to be all outliers) and reject the prior information. In this case, both likelihood and prior must have heavy tails. O'Hagan (1979) shows how the relative heaviness of tails can be chosen to achieve this kind of posterior behaviour.

7.29 Deciding how conflict should be resolved is potentially a powerful way of specifying prior tails. It can be seen as a use of imaginary observations, extending the ideas of **6.30**.

O'Hagan (1988) and O'Hagan and Le (1993) go further, discussing how tails of a multivariate prior distribution can be formulated to obtain different resolutions of conflicts between information relating to a number of parameters.

A hyperparameter representation

7.30 We can think of a class Γ of prior distributions as indexed by a hyperparameter ϕ, writing

$$\Gamma = \{f_\phi(\theta) : \phi \in \Phi\}. \tag{7.30}$$

Then we can formally treat $f_\phi(\theta)$ as a conditional distribution $f(\theta \mid \phi)$. A prior distribution $f(\phi)$ for ϕ would now enable us to give different prior probabilities to the distributions in Γ. Instead of saying only that Your true prior distribution is some member of Γ, You can now assert prior beliefs about which member of Γ the true $f(\theta)$ is. You could, for instance, represent a belief that $f(\theta)$ is more likely to be close to Your stated $f_0(\theta)$, in some sense, than to be further away at the edges of Γ.

Now in one sense this idea of putting a prior distribution over Γ ignores the original reason for introducing Γ and runs counter to this chapter's theme of studying sensitivity or robustness. For it allows You immediately to deduce a single prior distribution

$$f(\theta) = \int f(\theta \mid \phi) f(\phi)\, \mathrm{d}\phi, \tag{7.31}$$

which becomes *the* prior distribution, We similarly obtain a single posterior distribution from (7.31) and the likelihood $f(x \mid \theta)$,

$$f(\theta \mid x) = f(x \mid \theta) f(\theta) / f(x), \tag{7.32}$$

where

$$f(x) = \int f(x \mid \theta) f(\theta)\, \mathrm{d}\theta, \tag{7.33}$$

which becomes *the* posterior distribution. Inference derived from (7.32) could then be treated as *the* inference, and there is apparently no need to consider robustness.

7.31 A little algebra illustrates the relationship between these results and our earlier sensitivity analysis. The prior (7.31) is simply a weighted average of the prior distributions $f(\theta \mid \phi)$ in Γ, with weight function $f(\phi)$. We can similarly write the posterior (7.32) as a weighted average

$$f(\theta \mid x) = \int f(\theta \mid x, \phi) f(\phi \mid x)\, d\phi \tag{7.34}$$

with weight function

$$f(\phi \mid x) = f(x \mid \phi) f(\phi)/f(x), \tag{7.35}$$

where $f(\theta \mid x, \phi)$ is the posterior distribution given by the prior distribution $f(\theta \mid \phi)$. That is

$$f(\theta \mid x, \phi) = f(x \mid \theta) f(\theta \mid \phi)/f(x \mid \phi), \tag{7.36}$$

and

$$f(x \mid \phi) = \int f(x \mid \theta) f(\theta \mid \phi)\, d\theta. \tag{7.37}$$

Now if we consider inference in the form of an integration-based summary

$$\begin{aligned} E\{g(\theta) \mid x\} &= \int g(\theta) f(\theta \mid x)\, d\theta \\ &= \int \left\{ \int g(\theta) f(\theta \mid x, \phi)\, d\theta \right\} f(\phi \mid x)\, d\phi \\ &= \int E\{g(\theta) \mid x, \phi\} f(\phi \mid x)\, d\phi, \end{aligned} \tag{7.38}$$

assuming that we can reverse the order of integration. Then (7.38) shows *the* posterior inference as a weighted average of the inferences $E\{g(\theta) \mid x, \phi\}$ deriving from the various priors $f(\theta \mid \phi)$ in Γ, via the posterior weight function $f(\phi \mid x)$.

In studying the sensitivity of inference to variation of the prior distribution over Γ, we were interested in the maximum and minimum values of $E\{g(\theta) \mid x, \phi\}$. By assuming a prior distribution $f(\theta)$ over Γ we obtain instead an averaging of the possible inferences $E\{g(\theta) \mid x, \phi\}$ and lose the consideration of sensitivity completely.

7.32 However, something is gained from this representation because we see that the prior weight function is modified by Bayes' theorem in (7.35) to a posterior weight function which makes use of the data to indicate which prior distributions $f(\theta \mid \phi)$ in Γ are most likely. The role of the likelihood here is played by $f(x \mid \phi)$, equation (7.37), which is the predictive distribution for the data derived from the prior $f(\theta \mid \phi)$. We have argued in **7.19** and **7.20** that $f(x \mid \phi)$ represents the degree of agreement between the likelihood $f(x \mid \theta)$ and the prior $f(\theta \mid \phi)$, and so measures how well the prior agrees with the data. We have also seen that sensitivity is to be expected when the stated prior $f_0(\theta) = f(\theta \mid \phi_0)$ conflicts with the likelihood to produce a low predictive density $f(x \mid \phi_0)$. If there are other distributions $f(\theta \mid \phi)$ in Γ which agree more closely with the data then we can expect the inference $E\{g(\theta) \mid x\}$ in (7.38) to differ from the inference $E\{g(\theta) \mid x, \phi_0\}$ derived from the stated prior.

7.33 It is reasonable to argue that, at least in principle, You do have beliefs about which $f(\theta \mid \phi)$ in Γ are more or less reasonable representations of your true prior distribution for θ, which could be stated as a prior distribution $f(\phi)$ for ϕ. In practice, for any reasonable class Γ it will be very difficult to formulate such a distribution. If we allow $f(\phi)$ to be any distribution in some class Γ_ϕ we could examine sensitivity of posterior inference to varying $f(\phi)$ over Γ_ϕ. In particular, if we allowed Γ_ϕ to be the class of all possible distributions then the resulting class of distributions $f(\theta)$ given by (7.34) is just Γ and we obtain exactly the same range of inferences. There is little to be gained in practice, therefore, by using the hyperparameter representation. However, the mathematics of **7.30** can be adapted to a variety of other analyses.

Bayes factors for prior distributions

7.34 If Γ contains only two distributions $f_0(\theta) = f(\theta \mid \phi_0)$ and $f_1(\theta) = f(\theta \mid \phi_1)$, let $p = f(\phi_1) = P(\phi = \phi_1)$ and $1 - p = f(\phi_0) = P(\phi = \phi_0)$. Then (7.31) and (7.34) can be rewritten as a prior distribution

$$f(\theta) = (1-p)f_0(\theta) + pf_1(\theta), \tag{7.39}$$

and a posterior distribution

$$f(\theta \mid x) = (1-p^\star)f_0(\theta \mid x) + p^\star f_1(\theta \mid x), \tag{7.40}$$

where $p^\star$ is the posterior probability

$$p^\star = P(\phi = \phi_1 \mid x) = f(\phi_1 \mid x) = p\, m_1(x)/f(x), \tag{7.41}$$

$$m_0(x) = f(x \mid \phi_0) = \int f(x \mid \theta) f_0(\theta)\, d\theta,$$

$$m_1(x) = f(x \mid \phi_1) = \int f(x \mid \theta) f_1(\theta)\, d\theta,$$

and

$$f(x) = (1-p)m_0(x) + pm_1(x). \tag{7.42}$$

These formulae are equivalent to those found in **7.7** for epsilon contamination, if we replace p by ϵ, $p^\star$ by $\epsilon^\star$ and $q(\theta)$ by $f_1(\theta)$. Equation (7.10) is (7.41) with (7.42) substituted for $f(x)$. We can therefore think of the ϵ in **7.7** as a prior probability for the contaminating distribution $q(\theta)$ and $\epsilon^\star$ as the corresponding posterior probability.

7.35 Another way to write (7.41) is

$$\frac{p^\star}{1-p^\star} = \frac{p}{1-p}\,\frac{m_1(x)}{m_0(x)}\,. \tag{7.43}$$

The left hand side is the posterior odds against $\phi = \phi_0$ (and in favour of $\phi = \phi_1$). The first term on the right hand side is the corresponding prior odds. The ratio $m_1(x)/m_0(x)$ is called the Bayes factor against $\phi = \phi_0$ or the Bayes factor in favour of $\phi = \phi_1$. A Bayes factor greater than one means that the posterior odds against $\phi = \phi_0$ is greater than the prior odds, and hence $P(\phi = \phi_0 \mid x) < P(\phi = \phi_0)$. Or $P(\phi = \phi_1 \mid x) > P(\phi = \phi_1)$. A Bayes factor less than one against ϕ_0 is a Bayes factor greater than one in favour of ϕ_0 and the above inequalities are reversed.

In fact we can interpret $m_i(x)$ as a likelihood, since $m_i(x) = f(x \mid \phi_i)$, and this appears formally as a likelihood in Bayes' theorem, equation (7.41). A Bayes factor in general is simply a ratio of likelihoods, expressing the relative likelihood for one value of a parameter against another. In this case the parameter is ϕ and we are comparing the two values ϕ_0 and ϕ_1. Bayes factors are helpful in assessing the relative weight of evidence in the data for one value of a parameter against another. The ratio of posterior probabilities for the two parameter values is their prior probability ratio multiplied by the Bayes factor. A Bayes factor far from 1, either very large or very small, represents very strong evidence in the data for one parameter value against the other. It will lead to the posterior probability for the favoured parameter value being higher than for the other value unless the ratio of prior probabilities is very strongly in the opposite direction.

Bayes factors for comparing models

7.36 The idea of a Bayes factor arises more often in comparing models rather than comparing prior distributions. Suppose that two models are proposed for data x. In the ith model the likelihood is expressed as $f_i(x \mid \theta_i)$ in terms of parameters θ_i having prior distribution $f_i(\theta_i)$, $i = 1, 2$. (As discussed in **7.4**, we allow the parametrization to change when we change the likelihood.) Then the marginal distribution of x under model i is

$$m_i(x) = \int f_i(x \mid \theta_i) f_i(\theta_i) \, d\theta_i \,, \tag{7.44}$$

and we can define the Bayes factor for model 1 against model 2 as $m_1(x)/m_2(x)$. If You believe that one of these models correctly represents the process actually generating the data x, and Your prior probability is p that the correct model is model 1, then Your prior *odds* on model 1 is the ratio $p/(1-p)$. The corresponding posterior odds on model 1 will be the prior odds multiplied by the Bayes factor. Your probability for model 1 will increase if the Bayes factor exceeds one. Broadly speaking this is when model 1 fits the data x better than model 2.

The remainder of this chapter is concerned with comparing models, and in particular with the role of Bayes factors in measuring the relative fit of the models to the data, or in measuring the weight of evidence of the data for one model against another. Mosteller and Wallace (1984) present a substantial application of Bayes factors to a problem of determining authorship.

7.37 The link between Bayes factors and sensitivity is less direct now than in comparing prior distributions. We can of course consider a class $\mathcal{M}$ of models, where a model $m \in \mathcal{M}$ consists of a likelihood $f_m(x \mid \theta_m)$ for the data in terms of parameters θ_m having prior distribution $f_m(\theta_m)$. We could consider sensitivity of inference to the choice of model as m varies over the class $\mathcal{M}$, but what inference? We cannot in general think of the required inference now as a summary of the posterior distribution, because as we change the model the parameters θ_m of the posterior distribution $f_m(\theta_m \mid x)$ may change. Then posterior inferences from different models just cannot be compared and we cannot talk about the range of possible inferences. If You are uncertain about the model to the extent of being uncertain what parameters are appropriate to describe it, the required inference cannot be a summary of the posterior distribution.

Instead, it is appropriate to work with predictive inference (see **3.58**) for some other observations y which will be observed at some future time. Each model $m \in \mathcal{M}$ then includes also a distribution $f_m(y \mid \theta_m, x)$ for y, and we can deduce a predictive distribution

$$\begin{aligned} f_m(y \mid x) &= \int f_m(y \mid \theta_m, x) f_m(\theta_m \mid x)\, \mathrm{d}\theta_m \\ &\propto \int f_m(y \mid \theta_m, x) f_m(x \mid \theta_m) f_m(\theta_m)\, \mathrm{d}\theta_m \end{aligned} \tag{7.45}$$

for y given the data x. The predictive distributions (7.45) are now comparable across models, and we could consider how much a specific predictive inference about y varies as the model varies over $\mathcal{M}$, as a measure of sensitivity of that inference to mis-specification of the model.

Robustification of the model

7.38 Consider, however, the case where You believe that the data $x_1, x_2, \ldots, x_n$ are independently and identically distributed either as $N(\theta, 1)$ or $N(\theta, 2)$. Here there are two possible models with the *same* parameters θ. In this case it would clearly be unnecessary to think in terms of predictive inference, since in both models it is appropriate to be interested in inference about their common parameter θ. We therefore consider the case in which every model m in the set $\mathcal{M}$ employs the same parameters θ, i.e. $\theta_m = \theta$ for all $m \in \mathcal{M}$. The models are distinguished only by giving different likelihoods $f_m(x \mid \theta)$ and possibly different prior distributions $f_m(\theta)$.

An alternative formulation for this case is to treat the model as an extra parameter, which we will denote by ϕ. Considering a particular model m corresponds to setting $\phi = m$, so $f_m(x \mid \theta) = f(x \mid \theta, \phi = m)$ and $f_m(\theta) = f(\theta \mid \phi = m)$. So the set of likelihoods for the various models combine to form a single likelihood $f(x \mid \theta, \phi)$ for the augmented parameter (θ, ϕ). And the set of prior distributions form the conditional prior distributions $f(\theta \mid \phi)$. To complete this representation we need the marginal prior distribution $f(\phi)$ for the nuisance parameter ϕ, which in the original formulation simply gives prior probabilities for the various models.

In the case of the data distributed as either $N(\theta, 1)$ or $N(\theta, 2)$ we could introduce a parameter ϕ taking possible values 1 and 2. We then have parameters θ and ϕ, and a single model in which the data are distributed as $N(\theta, \phi)$. The prior probabilities for the models give the marginal prior distribution $P(\phi = 1) = 1 - P(\phi = 2)$.

7.39 Corresponding to this analysis is a simple way to study sensitivity of inferences to errors in specification of the likelihood. We simply introduce an extra parameter ϕ, such that the stated model $f_0(x \mid \theta)$ is the distribution $f(x \mid \theta, \phi = \phi_0)$ arising when ϕ takes the particular value ϕ_0.

Example 7.14
If the stated model supposes that observations x_i are distributed as $N(\mu, \sigma^2)$, with parameters $\theta = (\mu, \sigma^2)$, we can consider this as the case $\phi = 2$ of the more general exponential power family

$$f(x_i \mid \mu, \sigma^2, \phi) = \frac{1}{\Gamma(1+\phi^{-1})2^{1+\phi^{1}}}\sigma^{-1} \exp\left\{-\frac{1}{2}\left|\frac{x_i - \mu}{\sigma}\right|^{\phi}\right\}. \tag{7.46}$$

Box and Tiao (1962) discuss Bayesian analysis using this family of distributions.

7.40 The sensitivity analysis would now let ϕ vary over some neighbourhood of ϕ_0, looking for the maximum amd minimum values of a given inference about θ. In effect, this means looking at the range of inferences derived from the conditional posterior distributions $f(\theta \mid x, \phi) \propto f(\theta \mid \phi) f(x \mid \theta, \phi)$. If, however, we introduce a prior distribution $f(\phi)$ we can derive the marginal posterior distribution $f(\theta \mid x)$ for the parameters of interest and compute a single inference from that. This approach is equivalent to the hyperparameter representation of **7.30** which places a prior distribution over the class Γ of priors.

Box (1980) calls this process of introducing an extra parameter to the model *robustification.* It is a particularly Bayesian device, replacing an implicit modelling assumption that $\phi = \phi_0$ with a prior distribution for ϕ, representing some uncertainty over whether ϕ is equal to (or close to) ϕ_0. Classical inference offers no such intermediate state of knowledge about a model: one must either make the assumption of a fixed $\phi = \phi_0$ or treat ϕ as completely unknown.

Nested models

7.41 Having introduced the parameter ϕ it is interesting to ask whether the original model would nevertheless have been accurate. Then we wish to compare the original model having likelihood $f_0(x \mid \theta) = f(x \mid \theta, \phi = \phi_0)$ with the more general, robustified model with likelihood $f(x \mid \theta, \phi)$. The first model's parameters θ are now a subset of the parameters (θ, ϕ) of the second model, and we are said to be comparing two *nested* models. In general, nested models arise when one model is a special case of the other, which often allows it to be expressed using fewer parameters.

The prior distribution for the first model is a distribution $f_0(\theta)$, while for the second model we have a distribution $f_1(\theta, \phi)$ for (θ, ϕ). For a full specification of prior beliefs we also need a prior probability π_0 for the first model and $\pi_1 = 1 - \pi_0$ for the second. Notice, however, that π_0 can also be interpreted as a prior probability that $\phi = \phi_0$. Similarly, $f_0(\theta)$ is the conditional distribution of θ given $\phi = \phi_0$, while for all $\phi \neq \phi_0$ that conditional distribution would be $f_1(\theta \mid \phi)$ derived from the second prior distribution. Therefore if we write $f_1(\theta, \phi)$ as $f_1(\theta \mid \phi) f_1(\phi)$ the prior distribution for ϕ consists of a discrete probability π_0 that $\phi = \phi_0$, together with a continuous density $\pi_1 f_1(\phi)$ for all $\phi \neq \phi_0$. (The total probability in this distribution is of course one, since the total probability that $\phi \neq \phi_0$ is $\int \pi_1 f_1(\phi) d\phi = \pi_1 = 1 - \pi_0$.) This rather strange prior belief about ϕ is a necessary consequence of comparing nested models, since we are implicitly giving a non-zero prior probability that the special case $\phi = \phi_0$ holds.

Believing that there is a non-zero probability π_0 that $\phi = \phi_0$ inevitably creates a discontinuity in the marginal distribution of ϕ, between the density function $f_1(\phi)$ for $\phi \neq \phi_0$ and the discrete probability mass π_0. Nevertheless it will generally be reasonable in practice for the conditional density of θ given ϕ to be continuous in ϕ at ϕ_0. That is, $f_0(\theta)$ should be the limit of $f_1(\theta \mid \phi)$ as $\phi \to \phi_0$.

7.42 The prior odds ratio in favour of the first model is π_0/π_1. Posterior odds will be given by multiplying by the Bayes factor $f_0(x)/f_1(x)$, where

$$f_0(x) = \int f_0(x \mid \theta) f_0(\theta)\, \mathrm{d}\theta, \tag{7.47}$$

$$f_1(x) = \int\int f(x \mid \theta, \phi) f_1(\theta, \phi)\, \mathrm{d}\theta\, \mathrm{d}\phi. \tag{7.48}$$

The Bayes factor describes the weight of evidence from the data in favour of the first model, and can be calculated without specifying the prior probabilities π_0 and π_1.

Now since $f(\theta \mid \phi) = f_0(\theta)$ if $\phi = \phi_0$ and otherwise $f(\theta \mid \phi) = f_1(\theta \mid \phi)$, we can write (7.47) and (7.48) as

$$f_0(x) = \int f(x \mid \theta, \phi = \phi_0) f(\theta \mid \phi = \phi_0)\, \mathrm{d}\theta = f(x \mid \phi = \phi_0), \tag{7.49}$$

$$f_1(x) = \int \left\{ \int f(x \mid \theta, \phi) f(\theta \mid \phi)\, \mathrm{d}\theta \right\} f_1(\phi)\, \mathrm{d}\phi = \int f(x \mid \phi) f_1(\phi)\, \mathrm{d}\phi. \tag{7.50}$$

That is, the numerator of the Bayes factor is the value of $f(x \mid \phi)$ at $\phi = \phi_0$, while the denominator is a weighted average of the values of $f(x \mid \phi)$ for $\phi \neq \phi_0$, weighted by the prior distribution $f_1(\phi)$ for ϕ under the second model.

Example 7.15

Observations $x_1, x_2, \ldots, x_n$ are independently and identically distributed as $N(\theta, \phi)$. In the first model we suppose that the variance is known, $\phi = \phi_0$. In the second model both θ and ϕ are unknown and we assume a prior distribution of the conjugate form (Example 6.6)

$$f_1(\theta, \phi) \propto \phi^{-t-3/2} \exp\{-d(\theta - m)^2/(2\phi) - b\phi^{-1}\} \tag{7.51}$$

in which the conditional distribution of θ given ϕ is $N(m, \phi/d)$ and the marginal distribution of ϕ is

$$f_1(\phi) \propto \phi^{-t-1} \exp(-b\phi^{-1}). \tag{7.52}$$

The prior distribution of θ under the first model, $f_0(\theta)$, is specified to be $N(m, \phi_0/d)$ to reflect the fact that this is just the case $\phi = \phi_0$ of the more general model.

Then

$$f(x \mid \phi) = \int f(x \mid \theta, \phi) f(\theta \mid \phi)\, \mathrm{d}\theta$$

$$= \int \left[\prod_{i=1}^{n} (2\pi\phi)^{-1/2} \exp\{-(x_i - \theta)^2/(2\phi)\} \right] \left[(2\pi\phi/d)^{-1/2} \exp\{-d(\theta - m)^2/(2\phi) \right] \mathrm{d}\theta.$$

Simple manipulation gives

$$f(x \mid \phi) = (2\pi\phi)^{-n/2}(n+d)^{-1/2}\exp\{-y/(2\phi)\}, \tag{7.53}$$

where

$$y = \sum_{i=1}^{n}(x_i - \bar{x})^2 + nd(\bar{x} - m)^2/(n+d).$$

Setting $\phi = \phi_0$ gives the numerator (7.49) of the Bayes factor. The denominator (7.50) requires us to integrate the product of (7.52) and (7.53) with respect to ϕ, yielding

$$f_1(x) = (2\pi)^{-n/2}(n+d)^{-1/2}(b+y/2)^{-t-n/2}\Gamma(t)^{-1}\Gamma(t+n/2)b^t.$$

Therefore the Bayes factor is

$$f_0(x)/f_1(x) = b^{-t}(b+y/2)^{t+n/2}\phi_0^{-n/2}\Gamma(t)\Gamma(t+n/2)^{-1}\exp\{-y/(2\phi_0)\}. \tag{7.54}$$

Sensitivity of the Bayes factor

7.43 Difficulties arise with nested models if $f_1(\phi)$ is specified to represent very weak prior information about ϕ. We cannot, for instance, just specify a uniform distribution by $f_1(\phi) \propto 1$, because a proportionality constant in $f_1(\phi)$ does not cancel out anywhere. If the range of possible values for ϕ is unbounded then the uniform prior distribution $f_1(\phi) \propto 1$ is improper, so the proportionality constant does not actually exist.

Suppose that ϕ is a scalar parameter on $(-\infty, \infty)$, and we try to specify the uniform prior distribution as a limit of proper uniform distributions

$$f_1(\phi) = (2c)^{-1} \quad \text{for} \quad -c \leqslant \phi \leqslant c$$

and $f_1(\phi) = 0$ otherwise, letting c tend to infinity. Then (7.50) becomes

$$f_1(x) = (2c)^{-1}\int_{-c}^{c} f(x \mid \phi)\,\mathrm{d}\phi. \tag{7.55}$$

Now for most problems $f(x \mid \phi)$ will tend to zero as $\phi \to \pm\infty$ such that the limit of the integral in (7.55) is finite. Then letting c go to infinity means $f_1(x) \to 0$ and the Bayes factor tends to infinity. This is the problem discussed in **3.33**.

Example 7.16

Following Example 7.15, the usual improper prior distribution $f_1(\phi) \propto \phi^{-1}$, representing weak prior information about a variance, may be obtained by setting $b = t = 0$. Then the Bayes factor (7.54) tends to infinity because $\Gamma(t) \to \infty$. The uniform prior distribution corresponds to $b = 0, t = 1$, and again the Bayes factor becomes infinite because $b^{-t} \to \infty$.

7.44 This behaviour arises because conventional representations of weak prior information about an unbounded parameter ϕ usually have the effect of giving zero prior probability to ϕ lying in any finite region. Since the data generally result in a likelihood which is negligible outside a finite region, there is inevitably some kind of conflict generated. Specifically zero prior probability is given to the event that $f(x \mid \phi)$ exceeds $\epsilon f(x \mid \phi_0)$ for arbitrarily small $\epsilon > 0$ (and regardless of x). So the averaging of $f(x \mid \phi)$ in (7.50) must produce a result $f_1(x)$ less than $\epsilon f(x \mid \phi_0)$ and hence a Bayes factor greater than ϵ^{-1}.

7.45 This is one aspect of a general problem of sensitivity of Bayes factors to perturbations of the prior distribution. The fact that an improper prior distribution produces an infinite Bayes factor is not in itself the heart of the problem, since we do not believe that genuine beliefs could ever be represented by an improper distribution. Any proper prior distribution will yield a finite Bayes factor (assuming that the marginal likelihood $f(x \mid \phi)$ is positive and bounded). The message of **7.43** appears to be simply that we cannot use an improper prior distribution, as a convenient approximation to weak prior information, for a problem of comparing models. There is, however, a deeper problem than this.

When prior information is weak it is not easy to specify a prior distribution with any accuracy. The Bayes factor can then be very sensitive to the choice of prior distribution. To take a very simplistic example, we could give ϕ a proper uniform distribution over the finite range $[-c, c]$. Then as shown in **7.43** if c is large, representing weak information, the Bayes factor will be of order c. Changing from a uniform distribution over $[-10, 10]$ to a uniform distribution over $[-100, 100]$ multiplies the Bayes factor by 10, and yet when these distributions are being proposed to represent weak prior information You may not be willing to say which more accurately represents Your prior beliefs. In general, when Bayesian inference is not formally possible using an improper prior distribution the problem is not avoided simply by assuming a uniform prior distribution over some large but finite range. The prior distribution becomes proper, but inference can depend very strongly on how wide a range is used.

Example 7.17

A more realistic example of sensitivity can be obtained from Example 7.15. The prior mean of ϕ in (7.52) is $b/(t-2)$. For any given b we can choose t to obtain any required prior mean. If b is small, then the prior variance will be large, representing weak prior information, but in fact the variance is infinite for $t < 3$. So all sufficiently small values of b (and t correspondingly between 2 and 3) could represent weak prior information in this sense. Yet the term b^{-t} in (7.54) shows that the Bayes factor will then be highly sensitive to the particular choice of b.

7.46 Sensitivity will also be a problem even when prior information about ϕ is not particularly weak but the data are strong. Strong data will typically mean that the likelihood $f(x \mid \phi)$ is negligible outside a relatively small neighbourhood of its maximum $f(x \mid \hat{\phi})$ at $\phi = \hat{\phi}$, say. In particular, if the data are much stronger than the prior information then $f_1(\phi)$ will vary little over the same region. So the integral (7.50) will be approximated by $f_1(\hat{\phi}) \int_{-\infty}^{\infty} f(x \mid \phi) \mathrm{d}\phi$. Therefore any perturbation of the prior distribution $f_1(\phi)$ that alters its value $f_1(\hat{\phi})$ at $\phi = \hat{\phi}$ will result in a proportionately identical change in the Bayes factor.

This is in direct contrast to the argument in **3.26** which says that as the amount of data increases the prior information is overwhelmed by the data and becomes irrelevant. That argument was concerned with inference about θ. The posterior distribution of θ does not depend on the absolute values of the prior density, since proportionality constants cancel out in Bayes' theorem, but only on relative values. When the data are strong the fact that the prior density varies little over the range of θ values for which the likelihood is non-negligible results in the prior information becoming irrelevant. So for the basic problem, sensitivity to the prior distribution *reduces* in general as we collect more data.

For the problem of comparing models, however, no matter how strong the data are, the prior distribution remains important. We have seen, in fact, that ultimately the value $f_1(\hat{\phi})$ of the prior density at the maximum likelihood value $\hat{\phi}$ determines the Bayes factor.

Example 7.18

Let Your true prior distribution $f_1(\phi)$ be allowed to vary over a density bounded class Γ (see Example 7.4) containing all distributions satisfying

$$k^{-1}f_0(\phi) \leqslant f_1(\phi) \leqslant kf_0(\phi),$$

where $f_0(\phi)$ might be Your elicited prior distribution and $k > 1$ determines the size of the neighbourhood Γ. Then for suitably strong data, the Bayes factor (for comparing the general model with the initially assumed model having $\phi = \phi_0$) will vary in the same way, the ratio of the maximum and minimum Bayes factors being also k^2.

7.47 A different aspect of this problem arises if the ideas of the earlier part of this chapter are applied with improper priors. That is, any class Γ of priors which includes improper prior distributions will typically admit arbitrarily great variation of posterior inferences, so that robustness cannot be achieved with such a class. Wasserman (1992a) gives a thorough analysis of such behaviour.

Asymptotic behaviour

7.48 We now develop the argument of **7.46** rather more generally. Consider comparing two arbitrary models as in **7.36**, where the Bayes factor in favour of model 1 against model 2 is $m_1(x)/m_2(x)$, and where $m_i(x)$ is the marginal distribution of the data x under model i,

$$m_i(x) = \int f_i(x \mid \boldsymbol{\theta}_i) f_i(\boldsymbol{\theta}_i)\, \mathrm{d}\boldsymbol{\theta}_i. \tag{7.56}$$

Let $\boldsymbol{\theta}_i$ be a vector parameter of p_i elements. Now suppose that $x = (x_1, x_2, \ldots, x_n)$ and $x_1, x_2, \ldots, x_n$ are identically and independently distributed under model i with common density g_i. That is

$$f_i(x \mid \boldsymbol{\theta}_i) = \prod_{j=1}^{n} g_i(x_j \mid \boldsymbol{\theta}_i). \tag{7.57}$$

We now expand the likelihood (7.57) as in **3.24**. Let $\hat{\boldsymbol{\theta}}_i$ be the maximum likelihood estimate of $\boldsymbol{\theta}_i$ in model i, so that $\hat{\boldsymbol{\theta}}_i$ maximizes (7.57) for given x. Expanding the logarithm in a Taylor series around $\boldsymbol{\theta}_i = \hat{\boldsymbol{\theta}}_i$,

$$\log f_i(x \mid \boldsymbol{\theta}_i) = \log L_i - (n/2)(\boldsymbol{\theta}_i - \hat{\boldsymbol{\theta}}_i)' \mathbf{V}_i^{-1} (\boldsymbol{\theta}_i - \hat{\boldsymbol{\theta}}_i) + R,$$

where $L_i = \log f_i(x \mid \hat{\boldsymbol{\theta}}_i)$ is the logarithm of the maximized likelihood, $n\mathbf{V}_i$ is the modal dispersion matrix (defined in **2.15** as minus the inverse of the matrix of second order partial derivatives of $\log f_i(x \mid \boldsymbol{\theta}_i)$ evaluated at $\boldsymbol{\theta}_i = \hat{\boldsymbol{\theta}}_i$) and R is the remainder term involving third order derivatives. For $|\boldsymbol{\theta}_i - \hat{\boldsymbol{\theta}}_i|$ of order $n^{-1/2}$, the remainder term is of order $n^{-1/2}$ and can be ignored for large n (and for larger $|\boldsymbol{\theta}_i - \hat{\boldsymbol{\theta}}_i|$ the likelihood $f_i(x \mid \boldsymbol{\theta}_i)$ is negligible). But for $|\boldsymbol{\theta}_i - \hat{\boldsymbol{\theta}}_i|$ of order $n^{-1/2}$, $f_i(\boldsymbol{\theta}_i)$ varies slowly and can be approximated by the constant $f_i(\hat{\boldsymbol{\theta}}_i)$.

7.49 In **3.23** and **3.24** these results were used to show that the posterior distribution is asymptotically normal. Also, since the constant $f_i(\hat{\boldsymbol{\theta}}_i)$ cancels out in Bayes' theorem, the asymptotic posterior distribution is independent of the prior distribution. This implies that the more data we have, the more posterior inference can be expected to be robust to perturbation of the prior distribution. The same is *not* true of model comparison, since $f_i(\hat{\boldsymbol{\theta}}_i)$ remains as a constant in (7.56). In fact we have the approximation

$$\begin{aligned} m_i(x) &\approx f_i(\hat{\boldsymbol{\theta}}_i)L_i \int \exp\{-(n/2)(\boldsymbol{\theta}_i - \hat{\boldsymbol{\theta}}_i)'\mathbf{V}_i^{-1}(\boldsymbol{\theta}_i - \hat{\boldsymbol{\theta}}_i)\}\, d\boldsymbol{\theta}_i \\ &= f_i(\hat{\boldsymbol{\theta}}_i)L_i(2\pi)^{p_i/2}|n^{-1}\mathbf{V}_i|^{1/2} \\ &= f_i(\hat{\boldsymbol{\theta}}_i)L_i(2\pi)^{p_i/2}n^{-p_i/2}|V_i|^{1/2}. \end{aligned} \tag{7.58}$$

In particular, as noted already, the Bayes factor $m_1(x)/m_2(x)$ will depend asymptotically on the ratio $f_1(\hat{\boldsymbol{\theta}}_1)/f_2(\hat{\boldsymbol{\theta}}_2)$ of prior densities under the two models, and will be sensitive to variation in those prior densities no matter how much data we have.

7.50 It follows that

$$-2\log(m_1(x)/m_2(x)) = -2\log l + (p_1 - p_2)\log n + a, \tag{7.59}$$

where $l = L_1/L_2$ so that $-2\log l$ is the classical likelihood ratio test statistic (see A**23.7**) and a is $O(1)$. This result was obtained for quite general models by Poskitt (1987), who also applied a decision-theoretic approach to model comparison. Ignoring a (or setting $a = 0$) produces the Schwarz (1978) test criterion $-2\log l + (p_1 - p_2)\log n$. The Schwarz criterion, like Akaike's Information Criterion (Akaike, 1973) $-2\log l + 2(p_1 - p_2)$, adjusts the classical likelihood ratio criterion to favour more strongly the model with fewer parameters. Akaike's Information Criterion was proposed to counteract the tendency of classical tests always to favour the more complex model if the amount of data is large. The adjustment of $2(p_1 - p_2)$, it was argued, would favour the simpler model provided it fitted the data almost as well as the more complex model. The approach accords with the principle of Occam's Razor, that models should not be unnecessarily complex.

Schwarz (1978) shows that a Bayesian argument supports an even more radical adjustment $(p_1 - p_2)\log n$, which increases with n, and this is the result (7.59). A key implication is that classical comparison of models using the likelihood ratio criterion $-2\log l$ can produce very different conclusions from the Bayesian approach.

7.51 To explore this fact, suppose again that we have nested models with $\boldsymbol{\theta}_1 = \boldsymbol{\theta}$ and $\boldsymbol{\theta}_2 = (\boldsymbol{\theta}, \boldsymbol{\phi})$. Then p_1 is the number of elements of $\boldsymbol{\theta}$ and $p_2 - p_1 = q$ is the number of elements of $\boldsymbol{\phi}$. Then the classical theory, A**23.7**, asserts that the asymptotic distribution of

$-2\log l$ given that model 1 is true, $\boldsymbol{\phi} = \boldsymbol{\phi}_0$, is χ_q^2. Now it is clear from (7.59) that for large n the data can be such that $-2\log l$ is very large, and so the classical test will strongly favour model 2, and yet the Bayes factor $\exp\{(2\log l + q\log n - a)/2\}$ will be very large, strongly favouring model 1. For *any* prior distribution (which governs the constant a and the prior odds ratio), data giving a fixed classical significance level (i.e. fixed l) yield an increasing posterior probability for model 1 as n increases. Therefore, from the Bayesian point of view, the classical significance level has no absolute meaning, but is dependent on the sample size. For further discussion of this point see Berger and Delampady (1987), Berger and Sellke (1987), Delampady (1989) and Berger and Mortera (1991).

The fact that classical testing tends to favour the more complex model is clear from this analysis, as is the fact that Akaike's Information Criterion does not fully correct that tendency.

7.52 We can examine the different approaches further by using the asymptotic distribution of $-2\log l$ derived in A**23.7**. It is shown there that $-2\log l$ has an approximate non-central χ^2 distribution with q degrees of freedom and non-centrality parameter

$$\lambda = n(\boldsymbol{\phi} - \boldsymbol{\phi}_0)'\mathbf{V}_{\phi}^{-1}(\boldsymbol{\phi} - \boldsymbol{\phi}_0),$$

where $\mathbf{V}_{\phi}$ derives from the information matrix of a single observation. The expectation of $-2\log l$ is therefore asymptotically $q + \lambda$, so the expectation of (7.59) is

$$E\{-2\log(m_1(x)/m_2(x))\} = n(\boldsymbol{\phi} - \boldsymbol{\phi}_0)'\mathbf{V}_{\phi}^{-1}(\boldsymbol{\phi} - \boldsymbol{\phi}_0) - q\log n + O(1). \qquad (7.60)$$

Now if $\boldsymbol{\phi} \neq \boldsymbol{\phi}_0$, so that model 2 is true, (7.60) tends to infinity as $n \to \infty$, the Bayes factor in favour of model 1 tends to 0, and so the posterior probability of model 2 tends almost surely to one regardless of the prior odds. Conversely, if $\boldsymbol{\phi} = \boldsymbol{\phi}_0$ so that model 1 is true, (7.60) tends to minus infinity and the posterior probability of model 1 tends to one regardless of the prior odds.

Therefore the Bayesian approach is consistent: as the number of observations tends to infinity, the probability of selecting the correct model tends to one. It is easy to see that this is not true of the classical analysis, with or without the Akaike Information Criterion. The classical likelihood ratio criterion $-2\log l$ corresponds to ignoring the $q\log n$ term in (7.60). If this were taken as the Bayes factor, the posterior probability that $\phi = \phi_0$ does not tend to one, even when this is true. Since Akaike's Information Criterion is independent of n it does not change this behaviour. The classical approach does not, of course, employ the Bayes factor, but an analogous result holds. If $\phi = \phi_0$ then, no matter how large n is, a hypothesis test with size α wrongly rejects the null hypothesis that $\phi = \phi_0$ with probability α. The classical method as traditionally practised cannot consistently *accept* the null hypothesis.

From a strictly non-Bayesian viewpoint one can in fact argue that α should decrease as n increases. A very large amount of data can produce a very powerful test, so justifying a smaller α. This is now seen to accord with the Bayesian approach, and if α tends to zero can achieve consistent choice of models. However, it is by no means clear from a classical viewpoint *how* α should decrease with sample size, whereas the Bayesian approach deals with the problem automatically and more naturally.

7.53 Ultimately the Bayes factor selects the correct model, but if $|\phi - \phi_0|$ is small then (7.60) can be negative for relatively large n even though it tends to infinity as $n \to \infty$, so that the Bayes factor then favours the wrong model. Smith and Spiegelhalter (1980) argue that this behaviour is consistent with Occam's Razor, since the Bayes factor favours the simpler model if the discrepancy between the true value of ϕ and the assumption that $\phi = \phi_0$ is small enough to be unimportant.

Further classical discussion of these matters is given in A**28.27**.

Approaches to model comparison with improper priors

7.54 The difficulty of defining a Bayes factor when we try to represent weak prior information by an improper prior distribution was discussed in **7.43**. We now develop some approaches to solving this problem. Consider comparing two arbitrary models by the Bayes factor $B = m_1(x)/m_2(x)$ where $m_i(x)$ is given by (7.56). Suppose we wish the prior distribution under model 2 to be expressed in the improper form

$$f_2(\boldsymbol{\theta}_2) \propto g_2(\boldsymbol{\theta}_2),$$

where $\int g_2(\boldsymbol{\theta}_2)\mathrm{d}\boldsymbol{\theta}_2$ does not exist. We could write

$$f_2(\boldsymbol{\theta}_2) = c_2 g_2(\boldsymbol{\theta}_2), \tag{7.61}$$

where c_2 is an undefined (and strictly non-existent) normalizing constant. Then the Bayes factor becomes

$$B = \frac{1}{c_2}\,\frac{\int f_1(x \mid \boldsymbol{\theta}_1) f_1(\boldsymbol{\theta}_1)\,\mathrm{d}\boldsymbol{\theta}_1}{\int f_2(x \mid \boldsymbol{\theta}_2) g_2(\boldsymbol{\theta}_2)\,\mathrm{d}\boldsymbol{\theta}_2} \tag{7.62}$$

and depends on the unspecified c_2.

Similarly, we could have an improper prior distribution under model 1 expressed as $f_1(\boldsymbol{\theta}_1) = c_1 g_1(\boldsymbol{\theta}_1)$, so that

$$B = \frac{c_1}{c_2}\cdot\frac{\int f_1(x \mid \boldsymbol{\theta}_1) g_1(\boldsymbol{\theta}_1)\,\mathrm{d}\boldsymbol{\theta}_1}{\int f_2(x \mid \boldsymbol{\theta}_2) g_2(\boldsymbol{\theta}_2)\,\mathrm{d}\boldsymbol{\theta}_2} \tag{7.63}$$

depending on the unspecified ratio c_1/c_2.

7.55 One way to specify c_2 or c_1/c_2 is to use the device of imaginary observations, **6.30**. If You can imagine some data y such that You feel able to specify the value that the Bayes factor would have if You were to observe y, then this given value of B will determine c_2 or c_1/c_2. An obvious choice would be to imagine data y which would not in Your opinion favour either model 1 or model 2. Then the Bayes factor from data y will be fixed at one, so that (7.62) yields

$$c_2 = \frac{\int f_1(y \mid \boldsymbol{\theta}_1) f_1(\boldsymbol{\theta}_1)\,\mathrm{d}\boldsymbol{\theta}_1}{\int f_2(y \mid \boldsymbol{\theta}_2) g_2(\boldsymbol{\theta}_2)\,\mathrm{d}\boldsymbol{\theta}_2}, \tag{7.64}$$

or from (7.63)

$$\frac{c_1}{c_2} = \frac{\int f_2(y \mid \boldsymbol{\theta}_2) g_2(\boldsymbol{\theta}_2)\,\mathrm{d}\boldsymbol{\theta}_2}{\int f_1(y \mid \boldsymbol{\theta}_1) g_1(\boldsymbol{\theta}_1)\,\mathrm{d}\boldsymbol{\theta}_1}. \tag{7.65}$$

7.56 Spiegelhalter and Smith (1982) present the following general argument for identifying a suitable y. First, let y arise from the smallest possible experiment that will provide proper posterior distributions under both models. This is therefore the smallest amount of data for which both numerator and denominator of (7.64) or (7.65) are actually defined.

The next step is to set the actual value of y so as to maximize the Bayes factor (assuming that model 1 is the simpler model), and set this maximum value to one. That is, we apply (7.64) with y chosen to maximize c_2, or (7.65) with y chosen to minimize c_1/c_2. Spiegelhalter and Smith argue that with y chosen to give maximum support to model 1 the Bayes factor should actually be greater than one, but since y arises from a minimal experiment it should only be possible to give minimal support to the simpler model. Therefore it is reasonable to approximate the maximum value of B by one.

Example 7.19

Model 1 states that $x_1, x_2, x_3, \ldots,$ are iid Poisson random variables with mean 1, whereas model 2 also states that they are iid Poisson random variables but with unknown mean $\theta > 0$. Imagine a single observation y. Model 1 has no parameters and we simply have

$$m_1(y) = \mathrm{e}^{-1}/y!,$$

the Poisson distribution with mean 1. Under model 2 we wish to have the improper prior distribution $f_2(\theta) \propto \theta^{-t}$. (This is improper for any t). Then

$$m_2(y) = \int \theta^y \mathrm{e}^{-\theta}(y!)^{-1} c_2 \theta^{-t}\, \mathrm{d}\theta$$

Provided $t < 1$, this is proper for all $y = 0, 1, 2, \ldots$. The Bayes factor for model 1 against model 2 is therefore

$$B = \mathrm{e}^{-1}/\{c_2 \Gamma(y + 1 - t)\}. \tag{7.66}$$

We now choose y to maximize B, i.e. to minimize $\Gamma(y+1-t)$. Since $\Gamma(y+2-t)/\Gamma(y+1-t) = y + 1 - t$, the minimum occurs when $y = 0$ if $t \leqslant 0$, or when $y = 1$ if $0 \leqslant t < 1$. Setting the minimized value of B to one gives, for $0 \leqslant t < 1$,

$$c_2 = \{\mathrm{e}\Gamma(2 - t)\}^{-1}.$$

Thus, if the improper uniform prior $f_2(\theta) \propto 1$ is used, we should let $f_2(\theta) = c_2$ where $c_2 = \mathrm{e}^{-1}$. If we use instead $f_2(\theta) = c_2\theta^{-1/2}$ then $c_2 = \{\mathrm{e}\Gamma(3/2)\}^{-1} = 2\mathrm{e}^{-1}/\sqrt{\pi}$.

If we wish to use $f_2(\theta) = c_2\theta^{-1}$ then a single observation y provides a proper Bayes factor provided $y > 0$. Then (7.66) is minimized by $y = 1$ or 2 and we again have $c_2 = \mathrm{e}^{-1}$.

Having determined c_2, we can proceed to compute the Bayes factor (7.62) arising from the actual data x (as opposed to the imaginary observation y).

Example 7.20

Model 1 states that $x_1, x_2, x_3, \ldots,$ are independent $N(0, 1)$ random variables, while model 2 states that they are independent $N(\theta, 1)$ with unknown mean θ. A uniform prior distribution $f_2(\theta) = c_2$ is proposed under model 2, and again the minimal experiment is a single observation y.

Under model 1 we simply have

$$m_1(y) = (2\pi)^{-1/2} \exp(-y^2/2),$$

and under model 2,

$$m_2(y) = \int (2\pi)^{-1/2} \exp\{-(y-\theta)^2/2\} c_2 \, d\theta = c_2.$$

So the Bayes factor is $c_2^{-1}(2\pi)^{-1/2}\exp(-y^2/2)$ and is maximized at $y = 0$ (which obviously does give maximal support to the simpler model). Equating this maximum to one gives $c_2 = (2\pi)^{1/2}$.

7.57 Spiegelhalter and Smith (1982) proposed this method in the context of nested linear models, and we will consider that case in **9.35**.

Partial Bayes factors

7.58 A different solution to the problem of weak prior information is to reserve some of the data as a 'training sample' to provide improved 'prior' information, and use only the remainder of the data for model comparison. This suggestion was made by Lempers (1971, section 5.3). We therefore divide the data x into $x = (y, z)$, where y is the training sample and z is the comparison sample. First use y to produce posterior distributions

$$f_i(\boldsymbol{\theta}_i \mid y) = f_i(\boldsymbol{\theta}_i) f_i(y \mid \boldsymbol{\theta}_i) / f_i(y), \tag{7.67}$$

under the various models, where the normalizing constant $f_i(y)$ is the marginal density

$$f_i(y) = \int f_i(\boldsymbol{\theta}_i) f_i(y \mid \boldsymbol{\theta}_i) \, d\boldsymbol{\theta}_i.$$

A Bayes factor $B_y = f_1(y)/f_2(y)$, for comparing models using data y, would suffer from the related sensitivity problems of **7.43** to **7.46** if either or both of the prior distributions $f_i(\boldsymbol{\theta}_i)$ were improper. However, we do not use y for model comparison. Its role is to produce the posterior (7.67), which is then used as a prior distribution (in the usual way, see **3.5**) when receiving the remainder of the data, z. We use z for model comparison, so that the Bayes factor is now

$$B_{z \mid y} = m_1(z \mid y)/m_2(z \mid y), \tag{7.68}$$

where

$$m_i(z \mid y) = \int f_i(z \mid y, \boldsymbol{\theta}_i) f_i(\boldsymbol{\theta}_i \mid y) \, d\boldsymbol{\theta}_i. \tag{7.69}$$

Combining (7.67) and (7.69), we see that

$$B_{z \mid y} = B_x / B_y, \tag{7.70}$$

where B_x is the full Bayes factor $B_x = m_1(x)/m_2(x)$ based on the full data x, while B_y we have defined as $f_1(y)/f_2(y)$, which is the Bayes factor based only on data y. (7.70) is in fact just another expression of the result $B_x = B_y B_{z \mid y}$, which shows how Bayes factors may be computed sequentially. We will call $B_{z \mid y}$ a partial Bayes factor because it can be viewed as only one component of the full Bayes factor B_x.

7.59 If prior information in either or both of the models is weak and is expressed as an improper prior distribution, then the term c_2^{-1} or c_1/c_2 occurring in the expressions (7.62) and (7.63) for the full Bayes factor B_x will also occur in B_y. They will therefore cancel in (7.70) leaving the partial Bayes factor unaffected by these arbitrary quantities. The partial Bayes factor therefore avoids the improper prior problems of **7.43** and **7.54**.

7.60 Although partial Bayes factors make it possible to compare models based on improper prior distributions, there is the problem of how to divide the full data x into the two parts y and z. Berger and Pericchi (1993) propose that y should be a minimal experiment as in **7.50**, i.e. a part of the data containing the smallest number of observations such that posterior distributions $f_i(\boldsymbol{\theta}_i \mid y)$ are properly defined (and therefore B_y exists). However, if n_1 observations suffice to produce proper posterior distributions then every one of the $\binom{n}{n_1}$ possible selections of n_1 from the n observations in the full data x could serve as y. Berger and Pericchi call the average of all the resulting partial Bayes factors the *intrinsic Bayes factor*. If the partial Bayes factor $B_{z\mid y}$ is robust, in the sense of not varying so much with varying choices of y as to produce large changes in the posterior probabilities of the two models, then the choice of y does not matter and the average value is a sensible choice. If, however, $B_{z\mid y}$ is sensitive to y then the partial Bayes factor approach has not really solved the problem of model comparison with improper prior distributions.

7.61 We now apply the asymptotic argument of **7.48** to **7.50** to partial Bayes factors. If, as the sample size n becomes large, the number of observations in the training sample y is fixed, then the same asymptotics apply. We can change the prior $f_i(\boldsymbol{\theta}_i)$ to $f_i(\boldsymbol{\theta}_i \mid y)$ and the data to z. Then as the number of observations in z becomes large we have a result like (7.59); partial Bayes factors with a fixed size training sample will consistently choose the correct model. However, O'Hagan (1993) argues that they also retain the sensitivity problems of **7.46**.

O'Hagan (1991a) considers instead the case where the numbers of observations in both y and z increase. If y contains bn observations and z contains $(1-b)n$ observations, then (7.58) applies for the full data x, while for the training sample y we have analogously

$$f_i(y) \approx f_i(\boldsymbol{\theta}_i^\star) L_i^\star (2\pi)^{p_i/2} b^{-p_i/2} n^{-p_i/2} |\mathbf{V}_i^\star|^{1/2},$$

where $\boldsymbol{\theta}_i^\star$ is the maximum likelihood estimate of $\boldsymbol{\theta}_i$ under model i from data y, $L_i^\star = f_i(y \mid \boldsymbol{\theta}_i^\star)$ is the corresponding maximized likelihood, and $bn\mathbf{V}_i^\star$ is the corresponding modal dispersion matrix. Now as n tends to infinity $\boldsymbol{\theta}_i^\star$ and $\hat{\boldsymbol{\theta}}_i$ will both tend to the same limit (which is the true value of $\boldsymbol{\theta}_i$ if the data really follow model i). $\mathbf{V}_i^\star$ and $\mathbf{V}_i$ will also tend to the same limit, which is the Fisher information matrix for a single observation. Asymptotically we also have $L_i^\star \approx L_i^b$. Therefore (7.59) is replaced by

$$\begin{aligned} -2\log B_{z\mid y} &= -2\{\log m_1(x) + \log f_2(y) - \log m_2(x) - \log f_1(y)\} \\ &\approx (1-b)\{-2\log l + (p_1 - p_2)c(b)\} \end{aligned} \quad (7.71)$$

where

$$c(b) = (-\log b)/(1 - b). \tag{7.72}$$

The factor $(1 - b)$ in (7.71) results from the fact that y, representing a proportion b of the data, has not been used for model comparison. The multiplier $c(b)$ replaces the term $\log n$ in (7.59) as the correction to favour the simpler model, and no longer depends on n. Whereas $\log n$ corresponds to the Schwarz criterion, $c(b)$ is more like the fixed multiplier 2 used in Akaike's Information Criterion. In fact (7.72) produces $c(b) = 2$ if $b = 0.203$, so the AIC corresponds to a partial Bayes factor in which one fifth of the data are applied as a training sample and four fifths are used for model comparison. In his numerical example, Lempers (1971, chapter 6) uses $b = 0.5$, corresponding to $c(b) = 1.386$, but such a large b entails a considerable loss of information through the factor $(1 - b)$ in (7.71).

Now from (7.71), if a fixed proportion of the data are used for the training sample y, rather than a fixed number of observations, then the partial Bayes factor has different asymptotic behaviour from that of the full Bayes factor. In particular, it will not be consistent in the way described in **7.52**. Further discussion of the effect of different correction terms $c(b)$ is given in Smith and Spiegelhalter (1980) and Aitkin (1991).

7.62 According to the above asymptotic development, the likelihood for y behaves like the full likelihood raised to the power b. We can incorporate this idea as an alternative formulation of the Bayes factor

$$B_b = m_1^\star(x)/m_2^\star(x), \tag{7.73}$$

where

$$m_i^\star(x) = \frac{\int f_i(x \mid \boldsymbol{\theta}_i) f_i(\boldsymbol{\theta}_i)\,\mathrm{d}\boldsymbol{\theta}_i}{\int \{f_i(x \mid \boldsymbol{\theta}_i)\}^b f_i(\boldsymbol{\theta}_i)\,\mathrm{d}\boldsymbol{\theta}_i}. \tag{7.74}$$

In (7.74) we have formally replaced $f_i(y \mid \boldsymbol{\theta}_i)$ by $f_i(x \mid \boldsymbol{\theta}_i)$ raised to the power b. We will call B_b a *fractional* Bayes factor, and note that it does not depend on a specific division of x into a training sample y and a comparison sample z. O'Hagan (1993) advocates the use of fractional Bayes factors to improve robustness of model comparisons.

7.63 If we replace (7.74) by a more general class of integral ratios I_a/I_b, where

$$I_a = \int \{f_i(x \mid \boldsymbol{\theta}_i)\}^a f_i(\boldsymbol{\theta}_i)\,\mathrm{d}\boldsymbol{\theta}_i,$$

then the fractional Bayes factor with fraction b corresponds to $a = 1$ and $0 < b < 1$, while the full Bayes factor is the case $a = 1, b = 0$. Aitkin (1991) suggests $a = 2, b = 1$, which he calls the posterior Bayes factor. He reasons that this simply corresponds to using all of x for model comparison but replacing the prior distributions $f_i(\boldsymbol{\theta}_i)$ by the posterior distributions $f_i(\boldsymbol{\theta} \mid x)$. In terms of our discussion it seems that x is being used twice, as a training sample and as a comparison sample, and the procedure is not apparently justifiable within the Bayesian framework. See the discussion following Aitkin's paper and the author's reply. Alternative Bayesian approaches to model comparison are given by Geisser and Eddy (1979), Pericchi (1984) and San Martini and Spezzaferri (1984), based on ideas of predicative distributions of entropy.

Relationship with classical likelihood ratios

7.64 Most standard classical hypothesis tests are likelihood ratio tests. In the case of two simple hypotheses, the Neyman–Pearson lemma (see A**21.10**) shows that the most powerful classical test rejects the null hypothesis that $\phi = \phi_0$ if the likelihood ratio $f(x \mid \phi_1)/f(x \mid \phi_0)$ is sufficiently high. In the notation of this chapter we think of $\phi = \phi_0$ and $\phi = \phi_1$ as indexing two different models. The fact that they are simple hypotheses corresponds to there being no further unknown parameters θ_0 and θ_1 within the two models. Then the marginal density $m_i(x)$ for x under model i reduces simply to $f(x \mid \phi_i)$, and the likelihood ratio is seen as a Bayes factor. Therefore for comparing simple hypotheses there is some agreement between classical and Bayesian approaches in the sense that both use the Bayes factor as a measure of the evidence in the data for or against a model.

More generally, however, the existence of further unknown parameters θ_i under model i corresponds to comparing compound hypotheses. Then the classical generalized likelihood ratio criterion is $f_1(x \mid \hat{\theta}_1)/f_0(x/\hat{\theta}_0)$, where $\hat{\theta}_i$ is the maximum likelihood estimate under model i. Instead of maximizing $f_i(x \mid \theta_i)$, the Bayes factor $m_1(x)/m_0(x)$ is defined through (7.44) in terms of an averaging of $f_i(x \mid \theta_i)$ with respect to the prior distribution $f_i(\theta_i)$. Classical tests based on generalized likelihood ratios may therefore be very different from Bayesian procedures, which are based on Bayes factors.

EXERCISES

7.1 Suppose that You have specified a prior density $f_0(\theta)$ which is the $N(0,1)$ density. You wish to consider robustness to this specification by allowing $f(\theta)$ to range over a set Γ containing $f_0(\theta)$. Can Γ contain the Cauchy density $f(\theta)=\pi^{-1}(1+\theta^2)^{-1}$ if it is defined (a) as a total variation class (7.2) for some ϵ, (b) as a density ratio class (7.5) for some k, (c) as an epsilon contamination class (7.7) for some ϵ and some set Q of unimodal density functions?

7.2 Let Γ be the set of all prior distributions with given mean $E(\theta)=m$. Suppose that $f(x\,|\,\theta)\to 0$ as $\theta\to\pm\infty$. Show that a prior distribution can be found in Γ such that $E(\theta\,|\,x)$ is arbitrarily close to any given θ_1 for which $f(x\,|\,\theta_1)>0$, regardless of m. (Note that Γ includes discrete distributions.) (Goutis, 1993)

7.3 Let x be the number of successes in n independent trials with probability θ of success in each trial. The prior distribution is assumed to lie in the ϵ-contamination class (7.7) where Q is the class of all prior distributions for $0\leqslant\theta\leqslant 1$ having the same mean and variance as $f_0(\theta)$. Prove that if $n=1$ or $n=2$, then for every possible x and for every $q(\theta)\in Q$, the weight $\epsilon^\star$ in (7.10) equals ϵ (and therefore $\Gamma^\star$ is an ϵ-contamination class).

7.4 Let the prior density $f(\theta)$ be in a set Γ of possible prior distributions. If v is the maximum posterior variance as $f(\theta)$ ranges over Γ, prove that v is a solution of the equation

$$0=\sup_{\Gamma}\Big[\left\{\textstyle\int\theta^2 f(\theta)f(x\,|\,\theta)\,\mathrm{d}\theta\right\}$$
$$\left\{\textstyle\int f(\theta)f(x\,|\,\theta)\,\mathrm{d}\theta\right\}-\left\{\textstyle\int\theta f(\theta)f(x\,|\,\theta)\,\mathrm{d}\theta\right\}^2-v\left\{\textstyle\int f(\theta)f(x\,|\,\theta)\,\mathrm{d}\theta\right\}^2\Big].$$

7.5 Let $x_1,x_2,\ldots,x_n$ be independently and identically distributed as $N(\theta,1)$ given θ, and let the stated prior $f_0(\theta)$ be the $N(0,1)$ density. Prove that the surprise measure $\{m_0(x)\}^{-1}\sup_\theta f(x\,|\,\theta)$ of **7.21** is of order $\sqrt{n}$ (with probability one) as $n\to\infty$. (Gustafson and Wasserman, 1993).

7.6 Data $x_1,x_2,\ldots,x_n$ are independently and identically distributed as $N(\theta,v)$ given θ. Under model $1, v=1$ and under model 2, $v=2$. In either model the prior distribution for θ is $N(m,w)$. Show that as $w\to\infty$ the Bayes factor for model 1 against model 2 tends to

$$2^{n/2}\exp\left\{-\textstyle\sum_{i=1}^n(x_i-\bar{x})^2/4\right\}.$$

Confirm that this tends with probability one to infinity if $v=1$ and to zero if $v=2$.

7.7 A single observation x has the Poisson f.f. with mean θ, i.e. $f(x\,|\,\theta)=\theta^x\mathrm{e}^{-\theta}/x!$. Under model 1, $\theta=1$. Under model 2, θ has the gamma density $f_2(\theta)=a^b\theta^{b-1}\mathrm{e}^{-a\theta}/\Gamma(b)$. Prove that the Bayes factor for model 1 against model 2 is maximized when x is the integer part of $a-b+2$. In the case when the prior mode equals one (which is reasonable since model 1 asserts that $\theta=1$), show that the Bayes factor becomes

$$\mathrm{e}^{-1}(a+1)^{a+x+1}a^{-(a+1)}\Gamma(a+1)/\Gamma(a+x+1)$$

and tends to infinity as $a\to 0$. What is the limiting form of $f_2(\theta)$ in this case?

7.8 Observations $x_1,x_2,\ldots,x_n$ are independently and identically distributed as $N(\theta,v)$ given θ. Under model 1, $\theta=0$. Under model 2, θ has the $N(0,w)$ prior distribution. Let $x=100\sqrt{v/n}$, so that a classical hypothesis test would reject model 1 at an extremely high level of significance. Show that the Bayes factor for model 1 against model 2 tends to infinity as $n\to\infty$, and hence that for sufficiently large sample size the Bayesian analysis gives a very high posterior probability to model 1.

7.9 Let x be the number of successes in n independent trials with probability θ of success in each trial, and suppose that $0<x<n$. Under model 1, $\theta=1/2$. Under model 2, θ has the improper prior density $f_2(\theta)\propto\theta^{-1}(1-\theta)^{-1}$. Show that the method of **7.56** leads to a

Bayes factor, in favour of model 1 against model 2, of $\{2^{n-2}B(x, n-x)\}^{-1}$. Prove also that the fractional Bayes factor (7.73) is $B_b = 2^{-n(1-b)}B(bx, b(n-x))/B(x, n-x)$. What difficulty would be encountered in trying to derive the intrinsic Bayes factor of **7.60** by averaging all the partial Bayes factors from minimal training samples?

CHAPTER 8

COMPUTATION

The importance of computational techniques

8.1 Practical implementation of Bayesian methods usually requires substantial computation. This computational requirement is essentially to calculate summaries of the posterior distribution. The Bayesian method, as outlined in Chapter 1, consists of combining the prior distribution and likelihood to derive the posterior distribution by Bayes' theorem. So once the model has been determined, which determines the likelihood and the nature of the unknown parameters, and once a prior distribution for those parameters has been specified, then we can immediately write the posterior distribution down as proportional to the product of prior and likelihood. Now, as stressed in Chapter 2, the crucial task is to calculate relevant summaries of the posterior distribution, to express the posterior information in a usable form, and to serve as formal inferences if appropriate. It is in the task of summarizing that computation is typically needed. In many examples in other chapters we have worked with prior distributions and likelihoods of sufficiently convenient forms to enable the necessary results to be obtained by straightforward mathematics. In practice, however, a statistician needs often to work with much more complex models, and a conscientious Bayesian must specify prior distributions to reflect accurately the available prior information, which can also lead to complex modelling. The combination of likelihood and prior, furthermore, will generally produce a posterior distribution too complex for mathematical summarization, even if the two constituents separately are sufficiently simple. The Bayesian statistician therefore needs, on a routine basis, general computational tools to calculate a variety of summaries from posterior distributions that are mathematically complex, and also often high-dimensional. Some of the most important computational methods are dealt with in this chapter.

8.2 We are concerned with summarizing

$$g(\theta) = f(\theta)f(x \mid \theta), \tag{8.1}$$

the product of prior distribution and likelihood. $g(\theta)$ is not the posterior distribution but proportional to it: $g(\theta) = kf(\theta \mid x)$, where the normalizing constant $k = \{\int f(\theta)f(x \mid \theta)\mathrm{d}\theta\}^{-1}$ is typically unknown because the integral cannot be carried out analytically. Problems for which computational methods are needed to calculate summaries are almost invariably ones for which this integral cannot be done. In a sense, then, the first computational requirement is to do this integral numerically, and hence calculate the normalizing constant k. In practice, as we shall see, many of the computational techniques do not explicitly require k, but compute summaries of $f(\theta \mid x)$ directly from $g(\theta)$.

As in Chapter 2, $g(\theta)$ could generally be some other distribution than proportional to the posterior distribution, since many techniques do not explicitly make use of (8.1). We could therefore use them, for instance, to compute summaries of the prior distribution. This could be useful in specifying a prior distribution to match some stated prior summaries, to confirm that a particular prior distribution has the required summaries, or to calculate other summaries to feed back as part of checking the validity of a chosen distribution (see **6.10**). g could also be (proportional to) a predictive distribution $f(y \mid x)$ for some future data y.

Reducing dimensionality

8.3 The power of computers is continually increasing, but there will always be problems too large and complex to be solved numerically in a reasonable amount of time. As computing power and technique improves to solve problems previously beyond our reach, so our attention shifts to still more difficult problems. Indeed, our ability to grasp and formulate more and more complex problems also grows with time.

In summarizing a posterior distribution its dimensionality is the principal factor determining the scale of computation required. By dimensionality we mean simply the number of parameters. A model in which nk observations x_{ij} are independently distributed as $N(\mu_i, \sigma^2)$, for $i = 1, 2, \ldots, k$ and $j = 1, 2, \ldots, n$, has $k + 1$ parameters $\mu_1, \mu_2, \ldots, \mu_k$ and σ^2. Therefore the joint posterior distribution is $(k + 1)$-dimensional. The computing time needed to calculate many kinds of summary increases rapidly with the number of dimensions. Statistical methodology is limited in practice partly by our inability to understand and model complex practical problems, and partly by our inability to compute the summaries needed to analyse the data. Fortunately, even if we cannot obtain a certain summary of a p-dimensional $g(\theta)$ analytically we may not need to tackle it as a p-dimensional computing problem. We may be able to obtain a partial solution mathematically which reduces the dimensionality of the computing problem.

8.4 The key to reducing dimensionality is to be able integrate the joint posterior with respect to some elements of θ, leaving the marginal distribution of the remainder. Let $\theta = (\phi, \psi)$ and suppose that we can write

$$g(\theta) = f(\phi \mid \psi) g_1(\psi), \tag{8.2}$$

such that $f(\phi \mid \psi)$ is the corresponding conditional distribution satisfying $\int f(\phi \mid \psi) d\phi = 1$ for all ψ. If g is given by (8.1) so that $g(\theta) = kf(\theta \mid x)$, then $f(\phi \mid \psi)$ is the conditional posterior distribution $f(\phi \mid \psi, x)$, and $g_1(\psi) = kf(\psi \mid x)$ is proportional to the marginal posterior for ψ with the same proportionality constant k. Now if we can find $f(\phi \mid \psi)$ to satisfy (8.2) we must at least understand this conditional distribution well enough to know its proportionality constant to ensure that $f(\phi \mid \psi)$ integrates to one. Typically we will then also be able to derive a number of useful summaries of $f(\phi \mid \psi)$ analytically. It may be possible to use these to express the required summaries of $g(\theta)$ in terms of summaries of $g_1(\psi)$. To summarize $g_1(\psi)$ is a lower dimensional problem.

One important way of achieving this is through conditional conjugacy (**6.33**).

Example 8.1
In Example 6.9 the distribution (6.6) is considered, which we now write as

$$g(\theta) = (\sigma^2)^{-t} \exp[-(\mu - m_1)^2/(2v) - \{b + d(\mu - m_2)^2\}/(2\sigma^2)].$$

Although this is only a two-dimensional distribution it is not easy to obtain summaries by analytical means. However, Example 6.9 points out that it can be integrated with respect to either μ or σ^2, the first of which corresponds to $f(\mu \mid \sigma^2)$ being a normal distribution and

$$g_1(\sigma^2) = (\sigma^2)^{-t}(v^{-1} + d\sigma^{-2})^{-1/2} \exp\{-b/(2\sigma^2) - (v + d^{-1}\sigma^2)^{-1}(m_1 - m_2)^2/2\}. \tag{8.3}$$

The normal distribution is, of course, very well understood so that many useful summaries of $f(\mu \mid \sigma^2)$ are available analytically. Almost any required summary of the two-dimensional $g(\theta)$ could thereby be expressed in terms of summaries of the one-dimensional $g_1(\sigma^2)$.

8.5 An important class of summaries for which this approach is useful are the integration-based summaries. The posterior expectation of a function $h(\theta)$ is

$$E\{h(\theta) \mid x\} = E[E\{h(\theta) \mid x, \psi\} \mid x]. \tag{8.4}$$

The inner expectation is

$$E\{h(\theta) \mid x, \psi\} = \int h(\theta) f(\phi \mid \psi) \mathrm{d}\phi,$$

which we assume is available as a known function $h_1(\psi)$ because of $f(\phi \mid \psi)$ being well understood. Then

$$\begin{aligned} E\{h(\theta) \mid x\} &= E\{h_1(\psi) \mid x\} = k^{-1} \int h_1(\psi) g_1(\psi)\, \mathrm{d}\psi \\ &= \frac{\int h_1(\psi) g_1(\psi)\, \mathrm{d}\psi}{\int g_1(\psi)\, \mathrm{d}\psi} \end{aligned} \tag{8.5}$$

requires only computations on the lower dimensional $g_1(\psi)$.

Example 8.2
Following Example 8.1, the normal distribution of μ given σ^2 has mean

$$h_1(\sigma^2) = (v^{-1} + d\sigma^{-2})^{-1}(v^{-1}m_1 + d\sigma^{-2}m_2).$$

Therefore the (posterior) expectation of μ can be found by calculating the expectation of this function $h_1(\sigma^2)$ under the marginal distribution (8.3). In (8.5) we see this as a ratio of one-dimensional integrals, which can be calculated numerically by techniques described later in this chapter.

8.6 Calculation of other kinds of summary, such as finding modes, can benefit from the same dimensionality reduction. Rather than general formulations like that in **8.5**, we need to consider each case and technique separately. Examples will be found in **8.31** and **8.76**.

Normal approximations

8.7 We first consider a number of relatively quick calculations which can be used to give approximations to summaries. The simplest approach is to approximate the posterior distribution by its asymptotic normal form, as described in **3.24**. In terms of a general $g(\boldsymbol{\theta})$,

$$g(\boldsymbol{\theta}) \approx g(\mathbf{m}) \exp\{-(\boldsymbol{\theta} - \mathbf{m})'\mathbf{V}^{-1}(\boldsymbol{\theta} - \mathbf{m})/2\}, \tag{8.6}$$

where $\mathbf{m}$ is the mode, i.e. the value of $\boldsymbol{\theta}$ maximizing $g(\boldsymbol{\theta})$, and $\mathbf{V}$ is the modal dispersion matrix, i.e. $-\mathbf{V}^{-1}$ is the matrix of second derivatives of $\log g(\boldsymbol{\theta})$ evaluated at $\boldsymbol{\theta} = \mathbf{m}$. This approximation is simply the Taylor series expansion of $\log g(\boldsymbol{\theta})$ around $\boldsymbol{\theta} = \mathbf{m}$ with remainder terms involving third and higher derivatives ignored.

In the case of (8.1) where $g(\boldsymbol{\theta})$ is proportional to the posterior density $f(\boldsymbol{\theta} \mid x)$, $\mathbf{m}$ is the posterior mode and $\mathbf{V}$ the posterior modal dispersion matrix. Our approximation is that the posterior distribution is $N(\mathbf{m}, \mathbf{V})$, and in particular this gives an approximation to the normalizing constant via the integral approximation

$$\int g(\boldsymbol{\theta})\, \mathrm{d}\boldsymbol{\theta} \approx g(\mathbf{m})(2\pi)^{p/2}|\mathbf{V}|^{1/2}, \tag{8.7}$$

where p is the number of elements in $\boldsymbol{\theta}$. We immediately have approximations for any summaries of interest by using the corresponding summaries of the $N(\mathbf{m}, \mathbf{V})$ distribution. For instance, the posterior mean is approximated by the mode $\mathbf{m}$.

8.8 We know from **3.24** that this approximation will be accurate if the data x consist of a suitably large number of observations. In general we can expect it to be poor if the distribution $g(\boldsymbol{\theta})$ does not have a single dominant mode or has marked skewness. The approximation obviously ignores any local maximum of $g(\boldsymbol{\theta})$ other than $\mathbf{m}$, and ignores skewness by only taking the Taylor series expansion to second order derivatives.

In order to compute the approximation (8.6) we must first compute $\mathbf{m}$ and $\mathbf{V}$. These in turn require differentiation of $g(\boldsymbol{\theta})$. In most problems, even if we cannot integrate $g(\boldsymbol{\theta})$ analytically we can almost always differentiate. Methods of finding $\mathbf{m}$ are considered in **8.27** to **8.34**, both when derivatives are available and when they are not. If $\mathbf{V}$ cannot be found by analytical differentiation, numerical differentiation of $\log g(\boldsymbol{\theta})$ at $\boldsymbol{\theta} = \mathbf{m}$ can be used (see for example Thisted, 1988) or it could be approximated using a contour around $\boldsymbol{\theta} = \mathbf{m}$ as described in **2.15**.

8.9 For a multimodal g we can approximate by a mixture of normals. If modes are at $\mathbf{m}_1, \mathbf{m}_2, \ldots, \mathbf{m}_d$ with corresponding modal dispersion matrices $\mathbf{V}_1, \mathbf{V}_2, \ldots, \mathbf{V}_d$ then

$$g(\boldsymbol{\theta}) \approx \sum_{i=1}^{d} g(\mathbf{m}_i) \exp\{-(\boldsymbol{\theta} - \mathbf{m}_i)'\mathbf{V}_i^{-1}(\boldsymbol{\theta} - \mathbf{m}_i)/2\}. \tag{8.8}$$

This approximation will be good if the modes are well separated so that $(\mathbf{m}_j - \mathbf{m}_i)'\mathbf{V}_i^{-1}(\mathbf{m}_j - \mathbf{m}_i)$ is sufficiently large (e.g. greater than 8) for all $i \neq j$. The integral of g is then approximated by $\sum_{i=1}^{d} s_i$, where

$$s_i = (2\pi)^{p/2} g(\mathbf{m}_i)|\mathbf{V}_i|^{1/2}$$

is the measure (2.16) of content of each mode. If g is proportional to a posterior density, so that the $\mathbf{m}_i$s are posterior modes and the $\mathbf{V}_i$s posterior modal dispersion matrices, then (8.8) corresponds to approximating $f(\boldsymbol{\theta} \mid x)$ by a mixture of normal distributions $N(\mathbf{m}_i, \mathbf{V}_i)$ with weights $w_i = s_i / \sum_{i=1}^{d} s_i$.

Summaries can be derived from this approximation as before. For instance the posterior mean vector is approximated by $\bar{\mathbf{m}} = \sum_{i=1}^{d} w_i \mathbf{m}_i$ and the posterior variance matrix by

$$\sum_{i=1}^{d} w_i(\mathbf{V}_i + \mathbf{m}_i\mathbf{m}_i') - \bar{\mathbf{m}}\bar{\mathbf{m}}' = \sum_{i=1}^{d} w_i \mathbf{V}_i + \sum_{i=1}^{d} w_i(\mathbf{m}_i - \bar{\mathbf{m}})(\mathbf{m}_i - \bar{\mathbf{m}})'.$$

Approximations using third derivatives

8.10 We can take some account of skewness in g by using higher derivatives. Expanding $\log g(\boldsymbol{\theta})$ about $\boldsymbol{\theta} = \mathbf{m}$ and retaining also the terms involving third derivatives,

$$g(\boldsymbol{\theta}) \approx g(\mathbf{m}) \exp\{-(\boldsymbol{\theta} - \mathbf{m})'\mathbf{V}^{-1}(\boldsymbol{\theta} - \mathbf{m})/2 + t(\boldsymbol{\theta})/6\}, \tag{8.9}$$

where

$$\begin{aligned} t(\boldsymbol{\theta}) &= \sum_{i,j,k} g^{ijk}(\theta_i - m_i)(\theta_j - m_j)(\theta_k - m_k), \qquad (8.10)\\ &= \sum_{i} g^{iii}(\theta_i - m_i)^3 + 3\sum_{i \neq j} g^{iij}(\theta_i - m_i)^2(\theta_j - m_j)\\ &\quad + 6 \sum_{i>j>k} g^{ijk}(\theta_i - m_i)(\theta_j - m_j)(\theta_k - m_k), \end{aligned}$$

θ_i and m_i are the ith elements of $\boldsymbol{\theta}$ and $\mathbf{m}$, and

$$g^{ijk} = \partial^3 \log g(\boldsymbol{\theta})/(\partial\theta_i \partial\theta_j \partial\theta_k).$$

The principal difficulty with (8.9) is that for $\boldsymbol{\theta}$ far enough from $\mathbf{m}$ the cubic terms in $t(\boldsymbol{\theta})$ will dominate and the exponent will tend to $\pm\infty$. Therefore the approximation becomes very bad in the tails. It will be good for some summaries, particularly those depending only on the behaviour of $g(\boldsymbol{\theta})$ near its mode. For $\boldsymbol{\theta}$ close to $\mathbf{m}$, (8.9) will typically be a more accurate approximation than (8.6). However, all integration-based summaries fail because the integral of (8.9) diverges.

8.11 For $\boldsymbol{\theta}$ close to $\mathbf{m}$, where (8.9) is best, we can make a further approximation using $\exp\{t(\boldsymbol{\theta})/6\} \approx 1 + t(\boldsymbol{\theta})/6$. Thus

$$g(\boldsymbol{\theta}) \approx g(\mathbf{m})\{1 + t(\boldsymbol{\theta})/6\} \exp\{-(\boldsymbol{\theta} - \mathbf{m})'\mathbf{V}^{-1}(\boldsymbol{\theta} - \mathbf{m})/2\}. \tag{8.11}$$

This approximation has similar accuracy to (8.9) for $\boldsymbol{\theta}$ near $\mathbf{m}$ but otherwise will be much superior in the tails, because now the term $\{1 + t(\boldsymbol{\theta})/6\}$ is dominated by the exponential for large $\boldsymbol{\theta}$. It has a different disadvantage in that it is not positive for all $\boldsymbol{\theta}$. For $\boldsymbol{\theta}$ sufficiently far from $\mathbf{m}$ the term $\{1 + t(\boldsymbol{\theta})/6\}$ may be negative. However, if the derivatives g^{ijk} are not too large this will occur only for $\boldsymbol{\theta} - \mathbf{m}$ so large that the exponential term is very small. Then although (8.11) may become negative it only becomes 'slightly negative' and this behaviour will cause no problems with most summaries. In particular, not only does the integral of (8.11) exist, but it is equal to the integral (8.7) of (8.6) since third order central moments of the normal distribution are zero.

8.12 The approximation (8.11) is analogous to multivariate versions of the Gram–Charlier and Edgeworth expansions, see **A6.17** to **A6.20**. However, the coefficients g^{ijk} in (8.10) are expressed in terms of derivatives rather than moments. The expansion could be taken further to include fourth or higher derivatives. Including fourth derivatives, for instance, will take account of kurtosis. Some other techniques of Chapter A6 may also be adaptable to approximating posterior distributions.

Approximating expectations

8.13 An integration-based summary can be expressed as a ratio of integrals

$$E\{h(\boldsymbol{\theta})\} = \frac{\int h(\boldsymbol{\theta})g(\boldsymbol{\theta})\,\mathrm{d}\boldsymbol{\theta}}{\int g(\boldsymbol{\theta})\,\mathrm{d}\boldsymbol{\theta}}. \tag{8.12}$$

If $g(\boldsymbol{\theta})$ is proportional to the posterior density, (8.1), then (8.12) is the posterior expectation $E\{h(\boldsymbol{\theta})\,|\,x\}$. The various approximations given previously for $g(\boldsymbol{\theta})$ can be used to approximate (8.12). From (8.6) we simply obtain the expectation of $h(\boldsymbol{\theta})$ with respect to a $N(\mathbf{m}, \mathbf{V})$ distribution. (8.8) gives

$$E\{h(\boldsymbol{\theta})\,|\,x\} \approx \sum_{i=1}^{d} w_i h_i,$$

where the w_is are the weights $s_i/\sum_{i=1}^{d} s_i$ and the h_is are expectations of $h(\boldsymbol{\theta})$ with respect to the $N(\mathbf{m}_i, \mathbf{V}_i)$ distributions. We cannot use (8.9) for integration-based summaries, but (8.11) yields

$$E\{h(\boldsymbol{\theta})\} = E^{\star}\{h(\boldsymbol{\theta})\} + E^{\star}\{h(\boldsymbol{\theta})t(\boldsymbol{\theta})\}/6, \tag{8.13}$$

where $E^{\star}$ denotes expectation with respect to $N(\mathbf{m}, \mathbf{V})$. The first term of (8.13) is the result of using (8.6) and the second is a correction for skewness.

Example 8.3

If θ is a scalar parameter, then (8.10) simplifies to

$$t(\theta) = g'''(\theta - m)^3,$$

where g''' is the third derivative of $\log g(\theta)$ at $\theta = m$. Then with $h(\theta) = \theta$ we find the following approximation to the mean,

$$E(\theta) \approx m + g''' E^{\star}(\theta - m)^4/6 = m + g''' v^2/2.$$

(In this case the modal dispersion matrix reduces to the scalar $v = (-g'')^{-1}$, where g'' is the second derivative of $\log g(\theta)$ at $\theta = m$.) More generally, if $\boldsymbol{\theta}$ is a vector and $h(\boldsymbol{\theta}) = \theta_s$,

$$E(\theta_s) \approx m_s + \sum_{i,j,k} g^{ijk} E^{\star}\{(\theta_i - m_i)(\theta_j - m_j)(\theta_k - m_k)(\theta_s - m_s)\}/6,$$

with $E^\star$ defined as before. From Anderson (1958, section 2.6) this fourth order central moment of the normal distribution is $v_{ij}v_{ks} + v_{ik}v_{js} + v_{is}v_{jk}$, where v_{ij} is the (i, j)th element of $\mathbf{V}$. Therefore

$$E(\theta_s) \approx m_s + \sum_{i,j,k} g^{ijk} v_{ij} v_{ks}/2. \tag{8.14}$$

This result is given by Lindley (1980).

8.14 For some forms of $h(\boldsymbol{\theta})$ it may not be practical to use these approximations because $E^\star\{h(\boldsymbol{\theta})\}$ is not known. Functions whose expectations under a normal distribution are straightforward to derive include polynomials, exponentials and trigonometric functions. But an example of a function whose expectation cannot be derived simply is $h(\boldsymbol{\theta}) = (1 + \theta_s^2)^{-1}$. In such a case we can expand $h(\boldsymbol{\theta})$ also in a Taylor series about $\boldsymbol{\theta} = \mathbf{m}$, giving

$$\begin{aligned} E^\star\{h(\boldsymbol{\theta})\} &\approx E^\star\{h(\mathbf{m}) + \mathbf{h}'(\boldsymbol{\theta} - \mathbf{m}) + (\boldsymbol{\theta} - \mathbf{m})'\mathbf{H}(\boldsymbol{\theta} - \mathbf{m})/2 + \ldots\} \\ &\approx h(\mathbf{m}) + tr\mathbf{HV}/2 \end{aligned} \tag{8.15}$$

where $\mathbf{h}$ is the vector of first derivatives of $h(\boldsymbol{\theta})$ and $\mathbf{H}$ the matrix of second derivatives, both evaluated at $\boldsymbol{\theta} = \mathbf{m}$. Since fifth order moments of a normal distribution are zero, inserting the same second-order expansion of $h(\boldsymbol{\theta})$ in (8.13) gives

$$E\{h(\boldsymbol{\theta})\} \approx h(\mathbf{m}) + tr\mathbf{HV}/2 + \mathbf{h}'\mathbf{b}, \tag{8.16}$$

where the sth element of $\mathbf{b}$ is the last term in (8.14), i.e.

$$b_s = \sum_{i,j,k} g^{ijk} v_{ij} v_{ks}/2.$$

This approximation was also derived by Lindley (1980).

Example 8.4

With $h(\boldsymbol{\theta}) = (1 + \theta_s^2)^{-1}$ we find that the sth element of $\mathbf{h}$ is $-2m_s(1 + m_s^2)^{-2}$ and the remaining elements are zero. The (s, s)th element of $\mathbf{H}$ is $(6m_s^2 - 2)(1 + m_s^2)^{-3}$ and the rest are zero. Applying (8.16) we find

$$E\{(1 + \theta_s^2)^{-1}\} \approx (1 + m_s^2)^{-1} + v_{ss}(3m_s^2 - 1)(1 + m_s^2)^{-3} - m_s(1 + m_s^2)^{-2} \sum_{i,j,k} g^{ijk} v_{ij} v_{ks}.$$

8.15 Instead of applying the same approximation for $g(\boldsymbol{\theta})$ in both numerator and denominator of (8.12), we can approximate the product $g^\star(\boldsymbol{\theta}) = h(\boldsymbol{\theta})g(\boldsymbol{\theta})$ in the numerator by the same methods used to approximate $g(\boldsymbol{\theta})$. Since these methods are based on expanding the logarithm, we now require $h(\boldsymbol{\theta})$ to be positive for all $\boldsymbol{\theta}$. Then let $\mathbf{m}^\star$ be the mode of $g^\star$ and $\mathbf{V}^\star$ be its modal disprsion matrix, and apply the basic normal approximation of **8.7**. Then (8.7) gives

$$E\{h(\boldsymbol{\theta})\} \approx \frac{g^\star(\mathbf{m}^\star)|\mathbf{V}^\star|^{1/2}}{g(\mathbf{m})|\mathbf{V}|^{1/2}}, \tag{8.17}$$

an approximation advocated by Tierney and Kadane (1986).

The approximation (8.11), using third derivatives, gives the same result. More accuracy might be achieved, however, by using fourth derivatives.

If $h(\boldsymbol{\theta})$ is not a positive function, we can use (8.17) to approximate $E\{h(\boldsymbol{\theta}) + a\}$, where a is a sufficiently large constant so that for every $\boldsymbol{\theta}$ value for which $h(\boldsymbol{\theta}) + a$ is negative $g(\boldsymbol{\theta})$ is very small. Then we simply use $E\{h(\boldsymbol{\theta})\} = E\{h(\boldsymbol{\theta}) + a\} - a$.

Approximating marginal densities

8.16 Suppose we wish to compute

$$g_1(\boldsymbol{\theta}_1) = \int g(\boldsymbol{\theta})\,\mathrm{d}\boldsymbol{\theta}_2, \tag{8.18}$$

where, $\boldsymbol{\theta}_1$ and $\boldsymbol{\theta}_2$ are subvectors of $\boldsymbol{\theta}$. If, for instance, $g(\boldsymbol{\theta})$ is proportional to the joint posterior density of $\boldsymbol{\theta}_1$ and $\boldsymbol{\theta}_2$ then $g_1(\boldsymbol{\theta}_1)$ is proportional to the marginal posterior density of $\boldsymbol{\theta}_1$. The basic normal approximation (8.6) implies that the marginal distribution of $\boldsymbol{\theta}_1$ is approximately $N(\mathbf{m}_1, \mathbf{V}_{11})$, where $\mathbf{m}_1$ and $\mathbf{V}_{11}$ are the appropriate subvector of $\mathbf{m}$ and submatrix of $\mathbf{V}$, respectively.

An alternative is to apply the normal approximation separately for each $\boldsymbol{\theta}_1$, regarding $g(\boldsymbol{\theta})$ as a function of $\boldsymbol{\theta}_2$ for each fixed $\boldsymbol{\theta}_1$. Then we let $\mathbf{m}_2(\boldsymbol{\theta}_1)$ be the value of $\boldsymbol{\theta}_2$ maximizing $g(\boldsymbol{\theta})$ for given $\boldsymbol{\theta}_1$. The corresponding modal dispersion matrix $\mathbf{V}_{22}(\boldsymbol{\theta}_1)$ is then minus the inverse of the $p_2 \times p_2$ matrix of second derivatives of $\log g(\boldsymbol{\theta})$ with respect to the p_2 elements of $\boldsymbol{\theta}_2$, evaluated at $\boldsymbol{\theta}_2 = \mathbf{m}_2(\theta_1)$. This yields the approximation

$$g_1(\boldsymbol{\theta}_1) \approx g(\boldsymbol{\theta}_1, \mathbf{m}_2(\boldsymbol{\theta}_1))(2\pi)^{p_2/2}|\mathbf{V}_{22}(\theta_1)|^{1/2}. \tag{8.19}$$

8.17 It may sometimes be possible to derive $\mathbf{m}_2(\theta_1)$ and $\mathbf{V}_{22}(\theta_1)$ analytically as functions of $\boldsymbol{\theta}_1$, but typically if this is possible then we can also derive $g_1(\boldsymbol{\theta}_1)$ analytically without the need for approximation. Conditional conjugacy is a case where this applies. Note also that if $\mathbf{m}_2(\boldsymbol{\theta}_1)$ can be found analytically we will generally be able to write down formulae for second derivatives at $\boldsymbol{\theta}_2 = \mathbf{m}_2(\boldsymbol{\theta}_1)$, but analytically finding the determinant of this matrix to obtain $|\mathbf{V}_{22}(\boldsymbol{\theta}_1)|$ will often be impractical for $p_1 = p - p_2$ greater than two.

Therefore in practice (8.19) will generally require a separate computational exercise for each $\boldsymbol{\theta}_1$, either a maximization to calculate $\mathbf{m}_2(\boldsymbol{\theta}_1)$ or a determinant calculation for $|\mathbf{V}_{22}(\boldsymbol{\theta})|$. The scale of computation involved may prohibit the use of (8.19) as an intermediate step towards computing summaries of the marginal distribution of $\boldsymbol{\theta}_1$. However, a plot of the marginal density of a scalar θ_1 or a contour plot for a two-dimensional $\boldsymbol{\theta}_1$ are useful summaries in themselves. They can be approximated by computing (8.19) for a selection of $\boldsymbol{\theta}_1$ values and then interpolating between these points.

Asymptotic accuracy of approximations

8.18 The normal approximation corresponds to the asymptotic posterior distribution derived in Chapter 3, for the situation in which the quantity of data tends to infinity. One guide to the relative accuracies of the various approximations is to consider their behaviour as the number of observations, n, tends to infinity. We assume now equation (8.1), so that $g(\boldsymbol{\theta})$ is proportional to a posterior density $f(\boldsymbol{\theta} \mid x)$. The loglikelihood $\log f(x \mid \boldsymbol{\theta})$ is a sum of n terms and so all its derivatives are of order n. However, this means that the modal dispersion matrix $\mathbf{V}$ is of order n^{-1}. Therefore posterior standard deviations are of order $n^{-1/2}$ and the approximation is required primarily over a region where the elements of $\boldsymbol{\theta} - \mathbf{m}$ are of order $n^{-1/2}$. Outside such a region, $g(\boldsymbol{\theta})$ is very small. In considering the accuracy of the approximations over such a region, we regard $\boldsymbol{\theta} - \mathbf{m}$ as $O(n^{-1/2})$. Derivatives of the loglikelihood are $O(n)$.

The simple normal approximation (8.6) includes terms of $O(1)$. The first neglected term is $t(\boldsymbol{\theta})/6$ in (8.9) which is of $O(n^{-1/2})$. Therefore the error in (8.6) is $O(n^{-1/2})$. Notice, however, that the error in the integral approximation (8.7) is $O(n^{-1})$ because this is also the integral of (8.9) whose error is $O(n^{-1})$. Although (8.15) contains the $O(n^{-1})$ term $tr\mathbf{HV}/2$, its error is also $O(n^{-1})$ because it omits the $O(n^{-1})$ term $\mathbf{h}'\mathbf{b}$ in (8.16). The simpler approximation $E\{h(\boldsymbol{\theta})\} \approx h(\mathbf{m})$ also has an error of $O(n^{-1})$ and is called the first order Laplace approximation by Kass et al. (1988). The first neglected terms in (8.16) are $O(n^{-2})$.

8.19 Tierney and Kadane (1986) point out that the error in (8.17) is also $O(n^{-2})$ because $O(n^{-1})$ terms in both numerator and denominator cancel. Therefore this method has the same asymptotic accuracy as the more complicated-looking (8.16). Discussion of the relative merits of the two approximations, particularly in the context of interactive computing, is given by Kass et al. (1988). Extensions of (8.15) to various more complex situations are given by Tierney et al. (1989) and Bagchi and Kadane (1991).

8.20 For approximating marginal distributions the simple normal approximation $N(\mathbf{m}_1, \mathbf{V}_{11})$ has an error of $O(n^{-1/2})$ in the same way as (8.6). Tierney and Kadane (1986) show that (8.19) is accurate to $O(n^{-3/2})$ over the same region, i.e. for $\boldsymbol{\theta}_1 - \mathbf{m}_1$ of order $n^{-1/2}$. Furthermore, the error of (8.19) is $O(n^{-1})$ for any bounded (i.e. $O(1)$) region. Therefore this approximation is accurate into the tails of $g_1(\boldsymbol{\theta}_1)$, which can be an important property.

Expansion about the MLE

8.21 These considerations show only the relative accuracy of approximations in large samples. In moderate or small samples there is no guarantee that an $O(n^{-2})$ term will be smaller than an $O(n^{-1})$ term. Furthermore, approximations which are equivalent asymptotically may be very different in moderate or small samples. To illustrate this last point, note that in the case of $g(\boldsymbol{\theta})$ proportional to a posterior distribution we could approximate $f(x \mid \boldsymbol{\theta})$ by any of the above methods. Then $\mathbf{m}$ becomes the value of $\boldsymbol{\theta}$ maximizing $f(x \mid \boldsymbol{\theta})$, i.e. the maximum likelihood estimate (MLE). $\mathbf{V}$ is minus the inverse of the hessian matrix of $f(x \mid \boldsymbol{\theta})$ at the MLE $\mathbf{m}$, which is sometimes called the *observed information matrix*. Then

$$g(\boldsymbol{\theta}) \approx f(\boldsymbol{\theta}) f(\mathbf{x} \mid \mathbf{m}) \exp\{-(\boldsymbol{\theta} - \mathbf{m})'\mathbf{V}^{-1}(\boldsymbol{\theta} - \mathbf{m})/2\}. \tag{8.20}$$

However, given a large sample the likelihood is negligible outside an $O(n^{-1/2})$ region around the MLE, and to the same order of accuracy we can replace $f(\boldsymbol{\theta})$ simply by $f(\mathbf{m})$, the value of the prior density at $\boldsymbol{\theta} = \mathbf{m}$. Then (8.20) reduces to (8.6) except that

m is now the MLE and **V** is based only on derivatives of the loglikelihood. We can similarly replace all the other approximations by versions based on approximating the likelihood $f(x \mid \boldsymbol{\theta})$ rather than the whole of $g(\boldsymbol{\theta})$. We can expand the prior $f(\boldsymbol{\theta})$ separately in the same way as $h(\boldsymbol{\theta})$ where appropriate. Since $f(\boldsymbol{\theta})$ is a constant as we increase the amount of data, and is ultimately overwhelmed by the likelihood, the approximations are asymptotically equivalent to the original formulae obtained by approximating $g(\boldsymbol{\theta})$ as a whole. Nevertheless, in moderate or small samples the original formulae will generally be better. This conclusion is supported by Lindley (1980).

Reparametrization

8.22 An important device in many computational methods is reparametrization. If we have a distribution $g(\boldsymbol{\theta})$, then a one-to-one transformation from $\boldsymbol{\theta}$ to $\boldsymbol{\phi} = \mathbf{t}(\boldsymbol{\theta})$ produces a new distribution

$$g_{\phi}(\boldsymbol{\phi}) = \|\mathbf{J}(\boldsymbol{\phi})\| g(\mathbf{t}^{-1}(\boldsymbol{\phi})),$$

where $\mathbf{t}^{-1}(\boldsymbol{\phi})$ is the inverse transformation and $\mathbf{J}(\boldsymbol{\phi})$ is the Jacobian matrix of derivatives of $\mathbf{t}^{-1}(\boldsymbol{\phi})$ with respect to $\boldsymbol{\phi}$. $\|\mathbf{J}(\boldsymbol{\phi})\|$ denotes the absolute value of the determinant of $\mathbf{J}(\boldsymbol{\phi})$. If a transformation can be found such that g_{ϕ} is more accurately approximated by a normal distribution than g, then all of the approximations considered in **8.7** to **8.16** will be improved. Summaries of $\boldsymbol{\theta}$ can be obtained via summaries of $\boldsymbol{\phi}$, and in particular $E\{h(\boldsymbol{\theta})\}$ may be approximated by formulae for $E\{h(\mathbf{t}^{-1}(\boldsymbol{\phi}))\}$.

8.23 In general it is very difficult to theorize about an appropriate normalizing transformation from p-dimensional $\boldsymbol{\theta}$ to p-dimensional $\boldsymbol{\phi}$, but some progress can be made by considering transformations one dimension at a time. Let $\mathbf{t}(\boldsymbol{\theta})$ be such that the ith element is of the form $\phi_i = t_i(\boldsymbol{\theta}) = t_i(\theta_i)$. That is, each θ_i is transformed separately to ϕ_i. Then the Jacobian matrix is diagonal and its determinant reduces to

$$|\mathbf{J}(\boldsymbol{\theta})| = \prod_{i=1}^{p} dt_i^{-1}(\phi_i)/d\phi_i.$$

In suggesting suitable one-dimensional transformations to try to improve the normal approximation, one obvious consideration is the range of possible values of θ_i. The normal approximation assumes that each θ_i can range from $-\infty$ to ∞. If, for instance, θ_i must be positive we can expect the basic approximation (8.6) to be poor if it gives θ_i a mean m_i that is not large relative to its standard deviation $\sqrt{v_{ii}}$. In this case, letting $\phi_i = \log \theta_i$ removes the restriction, allowing ϕ_i to range over $(-\infty, \infty)$. Also, in most practical problems the distribution of a parameter constrained to be positive will be positively skewed. The log transformation will then have the second benefit of reducing skewness, so making the normal approximation even more accurate. It certainly works well for standard distributions: if θ_i has a gamma, chi-square or F distribution then $\phi_i = \log \theta_i$ is well-known as a transformation to approximate normality, and if θ_i has a log-normal distribution the transformation is obviously exact!

Similarly, the transformation $\phi_i = \log\{\theta_i/(1 - \theta_i)\}$, when θ_i is constrained to take values in $(0, 1)$, will often induce approximate normality in ϕ_i.

8.24 Hills and Smith (1993) proposed a diagnostic plot for assessing normality, and for suggesting suitable transformations. It is based on the Bates and Watts (1988) t-plot for assessing approximate normality of a likelihood. Consider first a scalar parameter θ. If $g(\theta)$ is exactly normal then

$$g(\theta) = g(m)\exp\{-(\theta - m)^2/(2v)\}, \tag{8.21}$$

for some m and v. Define

$$T(\theta) = \text{sign}(\theta - m)\{\log g(m) - \log g(\theta)\}^{1/2}. \tag{8.22}$$

Then if (8.21) holds exactly for some v we have

$$T(\theta) = (2v)^{-1/2}(\theta - m),$$

so that $T(\theta)$ is a linear function of θ. Hills and Smith (1993) suggest first finding the mode m of $g(\theta)$, calculating (8.22) for a range of values of θ, and plotting the results against θ. They call the plot of $T(\theta)$ against θ the Bayesian t-plot. If the plot is roughly linear the normal approximation will be acceptable. If not, the form of the plot may suggest a suitable transformation.

Example 8.5

If $g(\theta)$ is proportional to a log-normal distribution, so that $\log\theta$ has the $N(m^\star, v^\star)$ distribution, then we find $m = \exp(m^\star - v^\star)$ and

$$T(\theta) = (2v^\star)^{-1/2}(\log\theta + v^\star - m^\star),$$

which is linear in $\log\theta$. If, alternatively, $g(\theta)$ is proportional to a gamma distribution with

$$g(\theta) = \theta^{a-1}\exp(-b\theta)$$

we have $m = (a-1)/b$ and a much more complicated expression for $T(\theta)$. However, Figure 8.1 shows t-plots for the lognormal distribution with $m^\star = 0$, $v^\star = 0.3$ and the gamma distribution with $a = 5$, $b = 4$. The plots have similar shapes, suggesting again that a log transformation (which produces exact normality in the lognormal case) will convert a gamma distribution to approximate normality. Figure 8.2 shows the t-plot in the gamma case after the transformation $\phi = \log\theta$, and we see it is more nearly linear over the same range. However, Figure 8.1 shows that the gamma t-plot curves rather less than the lognormal, and Figure 8.2 shows a corresponding slight over-correction from using a logarithmic transformation.

8.25 Extending the method now to p-dimensional $\boldsymbol{\theta}$, let $g^\star(\theta_i)$ be the result of maximizing $g(\boldsymbol{\theta})$ with respect to $\theta_1, \theta_2, \ldots, \theta_{i-1}, \theta_{i+1}, \theta_{i+2}, \ldots, \theta_p$, as a function of θ_i. Hills and Smith (1993) then propose performing a t-plot on $g^\star$. This is realistic in practice if p is small or the maximizations can be done analytically. Otherwise the process of maximizing over $p-1$ variables for each of many different values of θ_i may be computationally demanding. A simpler alternative is to let $g^\star(\theta_i)$ be the result of setting θ_j for $j \neq i$ to arbitrary values (e.g. their modal values m_j) instead of maximizing over them. Whereas

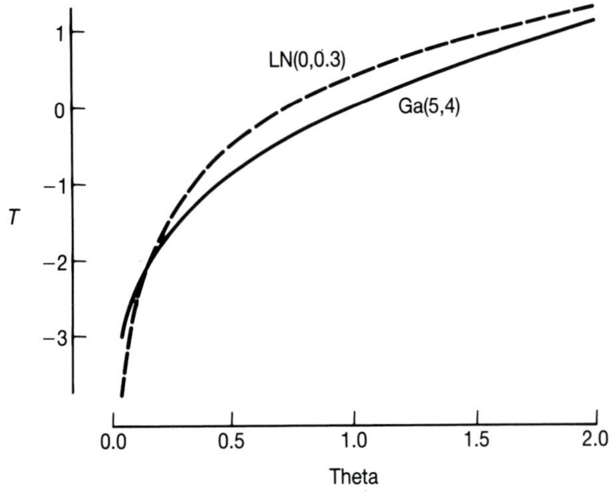

Fig. 8.1 Bayesian *t*-plots for lognormal and gamma

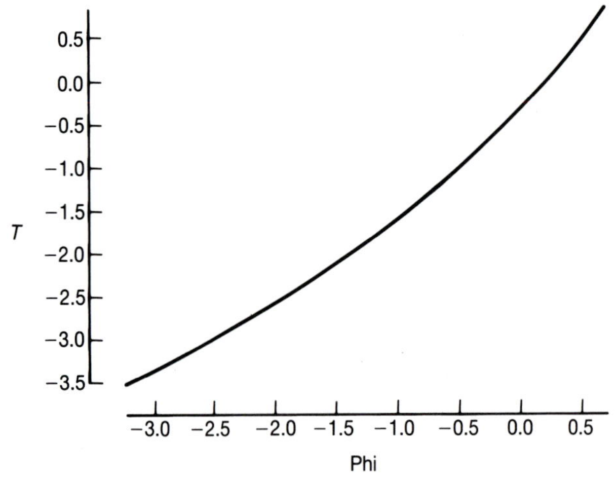

Fig. 8.2 Bayesian *t*-plot for logtransformed gamma

the Hills and Smith t-plot is similar to checking the marginal distribution of θ_i for normality, this will check its conditional distribution given the set values of the other θ_js. The conditional t-plot is quick to compute, and for any given θ_i it would be easy to plot conditional t-plots for a variety of different set values for the other θ_js. If we do not obtain similar shapes in these plots it is doubtful whether transformation of individual θ_is will succeed in producing approximate normality of the joint distribution.

8.26 Kass and Slate (1992) present a variety of non-graphical diagnostics for non-normality of $g(\boldsymbol{\theta})$, but the diagnostics themselves do not suggest suitable transformations.

Achcar and Smith (1990) discuss the effect of reparametrization on the approximations of **8.7** to **8.16**.

Optimization

8.27 All the methods we have discussed so far involve finding the mode of $g(\boldsymbol{\theta})$, which is itself generally a computational problem. A considerable literature exists on numerical methods for maximizing or minimizing (collectively known as optimizing) functions of many variables. Many of these are appropriate for finding modes of a posterior density. Indeed, most posterior densities will be sufficiently well-behaved functions for quite simple optimization methods to work very effectively.

8.28 One of the best known optimization techniques is the Newton–Raphson method. Starting from an initial guess $\boldsymbol{\theta}_0$, the method produces a series of new points $\boldsymbol{\theta}_1, \boldsymbol{\theta}_2, \boldsymbol{\theta}_3, \ldots$, with $\boldsymbol{\theta}_{n+1}$ defined by

$$\boldsymbol{\theta}_{n+1} = \boldsymbol{\theta}_n - \mathbf{G}(\boldsymbol{\theta}_n)^{-1}\mathbf{g}(\boldsymbol{\theta}_n), \tag{8.23}$$

where $\mathbf{g}(\boldsymbol{\theta})$ is the vector of first derivatives, i.e. with ith element $\partial g(\boldsymbol{\theta})/\partial\theta_i$, and $\mathbf{G}(\boldsymbol{\theta})$ is the Hessian matrix with (i, j)th element $\partial^2 g(\boldsymbol{\theta})/\partial\theta_i\partial\theta_j$. (8.23) is derived simply by expanding $g(\boldsymbol{\theta})$ in a Taylor series about $\boldsymbol{\theta} = \boldsymbol{\theta}_n$, as far as the second derivative terms. $\boldsymbol{\theta}_n$ converges rapidly to the mode $\mathbf{m}$ provided the starting point $\boldsymbol{\theta}_0$ is sufficiently close to a maximum or minimum $\mathbf{m}$ of $g(\boldsymbol{\theta})$. In fact, convergence to $\mathbf{m}$ is guaranteed if $\boldsymbol{\theta}_0$ is within the inflexion boundary (see **2.13**) around $\mathbf{m}$.

In general, both for Newton–Raphson and other methods, it is better to define the function to be maximized as $\log g(\boldsymbol{\theta})$ rather than $g(\boldsymbol{\theta})$. This is because in the case of a normal distribution the expansion of $\log g(\boldsymbol{\theta})$ to second derivatives is exact, so (8.23) converges in one step: $\boldsymbol{\theta}_1 = \mathbf{m}$. This is true for *any* starting point $\boldsymbol{\theta}_0$. If $g(\boldsymbol{\theta})$ is approximately normal, then maximizing $\log g(\boldsymbol{\theta})$ by Newton–Raphson should converge rapidly. If, as for the normal distribution, $\log g(\boldsymbol{\theta})$ is convex (so that its inflexion boundary is effectively at infinity), convergence is guaranteed for any $\boldsymbol{\theta}_0$. Even for distributions, such as those with tails like t distributions, which are not log-convex, the region of convergence using $\log g(\boldsymbol{\theta})$ is much wider than with $g(\boldsymbol{\theta})$. Reparametrization of $\boldsymbol{\theta}$ to improve the approximation of normality will obviously have the extra benefit of making maximization far more efficient.

8.29 Applying Newton–Raphson requires both first and second order derivatives of $\log g(\boldsymbol{\theta})$. Although methods exist for numerically approximating derivatives of a function, they are generally rather unreliable. Errors in calculating the Hessian matrix $\mathbf{G}(\boldsymbol{\theta}_n)$ can make the Newton–Raphson procedure unstable. Therefore if $\mathbf{G}(\boldsymbol{\theta})$ cannot be obtained analytically as an explicit function of $\boldsymbol{\theta}$ it is better to use alternative methods.

8.30 If the gradient vector $\mathbf{g}(\boldsymbol{\theta})$ is available analytically, then we can apply a method of steepest ascent. Given a current point $\boldsymbol{\theta}_n$ the gradient $\mathbf{g}(\boldsymbol{\theta}_n)$ determines the direction of steepest ascent, and we consider the set of points $\boldsymbol{\theta}_n + \alpha\mathbf{g}(\boldsymbol{\theta}_n)$ for $\alpha > 0$. Then $\boldsymbol{\theta}_{n+1}$ is defined as the point in this set for which $g(\boldsymbol{\theta})$ is maximized. That is, we carry out a one-dimensional maximization with respect to α. The one-dimensional maximizations can be carried out by any suitable method (including Newton–Raphson, which is generally stable in one dimension even if the second derivative is approximated numerically).

Various refinements of this simple gradient method are available, and there are other techniques suitable for the case where not even first derivatives of $\log g(\boldsymbol{\theta})$ are available analytically. Some authoritative treatments may be found in Kennedy and Gentle (1992) and Thisted (1988). Rather than attempt to describe a large number of such methods here, we end this discussion of optimization with some remarks on dimensionality reduction.

8.31 In **8.4** we considered dimesionality reduction by integrating out some parameters ψ analytically, but for optimization we need to maximize over ψ. So let $\theta = (\phi, \psi)$ and suppose that we can find

$$g_\phi(\phi) = \max_\psi g(\phi, \psi) = g(\phi, \hat{\psi}(\phi)). \tag{8.24}$$

Thus, $g_\phi(\phi)$ is the result of reducing dimensionality by maximizing with respect to ψ, and for given ϕ that maximum occurs at $\psi = \hat{\psi}(\phi)$. Then to maximize g we need only solve the lower-dimensional problem of maximizing g_ϕ. If this maximum is

$$g_\phi(\hat{\phi}) = \max_\phi g_\phi(\phi),$$

then $g_\phi(\hat{\phi})$ is also the maximum of g,

$$g_\phi(\hat{\phi}) = \max_{\phi,\psi} g(\phi, \psi) = g(\hat{\phi}, \hat{\psi}(\hat{\phi}),$$

with the maximum occurring at $(\phi, \psi) = (\hat{\phi}, \hat{\psi}(\hat{\phi}))$.

Lindley–Smith iteration

8.32 A variant on this is an iterative method due to Lindley and Smith (1972) which requires not only $\hat{\psi}(\phi)$ to be available as in (8.24) but also $\hat{\phi}(\psi)$. That is, we can maximize $g(\theta)$ analytically as a function of ψ for given ϕ, $\hat{\psi}(\phi)$, and also as a function of ϕ for given ψ, $\hat{\phi}(\psi)$. This will typically be possible when we have the kind of double conditional conjugacy seen in Example 6.9. We now start with an initial point ϕ_0 and proceed to compute the sequence $\psi_1, \phi_1, \psi_2, \phi_2, \psi_3, \phi_3, \ldots$ defined by

$$\psi_{n+1} = \hat{\psi}(\phi_n), \qquad \phi_{n+1} = \hat{\phi}(\psi_{n+1}). \tag{8.25}$$

This is a hill-climbing method, in the sense that the sequence of corresponding values of g, i.e. $g(\phi_0, \psi_1), g(\phi_1, \psi_1), g(\phi_1, \psi_2), g(\phi_2, \psi_2), \ldots$, are guaranteed to increase and the method will always converge to a mode. It is rather like the steepest ascent method of **8.30**, but instead of moving each time in the direction of steepest ascent we move alternately in the ϕ and ψ directions. Because these are not necessarily good directions for climbing $g(\theta)$ rapidly, the Lindley–Smith method may converge very slowly.

Example 8.6

Let (ϕ, ψ) have the bivariate normal density

$$g(\phi, \psi) = \exp\{-(\phi^2 + \psi^2 - 2r\phi\psi)/2\}$$

with zero means, unit variances and correlation r. Then $\hat{\psi}(\phi) = r\phi$ and $\hat{\phi}(\psi) = r\psi$. Therefore the result of the two steps (8.25) is that $\phi_{n+1} = r^2\phi_n$ (and $\psi_{n+1} = r^2\psi_n$). If the correlation is sufficiently close to ± 1 the iteration converges very slowly towards the mode at $(0, 0)$.

8.33 Example 8.6 shows that Lindley–Smith iteration can be slow even if $g(\theta)$ is exactly normal. The problem is correlation, and the method will generally converge slowly whenever there is high correlation in some form between ϕ and ψ. We shall return to methods for dealing with correlation in **8.41–42** and **8.67** to **8.68**.

8.34 The method is easy to generalize to more than two groups of parameters. If θ is partitioned by $\theta = (\theta_1, \theta_2, \ldots, \theta_m)$ into m groups of parameters, we begin with a starting point $\theta^0 = (\theta_1^0, \theta_2^0, \ldots, \theta_m^0)$ and at the nth iteration define θ^{n+1} by

$$\begin{aligned}
\theta_1^{n+1} &= \hat{\theta}_1(\theta_2^n, \theta_3^n, \ldots, \theta_m^n), \\
\theta_2^{n+1} &= \hat{\theta}_2(\theta_1^{n+1}, \theta_3^n, \theta_4^n, \ldots, \theta_m^n), \\
&\cdots \\
\theta_m^{n+1} &= \hat{\theta}_m(\theta_1^{n+1}, \theta_2^{n+1}, \ldots, \theta_{m-1}^{n+1}).
\end{aligned} \tag{8.26}$$

To apply this method we need to determine the functions $\hat{\theta}_i$ which give the maximum of $g(\theta)$ with respect to θ_i, conditional on fixed values of all the other elements of θ. Taking this approach to the extreme we could let $m = p$, the number of elements in θ, so that each θ_i is a scalar.

Quadrature

8.35 Numerical integration, also known as quadrature, is another problem with an extensive literature, some of which is directly applicable to summarizing distributions. Integration is needed to find the normalizing constant for $g(\theta)$, and integration-based summaries are ratios of integrals as in (8.12). Consider first a general one-dimensional integral

$$I = \int_a^b f(x)\mathrm{d}x. \tag{8.27}$$

Quadrature methods essentially approximate I by calculating f at a number of points $x_1, x_2, \ldots, x_n$ and applying some formula to the resulting values $f(x_1), f(x_2), \ldots, f(x_n)$. The simplest and by far the most common form is a weighted average

$$\hat{I} = \sum_{i=1}^{n} w_i f(x_i). \tag{8.28}$$

Different quadrature rules are distinguished by using different sets of design points $x_1, x_2, \ldots, x_n \in [a, b]$, and/or different sets of weights $w_1, w_2, \ldots, w_n$.

8.36 The simplest of all quadrature rules divides the interval $[a, b]$ into n equal parts, evaluates $f(x)$ at the midpoint of each subinterval, and then applies equal weights. Thus (8.28) becomes

$$\hat{I} = n^{-1}(b-a)\sum_{i=1}^{n} f\{a + (2i-1)(b-a)/(2n)\}. \tag{8.29}$$

This approximates (8.27) by the sum of areas of rectangles with base $n^{-1}(b-a)$ and height equal to the value of $f(x)$ at the midpoint of the base. For large n we have a sum of many thin rectangles whose area closely approximates I in exactly the way that integrals are often introduced in elementary calculus classes.

For one-dimensional integrals, acceptable accuracy can generally be achieved with $n = 100$ evaluations of f. Unless f itself is so complex that computing its value at a single point x_i is a substantial task, this approach is perfectly feasible by computer. And if greater accuracy is needed, n could be increased to thousands of points.

In the case of the expectation (8.12) and a scalar θ,

$$E\{h(\theta)\} \approx \sum_{i=1}^{n} h(\theta_i)g(\theta_i) / \sum_{i=1}^{n} g(\theta_i)$$

with the θ_is as in (8.29). A problem arises here in that the integrals in (8.12) will not usually be over a finite interval $[a, b]$. They are in principle over the whole line $(-\infty, \infty)$ and will only reduce to a finite interval if θ itself is confined to an interval (so that $g(\theta) = 0$ for $\theta \notin [a, b]$). In practice, we can simply set very wide limits $[a, b]$ so that $g(\theta)$ and $h(\theta)g(\theta)$ are negligible outside those limits.

8.37 We can take advantage of approximate normality of $g(\theta)$ by using Gauss–Hermite quadrature rules. Gauss–Hermite quadrature requires $a = -\infty$ and $b = \infty$, which is exactly the case required for the expectation (8.12) of an unconstrained parameter θ. As mentioned in **8.23**, normality of $g(\theta)$ is really only reasonable in this case also. The n-point Gauss–Hermite rule chooses design points $x_1, x_2, \ldots, x_n$ and weights $w_1, w_2, \ldots, w_n$ such that the approximation (8.28) yields the exact integral I if $\exp(x^2/2)f(x)$ is a polynomial of order $2n - 1$ or less. Therefore the approximation will be good if $\exp(x^2/2)f(x)$ can be closely approximated by a polynomial of order $2n - 1$. Since as n increases any arbitrary function can be approximated arbitrarily closely by a polynomial of order $2n-1$, Gauss–Hermite quadrature achieves greater and greater accuracy as n increases. Tables of design points and weights can be found in Abramowitz and Stegun (1965) and Salzer et al. (1952).

8.38 To apply Gauss–Hermite quadrature to the expectation $E\{h(\theta)\}$ for scalar θ, note that the denominator $\int g(\theta)d\theta$ will be integrated exactly if it is proportional to $\exp(-x^2/2)$, i.e. θ has the standard normal distribution. We need to adapt this to the case where θ is believed to be approximately $N(m, v)$. This can be achieved by a transformation from θ to $x = v^{-1/2}(\theta - m)$. Then we have the formula

$$\int g(\theta)\,d\theta \approx v^{1/2}\sum_{i=1}^{n} w_i g(m + v^{1/2}x_i), \tag{8.30}$$

where the x_is and w_is are the standard Gauss–Hermite design points and weights. Values of m and v can be chosen as usual to equal the mode and modal dispersion summaries. Sufficiently accurate integration is generally achievable with as few as $n = 8$ points.

If $h(\theta)$ is a polynomial, or well approximated by a polynomial, then the numerator of (8.12) can be integrated to a similar accuracy using the same points and weights, resulting in

$$E\{h(\theta)\} \approx \sum_{i=1}^{n} w_i h(\theta_i) g(\theta_i) / \sum_{i=1}^{n} w_i g(\theta_i). \tag{8.31}$$

where $\theta_i = m + v^{1/2} x_i$ as in (8.30). An alternative to (8.31) is to approximate the numerator by calculating new m and v values for the product $h(\theta)g(\theta)$, as in **8.15**. The approximation (8.6) with $p = 1$ is (8.30) with $n = 1$, since the 1-point Gauss–Hermite rule has $x_1 = 0$ and $w_1 = (2\pi)^{1/2}$. (8.17) corresponds to applying 1-point quadrature in numerator and denominator separately. The corrections for third derivatives as in (8.16) will have accuracy similar to 2-point Gauss–Hermite quadrature.

Cartesian product rules

8.39 We now turn to quadrature in two dimensions. The integral of $g(\boldsymbol{\theta})$ can be written as repeated one-dimensional integrals

$$\begin{aligned} \int g(\boldsymbol{\theta})\, \mathrm{d}\boldsymbol{\theta} &= \int \left\{ \int g(\phi, \psi)\, \mathrm{d}\psi \right\} \mathrm{d}\phi \\ &= \int g^\star(\phi)\, \mathrm{d}\phi, \end{aligned} \tag{8.32}$$

where ϕ and ψ are scalars and $g^\star(\phi)$ is thereby defined as $\int g(\phi, \psi) \mathrm{d}\psi$. We can calculate (8.32) by quadrature as $\sum w_i g^\star(\phi_i)$ for suitable weights w_i and points ϕ_i, but this requires $g^\star(\phi_i)$ to be known. Since $g^\star(\phi)$ is an integral, approximate each $g^\star(\phi_i)$ by a quadrature formula $\sum u_j g(\phi_i, \psi_j)$ using weights u_j and design points ψ_j. Putting the two stages together gives

$$\int g(\boldsymbol{\theta})\, \mathrm{d}\boldsymbol{\theta} \approx \sum_{i=1}^{n} \sum_{j=1}^{m} w_i u_j g(\phi_i, \psi_j). \tag{8.33}$$

We can think of this as a single two-dimensional rule with nm design points (ϕ_i, ψ_j) and weights $w_i u_j$. This is called the *Cartesian product* of the two one-dimensional rules.

8.40 It is simple now to generalize to p dimensions. Take p one-dimensional rules with $n_1, n_2, \ldots, n_p$ points θ_{ij} and weights w_{ij} $(i = 1, 2, \ldots, p;\ j = 1, 2, \ldots, m_i)$. Then the Cartesian product rule has $\prod_{i=1}^{p} n_i$ design points forming a $n_1 \times n_2 \times \ldots \times n_p$ grid in the p-dimensional $\boldsymbol{\theta}$ space. A typical point is $(\theta_{1j_1}, \theta_{2j_2}, \ldots, \theta_{pj_p})$ with weight $w_{1j_1} w_{2j_2} \ldots w_{pj_p}$.

Unless there is some reason to think that more points are needed to integrate adequately in some dimensions than in others, it is usual to make the p component rules identical. An n-point rule in one dimension then becomes an n^p-point *Cartesian power rule* in p-dimensions. The major difficulty with these rules is that unless p is sufficiently small the number of function evaluations, $\prod_{i=1}^{p} n_i$ or n^p, becomes impractically large. Even with Gauss–Hermite rules of 3 or 4 points in each dimension, and a powerful computer, it is not feasible to carry out quadrature in this way in more than about ten dimensions. Some high-dimensional quadrature rules with fewer points than product rules have been proposed; see Davis and Rabinowitz (1984). Some experience with such rules in Bayesian problems is reported by Shaw (1988).

Iterative scaling

8.41 The Gauss–Hermite product rules will integrate $g(\boldsymbol{\theta})$ exactly if $\exp(\boldsymbol{\theta}'\boldsymbol{\theta}/2)g(\boldsymbol{\theta})$ is a polynomial of order up to $2n_i - 1$ in θ_i, $i = 1, 2, \ldots, p$. They will certainly produce accurate integration if $g(\boldsymbol{\theta})$ is approximately proportional to the p-dimensional standard multivariate normal distribution $N(\mathbf{0}, \mathbf{I})$. Therefore it is important for this approach to consider reparametrization to improve normality, as in **8.22** to **8.25**. However, we also wish to achieve accurate integration when $g(\boldsymbol{\theta})$ approximates to a general $N(\mathbf{m}, \mathbf{V})$ density rather than the standard $N(\mathbf{0}, \mathbf{I})$. The scaling of (8.30) can be applied in each dimension so that the x variables have zero mean and unit variance, but in general they will still be correlated. Gauss–Hermite product rules can be highly inaccurate when there is correlation.

The alternative is to apply a more general linear transformation

$$\mathbf{x} = \mathbf{L}(\boldsymbol{\theta} - \mathbf{m}), \tag{8.34}$$

where the $p \times p$ matrix $\mathbf{L}$ satisfies

$$\mathbf{L}'\mathbf{L} = \mathbf{V}^{-1}. \tag{8.35}$$

Then if $\boldsymbol{\theta}$ is approximately $N(\mathbf{m}, \mathbf{V})$, $\mathbf{x}$ will be approximately $N(\mathbf{0}, \mathbf{I})$. Then (8.30) becomes

$$\int g(\boldsymbol{\theta})\,\mathrm{d}\boldsymbol{\theta} \approx |\mathbf{V}|^{1/2} \sum_{i=1}^{N} w_i g(\mathbf{m} + \mathbf{L}^{-1}\mathbf{x}_i), \tag{8.36}$$

where the $\mathbf{x}_i$s are the $N = n^p$ (or $\prod_{i=1}^{p} n_i$) design points of the Gauss–Hermite product rule and the w_is are the corresponding weights. $|\mathbf{V}|^{1/2}$ can be replaced by $|\mathbf{L}|^{-1}$, and a computationally convenient solution to (8.35) is the Cholesky decomposition (see Householder, 1964, or Thisted, 1988).

8.42 With $\mathbf{m}$ set equal to the mode of $g(\boldsymbol{\theta})$ and $\mathbf{V}$ to the modal dispersion matrix, this Gauss–Hermite product method is computationally straightforward and may achieve reasonable accuracy. But it can be sensitive to the choice of $\mathbf{m}$ and $\mathbf{V}$. Naylor and Smith (1982) advocate an iterative scaling approach. A first application of the Gauss–Hermite quadrature is used to approximate the mean vector and variance matrix, using (8.12) with $h(\boldsymbol{\theta}) = \theta_i$, $h(\boldsymbol{\theta}) = \theta_i^2$ and $h(\boldsymbol{\theta}) = \theta_i\theta_j$. They then replace $\mathbf{m}$ and $\mathbf{V}$ with these new values and perform a second Gauss–Hermite quadrature. The new $\mathbf{m}$ and $\mathbf{V}$ will imply function evaluation at new points in (8.36) and this quadrature is again used to approximate the mean vector and variance matrix. These new values then take the place of $\mathbf{m}$ and $\mathbf{V}$. The Gauss–Hermite quadrature is repeated until the calculated means, variances and covariances converge. Naylor and Smith (1982) suggest gradually increasing the number of grid points starting from, say, 3^p or 4^p point quadrature. When convergence is reached on a n^p point grid, increase to $(n+1)^p$ points. The procedure finally stops when there is negligible change between the converged estimates of means and variances on n^p points and $(n+1)^p$ points.

This iterative scaling approach ensures that the design points are shifted and scaled by $\mathbf{m}$ and $\mathbf{V}$ values that realistically reflect the mean and variance, so the scaled values should genuinely have zero means, unit variances and zero correlations. There is of course no guarantee that the Gauss–Hermite quadrature will be most accurate with the scaling chosen in this way, but we can show (as in Example 2.17) that the best multivariate normal approximation to $g(\boldsymbol{\theta})$ in the Kullback–Liebler sense is the normal distribution with mean vector and variance matrix agreeing with g. Further experiences with iterative scaling and Gauss–Hermite quadrature are reported by Smith et al. (1985).

Monte Carlo

8.43 A different approach to evaluating an integral evaluates $g(\boldsymbol{\theta})$ at *random* points, and is called *Monte Carlo* integration. Suppose that a series of points $\boldsymbol{\theta}_1, \boldsymbol{\theta}_2, \ldots, \boldsymbol{\theta}_n$ are drawn independently from the distribution with density $s(\boldsymbol{\theta})$. Now

$$I = \int g(\boldsymbol{\theta})\,\mathrm{d}\boldsymbol{\theta} = \int \{g(\boldsymbol{\theta})/s(\boldsymbol{\theta})\}s(\boldsymbol{\theta})\,\mathrm{d}\boldsymbol{\theta} = E_s\{g(\boldsymbol{\theta})/s(\boldsymbol{\theta})\}, \tag{8.37}$$

where E_s denotes expectation with respect to the distribution s. $s(\boldsymbol{\theta})$ must be positive wherever $g(\boldsymbol{\theta})$ is positive. Now the sample of points $\boldsymbol{\theta}_1, \boldsymbol{\theta}_2, \ldots, \boldsymbol{\theta}_n$ drawn independently from s gives a sample of values $g(\boldsymbol{\theta}_i)/s(\boldsymbol{\theta}_i)$ of the function $g(\boldsymbol{\theta})/s(\boldsymbol{\theta})$. We estimate the expectation (8.37) by the sample mean

$$\hat{I} = n^{-1}\sum_{i=1}^{n} g(\boldsymbol{\theta}_i)/s(\boldsymbol{\theta}_i). \tag{8.38}$$

According to classical statistical theory, $\hat{I}$ is an unbiased estimator of the integral I with variance

$$\operatorname{var}(\hat{I}) = n^{-1}\operatorname{var}_s\{g(\boldsymbol{\theta})/s(\boldsymbol{\theta})\}. \tag{8.39}$$

As n increases, $\hat{I}$ becomes a more and more accurate estimate of I. For large n, $\hat{I}$ is asymptotically normally distributed (by the central limit theorem) with mean I and variance (8.39). The variance can be estimated from the same sample, by $n^{-2}\sum_{i=1}^{n} g^2(\boldsymbol{\theta}_i)/s^2(\boldsymbol{\theta}_i) - n^{-1}\hat{I}^2$, and an asymptotic confidence interval derived.

Therefore, the Monte Carlo estimator (8.38) has an advantage over quadrature methods in that its accuracy can be assessed from the same data and confidence intervals given. The number of function evaluations, n, is not fixed, and can be increased until the estimated variance is as low as desired.

Of course, the theory behind Monte Carlo is entirely classical. A firm adherent of the Bayesian viewpoint would object to the use of Monte Carlo for evaluating posterior summaries purely because it is non-Bayesian. O'Hagan (1987a) gives more detailed criticism, and O'Hagan (1991b) develops an alternative, Bayesian approach to quadrature.

8.44 The Monte Carlo estimate of the expectation (8.12) is

$$E\{h(\boldsymbol{\theta})\} \approx \{\textstyle\sum_{i=1}^{n} h(\boldsymbol{\theta}_i)g(\boldsymbol{\theta}_i)/s(\boldsymbol{\theta}_i)\}/\{\textstyle\sum_{i=1}^{n} g(\boldsymbol{\theta}_i)/s(\boldsymbol{\theta}_i)\}. \tag{8.40}$$

As a classical statistical estimator, the ratio (8.40) is only asymptotically unbiased. Denoting the numerator and denominator of (8.40) by X and Y respectively, expanding $Y^{-1} = \{E(Y)+(Y-E(Y))\}^{-1} = E(Y)^{-1} - E(Y)^{-2}(Y-E(Y)) + E(Y)^{-3}(Y-E(Y))^2 - \dots$ and retaining only terms of order n^{-1} we obtain the asymptotic variance

$$\text{var}\left(\frac{X}{Y}\right) \approx \frac{\text{var}(X)}{E(Y)^2} - 2\frac{E(X)\text{cov}(X,Y)}{E(Y)^3} + \frac{E(X)^2\text{var}(Y)}{E(Y)^4}. \tag{8.41}$$

The variances and covariance can be estimated from the same sample, yielding an estimate of (8.41).

8.45 The variance (8.39) for the simple Monte Carlo estimator (8.38) depends on how accurately the sampling density $s(\boldsymbol{\theta})$ mimics $g(\boldsymbol{\theta})$. It will be zero if $g(\boldsymbol{\theta})/s(\boldsymbol{\theta})$ is a constant, but in that case $s(\boldsymbol{\theta})$ is a density $g(\boldsymbol{\theta})/\int g(\boldsymbol{\theta})\mathrm{d}\boldsymbol{\theta}$ and is exactly the distribution for $\boldsymbol{\theta}$ implied by g that we wish to summarize. If such an s were known numerical integration would not be necessary. We cannot expect to obtain a perfect sampling distribution in this way, but the objective is to obtain a density s such that (a) it is feasible to draw a sample from s, and (b) s is approximately proportional to g. An obvious choice for s is the normal approximation (8.6), but the following example shows that this can be a bad choice.

Example 8.7

Let $g(\theta) = (1+\theta^2)^{-c}$ and let $s(\theta)$ be the $N(m, v)$ density. Then

$$\begin{aligned} E_s\{g(\theta)^2/s(\theta)^2\} &= \int g(\theta)^2 s(\theta)^{-1}\,\mathrm{d}\theta \\ &= (2\pi v)^{1/2}\int_{-\infty}^{\infty}(1+\theta^2)^{-2c}\exp\{(\theta-m)^2/(2v)\}\,\mathrm{d}\theta, \end{aligned}$$

and this integral diverges for any c, m and v. Therefore $\text{var}_s\{g(\theta)/s(\theta)\}$ is infinite. If $g(\theta)$ is proportional to a t distribution with any degrees of freedom then setting $s(\theta)$ to any normal distribution is disastrous. The variance of $\hat{I}$ is infinite and convergence of $\hat{I}$ to I will not be achieved even as $n \to \infty$.

8.46 In order for Monte Carlo integration to converge, we must ensure a finite variance, which in effect means that the tails of $s(\theta)$ should not be too thin. We can derive a Cauchy-tailed sample $\boldsymbol{\theta}_1, \boldsymbol{\theta}_2, \dots$ by letting $\boldsymbol{\theta}_i = \mathbf{m} + z_i^{-1}(\boldsymbol{\phi}_i - \mathbf{m})$ where the $\boldsymbol{\phi}_i$s are a sample from the normal approximation $N(\mathbf{m}, \mathbf{V})$ and the z_is are independent $N(0,1)$ variables. The resulting density $s(\boldsymbol{\theta})$ is given in Exercise 8.4, and will have sufficiently heavy tails to achieve a finite variance for most practical problems.

A number of techniques can be applied to modify the basic Monte Carlo method, with a view to reducing the variance and so improving accuracy. Iterative scaling can be used to obtain an $s(\boldsymbol{\theta})$ with good location and scale. Hammersley and Handscomb

(1964) and Ripley (1987) provide a general overview of Monte Carlo methods. Techniques particularly relevant for Bayesian applications are presented by Kloek and van Dijk (1978), van Dijk and Kloek (1983, 1985), Stewart (1983), Geweke (1988, 1989).

8.47 If the number of dimensions is small, quadrature methods are much more efficient than Monte Carlo. In higher dimensions the number of function evaluations required by Cartesian product quadrature makes them impractical, but ultimately Monte Carlo has the same drawback of needing a sample size that escalates exponentially as we increase the number of dimensions. This is because $E_s\{g(\boldsymbol{\theta})^2/s(\boldsymbol{\theta})^2\}$ is a p-dimensional integral. If, for instance, the elements of $\boldsymbol{\theta}$ are independent in both $g(\boldsymbol{\theta})$ and $s(\boldsymbol{\theta})$, so that $g(\boldsymbol{\theta}) = \prod_{i=1}^p g_i(\theta_i)$ and $s(\boldsymbol{\theta}) = \prod_{i=1}^p s_i(\theta_i)$ then

$$\begin{aligned} E_s\{g(\boldsymbol{\theta})^2/s(\boldsymbol{\theta})^2\} &= \int g(\boldsymbol{\theta})^2 s(\boldsymbol{\theta})^{-1}\,\mathrm{d}\boldsymbol{\theta} \\ &= \prod_{i=1}^{p} \int g_i(\theta_i)^2 s_i(\theta_i)^{-1}\,\mathrm{d}\theta_i. \end{aligned}$$

This is a product of p one-dimensional integrals whose values depend on how accurately each $s_i(\theta_i)$ approximates $g_i(\theta_i)$. As the number of dimensions increases $E_s\{g(\boldsymbol{\theta})^2/s(\boldsymbol{\theta})^2\}$ will be of the order c^p, where c is the average (i.e. geometric mean) of these integrals. Essentially the same result applies generally, that if as the number p of dimensions increases the approximation $s(\boldsymbol{\theta})$ achieves a similar closeness to $g(\boldsymbol{\theta})$ in each dimension, then $E_s\{g(\boldsymbol{\theta})^2/s(\boldsymbol{\theta}^2)\}$ will be of the order of c^p for some c. (See Exercise 8.6). For large p, var$\{g(\boldsymbol{\theta})/s(\boldsymbol{\theta})\}$ will be dominated by this term, and so the sample size required to achieve a given accuracy with Monte Carlo integration ultimately tends to increase with p in the same exponential way as the number of function evaluations needed for Cartesian product quadrature rules.

This comparison is not entirely fair because the accuracy of Cartesian product rules with n^p points actually decreases with p in practice. Furthermore the factor c, by which the number of function evaluations required for Monte Carlo integration increases with each dimension, can be only slightly greater than unity for a well-chosen $s(\boldsymbol{\theta})$, whereas the number of points needed in Cartesian product rules increases by a factor of $n \geqslant 3$ for each dimension (where $n = 3$ corresponds to using only 3-point quadrature in each dimension). Therefore Monte Carlo is far more efficient in high-dimensional problems. Monte Carlo has been used effectively in problems of 10 to 20 dimensions, but in higher dimensions the sample size required will eventually become impractical.

Gibbs sampling

8.48 A group of methods known collectively as Markov Chain Monte Carlo (MCMC) methods may prove to be more powerful than either Monte Carlo integration or quadrature, and to be capable of handling very high-dimensional problems. The essence of these methods is that, at least in an asymptotic sense, they draw a sample directly from the normalized density $g(\boldsymbol{\theta}/\int g(\boldsymbol{\theta})\mathrm{d}\boldsymbol{\theta}$. This is despite the remark in **8.45** that it is not generally possible to sample from this distribution for the purposes of Monte Carlo integration, and that numerical integration would not be necessary if we could. The explanation of this apparent contradiction is that in Monte Carlo integration we have to know the density $s(\boldsymbol{\theta}) = g(\boldsymbol{\theta})/\int g(\boldsymbol{\theta})\mathrm{d}\boldsymbol{\theta}$ explicitly in order to use the estimator (8.38), and that does imply that we already know $I = \int g(\boldsymbol{\theta})\mathrm{d}\boldsymbol{\theta}$. MCMC methods, as we shall see, sample from the normalized density without knowing and without deriving the normalizing constant. Furthermore, a sample exactly from the normalized density is only achieved in an asymptotic sense.

Before considering how a sample from the normalized density may be used for summarization, we describe one of the most popular MCMC methods, known as Gibbs sampling.

8.49 Let $f(\boldsymbol{\theta}) = g(\boldsymbol{\theta})/\int g(\boldsymbol{\theta})\mathrm{d}\boldsymbol{\theta}$ be the normalized joint density. Let $g_i(\theta_i \mid \boldsymbol{\theta}_{(i)}) = g_i(\theta_i \mid \theta_1, \ldots, \theta_{i-1}, \theta_{i+1}, \theta_{i+2}, \ldots, \theta_p)$ be the function g regarded as a function of θ_i alone with the other components $\boldsymbol{\theta}_{(i)}$ of $\boldsymbol{\theta}$ fixed. In its normalized form $f_i(\theta_i \mid \boldsymbol{\theta}_{(i)}) = g_i(\theta_i \mid \boldsymbol{\theta}_{(i)})/\int g_i(\theta_i \mid \boldsymbol{\theta}_{(i)})\mathrm{d}\theta_i$, it is the conditional density of θ_i given $\boldsymbol{\theta}_{(i)}$, deriving from the joint density $f(\boldsymbol{\theta})$. Suppose that we are able to sample from each of these one-dimensional conditional distributions for any given values of $\boldsymbol{\theta}_{(i)}$. Then the Gibbs sampler consists of sampling from these distributions in a cyclical way.

We begin with an arbitrary starting point $\boldsymbol{\theta}^0 = (\theta_1^0, \theta_2^0, \ldots, \theta_p^0)$ and generate a series of random points $\boldsymbol{\theta}^1, \boldsymbol{\theta}^2, \boldsymbol{\theta}^3$, where $\boldsymbol{\theta}^{m+1}$ is derived from $\boldsymbol{\theta}^m$ in the following way.

$$\left.\begin{array}{lll}
\theta_1^{m+1} & \text{is drawn randomly from} & f_1(\theta_1 \mid \theta_2^m, \theta_3^m, \ldots, \theta_p^m) \\
\theta_2^{m+1} & \text{is drawn randomly from} & f_2(\theta_2 \mid \theta_1^{m+1}, \theta_3^m, \theta_4^m, \ldots, \theta_p^m) \\
\ldots & \ldots & \ldots \\
\theta_p^{m+1} & \text{is drawn randomly from} & f_p(\theta_p \mid \theta_1^{m+1}, \theta_2^{m+1}, \ldots, \theta_{p-1}^{m+1})
\end{array}\right\} \qquad (8.42)$$

(All the random draws in all the iterations are performed independently.) It is worth noting the similarity between the scheme (8.42) for Gibbs sampling and the scheme (8.26) for Lindley–Smith optimization. The Lindley–Smith method is deterministic, replacing each θ_i by the mode $\hat{\theta}_i$ of its relevant conditional distribution. Gibbs sampling is stochastic, replacing θ_i by a random draw from that conditional distribution. If we repeated the Gibbs sampler starting from the same $\boldsymbol{\theta}^0$ we would obtain a different sequence $\boldsymbol{\theta}^1, \boldsymbol{\theta}^2, \ldots$ each time. We can therefore think of $\boldsymbol{\theta}^m$ as a random variable whose distribution is determined by the starting point $\boldsymbol{\theta}^0$ and the distributions $f_i(\theta_i \mid \boldsymbol{\theta}_{(i)})$. We are interested in the distribution of $\boldsymbol{\theta}^m$.

8.50 $\boldsymbol{\theta}^0$ is fixed and $\boldsymbol{\theta}^1$ is obtained randomly by the procedure (8.42). Denote the resulting distribution of $\boldsymbol{\theta}^1$ by $f^\star(\boldsymbol{\theta}^1 \mid \boldsymbol{\theta}^0)$. (8.42) is now applied again with $\boldsymbol{\theta}^1$ fixed to obtain $\boldsymbol{\theta}^2$, and clearly the distribution of $\boldsymbol{\theta}^2$ given $\boldsymbol{\theta}^1$ is $f^\star(\boldsymbol{\theta}^2 \mid \boldsymbol{\theta}^1)$. The same distribution $f^\star$ applies to both steps, and to all steps of the iteration. $f^\star$ is determined by (8.42) and is a function of the conditional distributions $f_i(\theta_i \mid \boldsymbol{\theta}_{(i)})$. Another important property of the sequence $\boldsymbol{\theta}^0, \boldsymbol{\theta}^1, \boldsymbol{\theta}^2, \ldots$ is that the distribution of $\boldsymbol{\theta}^m$ given *all* previous values $\boldsymbol{\theta}^0, \boldsymbol{\theta}^1, \ldots, \boldsymbol{\theta}^{m-1}$ depends only on $\boldsymbol{\theta}^{m-1}$, and is just $f^\star(\boldsymbol{\theta}^m \mid \boldsymbol{\theta}^{m-1})$. In other words, $\boldsymbol{\theta}^m$ is independent of $\boldsymbol{\theta}^0, \boldsymbol{\theta}^1, \ldots, \boldsymbol{\theta}^{m-2}$ given $\boldsymbol{\theta}^{m-1}$. This is known as the Markov property and the sequence is called a *Markov* *chain.* (The Markov property of the exponential distribution in **A5.2** is related. The Markov property of the $\boldsymbol{\theta}_i$s says that the sequence has only a one-step memory).

We have a Markov chain with the same *transition* distribution $f^\star$ for every step, which constitutes a *homogeneous* Markov chain. Markov chains have been extensively studied; see, for instance, Feller (1968), Grimmett and Stirzaker (1992). One of the most important results is that under rather general conditions a homogeneous Markov chain converges, in the sense that the distribution of $\boldsymbol{\theta}^m$ given the starting point $\boldsymbol{\theta}^0$ converges as $m \to \infty$ to a limiting distribution f^∞ which is independent of $\boldsymbol{\theta}^0$ and depends only on the transition distribution $f^\star$.

8.51 The distribution whose convergence we are considering is $f(\boldsymbol{\theta}^m \mid \boldsymbol{\theta}^0)$. It is determined as follows. For $m = 1$, $f(\boldsymbol{\theta}^1 \mid \boldsymbol{\theta}^0)$ is just $f^\star(\boldsymbol{\theta}^1 \mid \boldsymbol{\theta}^0)$. For $m = 2$,

$$f(\boldsymbol{\theta}^2 \mid \boldsymbol{\theta}^0) = \int f(\boldsymbol{\theta}^2, \boldsymbol{\theta}^1 \mid \boldsymbol{\theta}^0)\,\mathrm{d}\boldsymbol{\theta}^1 = \int f(\boldsymbol{\theta}^2 \mid \boldsymbol{\theta}^1, \boldsymbol{\theta}^0) f(\boldsymbol{\theta}^1 \mid \boldsymbol{\theta}^0)\,\mathrm{d}\boldsymbol{\theta}^1$$
$$= \int f^\star(\boldsymbol{\theta}^2 \mid \boldsymbol{\theta}^1) f^\star(\boldsymbol{\theta}^1 \mid \boldsymbol{\theta}^0)\,\mathrm{d}\boldsymbol{\theta}^1$$

using the Markov property. In general we have the recurrence relation

$$f(\boldsymbol{\theta}^m \mid \boldsymbol{\theta}^0) = \int f^\star(\boldsymbol{\theta}^m \mid \boldsymbol{\theta}^{m-1}) f(\boldsymbol{\theta}^{m-1} \mid \boldsymbol{\theta}^0)\,\mathrm{d}\boldsymbol{\theta}^{m-1}. \tag{8.43}$$

Now if this converges to a distribution $f^\infty(\boldsymbol{\theta}^m)$ then (8.43) implies that f^∞ must satisfy the integral equation $f^\infty(\boldsymbol{\theta}^m) = \int f^\star(\boldsymbol{\theta}^m \mid \boldsymbol{\theta}^{m-1}) f^\infty(\boldsymbol{\theta}^{m-1})\,\mathrm{d}\boldsymbol{\theta}^{m-1}$ or, simplifying the notation,

$$f^\infty(\boldsymbol{\theta}) = \int f^\star(\boldsymbol{\theta} \mid \boldsymbol{\phi}) f^\infty(\boldsymbol{\phi})\,\mathrm{d}\boldsymbol{\phi}. \tag{8.44}$$

For given transition distribution $f^\star$ we can find the limiting distribution f^∞ of the chain by solving (8.44).

8.52 The first step of the Gibbs sampler determines the first element of $\boldsymbol{\theta}^{m+1}$ as a function of $\boldsymbol{\theta}^m$. Let $f_i^\infty(\theta_i) = \int f^\infty(\boldsymbol{\theta})\mathrm{d}\boldsymbol{\theta}_{(i)}$ be the limiting marginal distribution of θ_i, $f_{(i)}^\infty(\boldsymbol{\theta}_{(i)}) = \int f^\infty(\boldsymbol{\theta})\mathrm{d}\theta_i$ be the limiting distribution of $\boldsymbol{\theta}_{(i)}$, and $f_i^\infty(\theta_i \mid \boldsymbol{\theta}_{(i)}) = f^\infty(\boldsymbol{\theta})/f_{(i)}^\infty(\boldsymbol{\theta}_{(i)})$ be the limiting distribution of θ_i given $\boldsymbol{\theta}_{(i)}$. Then (8.44) and the first step of (8.42) imply

$$f_1^\infty(\theta_1) = \int f_1(\theta_1 \mid \boldsymbol{\theta}_{(1)}) f_{(1)}^\infty(\boldsymbol{\theta}_{(1)})\,\mathrm{d}\boldsymbol{\theta}_{(1)},$$

which can only be satisfied by $f_1^\infty(\theta_1 \mid \boldsymbol{\theta}_{(1)}) = f_1(\theta_1 \mid \boldsymbol{\theta}_{(1)})$. Repeating this argument for each step of (8.42) we have

$$f_i^\infty(\theta_i \mid \boldsymbol{\theta}_{(i)}) = f_i(\theta_i \mid \boldsymbol{\theta}_{(i)}), \tag{8.45}$$

for $i = 1, 2, \ldots, p$. Therefore the limiting distribution f^∞ has the same conditional distributions as the joint distribution $f(\boldsymbol{\theta})$ that we are interested in.

Subject to further mild conditions, the full set of conditional distributions $f_i(\theta_i \mid \boldsymbol{\theta}_{(i)})$ uniquely determine the joint distribution $f(\boldsymbol{\theta})$. Then the only possible solution of (8.45) is $f^\infty(\boldsymbol{\theta}) = f(\boldsymbol{\theta})$, and we have proved that the Gibbs sampler converges, regardless of the starting point $\boldsymbol{\theta}^0$, such that for large enough m the point $\boldsymbol{\theta}^m$ behaves like a random draw from the full normalized distribution $f(\boldsymbol{\theta}) = g(\boldsymbol{\theta})/\int g(\boldsymbol{\theta})\mathrm{d}\boldsymbol{\theta}$.

Conditions for Gibbs sampler convergence

8.53 We now briefly consider the conditions for the preceding argument to hold. There are two aspects to consider, whether the Gibbs sampler converges to a distribution f^∞ independent of $\boldsymbol{\theta}^0$, and whether conditional distributions uniquely determine the joint distribution f so that we can say $f^\infty = f$. The following rather contrived example shows that there are instances where the argument fails.

Example 8.8

Let $\boldsymbol{\theta} = (\theta_1, \theta_2)$ have the following joint distribution $f(\boldsymbol{\theta})$. Over the square $0 \leqslant \theta_1, \theta_2 \leqslant 1$, $f(\boldsymbol{\theta}) = c$ (and $0 < c < 1$). Over the square $2 \leqslant \theta_1, \theta_2 \leqslant 3$, $f(\boldsymbol{\theta}) = 1 - c$. Then if $\theta_2 \in [0, 1]$ the conditional distribution of θ_1 is uniform over $[0, 1]$, and if $\theta_2 \in [2, 3]$ the conditional distribution of θ_1 is uniform over $[2, 3]$. The conditional distribution of θ_2 given θ_1 is exactly the same, by symmetry. Now it is easy to see that the required convergence of the Gibbs sampler cannot happen. If the starting value $\boldsymbol{\theta}^0$ has $\theta_2^0 \in [0, 1]$ then the first step of (8.42) generates θ_1^1 uniformly from $[0, 1]$. Then since $\theta_1^1 \in [0, 1]$ the second step generates θ_2^1 uniformly from $[0, 1]$. Successive $\boldsymbol{\theta}^m$ will always be uniformly distributed over the unit square $[0, 1]^2$. If, however, the starting point $\boldsymbol{\theta}^0$ has $\theta_2^0 \in [2, 3]$ then $\boldsymbol{\theta}^m$ will be uniformly distributed over $[2, 3]^2$ for all m. We have convergence but the limiting distribution depends on the starting point.

This is also an example where conditional distributions do not determine the joint distribution, which is most clearly seen in the fact that c does not appear in the conditional distributions. From the conditionals we can deduce that the joint distribution is made up of two parts $[0, 1]^2$ and $[2, 3]^2$ with $\boldsymbol{\theta}$ uniformly distributed over each part. But we cannot determine the relative heights, c and $1 - c$, of $f(\boldsymbol{\theta})$ on those two parts.

8.54 The principal requirement for uniform convergence of a Markov chain is that the chain be irreducible and aperiodic. Precise definitions and conditions can be found in Roberts and Smith (1993), Schervish and Carlin (1992), but Example 8.8 shows the practical meaning of these conditions quite clearly. Let the set of possible values of $\boldsymbol{\theta}^m$ from starting point $\boldsymbol{\theta}^0$ be $C^m(\boldsymbol{\theta}^0)$. If there exist two possible starting values, say $\boldsymbol{\theta}^\star$ and $\boldsymbol{\theta}^{\star\star}$, such that $C^m(\boldsymbol{\theta}^\star) \cap C^m(\boldsymbol{\theta}^{\star\star}) = \phi$ for all m, then the same limiting distribution cannot be reached from both starting points. This occurs in Example 8.8, where there are two regions C_1 and C_2 such that the chain always stays in whichever region it starts in. Another way in which convergence may fail is if there are two disjoint regions C_1 and C_2 such that whenever $\boldsymbol{\theta}^{m-1} \in C_1$ then $\boldsymbol{\theta}^m \in C_2$, and $\boldsymbol{\theta}^{m-1} \in C_2$ implies $\boldsymbol{\theta}^m \in C_1$. Then the chain oscillates between C_1 and C_2, and if $\boldsymbol{\theta}^\star \in C_1$, $\boldsymbol{\theta}^{\star\star} \in C_2$ we again have $C^m(\boldsymbol{\theta}^\star) \cap C^m(\boldsymbol{\theta}^{\star\star}) = \phi$ for all m.

A Markov chain is irreducible if for every starting point $\boldsymbol{\theta}^0$ and set C such that $\int_C g(\boldsymbol{\theta}) d\boldsymbol{\theta} > 0$, there exists an m such that $P(\boldsymbol{\theta}^m \in C \mid \boldsymbol{\theta}^0) > 0$. So irreducibility ensures that the chain can reach all possible $\boldsymbol{\theta}$ values from any starting point. It is called aperiodic if it does not oscillate between two sets C_1, C_2, or cycle around a partition $C_1, C_2, \ldots, C_r$ of $r > 2$ disjoint sets.

These conditions are clearly linked to the requirement for conditional distributions to determine $f(\boldsymbol{\theta})$ uniquely. An irreducible and aperiodic chain converges to a unique limiting distribution f^{∞}. Yet if the conditional distributions did not uniquely determine f there would be more than one possible limiting distribution. In general, the convergence conditions will imply that $f(\boldsymbol{\theta})$ is uniquely determined by its conditional distributions.

Summarization with a Gibbs sample

8.55 There are a number of practical issues concerned with implementing the Gibbs sampler which we address later. Suppose now that we are able to apply it (or one of the other MCMC methods) in practice to obtain n independent draws $\boldsymbol{\theta}_1, \boldsymbol{\theta}_2, \ldots, \boldsymbol{\theta}_n$ from the normalized joint distribution $f(\boldsymbol{\theta})$. The problem of using this sample to calculate summaries of $f(\boldsymbol{\theta})$ is the very familiar one of making inferences about the underlying population distribution from a random sample. For instance, we can use the sample mean of the θ_1 values to approximate the population mean $E(\theta_1)$. More generally, for any function $h(\boldsymbol{\theta})$ the sample mean $n^{-1}\sum_{j=1}^{n} h(\boldsymbol{\theta}_j)$ is a classical unbiased estimator of $E\{h(\boldsymbol{\theta})\}$. Its variance is $n^{-1}\text{var}\,\{h(\boldsymbol{\theta})\}$, which can also be estimated from the same sample. The method therefore enjoys the same advantages as Monte Carlo integration, i.e. the accuracy of any estimated summary can be estimated from the same data, and can be made as small as we like by increasing the sample size. However, in contrast to the discussion in **8.47**, the sample size needed to achieve any desired accuracy does not increase with the dimensionality of $\boldsymbol{\theta}$. It is true in practice that the effort required to generate a sample of n increases with the number of dimensions, if only because the number of steps in each iteration (8.42) increases, and furthermore that convergence may be much slower in higher dimensions, see **8.65** to **8.68**. Nevertheless, Gibbs sampling in practice allows us to analyse very high-dimensional problems.

8.56 It was said in **2.3** that one of the most powerful summaries is a plot of a marginal density. From the ith elements of each of $\boldsymbol{\theta}_1, \boldsymbol{\theta}_2, \ldots, \boldsymbol{\theta}_n$ we have a sample of values of θ_i drawn from its marginal distribution $f_i(\theta_i)$. One way to estimate the marginal density f_i from this sample is to use general methods of *density estimation*, such as kernel density estimates; see Silverman (1986) and **10.24**. However, Gelfand and Smith (1990) propose an alternative method making use of the conditional densities $f_i(\theta_i \mid \boldsymbol{\theta}_{(i)})$. Let $\boldsymbol{\theta}_{j(i)}$ be $\boldsymbol{\theta}_j$ with the ith element removed. Then the Gelfand and Smith approximation is

$$f_i(\theta_i) \approx n^{-1} \sum_{j=1}^{n} f_i(\theta_i \mid \boldsymbol{\theta}_{j(i)}), \tag{8.46}$$

and is based on taking the $\boldsymbol{\theta}_{j(i)}$s as a sample from $f_{(i)}(\boldsymbol{\theta}_{(i)})$ and using $f_i(\theta_i) = \int f(\boldsymbol{\theta})\mathrm{d}\boldsymbol{\theta}_{(i)} = \int f_i(\theta_i \mid \boldsymbol{\theta}_{(i)}) f_{(i)}(\boldsymbol{\theta}_{(i)})\mathrm{d}\boldsymbol{\theta}_{(i)} = E\{f_i(\theta_i \mid \boldsymbol{\theta}_{(i)})\}$. (8.46) produces a smooth plot and because it makes use of the conditional distributions will estimate the tails of $f_i(\theta_i)$ better than more general density estimation techniques.

Sampling from the conditional distributions

8.57 In implementing the Gibbs sampler, an important question is how to make a random draw from $f_i(\theta_i \mid \boldsymbol{\theta}_{(i)})$. This conditional distribution is obtained by normalizing

$g_i(\theta_i \mid \boldsymbol{\theta}_{(i)})$, which in turn is just $g(\boldsymbol{\theta})$ regarded as a function of θ_i alone with $\boldsymbol{\theta}_{(i)}$ fixed. In looking at $g(\boldsymbol{\theta})$ as a function of θ_i we may recognize $f_i(\theta_i \mid \boldsymbol{\theta}_{(i)})$ as a standard distribution. Ways of drawing random samples from the common standard distributions have been extensively studied, and recommendations of the best available algorithms are given by Devroye (1986) and Ripley (1987).

Example 8.9
Consider the distribution

$$g(\mu, \sigma^2) = (\sigma^2)^{-t} \exp\left[-(\mu - m_1)^2/(2v) - \{b + d(\mu - m_2)^2\}/(2\sigma^2)\right]$$

first introduced in **6.20**. As a function of μ for fixed σ^2 it is clearly proportional to a normal distribution. As a function of σ^2 for fixed μ it is easy to show that σ^{-2} has a gamma distribution. Therefore, identifying θ_1 with μ and θ_2 with σ^2, the first step of (8.42) is to draw μ randomly from the appropriate normal distribution (whose parameters are calculated as functions of the current σ^2 value). Then the second and last step is to let σ^2 be the reciprocal of a random draw from the appropriate gamma distribution (with parameters calculated as functions of the current μ value drawn in the first step). Algorithms for sampling from both normal and gamma distributions are given by Devroye (1986).

8.58 If $g_i(\theta_i \mid \boldsymbol{\theta}_{(i)})$ is not recognized as proportional to a standard distribution, then it may also not be possible to integrate it analytically to obtain $f_i(\theta_i \mid \boldsymbol{\theta}_{(i)})$. We then require a method to draw from an arbitrary distribution g_i with unknown constant of proportionality. In discussing such methods we suppress the subscript i and simplify notation to consider sampling from an arbitrary $g(\theta)$, for scalar θ.

Rejection methods

8.59 Let s be a density from which we can conveniently sample, such as one of the standard distributions for which a well-established algorithm exists. And let the ratio $g(\theta)/s(\theta)$ be bounded. The supremum of $g(\theta)/s(\theta)$ may not be known analytically, but suppose that we can determine an upper bound $A \geqslant g(\theta)/s(\theta)$ for all θ. Now take two random draws, θ from the distribution $s(\theta)$, and y from the uniform distribution on [0.1]. Then if

$$A\, y \leqslant g(\theta)/s(\theta) \tag{8.47}$$

retain the draw θ, otherwise reject this θ, draw a fresh θ and y, and continue doing so until (8.47) is satisfied.

The joint density of θ and y from one pair of draws is $f(\theta, y) = s(\theta)$. The distribution of the value of θ that is retained is the conditional distribution of θ given (8.47). That is,

$$f(\theta \mid y \leqslant g(\theta)/\{As(\theta)\}) = \frac{\int_0^{g(\theta)/\{As(\theta)\}} f(\theta, y)\,\mathrm{d}y}{\int \int_0^{g(\theta)/\{As(\theta)\}} f(\theta, y)\,\mathrm{d}\theta\,\mathrm{d}y}$$

$$= \frac{s(\theta) \int_0^{g(\theta)/\{As(\theta)\}} \mathrm{d}y}{\int s(\theta) \left\{ \int_0^{g(\theta)/\{As(\theta)\}} \mathrm{d}y \right\} \mathrm{d}\theta}$$

$$= \frac{s(\theta)g(\theta)/\{As(\theta)\}}{\int s(\theta)g(\theta)/\{As(\theta)\}\,\mathrm{d}\theta} = \frac{g(\theta)}{\int g(\theta)\,\mathrm{d}\theta}. \qquad (8.48)$$

Therefore, this procedure yields a retained value θ whose distribution is precisely the $f(\theta) = g(\theta)/\int g(\theta)\mathrm{d}\theta$ from which we wish to sample. This is the basic rejection method.

Example 8.10

Let $f(\theta)$ be the truncated normal distribution given by $g(\theta) = \exp(-\theta^2/2)$ for $\theta \geqslant 0$, and $g(\theta) = 0$ for $\theta < 0$. Let $s(\theta)$ be the (untruncated) $N(0, 1)$ density $s(\theta) = (2\pi)^{-1/2}\exp(-\theta^2/2)$. Setting $A = (2\pi)^{1/2}$, (8.47) results in θ always being rejected if it is negative, and never otherwise. In general, applying the rejection method to a truncated distribution leads to the obvious algorithm; generate θ from the untruncated distribution and retain it if and only if it lies in the truncated range.

8.60 The probability of θ being accepted on any draw of (θ, y) in the basic rejection method is $P(y \leqslant g(\theta)/\{As(\theta)\})$, which is the denominator of (8.48), i.e. $A^{-1}\int g(\theta)\mathrm{d}\theta$. (Repeated use of the rejection method to draw from this distribution, therefore, would yield an estimate of the normalizing constant but a better estimator is given in Exercise 8.8.) The number of pairs (θ, y) generated before one is accepted is therefore geometrically distributed with mean $A/\int g(\theta)\mathrm{d}\theta$. This is greater than or equal to one, with equality if and only if the sampled distribution $s(\theta)$ equals the target distribution $f(\theta)$. Setting A no greater than the supremum of $g(\theta)/s(\theta)$ will obviously reduce rejection and increase the efficiency of the method. The other way to increase efficiency is to choose $s(\theta)$ to mimic $g(\theta)$ as closely as possible, so reducing $\sup g(\theta)/s(\theta)$.

Each pair of (θ, y) values generated requires computation of $g(\theta)/s(\theta)$ to apply the test (8.47), which can make the procedure much slower, particularly if many pairs are being rejected.

8.61 Gilks and Wild (1992) propose a method which combines a simple $s(\theta)$ for generating candidate θ values with a quick test that avoids the need to apply (8.47) for every generated (θ, y). The method is applicable when $g(\theta)$ is log-concave, i.e. $g'(\theta) = \mathrm{d}\log g(\theta)/\mathrm{d}\theta$ exists and is non-increasing in θ. For any set of f values $t_1 < t_2 < \ldots < t_r$, first evaluate $g(t_1), \ldots, g(t_r)$ and $g'(t_1), \ldots, g'(t_r)$. Since $g(\theta)$ is log-concave, we can bound $\log g(\theta)$ above by joining the tangents at $t_1, \ldots, t_r$, and below by joining the chords between $\log g(t_1), \ldots, \log g(t_r)$, as in Figure 8.3. $g(\theta)$ is thereby bounded above and below by piecewise exponential segments. The normalized upper bound is $s(\theta)$, and its normalizing constant is A. Since the bound is piecewise exponential, A is easily calculated and drawing

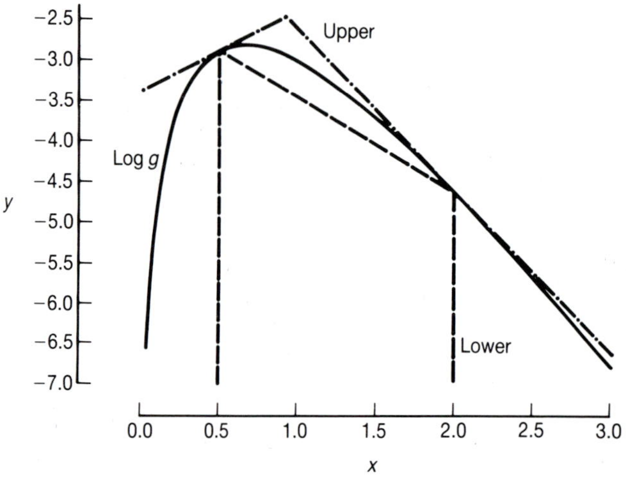

Fig. 8.3 Bounding a log-concave density

from $s(\theta)$ is also straightforward; see Gilks and Wild (1992), Wild and Gilks (1993) for detailed algorithms.

The lower bound provides a quick acceptance region without needing to compute $g(\theta)$. If the generated θ lies in the interval $[t_i, t_{i+1}]$ then θ can be accepted if

$$\log y \leqslant (t_{i+1} - t_i)^{-1}\{(\theta - t_i)\log g(t_{i+1}) + (t_{i+1} - \theta)\log g(t_i)\}$$
$$- \min\{\log g(t_{i+1}) + (\theta - t_i)h(t_i), \log g(t_{i+1}) + (\theta - t_{i+1})h(t_{i+1})\},$$

which only requires the precomputed $\log g(t_i)$ values. Only if this quick test fails does the proper test (8.47) need to be applied.

The method can be further improved by gradually refining the upper and lower bounds. Every time we need to apply (8.47) we compute a new $g(\theta)$ value, which can be added to the precomputed $g(t_i)$ values. Thus the lower bound becomes closer and closer to $g(\theta)$. If we also compute $g'(\theta)$ then we can also refine the upper bound so that it too becomes closer and closer to $g(\theta)$. This adaptive rejection method dramatically reduces the expected number of function evaluations required to obtain an accepted θ value.

8.62 The possibility of sampling from an arbitrary $g(\theta)$ by rejection sampling suggests that we could sample directly from $g(\boldsymbol{\theta})$ in this way. The argument of **8.59** applies also to this case, where $\boldsymbol{\theta}$ is sampled first from $s(\boldsymbol{\theta})$ and rejection applies as in (8.47) based on a uniformly distributed scalar y. We then obtain a sample direct from the normalized distribution $g(\boldsymbol{\theta})/\int g(\boldsymbol{\theta})d\boldsymbol{\theta}$ without needing Gibbs sampling at all. Whilst this is certainly possible it is hard to implement without obtaining enormous numbers of rejections. This requirement that $g(\boldsymbol{\theta})/s(\boldsymbol{\theta})$ be bounded is the same as that considered in **8.46** to ensure a finite variance for Monte Carlo integration, and is not difficult to achieve. A bound A can be found by numerical maximization of $g(\boldsymbol{\theta})/s(\boldsymbol{\theta})$, but it is difficult to obtain a sampling distribution $s(\boldsymbol{\theta})$ for which the acceptance rate $A^{-1}\int g(\boldsymbol{\theta})d\boldsymbol{\theta}$ is not small. Gilks and Wild adaptive rejection sampling can also be generalized to p dimensions if $g(\boldsymbol{\theta})$ is fully log-concave, but the upper bound is now made up of intersections of tangent planes, and is much more difficult to use as a sampling distribution $s(\boldsymbol{\theta})$.

Ratio-of-uniforms sampling

8.63 Let C_g be a subset of R^2 defined by

$$C_g = \{(u, v) : 0 < u \leqslant (g(v/u))^{1/2}\}, \tag{8.49}$$

and let the bivariate random variable (u, v) be uniformly distributed over C_g, i.e. $f(u, v) = k$ if $(u, v) \in C_g$, and otherwise $f(u, v) = 0$. Now define the random variable $\theta = v/u$. Then the joint density of θ and u is $f(\theta, u) = k\,u$ for $0 < u \leqslant (g(\theta))^{1/2}$ and zero otherwise. The marginal density of θ is

$$f(\theta) = \int_0^{\sqrt{g(\theta)}} f(\theta, u)\,\mathrm{d}u = k\,\left[u^2/2\right]_0^{\sqrt{g(\theta)}} = k\,g(\theta)/2. \tag{8.50}$$

Therefore $f(\theta) \propto g(\theta)$ and the normalizing constant must be $k = 2/\int g(\theta)\mathrm{d}\theta$. So if we can draw (u, v) uniformly over the region C_g, we can obtain a random draw θ from the normalized density $f(\theta)$ as a 'ratio-of-uniforms', $\theta = v/u$. This result is due to Kinderman and Monahan (1977).

8.64 One way to sample uniformly from C_g is to use rejection. We can generate (u, v) uniformly over the rectangle $0 < u \leqslant a$, $b \leqslant v \leqslant c$ very easily as independent uniform random variables over $(0, a]$ and $[b, c]$. We reject the pair (u, v) if it does not lie in C_g, continue sampling until $(u, v) \in C_g$ and then let $\theta = v/u$. This method works provided, of course, the rectangle contains C_g. While it is difficult to compute the boundary of C_g for any g, the values

$$a = \sup \sqrt{g(x)}, \qquad b = \inf x\sqrt{g(x)}, \qquad c = \sup x\sqrt{g(x)}$$

define the bounding rectangle. Any rectangle containing this minimal rectangle may be used, but will lead to a higher probability of rejection. In general, since the area of C_g is k^{-1}, the probability of acceptance is $\int g(\theta)\mathrm{d}\theta/\{2a(c-b)\}$.

Correlation and convergence

8.65 The second major question on implementation of the Gibbs sampler concerns how many iterations are required before convergence to the limiting distribution can be assumed. That is, how large must m be in order that the distribution of $\boldsymbol{\theta}^m$ given $\boldsymbol{\theta}^0$ is sufficiently close to the limiting distribution $f(\boldsymbol{\theta})$ for us to take it as a random draw from $f(\boldsymbol{\theta})$? Although the underlying theory is difficult, it is easy to see that the rate of convergence will depend on the degree of correlation between the θ_is.

If the θ_is are independent, then $f_i(\theta_i \mid \theta_{(i)})$ does not depend on $\theta_{(i)}$, and is in fact the marginal distribution $f_i(\theta_i)$ of θ_i. Then convergence is immediate. $\boldsymbol{\theta}^1, \boldsymbol{\theta}^2, \boldsymbol{\theta}^3, \ldots$ are independent draws from $f(\boldsymbol{\theta})$. Independence of the θ_is is therefore associated not only with immediate convergence but also with independence between successive iterations $\boldsymbol{\theta}^m$. Any correlation between the θ_is will tend both to slow convergence and to create correlation between $\boldsymbol{\theta}^m$s. The possible behaviour is illustrated by the following rather simplified examples.

Example 8.11

Let θ_1 and θ_2 take only the values 0 and 1. The joint distribution $f(\boldsymbol{\theta})$ is given by $P(\theta_1 = \theta_2 = 1) = P(\theta_1 = \theta_2 = 0) = p/2$ and $P(\theta_1 = 1, \theta_2 = 0) = P(\theta_1 = 0, \theta_2 = 1) = (1-p)/2$. The conditional distribution of θ_1 given θ_2 gives probability p to $\theta_1 = \theta_2$ and probability $1-p$ to $\theta_1 = 1-\theta_2$. The conditional distribution of θ_2 given θ_1 is the same. If the Gibbs sampler is applied with these conditional distributions, consider the probability distribution of θ_2 at the mth iteration.

$$\begin{aligned} P(\theta_2^m = 1 \mid \theta_2^{m-1} = 1) &= pP(\theta_1^m = 1 \mid \theta_2^{m-1} = 1) + (1-p)P(\theta_1^m = 0 \mid \theta_2^{m-1} = 1) \\ &= p^2 + (1-p)^2. \end{aligned} \tag{8.51}$$

Conversely $P(\theta_2^m = 1 \mid \theta_2^{m-1} = 0) = 2p(1-p)$. Now if p is close to 0 or 1, there is a high probability that $\boldsymbol{\theta}^m = \boldsymbol{\theta}^{m-1}$, and this is seen to be caused by the correspondingly high correlation between θ_1 and θ_2. Let $p_m = P(\theta_2^m = 1)$, and (8.51) implies the difference equation $p_m = ap_{m-1} + b$, where $a = p^2 + (1-p)^2 - 2p(1-p) = (2p-1)^2$ and $b = 2p(1-p)$. The general solution of this equation is

$$p_m = a^{m+1}p_0 + b(1-a)^{-1}(1-a^{m+1}). \tag{8.52}$$

As $m \to \infty$, $p_m \to b(1-a)^{-1} = 1/2$, which is the correct limiting probability $P(\theta_2 = 1)$. But if p is close to 0 or 1 then a is close to 1 and convergence to this limit is slow. If, for instance, $p = 0.999$ then $a = 0.996\,004$. After 100 iterations $p_{100} = 0.67p_0 + 0.165$, which is 0.165 or 0.835 depending on whether the starting point has $\theta_2^0 = 0$ or $\theta_2^0 = 1$. Even after 1000 iterations we find $p_{1000} = 0.491$ or 0.509 depending on the starting point.

A further practical problem is that the series may appear to have converged long before it has done. If p is close to 0 or 1, the Gibbs sampler will produce long sequences where $\boldsymbol{\theta}^m$ does not change at all, which may give the impression of convergence. Thus, if $p = 0.999$ and we start at $\theta_1^0 = \theta_2^0 = 0$, the number of iterations until $\boldsymbol{\theta}^m$ is *not* equal to $(0, 0)$ is geometrically distributed with mean $(1-p^2)^{-1} = 500$.

Example 8.12

Let C_p and D_p be p-dimensional cubes, $C_p = [0, c]^p$ and $D_p = [d, d+1]^p$. Let $g(\boldsymbol{\theta}) = a$ if $\boldsymbol{\theta} \in C_p \cap D_p^C$, $g(\boldsymbol{\theta}) = b$ if $\boldsymbol{\theta} \in D_p$, and $g(\boldsymbol{\theta}) = 0$ if $\boldsymbol{\theta} \notin C_p \cup D_p$. If $c < d$ we have two disjoint regions as in Example 8.9 (which corresponds to $p = 2$, $c = 1$, $d = 2$), but if $c > d$ the Gibbs sampler produces an irreducible and aperiodic chain and convergence is guaranteed. The conditional distributions are all of the following form. If $\boldsymbol{\theta}_{(i)} \in C_{p-1} \cap D_{p-1}^C$ then θ_i is uniformly distributed over $[0, c]$. If $\boldsymbol{\theta}_{(i)} \in C_{p-1}^C \cap D_{p-1}$ then θ_i is uniformly distributed over $[d, d+1]$. If $\boldsymbol{\theta}_{(i)} \in C_{p-1} \cap D_{p-1}$ then $f(\theta_i \mid \boldsymbol{\theta}_{(i)}) = a/(ad+b)$ for $0 \leqslant \theta_i < d$ and $f(\theta_i \mid \boldsymbol{\theta}_{(i)}) = b/(ad+b)$ for $d \leqslant \theta_i \leqslant d+1$. Now suppose that we start with $\boldsymbol{\theta}^0 \in C_p \cap D_p^C$. The Gibbs sampler will continue to generate all future θ_i uniformly in $[0, c]$ until $p-1$ successive values all fall in $[d, c]$. Then we have $\boldsymbol{\theta}_{(i)} \in C_{p-1} \cap D_{p-1}$ so that there is a probability $b(d+1-c)/(ad+b)$ that θ_i will be drawn in the range $[c, d+1]$. This will be the first time that $\boldsymbol{\theta}^m$ moves outside $C_p \cap D_p^C$. The expected waiting time until $p-1$ successive points are generated in $[d, c]$ is $(1-d/c)^{p-1}$, which will be small if p is large enough.

This example is like Example 8.11 in that the Gibbs sampler remains in one region for a long time, and will appear to converge immediately to a uniform distribution over that region. But eventually it will jump into the other region, and convergence to the true limiting distribution is in fact very slow. The Gibbs sampler can spend very many iterations stuck in a region of very low probability, as is the case when p is large but c is less than 1. Then the true joint probability that $\boldsymbol{\theta} \in C_p \cap D_p^C$ is very small, but if $\boldsymbol{\theta}^0$ is in this region it may be very many iterations before $\boldsymbol{\theta}^m$ can jump out. Even if $p-1$ successive values in $[d, c]$ gives the opportunity to jump out, if a is large the probability of that opportunity being taken is also small. Although this may seem a very contrived and unrealistic example, O'Hagan (1987b) shows that standard Bayesian analysis can produce a joint posterior distribution with essentially this kind of behaviour.

8.66 Gelfand and Smith (1990) point out that in general the Gibbs sampler converges geometrically. The rate of geometric convergence is defined in terms of the transition distribution $f^\star$, but is shown nicely in the simple case of Example 8.11. There, equation (8.52) shows dependence on the starting point declining with m in a geometric series. However, the examples have shown that in practice convergence may be difficult to diagnose. The Gibbs sampler can spend many consecutive iterations in a certain region, appearing to have converged, but may in fact slowly or suddenly break out into a wider area. Very long sequences are then needed if we are to gain an accurate appreciation of the relative probabilities of these regions.

8.67 One way to improve the performance of the Gibbs sampler is to reduce correlation. Reparametrization to reduce correlations is therefore potentially very valuable. Hills and Smith (1992) advocate using the iterative scaling of **8.42**. From a short initial sequence of points, the covariance matrix of $\boldsymbol{\theta}$ is estimated and a linear rescaling transforms $\boldsymbol{\theta}$ to parameters $\boldsymbol{\phi}$ with zero estimated correlations. The Gibbs sampler is then recast in terms of $\boldsymbol{\phi}$, another sequence of points generated and the process repeated. Once an appropriate parametrization is created by a small number of such rescaling steps, correlation between iterations of the Gibbs sampler should be small in the final parametrization, with the hope of more rapid convergence. One difficulty with this approach may be finding and sampling from the conditional distributions $f(\phi_i \mid \boldsymbol{\phi}_{(i)})$ under the new parametrization.

8.68 Another way to reduce correlation is to generalize (8.42) so that instead of dividing $\boldsymbol{\theta}$ into its scalar elements θ_i and having a step in each iteration for each θ_i, we allow θ_i to be a subvector $\boldsymbol{\theta}_i$ of $\boldsymbol{\theta}$. Then $f(\boldsymbol{\theta}_i \mid \boldsymbol{\theta}_{(i)})$ is a multivariate distribution. If the high correlations in the Gibbs sampler are caused primarily by correlations between elements within each subvector $\boldsymbol{\theta}_i$, and correlations between $\boldsymbol{\theta}_i$ and $\boldsymbol{\theta}_j$ $(j \neq i)$ are lower, then we can expect much more rapid convergence. The price to pay for this gain is the added complexity of sampling from the multivariate distributions $f(\boldsymbol{\theta}_i \mid \boldsymbol{\theta}_{(i)})$.

Parallel runs

8.69 Another aspect of correlation is the problem of obtaining a sample of n *independent* observations drawn from $f(\boldsymbol{\theta})$. There are three basic approaches. One is to use several parallel runs of the Gibbs sampler. Even if they were to start from the same $\boldsymbol{\theta}^0$, independent application of the Gibbs sampler produces different values of $\boldsymbol{\theta}^1$ in each run. Obviously, if there is high correlation between the θ_is, the $\boldsymbol{\theta}^1$ value in any run is highly dependent on $\boldsymbol{\theta}^0$, so that n runs do not produce n independent $\boldsymbol{\theta}^1$ values. However, dependence between the runs becomes negligible after a large number of iterations, and can be reduced by using different $\boldsymbol{\theta}^0$ starting points for each run. By the time n parallel runs have converged, we can treat the n resulting $\boldsymbol{\theta}^m$ values as a sample of n independent observations from $f(\boldsymbol{\theta})$.

An alternative to this approach is to use one run, but once it has converged to use observations t steps apart. That is, the sample is $\boldsymbol{\theta}^m, \boldsymbol{\theta}^{m+t}, \ldots, \boldsymbol{\theta}^{m+(n-1)t}$. If t is large enough, the correlations in this sample will be small. If correlation is high and convergence is slow then both t and m will need to be large, but Hills and Smith (1992) suggest that we will typically be able to have $t < m$, so that the single run is more efficient than multiple runs.

8.70 A final alternative is to use a single run and accept that correlation exists between the iterations. Consider estimating $E\{h(\boldsymbol{\theta})\}$ as in **8.55**. If we use a single run and select iterations t apart as above, we obtain the estimate $n^{-1}\sum_{i=1}^{n} h(\boldsymbol{\theta}^{m+(i-1)t}) = E_0$, say. We can also consider $t-1$ other estimators $E_s = n^{-1}\sum_{i=1}^{n} h(\boldsymbol{\theta}^{m+s+(i-1)t})$, $s = 1, 2. \ldots, t-1$. If the separation t is enough to produce essentially independent samples, then each E_s has variance $n^{-1}\text{var}\{h(\boldsymbol{\theta})\}$. But if we use the whole series of observations we have the estimator $E = (nt)^{-1}\sum_{i=1}^{nt} h(\boldsymbol{\theta}^{m+i-1}) = t^{-1}\sum_{s=0}^{t-1} E_s$. Then

$$\text{var}(E) = t^{-2}\sum_{s=0}^{t-1}\sum_{u=0}^{t-1}\text{cov}(E_s, E_u).$$

Since $\text{cov}(E_s, E_u) \leqslant \{\text{var}(E_s)\text{var}(E_u)\}^{1/2} = n^{-1}\text{var}\{h(\boldsymbol{\theta})\}$, we have $\text{var}(E) \leqslant n^{-1}\text{var}\{h(\boldsymbol{\theta})\}$. This proves that if we simply average nt consecutive observations after convergence, the result will be at least as good as taking n observations t apart.

Convergence diagnostics

8.71 The decision whether to use a single run or multiple parallel runs is related also to the ability to diagnose convergence. We have seen that in a single run the Gibbs sampler can spend long periods in a relatively small region, which can mislead one into believing that it has converged. With many parallel runs it is unlikely that all runs will be showing this behaviour together, and a safer view of convergence is possible.

Gelman and Rubin (1992) make a similar argument for using multiple runs. They suggest that the $\boldsymbol{\theta}^0$ starting values for the different series should be more widely dispersed, so that they make up an initial sample with larger variances than the true variances $\text{var}(\theta_i)$ implied by $f(\boldsymbol{\theta})$. Then we should not only observe convergence of each run but also convergence between runs, in the sense that the variance matrix of the $\boldsymbol{\theta}^m$ values in the different series should reduce and converge with m. They propose a formal convergence diagnostic based on this behaviour.

Alternative convergence diagnostics for a single run are described by Geweke (1992a), Raftery and Lewis (1992) and Roberts (1992). Limited experience has been gained in the application of formal convergence tests. It is common practice to adopt a pragmatic and informal approach combining several ideas. Thus, n runs may be used, with widely spaced starting points. The values of $\boldsymbol{\theta}^m, \boldsymbol{\theta}^{m-1}, \ldots, \boldsymbol{\theta}^{m-t}$ from all n runs produce a sample of nt points. Moments and marginal distributions are examined and plotted against m, either every observation, every tth observation or at even wider intervals (depending upon the apparent correlation between iterations). These may be assessed for convergence by eye or using formal tests.

Metropolis–Hastings algorithms

8.72 We have described the Gibbs sampler in detail, since it is the most widely used of the class of MCMC methods. Another useful and general MCMC method was described by Hastings (1970), generalizing the algorithm of Metropolis et al. (1953). A transition distribution $f^\star(\boldsymbol{\theta}^m \mid \boldsymbol{\theta}^{m-1})$ is constructed directly such that the resulting Markov chain has limiting distribution $f(\boldsymbol{\theta}) = g(\boldsymbol{\theta}) / \int g(\boldsymbol{\theta}) \mathrm{d}\boldsymbol{\theta}$. Let $q(\boldsymbol{\theta}^m \mid \boldsymbol{\theta}^{m-1})$ be any transition distribution, and let

$$p(\boldsymbol{\theta}^m, \boldsymbol{\theta}^{m-1}) = \min\left\{ \frac{g(\boldsymbol{\theta}^m) q(\boldsymbol{\theta}^{m-1} \mid \boldsymbol{\theta}^m)}{g(\boldsymbol{\theta}^{m-1}) q(\boldsymbol{\theta}^m \mid \boldsymbol{\theta}^{m-1})}, 1 \right\}. \tag{8.53}$$

(If $g(\boldsymbol{\theta}^{m-1}) q(\boldsymbol{\theta}^m \mid \boldsymbol{\theta}^{m-1}) = 0$ define $p(\boldsymbol{\theta}^m, \boldsymbol{\theta}^{m-1}) = 1$.)

The Metropolis–Hastings method first draws $\boldsymbol{\theta}^m$ from the transition distribution $q(\boldsymbol{\theta}^m \mid \boldsymbol{\theta}^{m-1})$, then with probability $p(\boldsymbol{\theta}^m, \boldsymbol{\theta}^{m-1})$ retains this value of $\boldsymbol{\theta}^m$, and otherwise simply sets $\boldsymbol{\theta}^m = \boldsymbol{\theta}^{m-1}$. Therefore the resulting transition distribution is defined by

$$f^\star(\boldsymbol{\theta}^m \mid \boldsymbol{\theta}^{m-1}) = q(\boldsymbol{\theta}^m \mid \boldsymbol{\theta}^{m-1}) p(\boldsymbol{\theta}^m, \boldsymbol{\theta}^{m-1}) \tag{8.54}$$

if $\boldsymbol{\theta}^m \neq \boldsymbol{\theta}^{m-1}$, together with a probability

$$P^\star(\boldsymbol{\theta}^m = \boldsymbol{\theta}^{m-1} \mid \boldsymbol{\theta}^{m-1}) = 1 - \int q(\boldsymbol{\theta} \mid \boldsymbol{\theta}^{m-1}) p(\boldsymbol{\theta}, \boldsymbol{\theta}^{m-1}) \, \mathrm{d}\boldsymbol{\theta} \tag{8.55}$$

of the chain staying in the same place. It is easy to verify from (8.53) and (8.55) that for $\boldsymbol{\theta}^m \neq \boldsymbol{\theta}^{m-1}$ we have $f(\boldsymbol{\theta}^{m-1}) f^\star(\boldsymbol{\theta}^m \mid \boldsymbol{\theta}^{m-1}) = f(\boldsymbol{\theta}^m) f^\star(\boldsymbol{\theta}^{m-1} \mid \boldsymbol{\theta}^m)$, and this is shown in Roberts and Smith (1992) to imply that the limiting distribution of the Markov chain is $f(\boldsymbol{\theta})$, subject to the original transition distribution q being irreducible and aperiodic.

8.73 Different choices of q lead to different versions of the general Metropolis–Hastings algorithm. A common choice is for $q(\boldsymbol{\theta}^m \mid \boldsymbol{\theta}^{m-1})$ to be a normal distribution with mean $\boldsymbol{\theta}^{m-1}$, or a heavier-tailed Student t form as in **8.46**. Because these distributions are symmetric, $q(\boldsymbol{\theta}^m \mid \boldsymbol{\theta}^{m-1}) = q(\boldsymbol{\theta}^{m-1} \mid \boldsymbol{\theta}^m)$, and (8.53) simplifies to $p(\boldsymbol{\theta}^m, \boldsymbol{\theta}^{m-1}) = \min\{g(\boldsymbol{\theta}^m)/g(\boldsymbol{\theta}^{m-1}), 1\}$. The transition from $\boldsymbol{\theta}^{m-1}$ to $\boldsymbol{\theta}^m$ is now particularly easy to interpret. First we set $\boldsymbol{\theta}^m = \boldsymbol{\theta}^{m-1} + \boldsymbol{\delta}$, where $\boldsymbol{\delta}$ is generated from a fixed, zero-mean, symmetric distribution. If the new $\boldsymbol{\theta}^m$ has higher value $g(\boldsymbol{\theta}^m)$ than $g(\boldsymbol{\theta}^{m-1})$ we definitely retain this $\boldsymbol{\theta}^m$, otherwise we retain it with probability $g(\boldsymbol{\theta}^m)/g(\boldsymbol{\theta}^{m-1})$. It is easy to see in general terms that this implies the chain will spend more time in regions of high $g(\boldsymbol{\theta})$.

8.74 Another variant sets $g(\boldsymbol{\theta}^m \mid \boldsymbol{\theta}^{m-1}) = s(\boldsymbol{\theta}^m)$, a fixed distribution independent of $\boldsymbol{\theta}^{m-1}$. Then (8.53) becomes $p(\boldsymbol{\theta}^m, \boldsymbol{\theta}^{m-1}) = \min\{r(\boldsymbol{\theta}^m)/r(\boldsymbol{\theta}^{m-1}), 1\}$, where $r(\boldsymbol{\theta}) = g(\boldsymbol{\theta})/s(\boldsymbol{\theta})$. This is related to the rejection method of sampling directly from $f(\boldsymbol{\theta})$, discussed in **8.62**. There the acceptance probability is proportional to $r(\boldsymbol{\theta})$, and with this Metropolis–Hastings method we can see, again in general terms, that the chain will more often go to points with high $r(\boldsymbol{\theta})$. With the rejection method a single $\boldsymbol{\theta}$ is generated, which is the first $\boldsymbol{\theta}$ which is not rejected, and as noted in **8.62** this can lead to very large numbers of rejections. The Metropolis–Hastings method generates a Markov chain $\boldsymbol{\theta}^0, \boldsymbol{\theta}^1, \boldsymbol{\theta}^2, \ldots$ whose limiting distribution is $f(\boldsymbol{\theta})$. Since the acceptance probability $p(\boldsymbol{\theta}^m, \boldsymbol{\theta}^{m-1})$ is generally relatively high, if convergence is rapid the Metropolis–Hastings approach is much more efficient.

Data augmentation

8.75 Of the many devices that have have been developed for applying MCMC methods to complex problems (see references in **8.76** below), the data augmentation method originally proposed by Tanner and Wong (1987) deserves mention here. The method is simply to notice that it may be possible to sample by MCMC methods much more simply and efficiently from a distribution $f(\boldsymbol{\theta}, \boldsymbol{\phi} \mid \mathbf{x})$ than from $f(\boldsymbol{\theta} \mid \mathbf{x})$. Here the augmentation parameter $\boldsymbol{\phi}$ can be anything, and applications of this approach rely on a certain amount of ingenuity in seeing a $\boldsymbol{\phi}$ that can be used. Clearly, if we can sample from $f(\boldsymbol{\theta}, \boldsymbol{\phi} \mid \mathbf{x})$, the required $f(\boldsymbol{\theta} \mid \mathbf{x})$ is simply a marginal distribution of the augmented distribution, and a sample from $f(\boldsymbol{\theta} \mid \mathbf{x})$ consists just of ignoring the $\boldsymbol{\phi}$ components of the $(\boldsymbol{\theta}, \boldsymbol{\phi})$ sample.

There are two quite general classes of problem in which data augmentation can be valuable.

(i) Missing data problems. In many planned experiments, the posterior distribution is easy to summarize (or at least to simulate from) provided all the planned data are available. In practice, some observations are often missing, resulting in an unbalanced experiment and a more complex posterior distribution. Letting $\mathbf{x}$ be the data actually received, and $\boldsymbol{\phi}$ be the missing data, $f(\boldsymbol{\theta} \mid \boldsymbol{\phi}, \mathbf{x})$ is the posterior distribution from the complete data, which we have supposed to be easy to sample from (directly or by MCMC). Then $f(\boldsymbol{\phi} \mid \boldsymbol{\theta}, \mathbf{x})$ is the sampling distribution for the missing data, and will also typically be very easy to sample from. Gibbs sampling is therefore very straightforward. At each iteration, a draw from $f(\boldsymbol{\theta} \mid \boldsymbol{\phi}, \mathbf{x})$ is made from the posterior distribution based on full data, using the current sampled values of the missing data $\boldsymbol{\phi}$, and then a draw from $f(\boldsymbol{\phi} \mid \boldsymbol{\theta}, \mathbf{x})$ samples new values for the missing data, based on the current value of $\boldsymbol{\theta}$. This was the application from which the name 'data augmentation' derives. The second step of sampling missing data forms the basis of the data *imputation* method of Rubin (1987).

(ii) Hierarchical models. In a hierarchical model, suppose that data $\mathbf{x}$ are related to parameters $\boldsymbol{\theta}$ through a likelihood $f(\mathbf{x} \mid \boldsymbol{\theta})$, the prior distribution is modelled in the first stage as $f(\theta \mid \boldsymbol{\phi})$ conditional on hyperparameters $\boldsymbol{\phi}$, and completed by $f(\boldsymbol{\phi})$. In many applications, the posterior distribution $f(\boldsymbol{\theta} \mid \mathbf{x})$ for $\boldsymbol{\theta}$ alone is complex, while $f(\boldsymbol{\theta}, \boldsymbol{\phi} \mid \mathbf{x})$ is much simpler.

The following is an example of a less obvious application.

Example 8.13

Data $x_1, x_2, \ldots, x_n$ are independently and identically distributed as $N(\theta, v)$ with unknown mean θ but known variance v. The prior distribution has the form

$$f(\theta) \propto \{1 + (\theta - m)^2\}^{-(c+1)/2}, \qquad (8.56)$$

so that $(\theta - m)\sqrt{c}$ has the standard Student t-distribution with c degrees of freedom. The combination of this with the normal likelihood produces an intractable posterior distribution. However, if we introduce a parameter ϕ such that the conditional prior distribution of θ given ϕ is $N(m, \phi^{-1})$ and then we assign to ϕ a χ^2_c prior distribution, it is easy to show that the marginal prior for θ in this formulation is (8.56). The conditional posterior distribution for θ given ϕ is now normal, and in general $f(\theta, \phi \mid \mathbf{x})$ can be sampled from easily and efficiently by Gibbs sampling. This device is employed in considerably more complex models by Carlin and Polson (1991) and Geweke (1992b). It can be seen as an application of data augmentatiion of a hierarchical model, but it is necessary first to see how (8.56) could be expressed in hierarchical form.

8.76 In the dimensionality reduction situation described in **8.4**, we wish to obtain appropriate summaries of $g(\theta) \propto f(\theta)$ where $\theta = (\phi, \psi)$, but the conditional distribution $f(\phi \mid \psi)$ is analytically simple and well-understood. Then

$$g_1(\psi) = \int g(\phi, \psi) \mathrm{d}\phi \propto f(\psi)$$

is obtainable analytically, and the suggestion is to use numerical methods to summarize $g_1(\psi)$ rather than $g(\theta)$, which should be quicker or more accurate because of the reduced dimentionality. Summaries of $f(\theta)$ are then deduced using known properties of $f(\phi \mid \psi)$. In this case, one might sample from $g_1(\psi)$ by Gibbs sampling or some other MCMC technique. Having obtained such a sample $\psi_1, \psi_2, \ldots, \psi_n$ one could apply (8.4) to estimate $E\{h(\theta)\}$ by

$$n^{-1} \sum_{i=1}^{n} h^\star(\psi_j),$$

where $h^\star(\psi) = E\{h(\theta) \mid \psi\}$ is derived analytically from $f(\phi \mid \psi)$. The approximation (8.46) relies on a similar idea.

If $h^\star(\psi)$ is not available in this way, dimensionality reduction can still be useful if $f(\phi \mid \psi)$ can be sampled from easily. First generate $\psi_1, \psi_2, \ldots, \psi_n$ as before, then generate $\phi_1, \phi_2, \ldots, \phi_n$ by sampling from $f(\phi \mid \psi)$. The advantage is that the process of convergence takes place in the reduced dimension problem of sampling from $f(\psi) \propto g_1(\psi)$, which will be quicker and can be more reliably diagnosed. Only after convergence in this reduced problem do we sample the ϕ_js.

In a sense this is the opposite of data augmentation. Indeed, there will be practical applications in which despite the availability of $g_1(\psi)$ analytically it is more efficient to augment and bring ϕ back. That is, despite the reduced dimensionality of $g_1(\psi)$, it is simpler and more efficient to sample from the original $g(\theta)$.

8.77 Smith and Roberts (1993) provide a review of techniques to apply MCMC methods in Bayesian computation. Many applications to medical problems are described by Gilks et al. (1993). Besag and Green (1993) present a substantial bibliography of MCMC techniques in image analysis and spatial statistics, and describe variations in the basic MCMC methods to achieve faster or more reliable convergence. These three articles, and the fifty pages of published discussion following them, provide excellent coverage of a rapidly expanding literature.

Overview

8.78 A variety of computational techniques are available for obtaining summaries of distributions. We conclude this chapter with some brief remarks about the strengths and weaknesses of the methods. The first remark is that approximation methods, quadrature, Monte Carlo integration and Markov Chain Monte Carlo are all primarily aimed at computing integration-based summaries. Chapter 2 makes it clear that many other kinds of summary are useful and worthy of attention. These typically rely on differentiation rather than integration, and in this chapter we have really only considered optimization to find modes, ignoring other differentiation-based summaries.

There are two main reasons for this emphasis on integration-based summaries. One is that they are analytically and computationally more difficult. We can usually differentiate analytically almost any $g(\boldsymbol{\theta})$, regardless of its complexity, whereas we can only integrate analytically in special cases where the integrals are known. The operations we need to carry out with the derivatives, like equating to zero to find a mode, may not be soluble analytically but are often simple to compute. The development of increasingly advanced quadrature, Monte Carlo and MCMC methods has been driven by the fact that numerical integration is a very heavy computational task. The other reason for emphasizing integration-based summaries is that moments, marginal distributions and probabilities of regions (as in highest density regions) have generally been regarded as some of the most important summaries.

8.79 However, these integration-based summaries are not adequate on their own. We have stressed in **2.8** that an understanding of shape is necessary before moments, or other measures of location, dispersion, skewness etc, can be properly interpreted. Furthermore, there is another reason for considering shape before applying many of the computational methods of this chapter, which is that an understanding of shape is important in determining the most efficient ways to implement the techniques.

Bimodality, skewness, heavy tails or high correlations are all indicators of where more sophisticated methods are needed. In any of these cases, simple normal approximations will be poor, quadrature rules will need careful application (and probably many more points), Monte Carlo integration with a symmetric unimodal $s(\boldsymbol{\theta})$ will have a large variance, or Gibbs sampling may converge slowly, become 'stuck' in one mode then flip to another or behave erratically in other ways. They indicate when and how reparametrization could be used to improve these methods.

8.80 Approximations are quick, and therefore preferable when great accuracy is not needed. Kass et al. (1988) show how they enable summarization to be done in an interactive, exploratory way. Having identified by such means the summaries which appear to be most interesting and informative, they can be calculated by more accurate, but slower, methods.

8.81 For integration-based summaries the choice of method seems to depend primarily on dimensionality. If $\boldsymbol{\theta}$ has only a few elements, or computation can be reduced analytically to few dimensions, quadrature methods are efficient and accurate. In higher dimensions, Monte Carlo integration is preferable. Wherever $\boldsymbol{\theta}$ has above 15 or 20 dimensions, or when the shape or complexity of $g(\boldsymbol{\theta})$ makes other methods ineffective, Gibbs sampling or other MCMC methods can often be applied successfully. They have the disadvantage of always leaving some uncertainty over whether the series has effectively converged, and hence uncertainty about how accurate the estimates are. Their great strengths are flexibility and efficiency in high-dimensional problems.

EXERCISES

8.1 Let x be the number of successes in n independent trials with probability θ of success in each trial, and let y similarly be the number of successes in m trials but with probability $c\theta$ of success, $c < 1$. x and y are independent given θ, and θ has the beta prior $f(\theta) \propto \theta^{a-1}(1-\theta)^{b-1}$.

(a) Prove that the posterior mode is the unique solution in $[0, 1]$ of the quadratic equation

$$c(n+m+a+b-2)\theta^2 - (n+cm+cx+y+a+b+ca-2-c)\theta + (x+y+a-1) = 0.$$

(b) Prove that the posterior mode can be expressed as

$$E(\theta \mid x, y) = \frac{\sum_{k=1}^{m-y}(-c)^k B(1+x+y+k, n-x)}{\sum_{k=1}^{m-y}(-c)^k B(x+y+k, n-x)}.$$

(c) In the case $a = b = 1$ (uniform prior distribution), $n = m = 5$, $x = 4$, $y = 3$, $c = 0.5$, calculate the mean and mode and plot the density. Calculate the approximation (8.13) to the mean, as described in Example 8.3. Also approximate the mean using the quadrature rule (8.29) using 5 and 20 subintervals.

8.2 Consider the t density $f(\theta) \propto g(\theta) = (1+\theta^2)^{-c}$. Prove that the approximation (8.7) underestimates $\int g(\theta)\,d\theta = B(c-1/2, 1/2)$ but that the approximation becomes exact as $c \to \infty$. What effect would introducing third derivatives as in (8.9) have?

8.3 A variant of Lindley–Smith iteration is obtained by replacing the second equation of (8.25) by $\phi_{n+1} = \hat{\phi}(\psi_n)$. Prove that this procedure will take twice as many iterations to converge.

8.4 Prove that the density function of each $\boldsymbol{\theta}_i$ produced by the procedure of **8.46** is

$$s(\boldsymbol{\theta}) = 2^{-1}\pi^{-(p+1)/2}|\mathbf{V}|^{-1/2}\Gamma\{(p+1)/2\}\{1+(\boldsymbol{\theta}-\mathbf{m})'\mathbf{V}^{-1}(\boldsymbol{\theta}-\mathbf{m})\}^{-(p+1)/2}$$

8.5 Let $g(\theta)$ be a normal density function and consider estimating $I = \int g(\theta)\,d\theta$ by Monte Carlo integration. The density $s(\theta)$ used for generating the Monte Carlo sample is also a normal density. Prove that the variance of the Monte Carlo estimator is infinite if the variance of $s(\theta)$ is less than or equal to half the variance of $g(\theta)$.

8.6 Let $g(\boldsymbol{\theta})$ be the p-dimensional $N(\mathbf{m}, \mathbf{V})$ density function. $I = \int g(\boldsymbol{\theta})\,d\boldsymbol{\theta}$ is to be estimated by Monte Carlo using the $N(\mathbf{m}, k\mathbf{V})$ distribution to generate the sample. Prove that the variance (8.39) of the Monte Carlo estimator is

$$n^{-1}\left\{\left(\frac{k^2}{2k-1}\right)^{p/2} - 1\right\}$$

which tends to infinity as $p \to \infty$ for any $k \neq 1$. (It is infinite if $k \leqslant 1/2$, as in Exercise 8.5).

8.7 Let θ_1 and θ_2 have a bivariate normal distribution with zero means, unit variances and correlation r. Consider sampling from this distribution using the Gibbs sampler (8.42). Given the starting value θ_2^0 for θ_2, prove that the distribution of the value θ_2^n of θ_2 after the nth iteration is $N(r^{2n}\theta_2^0, 1-r^{4n})$, and so convergence can depend critically upon r. Show also that Lindley–Smith iteration will converge very slowly if r is close to ± 1.

8.8 Let $\boldsymbol{\theta}^1, \boldsymbol{\theta}^2, \ldots, \boldsymbol{\theta}^n$ be a sample from the density $f(\boldsymbol{\theta}) \propto g(\boldsymbol{\theta})$ (obtained for instance by some Markov Chain Monte Carlo method). We wish to estimate the normalizing constant k, where $k^{-1} = \int g(\boldsymbol{\theta})\,d\boldsymbol{\theta}$. Let $s(\boldsymbol{\theta})$ be any proper density function such that $s(\boldsymbol{\theta}) = 0$ whenever $g(\boldsymbol{\theta}) = 0$. Prove that

$$\hat{k} = n^{-1}\sum_{i=1}^{n} s(\boldsymbol{\theta}^i)/g(\boldsymbol{\theta}^i)$$

is a classical unbiased estimator of k. By considering the variance of this estimator, show that $s(\boldsymbol{\theta})$ should be chosen to mimic $g(\boldsymbol{\theta})$, but with thinner tails.

8.9 Consider a continuous d.f. $F(x)$ with inverse function F^{-1}, i.e. if $F(x) = p$ then $F^{-1}(p) = x$. Let y be uniformly distributed over [0,1]. Prove that the random variable $x = F^{-1}(y)$ has the distribution with d.f. $F(x)$. Thereby show how to generate random draws from a Cauchy distribution given the availability of independent uniform random numbers.

CHAPTER 9

THE LINEAR MODEL

9.1 The Linear Model and its classical analysis are dealt with very fully in Volume 2. This chapter presents some corresponding Bayesian theory. Emphasis is on analysis of the normal linear model under various formulations of prior distributions.

The normal linear model

9.2 As in A**28.1**, we write the linear model in the form

$$\mathbf{y} = \mathbf{X}\boldsymbol{\beta} + \boldsymbol{\epsilon}, \tag{9.1}$$

where $\mathbf{y}$ is an $n \times 1$ vector of observations, $\mathbf{X}$ is an $n \times p$ matrix of known coefficients, $\boldsymbol{\beta}$ is a $p \times 1$ vector of parameters and $\boldsymbol{\epsilon}$ an $n \times 1$ vector of random errors. The elements of $\boldsymbol{\epsilon}$ are assumed to have zero mean, to be uncorrelated and to have common variance σ^2, which is an additional parameter.

If we further assume that the elements of $\boldsymbol{\epsilon}$ are jointly normally distributed then the model is described as the *normal linear model*. The model says simply that the conditional distribution of $\mathbf{y}$ given parameters $(\boldsymbol{\beta}, \sigma^2)$ is the multivariate normal distribution $N(\mathbf{X}\boldsymbol{\beta}, \sigma^2\mathbf{I})$, where as usual $\mathbf{I}$ denotes the $(n \times n)$ identity matrix. Therefore the likelihood becomes

$$f(\mathbf{y} \mid \boldsymbol{\beta}, \sigma^2) = (2\pi\sigma^2)^{-n/2} \exp\{-(\mathbf{y} - \mathbf{X}\boldsymbol{\beta})'(\mathbf{y} - \mathbf{X}\boldsymbol{\beta})/(2\sigma^2)\}. \tag{9.2}$$

9.3 We can write the quadratic form $(\mathbf{y} - \mathbf{X}\boldsymbol{\beta})'(\mathbf{y} - \mathbf{X}\boldsymbol{\beta})$ in the exponent of (9.2) in various ways. Just expanding,

$$(\mathbf{y} - \mathbf{X}\boldsymbol{\beta})'(\mathbf{y} - \mathbf{X}\boldsymbol{\beta}) = \boldsymbol{\beta}'\mathbf{X}'\mathbf{X}\boldsymbol{\beta} - \boldsymbol{\beta}'\mathbf{X}'\mathbf{y} - \mathbf{y}'\mathbf{X}\boldsymbol{\beta} + \mathbf{y}'\mathbf{y} \tag{9.3}$$

If the $p \times p$ matrix $\mathbf{X}'\mathbf{X}$ is non-singular we can go on to complete the square to give

$$(\mathbf{y} - \mathbf{X}\boldsymbol{\beta})'(\mathbf{y} - \mathbf{X}\boldsymbol{\beta}) = (\boldsymbol{\beta} - \hat{\boldsymbol{\beta}})'\mathbf{X}'\mathbf{X}(\boldsymbol{\beta} - \hat{\boldsymbol{\beta}}) + S \tag{9.4}$$

where $\hat{\boldsymbol{\beta}} = (\mathbf{X}'\mathbf{X})^{-1}\mathbf{X}'\mathbf{y}$ is the classical Maximum Likelihood or Least Squares estimator of $\boldsymbol{\beta}$, and $S = (\mathbf{y} - \mathbf{X}\hat{\boldsymbol{\beta}})'(\mathbf{y} - \mathbf{X}\hat{\boldsymbol{\beta}})$ is the residual sum of squares.

The normal-inverse-gamma distribution

9.4 From (9.2) and (9.4) the natural conjugate family of prior distributions has the form

$$f(\boldsymbol{\beta}, \sigma^2) \propto (\sigma^2)^{-(d+p+2)/2} \exp[-\{(\boldsymbol{\beta} - \mathbf{m})'\mathbf{V}^{-1}(\boldsymbol{\beta} - \mathbf{m}) + a\}/(2\sigma^2)], \tag{9.5}$$

with hyperparameters a, d, $\mathbf{m}$ and $\mathbf{V}$. Possible values for the hyperparameters include all those values for which (9.5) is a proper distribution. The reason for using the power $(d + p + 2)/2$ for σ^2 will soon become apparent.

9.5 We can first find the marginal prior distribution of σ^2 by integrating (9.5) with respect to $\boldsymbol{\beta}$.

$$\begin{aligned} f(\sigma^2) &= \int f(\boldsymbol{\beta}, \sigma^2)\,\mathrm{d}\boldsymbol{\beta} \\ &\propto (\sigma^2)^{-(d+p+2)/2} \exp\{-a/(2\sigma^2)\} |\sigma^2 \mathbf{V}|^{1/2} \\ &\propto (\sigma^2)^{-(d+2)/2} \exp\{-a/(2\sigma^2)\}. \end{aligned} \tag{9.6}$$

In doing this integration, the symmetric $p \times p$ matrix $\mathbf{V}^{-1}$ must be positive definite. Otherwise the integral diverges and (9.5) is not a proper distribution. A proper distribution must therefore have $\mathbf{V}$ positive definite.

The marginal distribution (9.6) is an *inverse-gamma* distribution, which is seen to be an appropriate name if we transform from σ^2 to $\phi = \sigma^{-2}$. Then (9.6) becomes

$$f(\phi) \propto \phi^{(d-2)/2} \exp(-a\phi/2). \tag{9.7}$$

This is a gamma distribution with hyperparameters $a/2$ and $d/2$. Alternatively, $a\phi$ has the chi-square distribution with d degrees of freedom. It is proper if both hyperparameters are positive. Therefore for (9.5) to be proper we require $a > 0$, $d > 0$ as well as positive definite $\mathbf{V}$.

We will therefore say that (9.6) is the inverse gamma distribution with hyperparameters a and d, and denote this by $IG(a, d)$. Its normalizing constant can be deduced from that of the gamma distribution (9.7), i.e.

$$f(\sigma^2) = \frac{(a/2)^{d/2}}{\Gamma(d/2)} (\sigma^2)^{-(d+2)/2} \exp\{-a/(2\sigma^2)\}. \tag{9.8}$$

9.6 Summaries of this distribution may be found directly or via summaries of ϕ. It is always unimodal with mode at $a/(d+2)$. Its mean is $E(\sigma^2) = a/(d-2)$ provided $d > 2$. If $2 \geqslant d > 0$ then the distribution is proper but the mean does not exist (and is effectively infinite). The mean is greater than the mode, reflecting the fact that the distribution is positively skewed. If $d > 4$, its variance is $\mathrm{var}\,(\sigma^2) = 2a^2/\{(d-2)^2(d-4)\}$.

9.7 In integrating out $\boldsymbol{\beta}$ in **9.5** we implicitly used the fact that the conditional distribution of $\boldsymbol{\beta}$ given σ^2 is $N(\mathbf{m}, \sigma^2\mathbf{V})$. In particular

$$E(\boldsymbol{\beta} \mid \sigma^2) = \mathbf{m}, \qquad \mathrm{var}\,(\boldsymbol{\beta} \mid \sigma^2) = \sigma^2 \mathbf{V}.$$

Therefore

$$E(\boldsymbol{\beta}) = E(E(\boldsymbol{\beta} \mid \sigma^2)) = \mathbf{m}, \tag{9.9}$$

$$\mathrm{var}\,(\boldsymbol{\beta}) = E(\mathrm{var}\,(\boldsymbol{\beta} \mid \sigma^2)) + \mathrm{var}\,(E(\boldsymbol{\beta} \mid \sigma^2)) = E(\sigma^2)\mathbf{V} = \{a/(d-2)\}\mathbf{V} \tag{9.10}$$

provided $d > 2$. Otherwise the variance of $\boldsymbol{\beta}$ is infinite.

9.8 From the two integrations in **9.5** we derive the normalizing constant of (9.5). The joint distribution becomes

$$f(\boldsymbol{\beta},\sigma^2)=\frac{(a/2)^{d/2}}{(2\pi)^{p/2}|\mathbf{V}|^{1/2}\Gamma(d/2)}(\sigma^2)^{-(d+p+2)/2}\exp[-\{(\boldsymbol{\beta}-\mathbf{m})'\mathbf{V}^{-1}(\boldsymbol{\beta}-\mathbf{m})+a\}/(2\sigma^2)]. \quad (9.11)$$

We will call this the *normal-inverse-gamma* distribution with hyperparameters a, d, $\mathbf{m}$ and $\mathbf{V}$, and denote it by $NIG(a,d,\mathbf{m},\mathbf{V})$.

9.9 To find the marginal distribution of $\boldsymbol{\beta}$ we integrate (9.11) with respect to σ^2. Notice that the conditional distribution of σ^2 given $\boldsymbol{\beta}$ is $IG((\boldsymbol{\beta}-\mathbf{m})'\mathbf{V}^{-1}(\boldsymbol{\beta}-\mathbf{m})+a, d+p)$. Therefore integrating using (9.8) we have

$$\begin{aligned}f(\boldsymbol{\beta})&=\frac{a^{d/2}\Gamma((d+p)/2)}{|\mathbf{V}|^{1/2}\pi^{p/2}\Gamma(d/2)}\{a+(\boldsymbol{\beta}-\mathbf{m})'\mathbf{V}^{-1}(\boldsymbol{\beta}-\mathbf{m})\}^{-(d+p)/2} \quad (9.12)\\ &\propto\{1+(\boldsymbol{\beta}-\mathbf{m})'(a\mathbf{V})^{-1}(\boldsymbol{\beta}-\mathbf{m})\}^{-(d+p)/2}.\end{aligned}$$

This is a generalization of the Student t distribution in **A16.10**. We will call it the (multivariate) t distribution with degrees of freedom d and hyperparameters $\mathbf{m}$ and $a\mathbf{V}$, and denote it by $t_d(\mathbf{m},a\mathbf{V})$.

The distribution is symmetric around $\mathbf{m}$, with mean and variance given by (9.9) and (9.10).

Conjugate analysis

9.10 Now suppose that the $NIG(a,d,\mathbf{m},\mathbf{V})$ distribution (9.5) or (9.11) is adopted as the prior distribution for $(\boldsymbol{\beta},\sigma^2)$. Combining with the likelihood (9.2) gives the posterior distribution

$$f(\boldsymbol{\beta},\sigma^2\,|\,\mathbf{y})\propto(\sigma^2)^{-(d+n+p+2)/2}\exp\{-Q/(2\sigma^2)\}, \quad (9.13)$$

where

$$\begin{aligned}Q&=(\mathbf{y}-\mathbf{X}\boldsymbol{\beta})'(\mathbf{y}-\mathbf{X}\boldsymbol{\beta})+(\boldsymbol{\beta}-\mathbf{m})'\mathbf{V}^{-1}(\boldsymbol{\beta}-\mathbf{m})+a \quad (9.14)\\ &=\boldsymbol{\beta}'(\mathbf{V}^{-1}+\mathbf{X}'\mathbf{X})\boldsymbol{\beta}-\boldsymbol{\beta}'(\mathbf{V}^{-1}\mathbf{m}+\mathbf{X}'\mathbf{y})-(\mathbf{m}'\mathbf{V}^{-1}+\mathbf{y}'\mathbf{X})\boldsymbol{\beta}+(\mathbf{m}'\mathbf{V}^{-1}\mathbf{m}+\mathbf{y}'\mathbf{y}+a)\\ &=(\boldsymbol{\beta}-\mathbf{m}^\star)'(\mathbf{V}^\star)^{-1}(\boldsymbol{\beta}-\mathbf{m}^\star)+\mathbf{a}^\star, \quad (9.15)\end{aligned}$$

and where

$$\mathbf{V}^\star=(\mathbf{V}^{-1}+\mathbf{X}'\mathbf{X})^{-1}, \quad (9.16)$$

$$\mathbf{m}^\star=(\mathbf{V}^{-1}+\mathbf{X}'\mathbf{X})^{-1}(\mathbf{V}^{-1}\mathbf{m}+\mathbf{X}'\mathbf{y}), \quad (9.17)$$

$$a^\star=a+\mathbf{m}'\mathbf{V}^{-1}\mathbf{m}+\mathbf{y}'\mathbf{y}-(\mathbf{m}^\star)'(\mathbf{V}^\star)^{-1}\mathbf{m}^\star. \quad (9.18)$$

Therefore letting $d^\star=d+n$ the posterior distribution of $(\boldsymbol{\beta},\sigma^2)$ is $NIG(a^\star,d^\star,\mathbf{m}^\star,\mathbf{V}^\star)$. Summaries of this distribution are therefore immediately given by the results of **9.5** to **9.9**, simply changing a to $a^\star$, d to $d^\star$, $\mathbf{m}$ to $\mathbf{m}^\star$ and $\mathbf{V}$ to $\mathbf{V}^\star$. We now consider these in a little more detail.

9.11 $E(\boldsymbol{\beta} \mid \mathbf{y}) = \mathbf{m}^\star$ is a posterior estimate of $\boldsymbol{\beta}$. In fact, since the posterior distribution of $\boldsymbol{\beta}$ is symmetric, this is also the posterior mode. If $\mathbf{X}'\mathbf{X}$ is non-singular, we can write

$$\mathbf{m}^\star = (\mathbf{V}^{-1} + \mathbf{X}'\mathbf{X})^{-1}(\mathbf{V}^{-1}\mathbf{m} + \mathbf{X}'\mathbf{X}\hat{\boldsymbol{\beta}}) = (\mathbf{I} - \mathbf{A})\mathbf{m} + \mathbf{A}\hat{\boldsymbol{\beta}}, \tag{9.19}$$

where $\mathbf{A} = (\mathbf{V}^{-1} + \mathbf{X}'\mathbf{X})^{-1}\mathbf{X}'\mathbf{X}$. (9.19) expresses the posterior estimate $\mathbf{m}^\star$ as a matrix-weighted average of the prior mean $\mathbf{m}$ and the classical estimate $\hat{\boldsymbol{\beta}}$, with weights $\mathbf{I} - \mathbf{A}$ and $\mathbf{A}$. If prior information is strong, the elements of $\mathbf{V}$ will be small, reflecting small prior variances for the elements of $\boldsymbol{\beta}$. Then $\mathbf{V}^{-1}$ will be large and $\mathbf{A}$ small, so that the posterior mean gives most weight to the prior mean. Conversely, if prior information is weak or the data substantial then most weight will be given to $\hat{\boldsymbol{\beta}}$.

If $\mathbf{X}'\mathbf{X}$ is singular there is no unique solution $\hat{\boldsymbol{\beta}}$ to the classical Least Squares equations (see A**19.13**). The posterior distribution is nevertheless proper with mean (9.19) as long as the prior distribution is proper ($\mathbf{V}$ is positive definite).

9.12 Posterior uncertainty about $\boldsymbol{\beta}$ is described in part by $\mathbf{V}^\star$. Thus $\text{var}(\boldsymbol{\beta} \mid \sigma^2, \mathbf{y}) = \sigma^2\mathbf{V}^\star$ and $\text{var}(\boldsymbol{\beta} \mid \mathbf{y}) = E(\sigma^2 \mid \mathbf{y})\mathbf{V}^\star$. $\mathbf{V}^\star$ also represents a combination of prior information and data. Now $(\mathbf{V}^\star)^{-1} = \mathbf{V}^{-1} + \mathbf{X}'\mathbf{X}$ is in some sense greater than $\mathbf{V}^{-1}$, so $\mathbf{V}^\star$ is 'smaller than' $\mathbf{V}$. The extra information from the data has reduced uncertainty about $\boldsymbol{\beta}$. We can make this vague argument precise as follows. Let $\phi = \mathbf{a}'\boldsymbol{\beta}$ be any linear combination of the elements of $\boldsymbol{\beta}$, $\mathbf{a} \neq \mathbf{0}$. Then

$$\begin{aligned}\sigma^{-2}\text{var}(\phi \mid \sigma^2, \mathbf{y}) &= \mathbf{a}'\mathbf{V}^\star\mathbf{a} = \mathbf{a}'(\mathbf{V}^{-1} + \mathbf{X}'\mathbf{X})^{-1}\mathbf{a} \\ &= \mathbf{a}'\{\mathbf{V} - \mathbf{V}\mathbf{X}'(\mathbf{I} + \mathbf{X}\mathbf{V}\mathbf{X}')^{-1}\mathbf{X}\mathbf{V}\}\mathbf{a} \\ &= \mathbf{a}'\mathbf{V}\mathbf{a} - (\mathbf{X}\mathbf{V}\mathbf{a})'(\mathbf{I} + \mathbf{X}\mathbf{V}\mathbf{X}')^{-1}(\mathbf{X}\mathbf{V}\mathbf{a}).\end{aligned} \tag{9.20}$$

Now $\mathbf{I} + \mathbf{X}\mathbf{V}\mathbf{X}'$ is positive definite, so $(\mathbf{I} + \mathbf{X}\mathbf{V}\mathbf{X}')^{-1}$ is also positive definite. Therefore $\text{var}(\phi \mid \sigma^2, \mathbf{y}) \leqslant \sigma^2\mathbf{a}'\mathbf{V}\mathbf{a} = \text{var}(\phi \mid \sigma^2)$. $\text{var}(\phi \mid \sigma^2, \mathbf{y})$ will be strictly less than $\text{var}(\phi \mid \sigma^2)$ for all σ^2, showing a real reduction in uncertainty about ϕ, if $\mathbf{X}\mathbf{V}\mathbf{a} \neq \mathbf{0}$. If $\mathbf{X}$ has full column rank, $\mathbf{X}\mathbf{V}\mathbf{a} = \mathbf{0}$ implies $\mathbf{a} = \mathbf{0}$, in which case the data $\mathbf{y}$ give a reduced variance for every linear function of $\boldsymbol{\beta}$. In particular the variance (given σ^2) of each element of $\boldsymbol{\beta}$ is reduced.

However, if the rank of $\mathbf{X}$ does not equal its number of columns, p, there will be some $\mathbf{a} \neq \mathbf{0}$ for which $\text{var}(\phi \mid \sigma^2, \mathbf{y}) = \text{var}(\phi \mid \sigma^2)$. The data provide no direct information about these functions of $\boldsymbol{\beta}$. This is the case of singular $\mathbf{X}'\mathbf{X}$, when classical Least Squares or Maximum Likelihood fail to produce a unique estimator $\hat{\boldsymbol{\beta}}$. With a proper prior distribution $\boldsymbol{\beta}$ has a proper posterior distribution with a unique posterior mean, but the absence of information in the data about certain aspects of $\boldsymbol{\beta}$ shows through in linear combinations $\mathbf{a}'\boldsymbol{\beta}$ for which $\mathbf{a}'\mathbf{V}^\star\mathbf{a} = \mathbf{a}'\mathbf{V}\mathbf{a}$. These are nonidentifiable in the sense of **3.15**.

Example 9.1

The simplest case of a linear model is when $p = 1$ and $\mathbf{X} = \mathbf{1}$, an $n \times 1$ vector of ones. $\boldsymbol{\beta}$ is a scalar, which we will denote by μ, and (9.1) reduces to $y_i = \mu + \epsilon_i$. Therefore the y_is are independent and identically distributed as $N(\mu, \sigma^2)$. Then $\mathbf{X}'\mathbf{X} = \mathbf{1}'\mathbf{1} = n$ and

$\hat{\boldsymbol{\beta}} = \hat{\mu} = n^{-1}\mathbf{1}'\mathbf{y} = n^{-1}\sum y_i = \bar{y}$ is the classical estimator of μ. In the prior distribution, $\mathbf{m}$ and $\mathbf{V}$ reduce to the scalars m and v. The posterior mean is

$$m^{\star} = (v^{-1} + n)^{-1}(v^{-1}m + n\bar{y}) = (1 - a)m + a\bar{y},$$

where $a = (v^{-1} + n)^{-1}n$ is the weight given to the data estimate $\bar{y}$. This weight is large if n is large (strong data) or v is large (weak prior). The posterior variance is $v^{\star}E(\sigma^2 \mid \mathbf{y})$ where $v^{\star} = (v^{-1} + n)^{-1}$.

Example 9.2

The simple regression model $y_i = \alpha + \beta x_i + \epsilon_i$ is a linear model with $p = 2$, $\boldsymbol{\beta} = (\alpha, \beta)'$ and $\mathbf{X} = (\mathbf{1}, \mathbf{x})$. Then

$$\mathbf{X}'\mathbf{X} = \begin{pmatrix} n & \sum x_i \\ \sum x_i & \sum x_i^2 \end{pmatrix}, \qquad \mathbf{X}'\mathbf{y} = \begin{pmatrix} \sum y_i \\ \sum x_i y_i \end{pmatrix}.$$

and $\hat{\boldsymbol{\beta}}$ is the usual least squares estimator for the simple regression model. The way in which this is modified by the prior information to give the posterior mean $\mathbf{m}^{\star}$ is no longer as simple as in Example 9.1. If the weight matrix $\mathbf{A}$ in (9.19) is diagonal, then the posterior mean of α will be a weighted average of its prior mean and the classical $\hat{\alpha}$, and the posterior mean of β will similarly be a weighted average of its prior mean and $\hat{\beta}$. But this will not generally be the case, and the posterior mean of either α or β will depend on both components of the prior mean $\mathbf{m}$ and both components of the classical estimate $\hat{\boldsymbol{\beta}}$.

9.13 Information about $\boldsymbol{\beta}$ is also obtained indirectly through $E(\sigma^2 \mid \mathbf{y})$ in the formula $\text{var}(\boldsymbol{\beta} \mid \mathbf{y}) = E(\sigma^2 \mid \mathbf{y})\mathbf{V}^{\star}$. If the data $\mathbf{y}$ suggest that σ^2 is smaller than its prior estimate so that $E(\sigma^2 \mid \mathbf{y}) < E(\sigma^2)$, then $\text{var}(\mathbf{a}'\boldsymbol{\beta} \mid \mathbf{y}) = E(\sigma^2 \mid \mathbf{y})\mathbf{a}'\mathbf{V}^{\star}\mathbf{a} \leqslant E(\sigma^2 \mid \mathbf{y})\mathbf{a}'\mathbf{V}\mathbf{a} < E(\sigma^2)\mathbf{a}'\mathbf{V}\mathbf{a} = \text{var}(\mathbf{a}'\boldsymbol{\beta})$. Then the posterior variance of every linear function of $\boldsymbol{\beta}$ is less than its prior variance. If, on the other hand, $E(\sigma^2 \mid \mathbf{y}) > E(\sigma^2)$ then posterior uncertainty about some elements or functions of $\boldsymbol{\beta}$ may increase despite the reduction in $\mathbf{V}^{\star}$. Since $E(E(\sigma^2 \mid \mathbf{y})) = E(\sigma^2)$, the data are not expected *a priori* either to increase or decrease the expectation of σ^2, and so $\mathbf{V}^{\star}$ is the main determinant of posterior uncertainty about $\boldsymbol{\beta}$.

If, therefore, we wish to design an experiment with the primary purpose of estimating $\phi = \mathbf{a}'\boldsymbol{\beta}$, we would do so by choosing $\mathbf{X}$ to maximize the term $(\mathbf{XVa})'(\mathbf{I} + \mathbf{XVX}')^{-1}(\mathbf{XVa})$ in (9.20).

Optimal design

9.14 Various other criteria for design of experiments have been proposed. In practice we may have the loose objective of 'obtaining best possible information about $\boldsymbol{\beta}$', and it may be difficult to be so specific as minimizing the posterior variance of a single linear function $\phi = \mathbf{a}'\boldsymbol{\beta}$. As an overall measure of quality of information about $\boldsymbol{\beta}$, the determinant of its posterior variance matrix, or 'generalized variance', is one possibility. We may therefore consider minimizing $|\mathbf{V}^{\star}|$, or equivalently maximizing $|\mathbf{V}^{\star}|^{-1} = |\mathbf{V}^{-1} + \mathbf{X}'\mathbf{X}|$. This approach has the drawback that $|\mathbf{V}^{\star}|$ can be made arbitrarily small, or even zero, by obtaining good information about any element or linear combination of $\boldsymbol{\beta}$, while posterior variances of other elements or combinations might still be large.

An alternative generalized criterion is to choose $\mathbf{X}$ to minimize the trace of $|\mathbf{V}^{\star}|$, which is the sum of the variances of the elements of $\boldsymbol{\beta}$. This criterion can only go to zero if *all* the variances go to zero, corresponding to perfect information about the whole of $\boldsymbol{\beta}$. On the other hand, this criterion is not invariant under linear reparametrization. If we let $\boldsymbol{\phi} = \mathbf{B}\boldsymbol{\beta}$, then the posterior variance matrix of $\boldsymbol{\phi}$ is $\mathbf{BV}^{\star}\mathbf{B}'$. If $\mathbf{B}$ is non-singular, $|\mathbf{BV}^{\star}\mathbf{B}'| = |\mathbf{B}|^2|\mathbf{V}^{\star}|$ and the criterion of minimizing $|\mathbf{V}^{\star}|$ is invariant in the sense that it does not matter whether we parametrize by $\boldsymbol{\beta}$ or $\boldsymbol{\phi}$. But minimizing $tr(\mathbf{BV}^{\star}\mathbf{B}') = tr(\mathbf{V}^{\star}\mathbf{L})$, where $\mathbf{L} = \mathbf{B}'\mathbf{B}$, is not the same as minimizing $tr\mathbf{V}^{\star}$. Implicitly, the trace criterion sets $\mathbf{L} = \mathbf{I}$ or may be generalized by using any other specific $\mathbf{L}$. (9.15) is the case $\mathbf{L} = \mathbf{aa}'$. Designs based on the trace and determinant criteria provide Bayesian alternatives to classical A-optimal and D-optimal designs, respectively. For references see **3.47**.

9.15 We now examine the posterior mean $E(\sigma^2 \mid \mathbf{y})$ of σ^2. Note that if $\mathbf{X}'\mathbf{X}$ is non-singular we can rewrite (9.18) after a little algebra as

$$a^{\star} = a + (n-p)\hat{\sigma}^2 + (\mathbf{m} - \hat{\boldsymbol{\beta}})'\{\mathbf{V} + (\mathbf{X}'\mathbf{X})^{-1}\}^{-1}(\mathbf{m} - \hat{\boldsymbol{\beta}}), \tag{9.21}$$

where $\hat{\sigma}^2$ is the classical unbiased estimator of σ^2 as derived in A**19.9**. That is $(n-p)\hat{\sigma}^2 = \mathbf{y}'\{\mathbf{I} - \mathbf{X}(\mathbf{X}'\mathbf{X})^{-1}\mathbf{X}'\}\mathbf{y}' = (\mathbf{y} - \mathbf{X}\hat{\boldsymbol{\beta}})'(\mathbf{y} - \mathbf{X}\hat{\boldsymbol{\beta}}) = S$, the residual sum of squares in classical theory. To interpret the last term in (9.21) we note the following.

$$E(\hat{\boldsymbol{\beta}} \mid \sigma^2) = E\{E(\hat{\boldsymbol{\beta}} \mid \boldsymbol{\beta}, \sigma^2) \mid \sigma^2\} = E(\boldsymbol{\beta} \mid \sigma^2) = \mathbf{m},$$

$$\begin{aligned}\operatorname{var}(\hat{\boldsymbol{\beta}} \mid \sigma^2) &= E\{\operatorname{var}(\hat{\boldsymbol{\beta}} \mid \boldsymbol{\beta}, \sigma^2) \mid \sigma^2\} + \operatorname{var}\{E(\hat{\boldsymbol{\beta}} \mid \boldsymbol{\beta}, \sigma^2) \mid \sigma^2\} \\ &= E\{\sigma^2(\mathbf{X}'\mathbf{X})^{-1} \mid \sigma^2\} + \operatorname{var}(\boldsymbol{\beta} \mid \sigma^2) = \sigma^2\{(\mathbf{X}'\mathbf{X})^{-1} + \mathbf{V}\}.\end{aligned}$$

$$\begin{aligned}\therefore E[(\mathbf{m} - \hat{\boldsymbol{\beta}})'&\{\mathbf{V} + (\mathbf{X}'\mathbf{X})^{-1}\}^{-1}(\mathbf{m} - \hat{\boldsymbol{\beta}}) \mid \sigma^2] \\ &= E[tr\{\mathbf{V} + (\mathbf{X}'\mathbf{X})^{-1}\}^{-1}(\mathbf{m} - \hat{\boldsymbol{\beta}})(\mathbf{m} - \hat{\boldsymbol{\beta}})' \mid \sigma^2] \\ &= tr[\{\mathbf{V} + (\mathbf{X}'\mathbf{X})^{-1}\}^{-1}E\{(\mathbf{m} - \hat{\boldsymbol{\beta}})(\mathbf{m} - \hat{\boldsymbol{\beta}})' \mid \sigma^2\}] \\ &= tr[\{\mathbf{V} + (\mathbf{X}'\mathbf{X})^{-1}\}^{-1}\sigma^2\{\mathbf{V} + (\mathbf{X}'\mathbf{X})^{-1}\}] \\ &= tr(\sigma^2\mathbf{I}) = p\sigma^2.\end{aligned} \tag{9.22}$$

Therefore

$$E(\sigma^2 \mid \mathbf{y}) = \frac{d-2}{d+n-2}E(\sigma^2) + \frac{n-p}{d+n-2}\hat{\sigma}^2 + \frac{p}{d+n-2}t \tag{9.23}$$

in which $t = p^{-1}(\mathbf{m} - \hat{\boldsymbol{\beta}})'\{\mathbf{V} + (\mathbf{X}'\mathbf{X})^{-1}\}^{-1}(\mathbf{m} - \boldsymbol{\beta})$, which from (9.22) has expectation σ^2. So the posterior mean of σ^2 is a weighted average of three estimates. The first is the prior mean, the second is the standard classical estimate, and the third is the estimate derived above. The third estimate arises from comparing the prior and classical estimators, $\mathbf{m}$ and $\hat{\boldsymbol{\beta}}$, for $\boldsymbol{\beta}$. Since the prior variance of $\boldsymbol{\beta}$ given σ^2 is $\sigma^2\mathbf{V}$, a large discrepancy between the two estimates of $\boldsymbol{\beta}$ suggests that the prior estimate may have been poor, which in turn suggests that σ^2 is large.

The relative weights given to the three estimates in (9.23), i.e. $d-2$, $n-p$ and p, reflect the strengths of these information sources. The strength of the prior information of σ^2 itself is shown by $d-2$, since for a fixed $E(\sigma^2)$ increasing d reduces $\operatorname{var}(\sigma^2)$. The strength of the classical estimate $\hat{\sigma}^2$ is denoted by the classical degrees of freedom, $n-p$. Finally, the third source of information about σ^2 comes from comparisons between $\mathbf{m}$ and $\hat{\boldsymbol{\beta}}$, and the weight p given to this is the number of dimensions in which these comparisons are made. As the number of observations increases, the weight for the classical estimate increases and $E(\sigma^2 \mid \mathbf{y}) \to \hat{\sigma}^2$.

Notice that essentially the same results apply if we look at the posterior mode rather than $E(\sigma^2 \mid \mathbf{y})$. The weight given to the prior mode is $d+2$ instead of $d-2$, but this is the only change. The posterior mode also tends to $\hat{\sigma}^2$ as $n \to \infty$.

9.16 $\text{var}(\sigma^2 \mid \mathbf{y}) = (a^\star)^2/\{(d^\star-2)^2(d^\star-4)\} = E(\sigma^2 \mid \mathbf{y})^2/(d^\star-4)$. So the relative variance $\text{var}(\sigma^2 \mid \mathbf{y})/E(\sigma^2 \mid \mathbf{y})^2$, which is the square of the posterior coefficient of variation of σ^2, is simply $(d^\star-4)^{-1}$. Since $d^\star > d$, the information from the data decreases the coefficient of variation, representing a reduction in uncertainty about σ^2 relative to its mean.

Weak prior information

9.17 We can represent weak prior information about $(\boldsymbol{\beta}, \sigma^2)$ within the conjugate family by letting prior variances tend to infinity. Letting the prior variances of elements of $\boldsymbol{\beta}$ tend to infinity results in $\mathbf{V}^{-1} \to \mathbf{0}$. Setting $\mathbf{V}^{-1} = \mathbf{0}$ in (9.5) produces

$$f(\boldsymbol{\beta}, \sigma^2) \propto (\sigma^2)^{-(d+p+2)/2} \exp\{-a/(2\sigma^2)\}. \tag{9.24}$$

In this expression, $\boldsymbol{\beta}$ has an improper uniform distribution and σ^2 has the $IG(a, d+p)$ distribution. The conventional improper prior distribution $f(\sigma^2) \propto \sigma^{-2}$, which is often recommended for positive parameters, is now obtained by setting $a=0$, $d=-p$. Then

$$f(\boldsymbol{\beta}, \sigma^2) \propto \sigma^{-2}. \tag{9.25}$$

This is not the only way in which weak prior information can be formulated. In particular, if we begin by looking at σ^2, its marginal prior distribution is $IG(a, d)$, which we can equate to the $f(\sigma^2) \propto \sigma^{-2}$ form by letting $a=0$, $d=0$. Then

$$f(\boldsymbol{\beta}, \sigma^2) \propto (\sigma^2)^{-(p+2)/2} \exp\{-(\boldsymbol{\beta}-\mathbf{m})'\mathbf{V}^{-1}(\boldsymbol{\beta}-\mathbf{m})/(2\sigma^2)\}. \tag{9.26}$$

If we now let $\mathbf{V}^{-1} \to \mathbf{0}$ we obtain the alternative form

$$f(\boldsymbol{\beta}, \sigma^2) \propto \sigma^{-(p+2)}. \tag{9.27}$$

This can be shown to be the Jeffreys prior distribution for the normal Linear Model.

These two representations both set $\mathbf{V}^{-1} = \mathbf{0}$ and $a=0$, but give alternative values of $-p$ and 0 to d. It is also reasonable to argue for the uniform prior $f(\sigma^2) \propto 1$ instead of $f(\sigma^2) \propto \sigma^{-2}$ giving values $d = -(p+2)$ or $d=-2$. We shall take the view, as in **4.35**, that if the different formulations lead to essentially the same posterior inference it clearly does not matter which we use. If different weak prior distributions lead to important differences in posterior inference it is necessary to think about prior information instead of adopting any standard formula.

9.18 We now consider the effect of weak prior information on the hyperparameters of the posterior $NIG(a^\star, d^\star, \mathbf{m}^\star, \mathbf{V}^\star)$ distribution. Letting $\mathbf{V}^{-1} \to \mathbf{0}$ results in $\mathbf{V}^\star = (\mathbf{X}'\mathbf{X})^{-1}$, $\mathbf{m}^\star = \hat{\boldsymbol{\beta}}$, $a^\star = a + (n-p)\hat{\sigma}^2$. Now it is necessary that $\mathbf{X}'\mathbf{X}$ be non-singular, otherwise the posterior distribution becomes improper. When prior information about $\boldsymbol{\beta}$ is very weak, the data must provide information about all elements and linear functions of $\boldsymbol{\beta}$.

The posterior distribution of $\boldsymbol{\beta}$ given σ^2 is now $N(\hat{\boldsymbol{\beta}}, \sigma^2(\mathbf{X}'\mathbf{X})^{-1})$, which is directly analogous to classical inference in which the estimator $\hat{\boldsymbol{\beta}}$ is normally distributed with mean $\boldsymbol{\beta}$ and variance $\sigma^2(\mathbf{X}'\mathbf{X})^{-1}$. The Bayesian estimate is the same as the classical estimate $\hat{\boldsymbol{\beta}}$, and its accuracy is described in the same terms, via a normal distribution with variance $\sigma^2(\mathbf{X}'\mathbf{X})^{-1}$. However, the prior distribution (9.24) clearly provides information about σ^2, and this is reflected in the posterior distribution of σ^2.

$$E(\sigma^2 \mid \mathbf{y}) = a^\star/(d^\star - 2) = \frac{d+p-2}{d+n-2}E(\sigma^2) + \frac{n-p}{d+n-2}\hat{\sigma}^2 \tag{9.28}$$

is a weighted average now of only two estimates. The third term comparing estimates of $\boldsymbol{\beta}$ disappears because there is no strength in the prior information about $\boldsymbol{\beta}$ and hence no comparison to make. The weight on the prior mean $E(\sigma^2)$ increases to $d+p-2$ instead. Strictly, (9.24) is improper as a joint distribution for $(\boldsymbol{\beta}, \sigma^2)$, but if we regard it as a proper $IG(a, d+p)$ marginal distribution for σ^2 times an improper distribution for $\boldsymbol{\beta}$ then we have the value $E(\sigma^2) = a/(d+p-2)$ used in (9.28).

9.19 Now letting $a \to 0$ gives $a^\star = (n-p)\hat{\sigma}^2$ and $E(\sigma^2 \mid \mathbf{y}) = \{(n-p)/(d+n-2)\}\hat{\sigma}^2$, which takes different values depending on which value we use for d in representing weak prior information. Setting $d = 2-p$ would lead to agreement between $E(\sigma^2 \mid \mathbf{y})$ and the classical estimate $\hat{\sigma}^2$, but $f(\boldsymbol{\beta}, \sigma^2) \propto \sigma^{-4}$ is not generally advocated for any other reasons. Since the posterior distribution of σ^2 is skew, $E(\sigma^2 \mid \mathbf{y})$ is not the only estimate which might be considered. The mode $\{(n-p)/(d+n+2)\}\hat{\sigma}^2$ or the median, which will lie between mean and mode, might also be used. Values $d = -p$ or $-(p+2)$ will therefore produce posterior estimates which are close to the classical $\hat{\sigma}^2$. The alternatives $d = 0$ or -2 will also produce similar results if n is much larger than p, but otherwise the different prior formulations (9.25) and (9.27) will not yield essentially the same posterior inferences.

Interval estimation

9.20 The discussion so far has concentrated on point estimates such as posterior means, together with posterior variances to indicate the strength of posterior information. Another useful form of inference is an interval estimate. Bayesian interval estimation in the form of highest density regions is considered in **2.50**, and we now develop highest posterior density regions for $\boldsymbol{\beta}$ and σ^2. In general, a highest posterior density interval for θ is a region C such that the posterior density for θ is higher at all points in C than at any point outside C. C is therefore bounded by a contour of the posterior density function. If the posterior probability that $\theta \in C$ is p, then C is made the smallest region amongst all those that contain θ with the same probability p. (Alternatively, p is higher than the probability that θ lies in any other region of the same size as C.)

9.21 Turning first to interval estimation for the entire $\boldsymbol{\beta}$ vector, the posterior density for $\boldsymbol{\beta}$ is the multivariate t distribution (9.12) but substituting $a^\star$ for a, $d^\star$ for d, $\mathbf{m}^\star$ for $\mathbf{m}$ and $\mathbf{V}^\star$ for $\mathbf{V}$. Contours of this density are values of $\boldsymbol{\beta}$ such that $f(\boldsymbol{\beta} \mid \mathbf{y})$ is constant, and are therefore the ellipsoids for which $(\boldsymbol{\beta}-\mathbf{m}^{\star})'(\mathbf{V}^{\star})^{-1}(\boldsymbol{\beta}-\mathbf{m}^{\star})$ is constant. A highest posterior density region for $\boldsymbol{\beta}$ therefore takes the form

$$C=\{\boldsymbol{\beta}:(\boldsymbol{\beta}-\mathbf{m}^{\star})'(\mathbf{V}^{\star})^{-1}(\boldsymbol{\beta}-\mathbf{m}^{\star})\leqslant c\}$$

Then

$$P(\boldsymbol{\beta}\in C\,|\,\mathbf{y})=P((\boldsymbol{\beta}-\mathbf{m}^{\star})'(\mathbf{V}^{\star})^{-1}(\boldsymbol{\beta}-\mathbf{m}^{\star})\leqslant c\,|\,\mathbf{y}) \tag{9.29}$$

and we wish to evaluate this probability and find c such that it equals some appropriate value, such as 0.9 or 0.99.

Now since the posterior distribution of $\boldsymbol{\beta}$ given σ^2 is $N(\mathbf{m}^{\star},\sigma^2\mathbf{V}^{\star})$, the posterior distribution of $\phi=\sigma^{-2}(\boldsymbol{\beta}-\mathbf{m}^{\star})'(\mathbf{V}^{\star})^{-1}(\boldsymbol{\beta}-\mathbf{m}^{\star})$ given σ^2 is χ^2_p, independent of σ^2 (see A**15.11** and A**15.21**). And since the posterior distribution of σ^2 is $IG(a^{\star},d^{\star})$, the relationship between IG and χ^2 distributions developed in **9.5** shows that $a^{\star}\sigma^{-2}$ is distributed as $\chi^2_{d^{\star}}$. Therefore $(p^{-1}\phi)/((d^{\star})^{-1}a^{\star}\sigma^{-2})=\{d^{\star}/(pa^{\star})\}(\boldsymbol{\beta}-\mathbf{m}^{\star})'(\mathbf{V}^{\star})^{-1}(\boldsymbol{\beta}-\mathbf{m}^{\star})$ has the F distribution with degrees of freedom p and $d^{\star}=d+n$. We can now solve (9.29), with $P(\boldsymbol{\beta}\in C\,|\,\mathbf{y})=P(F_{p,d+n}\leqslant(d^{\star}c)/(pa^{\star}))$.

We can therefore determine a highest posterior density region C with given probability $1-\alpha$ of containing $\boldsymbol{\beta}$. Letting F be the upper $100\alpha\%$ point of the $F_{p,d+n}$ distribution,

$$C=\{\boldsymbol{\beta}:(\boldsymbol{\beta}-\mathbf{m}^{\star})'(\mathbf{V}^{\star})^{-1}(\boldsymbol{\beta}-\mathbf{m}^{\star})\leqslant pa^{\star}F/d^{\star}\}. \tag{9.30}$$

The form of this region is the interior of an ellipsoid centred at $\mathbf{m}^{\star}$ and with shape matrix proportional to $\mathbf{V}^{\star}$. Its principal axes are therefore given by the principal components of $\mathbf{V}^{\star}$ (see **2.15**).

9.22 Now consider an arbitrary linear transformation $\boldsymbol{\Phi}=\mathbf{A}\boldsymbol{\beta}$, where $\mathbf{A}$ is an $r\times p$ matrix of rank r. Then the conditional posterior distribution of $\boldsymbol{\Phi}$ given σ^2 is $N(\mathbf{A}\mathbf{m}^{\star},\sigma^2\mathbf{A}\mathbf{V}^{\star}\mathbf{A}')$. The marginal posterior distribution of $\boldsymbol{\Phi}$ will therefore be $t_{d^{\star}}(\mathbf{A}\mathbf{m}^{\star},a^{\star}\mathbf{A}\mathbf{V}^{\star}\mathbf{A}')$. We can immediately deduce the highest posterior density region by analogy with (9.30):

$$C=\{\gamma:(\gamma-\mathbf{A}\mathbf{m}^{\star})'(\mathbf{A}\mathbf{V}^{\star}\mathbf{A}')^{-1}(\gamma-\mathbf{A}\mathbf{m}^{\star})\leqslant ra^{\star}F/d^{\star}\}, \tag{9.31}$$

where now F is the upper $100\alpha\%$ point of $F_{r,d+n}$. In particular, if $m_i^{\star}$ is the ith element of $\mathbf{m}^{\star}$ and $v_i^{\star}$ is the ith diagonal element of $\mathbf{V}^{\star}$, we have the following highest density region for an individual β_i.

$$C=\{\beta_i:(\beta_i-m_i^{\star})^2/v_i^{\star}\leqslant a^{\star}F/d^{\star}\}. \tag{9.32}$$

Now F is the upper $100\alpha\%$ point of $F_{1,d+n}$, so that $F=t^2$, where t is the upper $50\alpha\%$ point of the Student t distribution t_{d+n}. So (9.32) becomes a highest posterior density *interval*

$$\begin{aligned}C&=\{\beta_i:|\beta_i-m_i^{\star}|\leqslant(a^{\star}v_i^{\star}/d^{\star})^{1/2}t\}\\&=[m_i^{\star}-(a^{\star}v_i^{\star}/d^{\star})^{1/2}t,\ m_i^{\star}+(a^{\star}v_i^{\star}/d^{\star})^{1/2}t].\end{aligned} \tag{9.33}$$

This corresponds to the fact that the posterior distribution of β_i is $t_{d+n}(m_i^{\star},a^{\star}v_i^{\star})$ and hence $(\beta_i-m_i^{\star})(a^{\star}v_i^{\star}/d^{\star})^{-1/2}$ has the standard t_{d+n} distribution.

9.23 In the case of weak prior information, we have seen that $\mathbf{m}^\star = \hat{\boldsymbol{\beta}}$, $\mathbf{V}^\star = (\mathbf{X}'\mathbf{X})^{-1}$ and $a^\star = (n-p)\hat{\sigma}^2$. Then (9.27) becomes the interval $\hat{\beta}_i \pm \{(n-p)/(d+n)\}^{1/2} a_i^{1/2} \hat{\sigma} t$, where a_i is the ith diagonal element of $(\mathbf{X}'\mathbf{X})^{-1}$. If we also choose $d = -p$ to represent weak prior information, corresponding to the prior distribution (9.20), the interval is exactly equal to the standard classical confidence interval (A28.27). The more general ellipsoidal region (9.25) is also then a classical confidence region.

A different value of d will not exactly reproduce classical confidence regions, but if n is much larger than p the difference will be negligible.

Example 9.3

Following Example 9.1, the posterior marginal distribution of the population mean μ is a t distribution, and a highest posterior density interval for μ is, from (9.33)

$$m^\star \pm (a^\star v^\star / d^\star)^{1/2} t,$$

where t is an appropriate upper percentage point of the t_{d+n} distribution. In the case of weak prior information with $v^{-1} = 0$, $a = 0$, $d = -1$ (since $p = 1$), we have the standard classical confidence interval

$$\bar{y} \pm \hat{\sigma} n^{-1/2} t,$$

where t is a percentage point of the t_{n-1} distribution and $\hat{\sigma}^2 = (n-1)^{-1} \sum (y_i - \bar{y})^2$ is the usual unbiased estimator of σ^2.

Example 9.4

With the simple regression model of Example 9.2, we could use the general result (9.30) to construct an elliptical highest posterior density region for (α, β). Another use is to construct a highest posterior density interval using (9.31) for $\gamma = \alpha + \beta x$, which is the value of the regression line at the point x. If we write

$$\mathbf{m}^\star = \begin{pmatrix} a \\ b \end{pmatrix}, \quad \mathbf{V}^\star = \begin{pmatrix} v_a & c \\ c & v_b \end{pmatrix},$$

then (9.31) reduces (by a similar argument to the derivation of (9.33)) to the interval

$$a + bx \pm (a^\star v(x)/d^\star)^{1/2} t, \tag{9.34}$$

where

$$v(x) = v_a + 2cx + v_b x^2 \tag{9.35}$$

and t is an appropriate percentage point of t_{d+n}. Plotting the interval (9.34) as a function of x will yield hyperbolic limits for the regression line, as in Figure 9.1, where the posterior mean of the regression line $y = E(\alpha + \beta x \mid \mathbf{y}) = a + bx$ is also shown passing through the middle of the intervals.

The width of the interval is minimized at $x = c/v_b$ and becomes wider on either side of this minimum because of uncertainty about the true slope β of the line.

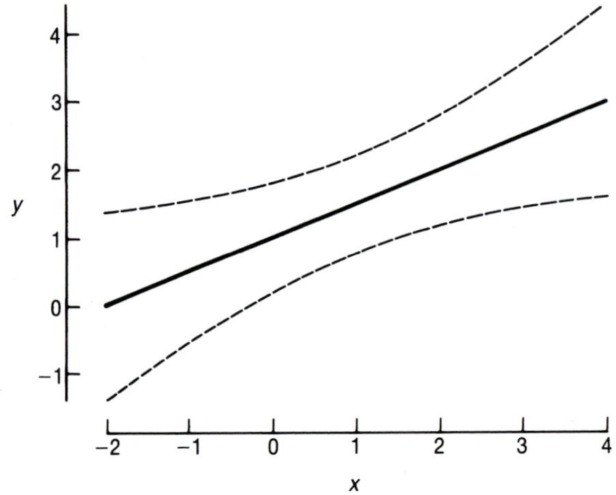

Fig. 9.1 Typical highest posterior density intervals for a simple regression line

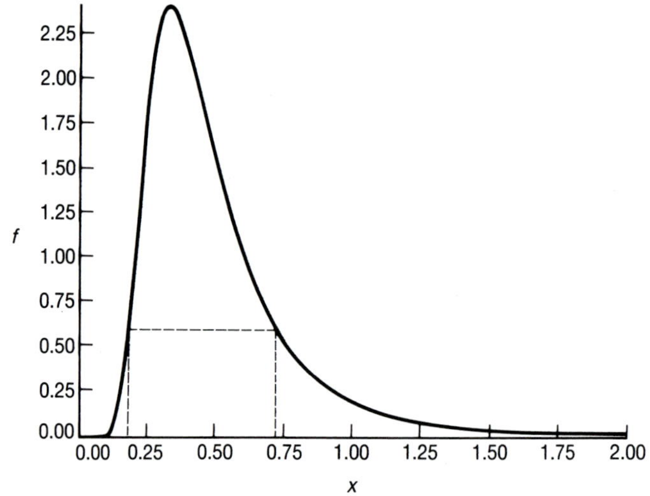

Fig. 9.2 Highest posterior density interval for $IG(4, 10)$

9.24 A highest posterior density interval for σ^2 is straightforward to construct. The posterior density is the IG density (9.6) with a changed to $a^\star$ and d changed to $d^\star$. The density is unimodal so the highest density region is an interval $[s_1, s_2]$ as in Figure 9.2 such that $f(\sigma^2 = s_1 \mid \mathbf{y}) = f(\sigma^2 = s_2 \mid \mathbf{y})$. However, the density is not symmetric and values of s_1 and s_2 to obtain an interval with given probability must be obtained numerically. Novick and Jackson (1974) give tables of these highest density regions.

In the case of weak prior information, $a = 0$, $d = -p$ gives $a^\star = (n-p)\hat{\sigma}^2$, $d^\star = n - p$. Then $(n-p)\hat{\sigma}^2\sigma^{-2}$ has a χ^2_{n-p} posterior distribution which agrees exactly with the classical distribution for $\hat{\sigma}^2$ (given σ^2). Therefore the highest posterior density interval is also a classical confidence interval. It does not, however, coincide with the usual confidence interval for this problem, because it is clear from Figure 9.1 that the probabilities in the two tails $(0, s_1)$ and (s_2, ∞) are not equal. The shortest confidence interval in Exercise A20.5 is similar to the highest posterior density interval, in being bounded by points of equal density, but with respect to the χ^2_{n-p+4} density.

Conditional distributions

9.25 We have derived the marginal distributions of $\boldsymbol{\beta}$ and σ^2, and in **9.21** the marginal distribution of an arbitrary linear transform $\gamma = \mathbf{A}\boldsymbol{\beta}$. We have also found conditional distributions for $\boldsymbol{\beta}$ given σ^2 and for σ^2 given $\boldsymbol{\beta}$, but now consider conditional distributions given partial specification of $\boldsymbol{\beta}$. First let $\boldsymbol{\beta}' = (\boldsymbol{\beta}_1', \boldsymbol{\beta}_2')$, and consider distributions conditional on $\boldsymbol{\beta}_2$. Suppose that $(\boldsymbol{\beta}, \sigma^2)$ has the $NIG(a, d, \mathbf{m}, \mathbf{V})$ distribution. Corresponding posterior distributions result if we change a to $a^\star$, d to $d^\star$, $\mathbf{m}$ to $\mathbf{m}^\star$ and $\mathbf{V}$ to $\mathbf{V}^\star$. If $\boldsymbol{\beta}_2$ has r elements write

$$\mathbf{m} = \begin{pmatrix} \mathbf{m}_1 \\ \mathbf{m}_2 \end{pmatrix}, \qquad \mathbf{V} = \begin{pmatrix} \mathbf{V}_{11} & \mathbf{V}_{12} \\ \mathbf{V}_{21} & \mathbf{V}_{22} \end{pmatrix},$$

where $\mathbf{m}_2$ is $r \times 1$ and $\mathbf{V}_{22}$ is $r \times r$. Now since $\boldsymbol{\beta}$ given σ^2 is distributed as $N(\mathbf{m}, \sigma^2\mathbf{V})$ we have the following distributions (using general results on multivariate normal distributions, as in A**15.4** and Exercise A15.1)

$$(\boldsymbol{\beta}_2 \mid \sigma^2): \quad N(\mathbf{m}_2, \sigma^2\mathbf{V}_{22}), \tag{9.36}$$

$$(\boldsymbol{\beta}_1 \mid \boldsymbol{\beta}_2, \sigma^2): \quad N(\mathbf{m}_{1.2}, \sigma^2\mathbf{V}_{11.2}), \tag{9.37}$$

where $\mathbf{m}_{1.2} = \mathbf{m}_1 + \mathbf{V}_{12}\mathbf{V}_{22}^{-1}(\boldsymbol{\beta}_2 - \mathbf{m}_2)$ and $\mathbf{V}_{11.2} = \mathbf{V}_{11} - \mathbf{V}_{12}\mathbf{V}_{22}^{-1}\mathbf{V}_{21}$.

From (9.36) and the $IG(a, d)$ marginal distribution of σ^2 we have the distribution

$$(\boldsymbol{\beta}_2, \sigma^2): \quad NIG(a, d, \mathbf{m}_2, \mathbf{V}_{22})$$

and hence

$$\boldsymbol{\beta}_2: \quad t_d(\mathbf{m}_2, \mathbf{V}_{22}),$$

$$(\sigma^2 \mid \boldsymbol{\beta}_2): \quad IG(a_2, d + r), \tag{9.38}$$

where

$$a_2 = a + (\boldsymbol{\beta}_2 - \mathbf{m}_2)'\mathbf{V}_{22}^{-1}(\boldsymbol{\beta}_2 - \mathbf{m}_2). \tag{9.39}$$

Now (9.37) and (9.38) together give

$$(\boldsymbol{\beta}_1, \sigma^2 \mid \boldsymbol{\beta}_2): \quad NIG(a_2, d + r, \mathbf{m}_{1.2}, \mathbf{V}_{11.2})$$

and finally

$$(\boldsymbol{\beta}_1 \mid \boldsymbol{\beta}_2): \quad t_{d+r}(\mathbf{m}_{1.2}, a_2\mathbf{V}_{11.2}) \tag{9.40}$$

The general linear hypothesis

9.26 Consider the hypothesis that $\mathbf{A}\boldsymbol{\beta} = \mathbf{c}$, where $\mathbf{A}$ is an $r \times p$ matrix of rank r and $\mathbf{c}$ is any $r \times 1$ vector of constants. Classical testing of this hypothesis is considered in

A23.25 to A23.29. We consider it from a Bayesian point of view in two ways. First suppose that the conjugate prior distribution $NIG(a, d, \mathbf{m}, \mathbf{V})$ is appropriate. This gives zero prior probability to the hypothesis, and so this prior distribution does not formally treat the hypothesis as having a positive probability of being true, and it is not sensible to ask for its posterior probability. Instead we ask more informally whether the posterior distribution suggests that $\mathbf{c}$ is a relatively probable or improbable value for $\boldsymbol{\gamma} = \mathbf{A}\boldsymbol{\beta}$. The posterior distribution of $\boldsymbol{\gamma}$ was found in **9.22** to be $t_{d^*}(\mathbf{Am}^*, a^*\mathbf{AV}^*\mathbf{A}')$. The ratio of the posterior density at $\boldsymbol{\gamma} = \mathbf{c}$ to its value at the mode $\boldsymbol{\gamma} = \mathbf{Am}^*$ is

$$\{1 + (\mathbf{c} - \mathbf{Am}^*)'(a^*\mathbf{AV}^*\mathbf{A}')^{-1}(\mathbf{c} - \mathbf{Am}^*)\}^{-(d^*+r)/2}$$

and if this is small we should regard $\mathbf{c}$ as a relatively implausible value for $\mathbf{A}\boldsymbol{\beta}$.

To quantify this further, note that the set of all values of $\boldsymbol{\gamma}$ having posterior density higher than at $\boldsymbol{\gamma} = \mathbf{c}$ is a highest posterior density region. Using the argument of **9.22**, the posterior probability that $\boldsymbol{\gamma}$ lies outside this set is

$$\begin{aligned} P\{f(\boldsymbol{\gamma} \mid \mathbf{y}) &\leqslant f(\boldsymbol{\gamma} = \mathbf{c} \mid \mathbf{y}) \mid \mathbf{y}\} \\ &= P\{(\boldsymbol{\gamma} - \mathbf{Am}^*)'(\mathbf{AV}^*\mathbf{A}')^{-1}(\boldsymbol{\gamma} - \mathbf{Am}^*) \geqslant (\mathbf{c} - \mathbf{Am}^*)'(\mathbf{AV}^*\mathbf{A}')^{-1}(\mathbf{c} - \mathbf{Am}^*) \mid \mathbf{y}\} \\ &= P\{F_{r,d+n} \geqslant d^*(\mathbf{c} - \mathbf{Am}^*)'(\mathbf{AV}^*\mathbf{A}')^{-1}(\mathbf{c} - \mathbf{Am}^*)/(ra^*)\}, \end{aligned} \quad (9.41)$$

which can be calculated from tables of the F distribution with r and $d + n$ degrees of freedom. If this is small we can regard the hypothesis $\mathbf{A}\boldsymbol{\beta} = \mathbf{c}$ as implausible in the sense that $\mathbf{A}\boldsymbol{\beta}$ has a small posterior probability of being so far from its posterior mean $\mathbf{Am}^*$. We can think of this as an informal Bayesian test of the general linear hypothesis. It is similar to the classical idea of rejecting a hypothesis if the hypothesized value does not lie in a confidence interval with sufficiently large confidence coefficient.

9.27 In the case of weak prior information we let $\mathbf{V}^{-1} = \mathbf{0}$ and $a = 0$, obtaining $\mathbf{V}^* = (\mathbf{X}'\mathbf{X})^{-1}$, $\mathbf{m}^* = \hat{\boldsymbol{\beta}}$ and $a^* = (n - p)\hat{\sigma}^2$ as discussed in **9.17**. Then (9.41) becomes

$$P\{f(\boldsymbol{\gamma} \mid \mathbf{y}) \leqslant f(\boldsymbol{\gamma} = \mathbf{c} \mid \mathbf{y}) \mid \mathbf{y}\} = P\{F_{r,d+n} \geqslant (d + n)F/(n + p)\}. \quad (9.42)$$

where

$$F = (\mathbf{c} - \mathbf{A}\hat{\boldsymbol{\beta}})'(\mathbf{A}(\mathbf{X}'\mathbf{X})^{-1}\mathbf{A}')^{-1}(\mathbf{c} - \mathbf{A}\hat{\boldsymbol{\beta}})/(r\hat{\sigma}^2)$$

is the standard classical test statistic for the general linear hypothesis (developed in a less explicit form in A23.28). Several possible values of d are discussed in **9.17**, all claiming to represent weak prior information in some sense. The case $d = -p$, corresponding to the improper prior distribution $f(\boldsymbol{\beta}, \sigma^2) \propto \sigma^{-2}$, makes (9.42) the classical observed significance probability, so that the informal Bayesian test then agrees exactly with the usual classical significance test.

9.28 An alternative Bayesian approach is to give a non-zero prior probability to the hypothesis, and apply the theory of **7.41** and **7.42** for comparing nested models. We would then wish to calculate the Bayes factor $B = f(\mathbf{y} \mid \boldsymbol{\gamma} = \mathbf{c})/f(\mathbf{y})$. We will consider this approach within a general treatment of Bayes factors for comparing linear models.

Bayes factors for linear models

9.29 Suppose that in addition to the original model (9.1) we have an alternative model $\mathbf{y} = \mathbf{X}_A\boldsymbol{\beta}_A + \boldsymbol{\epsilon}$, with $\boldsymbol{\epsilon}$ distributed as $N(\mathbf{0}, \sigma^2\mathbf{I})$ as before. The prior distribution for $(\boldsymbol{\beta}_A, \sigma^2)$ is $NIG(a, d, \mathbf{m}_A, \mathbf{V}_A)$. The two models make the same assumptions about the error term $\boldsymbol{\epsilon}$, including the same $IG(a, d)$ prior distribution for σ^2. They differ in the matrices $\mathbf{X}$ and $\mathbf{X}_A$ of coefficients, and so try to explain or predict the response variable $\mathbf{y}$ using different regressor variables. Accordingly, they have different parameter vectors $\boldsymbol{\beta}$ and $\boldsymbol{\beta}_A$.

The Bayes factor in favour of the alternative model is the ratio $B = f_A(\mathbf{y})/f(\mathbf{y})$ of the resulting marginal densities for $\mathbf{y}$ under the two models. The denominator is obtained as follows. From (9.2) and (9.11).

$$
\begin{aligned}
f(\mathbf{y}) &= \int\!\!\int f(\mathbf{y} \mid \boldsymbol{\beta}, \sigma^2) f(\boldsymbol{\beta}, \sigma^2)\, \mathrm{d}\boldsymbol{\beta}\, \mathrm{d}\sigma^2 \\
&= k \int\!\!\int (\sigma^2)^{-(d+n+p+2)/2} \exp\{-Q/(2\sigma^2)\}\, \mathrm{d}\boldsymbol{\beta}\, \mathrm{d}\sigma^2, \qquad (9.43)
\end{aligned}
$$

where

$$
k = \frac{(a/2)^{d/2}}{(2\pi)^{(n+p)/2} |\mathbf{V}|^{1/2} \Gamma(d/2)}
$$

and Q is given by (9.14). Now the equivalent expression (9.15) allows us to do the integration with respect to $\boldsymbol{\beta}$ in (9.43), to yield

$$
\begin{aligned}
f(\mathbf{y}) &= k|\mathbf{V}^\star|^{1/2} (2\pi)^{p/2} \int (\sigma^2)^{-(d^\star+2)/2} \exp\{-a^\star/(2\sigma^2)\}\, \mathrm{d}\sigma^2 \\
&= k|\mathbf{V}^\star|^{1/2} (2\pi)^{p/2} (a^\star/2)^{-d^\star/2} \Gamma(d^\star/2) \\
&= \frac{|\mathbf{V}^\star|^{1/2} a^{d/2} \Gamma(d^\star/2)}{|\mathbf{V}|^{1/2} \pi^{n/2} \Gamma(d/2)} (a^\star)^{-d^\star/2}. \qquad (9.44)
\end{aligned}
$$

Notice that $\mathbf{y}$ only appears in (9.44) through $a^\star$. The rest of the expression is the normalizing constant for $f(\mathbf{y})$.

9.30 The analogous expression for $f_A(\mathbf{y})$ adds subscript A to $\mathbf{V}$, $\mathbf{V}^\star$ and $\mathbf{a}^\star$, so that the Bayes factor is

$$
B = \frac{|\mathbf{V}|^{1/2} |\mathbf{V}_A^\star|^{1/2}}{|\mathbf{V}_A|^{1/2} |\mathbf{V}^\star|^{1/2}} \cdot \left(\frac{a^\star}{a_A^\star} \right)^{d^\star/2}. \qquad (9.45)
$$

The four determinants do not depend on the observed data $\mathbf{y}$, and are concerned with the relative strength of prior information and data information about the parameter vectors, as measured by $\mathbf{V}$, $\mathbf{V}_A$, $\mathbf{X}'\mathbf{X}$ and $\mathbf{X}_A'\mathbf{X}_A$. The term involving $\mathbf{y}$ is an increasing function of $a^\star/a_A^\star$, and so favours the alternative model if it leads to a smaller $a_A^\star$ than the original model's $a^\star$. Since $d^\star = d + n$ is the same in both models, the Bayes factor tends to favour the model producing the lower posterior estimate of σ^2. This is intuitively reasonable since σ^2 determines the magnitude of the errors $\boldsymbol{\epsilon} = \mathbf{y} - \mathbf{X}\boldsymbol{\beta}$ or $\boldsymbol{\epsilon} = \mathbf{y} - \mathbf{X}_A\boldsymbol{\beta}_A$ and so measures the lack of fit of the model to the data. An estimate such as $E(\sigma^2 \mid \mathbf{y}) = a^\star/(d^\star - 2)$ of σ^2 estimates this lack of fit.

9.31 We have two expressions for $a^\star$ in (9.18) and (9.21). Several others can be derived. For instance, (9.15) reduces to $Q = a^\star$ if $\boldsymbol{\beta} = \mathbf{m}^\star$, and making this substitution in (9.14) gives

$$a^\star = a + (\mathbf{y} - \mathbf{X}\mathbf{m}^\star)'(\mathbf{y} - \mathbf{X}\mathbf{m}^\star) + (\mathbf{m}^\star - \mathbf{m})'\mathbf{V}^{-1}(\mathbf{m}^\star - \mathbf{m}). \tag{9.46}$$

This is similar to (9.21) but here we have a Bayesian residual sum of squares $(\mathbf{y}-\mathbf{X}\mathbf{m}^\star)'(\mathbf{y}-\mathbf{X}\mathbf{m}^\star)$ in place of the classical $(n-p)\hat{\sigma}^2 = (\mathbf{y} - \mathbf{X}\hat{\boldsymbol{\beta}})'(\mathbf{y} - \mathbf{X}\hat{\boldsymbol{\beta}})$. The classical residual vector $\mathbf{y} - \mathbf{X}\hat{\boldsymbol{\beta}}$ is replaced by Bayesian residuals which are differences between the observations $\mathbf{y}$ and the posterior model fit $\mathbf{X}E(\boldsymbol{\beta}\,|\,\mathbf{y}) = \mathbf{X}\mathbf{m}^\star$. Similarly, the last term in (9.46) is a comparison between prior and posterior means, instead of the comparison between the prior mean and the classical estimate $\hat{\boldsymbol{\beta}}$ which appears in (9.21).

We obtain another expression for $a^\star$ by deriving $f(\mathbf{y})$ differently. The model (9.1) expresses $\mathbf{y}$ as a sum of $\mathbf{X}\boldsymbol{\beta}$ and $\boldsymbol{\epsilon}$, where $\boldsymbol{\epsilon}$ is distributed as $N(0, \sigma^2\mathbf{I})$ given σ^2, and $\boldsymbol{\beta}$ is distributed as $N(\mathbf{m}, \sigma^2\mathbf{V})$, also given σ^2. Therefore the distribution of $\mathbf{y}$ given σ^2 is $N(\mathbf{X}\mathbf{m}, \sigma^2(\mathbf{I} + \mathbf{X}\mathbf{V}\mathbf{X}'))$. The prior distribution of σ^2 being $IG(a, d)$, it follows that the joint distribution of $(\mathbf{y}, \sigma^2)$ is $NIG(a, d, \mathbf{X}\mathbf{m}, \mathbf{I} + \mathbf{X}\mathbf{V}\mathbf{X}')$. Therefore the marginal distribution of $\mathbf{y}$ is $t_d(\mathbf{X}\mathbf{m}, a(\mathbf{I} + \mathbf{X}\mathbf{V}\mathbf{X}'))$. By comparing (9.12) in this case with (9.44),

$$a^\star = a + (\mathbf{y} - \mathbf{X}\mathbf{m})'(\mathbf{I} + \mathbf{X}\mathbf{V}\mathbf{X}')^{-1}(\mathbf{y} - \mathbf{X}\mathbf{m}). \tag{9.47}$$

(We also obtain the matrix identity $|\mathbf{I} + \mathbf{X}\mathbf{V}\mathbf{X}'| = |\mathbf{V}||\mathbf{V}^{-1} + \mathbf{X}'\mathbf{X}|$).

Finally, using (9.17),

$$\mathbf{y} - \mathbf{X}\mathbf{m} = \mathbf{y} - \mathbf{X}\mathbf{m}^\star + \mathbf{X}(\mathbf{m}^\star - \mathbf{m}) = \mathbf{y} - \mathbf{X}\mathbf{m}^\star + \mathbf{X}(\mathbf{V}^{-1} + \mathbf{X}'\mathbf{X})^{-1}\mathbf{X}'(\mathbf{y} - \mathbf{X}\mathbf{m}),$$

so that

$$\mathbf{y} - \mathbf{X}\mathbf{m} = \{\mathbf{I} - \mathbf{X}(\mathbf{V}^{-1} + \mathbf{X}'\mathbf{X})^{-1}\mathbf{X}'\}^{-1}(\mathbf{y} - \mathbf{X}\mathbf{m}^\star) = (\mathbf{I} + \mathbf{X}\mathbf{V}\mathbf{X}')(\mathbf{y} - \mathbf{X}\mathbf{m}^\star).$$

Therefore

$$a^\star = a + (\mathbf{y} - \mathbf{X}\mathbf{m}^\star)(\mathbf{I} + \mathbf{X}\mathbf{V}\mathbf{X}')(\mathbf{y} - \mathbf{X}\mathbf{m}^\star). \tag{9.48}$$

9.32 The various expressions for $a^\star$ show various ways of looking at lack of fit of the model to the data. (9.21) looks at it in terms of the classical residuals and discrepancy between prior and classical estimates of $\boldsymbol{\beta}$. (9.46) expresses lack of fit in terms of the Bayesian residuals and discrepancy between prior and posterior estimates of $\boldsymbol{\beta}$. (9.47) uses a combined measure based on discrepancy between $\mathbf{y}$ and its prior estimate $\mathbf{X}\mathbf{m}$, and (9.48) shows that this can also be turned into a single measure involving the Bayesian residuals (but inflated by the use of $\mathbf{I} + \mathbf{X}\mathbf{V}\mathbf{X}'$ instead of $\mathbf{I}$).

9.33 If the alternative model is a special case of the original linear model then the models are nested. This arises when the columns of $\mathbf{X}_A$ are all linear combinations of the columns of $\mathbf{X}$, so that $\mathbf{X}_A = \mathbf{X}\mathbf{B}$, where $\mathbf{B}$ is a $p \times p_A$ matrix with $p_A < p$. Then the alternative model says $\mathbf{y} = \mathbf{X}\mathbf{B}\boldsymbol{\beta}_A + \boldsymbol{\epsilon}$, a special case of the original model in which $\boldsymbol{\beta} = \mathbf{B}\boldsymbol{\beta}_A$. Since $p_A < p$ we can define an $r \times p$ matrix $\mathbf{A}$, where $r = p - p_A$, such that $\mathbf{A}\mathbf{B} = \mathbf{0}$. Then the alternative model can be seen as asserting that $\mathbf{A}\boldsymbol{\beta} = \mathbf{0}$, a form of the general linear hypothesis. In fact the more general hypothesis $\mathbf{A}\boldsymbol{\beta} = \mathbf{c}$ corresponds to an alternative model of the form $\mathbf{y} = \mathbf{X}_A\boldsymbol{\beta}_A + \mathbf{X}\mathbf{d} + \boldsymbol{\epsilon}$, when the constant vector $\mathbf{d}$ is any solution of the equation $\mathbf{A}\mathbf{d} = \mathbf{c}$. This is the same alternative model if we simply redefine the response variable to be $\mathbf{y}_A = \mathbf{y} - \mathbf{X}\mathbf{d}$ instead of $\mathbf{y}$. We can go on to show by analogy to (9.39) that the general linear hypothesis results in an alternative model for which

$$a_A^\star = a^\star + (\mathbf{c} - \mathbf{A}\mathbf{m}^\star)'(\mathbf{A}\mathbf{V}^\star\mathbf{A}')^{-1}(\mathbf{c} - \mathbf{A}\mathbf{m}^\star),$$

and hence that the Bayes factor (9.45) depends on $\mathbf{y}$ through the same criterion as the informal Bayesian hypothesis test (9.41).

However, an extra complication is introduced by the fact that $\boldsymbol{\beta}$ and σ^2 are not independent in the natural conjugate prior distribution. If $(\boldsymbol{\beta}, \sigma^2)$ have the $NIG(a, d, \mathbf{m}, \mathbf{V})$ prior distribution we could follow the development of **9.25** to derive the conditional prior distribution of $(\boldsymbol{\beta}, \sigma^2)$ given the hypothesis $\mathbf{A}\boldsymbol{\beta} = \mathbf{c}$. This is a normal-inverse-gamma distribution, but its first two hyperparameters are not a and d. The conditioning information $\mathbf{A}\boldsymbol{\beta} = \mathbf{c}$ provides information about σ^2 and thereby changes its distribution. This is different from our assumption in **9.29** that the prior marginal distribution of σ^2 under both models is the same. Consequently we obtain a different result from (9.45) if we begin with a $NIG(a, d, \mathbf{m}, \mathbf{V})$ prior distribution and derive the Bayes factor $B = f(\mathbf{y} \mid \mathbf{A}\boldsymbol{\beta} = \mathbf{c})/f(\mathbf{y})$ as suggested in **7.42** for comparing nested models. This can be seen as an undesirable feature of the normal-inverse-gamma distribution, and the Bayes factor (9.45) seems preferable even in the case of nested models.

Bayes factors with weak prior information

9.34 Now consider weak prior information specified by $\mathbf{V}^{-1} = \mathbf{0}$ and $a = 0$, so that $\mathbf{V}^\star = (\mathbf{X}'\mathbf{X})^{-1}$ and $a^\star = (n-p)\hat{\sigma}^2$. For the alternative model we also let $\mathbf{V}_A^{-1} = \mathbf{0}$, yielding $\mathbf{V}_A^\star = (\mathbf{X}_A'\mathbf{X}_A)^{-1}$, and $a_A^\star = (n - p_A)\hat{\sigma}_A^2$ is the residual sum of squares under the alternative model. The Bayes factor (9.45) is indeterminate because of the terms $|\mathbf{V}|^{1/2}$ and $|\mathbf{V}_A|^{1/2}$, both of which are infinite. This is the difficulty noted in **7.43** of using Bayes factors with improper prior distributions. As in **7.54**, we will let their ratio be an undetermined constant c and write

$$B = c\frac{|\mathbf{X}'\mathbf{X}|^{1/2}}{|\mathbf{X}_A'\mathbf{X}_A|^{1/2}}\left(\frac{(n-p)\hat{\sigma}^2}{(n-p_A)\hat{\sigma}_A^2}\right)^{d^\star/2}. \tag{9.49}$$

This Bayes factor depends on the data through the ratio of residual sums of squares $\{(n-p)\hat{\sigma}^2\}/\{(n-p_A)\hat{\sigma}_A^2\}$, which is simply the classical likelihood ratio test statistic (A23.99).

In the case of nested models we can write (9.49) as

$$B = c\frac{|\mathbf{X}'\mathbf{X}|^{1/2}}{|\mathbf{X}_A'\mathbf{X}_A|^{1/2}}\left(1 + \frac{r}{n-p}F\right)^{-d^\star/2}, \tag{9.50}$$

where F is the classical test statistic

$$F = \frac{(n-p_A)\hat{\sigma}^2 - (n-p)\hat{\sigma}^2}{r\hat{\sigma}^2} \tag{9.51}$$

as derived in A**23.28**. A large value of F will lead to a small Bayes factor in favour of the alternative model, and so cause us to favour the original model. This is analogous to the classical procedure in which a large F causes the hypothesis, represented by the alternative model, to be rejected.

9.35 In order to use the Bayes factor (9.50) in the case of weak prior information we must specify the constant c. The approach of **7.56** was advocated by Spiegelhalter and Smith (1982) as a way of specifying c when comparing nested models.

The method requires You first to specify a minimal experiment such that proper posterior distributions are obtained under both the original and alternative models. In general, this will mean an experiment with $n = p + 1$ observations. This is certainly the case for the most commonly used weak prior distribution in which $d = -p$, since then the degrees of freedom $d^* = d + n$ of the posterior inverse-gamma distribution will not be positive for any smaller n. In general, $p + 1$ observations allow us to estimate the $p + 1$ components of the parameter vector $(\boldsymbol{\beta}, \sigma^2)$. Let $\mathbf{E}$ be the $\mathbf{X}$ matrix for this minimal experiment, under the original model, and let $\mathbf{E}_A$ be the corresponding $\mathbf{X}_A$ matrix for the alternative model.

We now choose the data arising from that hypothetical minimal experiment to maximize B. Since the data $\mathbf{y}$ only affect B through the test statistic F, choose $\mathbf{y}$ to obtain $F = 0$. Then equate the resulting value of B to one, and solve for c:

$$c = |\mathbf{E}_A'\mathbf{E}_A|^{1/2}/|\mathbf{E}'\mathbf{E}|^{1/2}. \tag{9.52}$$

We then insert this value of c into (9.50) to obtain the Spiegelhalter and Smith Bayes factor for the actual data.

Example 9.5

If $x_1, x_2, \ldots, x_n$ are identically and independently distributed as $N(\mu, \sigma^2)$ we have the simple linear model of Example 9.1 $\mathbf{X}'\mathbf{X} = n$, and for a minimal experiment we require two observations, so $\mathbf{E}'\mathbf{E} = 2$. A hypothesis that $\mu = c$ corresponds to the alternative model $y_i - c = \epsilon_i$, where there is no $\boldsymbol{\beta}_A$ vector or $\mathbf{X}_A$ matrix. The effect is to set $|\mathbf{X}_A'\mathbf{X}_A| = 1$ (or equivalently we can treat it as undefined and merge it with c). Therefore, from (9.52) we have $c = 1/\sqrt{2}$ and a Bayes factor of

$$B = (n/2)^{1/2}\{1 + F/(n-1)\}^{-d^*/2},$$

where in this case $F = n(\bar{y} - c)^2/\hat{\sigma}^2$ with $\hat{\sigma}^2 = (n-1)^{-1}\sum(y_i - \bar{y})^2$.

Example 9.6

If the observations follow the simple repression model $y_i = \alpha + \beta x_i + \epsilon_i$ of Example 9.2 then $|\mathbf{X}'\mathbf{X}| = \sum(x_i - \bar{x})^2$. A minimal experiment will have three observations, which we suppose have values e_i, e_2, e_3, of the regressor variable. If the alternative model is that $\beta = 0$, it reduces to independent $N(\alpha, \sigma^2)$ observations and $|\mathbf{X}_A'\mathbf{X}_A| = n$. (9.37) now gives $c = (n/\sum_{i=1}^3(e_i - \bar{e})^2)$, but this is not uniquely determined because there is not a unique minimal experiment.

9.36 In general there may be many possible experiments of minimal size. This is particularly true for regression models, where $\mathbf{X}$ depends on the values of the regressor variables. Spiegelhalter and Smith (1982) deal with regression problems by supposing that not just the observations $\mathbf{y}$ but also the design $\mathbf{X}$ of an experiment of minimal size $n = p+1$

should be chosen to maximize B, before then obtaining c by letting $B = 1$. This seems to run counter to their argument that we set $B = 1$ because a minimal experiment should not be able to provide more than negligible support for the alternative model. Choosing $\mathbf{X}$ to maximize B makes the experiment maximally informative. It is less defensible to set $B = 1$ when the data $\mathbf{y}$ are chosen to give maximal support to the alternative model in a maximally informative experiment, even if that experiment has a minimal number of observations. It might even be argued, conversely, that a truly minimal experiment should have $\mathbf{X}$ chosen to *minimize* B. The unfortunate consequence of that proposal would be to let $|E'E| \to 0$. (Although it must be strictly positive to obtain a proper posterior distribution, we can make it arbitrarily small.) Therefore $c \to \infty$ if we follow this idea. As the following example shows, there is typically no upper bound on $|\mathbf{X}'\mathbf{X}|$ either, so that Spiegelhalter and Smith's approach leads to the opposite extreme of $c = 0$. In general, their method appears unsuitable for problems where alternative minimal experiments give different possible values for c.

Example 9.7
Following Example 9.6, we can choose e_1, e_2 and e_3 to obtain an arbitrarily small value of $\sum(e_i - \bar{e})^2$, and have $c \to \infty$, if the minimal experiment is to be minimally informative. For it to be maximally informative, let $\sum(e_i - \bar{e})^2 \to \infty$ and hence $c \to 0$. Spiegelhalter and Smith (1982) actually proposed maximizing $\sum(e_i - \bar{e})^2$ subject to the constraint that $|e_i| \leqslant 1$ for $i = 1, 2, 3$. This gives a solution $\sum(e_i - \bar{e})^2 = 24/9$ and $c = (27/24)^{1/2}$, but the constraint seems to be completely arbitrary.

Fractional Bayes factors for linear models

9.37 An alternative approach presented in **7.62** is to use the fractional Bayes factor based on using a proportion b of the data as a training sample. We specify the prior distribution as $f(\boldsymbol{\beta}, \sigma^2) \propto g(\boldsymbol{\beta}, \sigma^2)$, where g is an improper prior distribution to represent weak prior information. We shall use $g(\boldsymbol{\beta}, \sigma^2) = \sigma^{-2}$ in both models, but the results are easily adapted to other powers of σ^2 corresponding to other values of the prior degrees of freedom parameter d. Under this approach we take as our definition of $f(\mathbf{y})$ the expression

$$f(\mathbf{y}) = \frac{\int\int f(\mathbf{y} \mid \boldsymbol{\beta}, \sigma^2) g(\boldsymbol{\beta}, \sigma^2)\, \mathrm{d}\boldsymbol{\beta}\, \mathrm{d}\sigma^2}{\int\int \{f(\mathbf{y} \mid \boldsymbol{\beta}, \sigma^2)\}^b g(\boldsymbol{\beta}, \sigma^2)\, \mathrm{d}\boldsymbol{\beta}\, \mathrm{d}\sigma^2}. \tag{9.53}$$

With $g(\boldsymbol{\beta}, \sigma^2) = \sigma^{-2}$, the denominator is

$$I_b = \int\int (2\pi\sigma^2)^{-nb/2} \exp\{-b(\mathbf{y} - \mathbf{X}\boldsymbol{\beta})'(\mathbf{y} - \mathbf{X}\boldsymbol{\beta})/(2\sigma^2)\}\sigma^{-2}\, \mathrm{d}\boldsymbol{\beta}\, \mathrm{d}\sigma^2.$$

Using (9.4) we have

$$I_b = (2\pi)^{-nb/2} \int (\sigma^2)^{-(nb/2)-1} \exp\{-b(n-p)\hat{\sigma}^2/(2\sigma^2)\} J_b\, \mathrm{d}\sigma^2, \tag{9.54}$$

where $(n-p)\hat{\sigma}^2 = (\mathbf{y} - \mathbf{X}\hat{\boldsymbol{\beta}})'(\mathbf{y} - \mathbf{X}\hat{\boldsymbol{\beta}})$ is the residual sum of squares as before, and

$$\begin{aligned} J_b &= \int \exp\{-b(\boldsymbol{\beta} - \hat{\boldsymbol{\beta}})'\mathbf{X}'\mathbf{X}(\boldsymbol{\beta} - \hat{\boldsymbol{\beta}})/(2\sigma^2)\}\, \mathrm{d}\boldsymbol{\beta} \\ &= (2\pi)^{p/2} |\sigma^2 b^{-1} (\mathbf{X}'\mathbf{X})^{-1}|^{1/2} = (2\pi)^{p/2} (\sigma^2)^{p/2} b^{-p/2} |\mathbf{X}'\mathbf{X}|^{-1/2}. \end{aligned}$$

Substituting into (9.54) gives

$$I_b = (2\pi)^{(p-nb)/2} b^{-p/2} |\mathbf{X}'\mathbf{X}|^{-1/2} \{b(n-p)\hat{\sigma}^2/2\}^{(p-nb)/2} \Gamma((nb-p)/2)$$
$$= \pi^{(p-nb)/2} b^{-nb/2} |\mathbf{X}'\mathbf{X}|^{-1/2} \{(n-p)\hat{\sigma}^2\}^{(p-nb)/2} \Gamma((nb-p)/2).$$

Then (9.53) is

$$f(\mathbf{y}) = I_1/I_b = \pi^{-n(1-b)/2} b^{nb/2} \{(n-p)\hat{\sigma}^2\}^{-n(1-b)/2} \Gamma((n-p)/2)/\Gamma((nb-p)/2). \qquad (9.55)$$

Under the alternative model we find the same expression for $f_A(\mathbf{y})$ except that p changes to p_A and $\hat{\sigma}^2$ to $\hat{\sigma}_A^2$. Therefore the fractional Bayes factor is

$$B_q = \frac{\Gamma((nb-p)/2)\Gamma((n-p_A)/2)}{\Gamma((nb-p_A)/2)\Gamma((n-p)/2)} \left(\frac{(n-p)\hat{\sigma}^2}{(n-p_A)\hat{\sigma}_A^2}\right)^{n(1-b)/2}. \qquad (9.56)$$

9.38 Comparing (9.56) with the full Bayes factor (9.49) we notice that the fractional Bayes factor does not have an indeterminate constant c, nor does it depend on the matrices $\mathbf{X}$ and $\mathbf{X}_A$. Instead, the constant term depends only on the number of observations, the numbers of parameters in the two models, and the proportion of the data to be regarded as a training sample. The other difference is in the power of the ratio of residual sums of squares. Pericchi (1984), from a very different approach, obtains a Bayes factor which is different from (9.56) but also does not involve $\mathbf{X}$ or $\mathbf{X}_A$. De Vos (1993) obtains a form of intrinsic Bayes factor with the same property.

Predictive inference

9.39 A common requirement for regression models is to predict the values of the response variable in some future observations. Suppose that You wish to predict a vector $\mathbf{y}_0$ of future observations, with values of the regressor variables $\mathbf{X}_0$. That is, in terms of the parameters $\boldsymbol{\beta}$ of the original model, the new observations are represented by the equation

$$\mathbf{y}_0 = \mathbf{X}_0\boldsymbol{\beta} + \boldsymbol{\epsilon}_0,$$

where $\boldsymbol{\epsilon}_0$ is a vector of new residuals, independently and identically distributed as $N(0, \sigma^2)$. The posterior predictive distribution for $\mathbf{y}_0$ can be derived in the same way as the prior predictive distribution of $\mathbf{y}$ in **9.29** or **9.31**. Following the latter approach, the posterior conditional distribution of $\mathbf{y}_0$ given σ^2 is $N(\mathbf{X}_0\mathbf{m}^\star, \sigma^2(\mathbf{I} + \mathbf{X}_0\mathbf{V}^\star\mathbf{X}_0'))$ and the posterior distribution of σ^2 is $IG(a^\star, d^\star)$. Hence the joint posterior distribution of $(\mathbf{y}_0, \sigma^2)$ is $NIG(a^\star, d^\star, \mathbf{X}_0\mathbf{m}^\star, \mathbf{I} + \mathbf{X}_0\mathbf{V}^\star\mathbf{X}_0')$ and the marginal posterior distribution of $\mathbf{y}_0$ is $t_{d^\star}(\mathbf{X}_0\mathbf{m}^\star, a^\star(\mathbf{I} + \mathbf{X}_0\mathbf{V}^\star\mathbf{X}_0'))$.

Example 9.8

For a simple regression model $y_i = \alpha + \beta x_i + \epsilon_i$ consider a single future observation y_0 when the regressor variable takes the value x_0, so that $\mathbf{X}_0 = (1, x_0)$. Writing

$$\mathbf{m}^\star = \begin{pmatrix} a \\ b \end{pmatrix}, \qquad \mathbf{V}^\star = \begin{pmatrix} v_a & c \\ c & v_b \end{pmatrix}$$

as in Example 9.4, $\mathbf{X}_0\mathbf{m}^\star = a + bx_0$ and $\mathbf{I} + \mathbf{X}_0\mathbf{V}^\star\mathbf{X}_0' = 1 + v_a + 2cx_0 + v_b x_0^2 = 1 + v(x_0)$, where $v(x)$ is as in (9.37). Then a highest posterior density interval for y_0 is

$$a + bx_0 \pm \{a^\star(1 + v(x_0))/d^\star\}^{1/2}t,$$

where t is an appropriate percentage point of the t_{d+n} distribution. This differs from the result (9.36) in Example 9.4 only by changing $v(x)$ to $1 + v(x)$. Whereas in Example 9.4 we were making posterior inference about the value of the regression line at a value x (or x_0) of the regressor variable, here we are making predictive inference about a new observation at that point, which is the regression line plus a further random error ϵ_0. Hence the increased variance and wider highest posterior density intervals in this case. The distinction is equivalent to that between the classical confidence intervals (A28.29) and (A28.33).

Ridge regression

9.40 If we take a normal-inverse-gamma prior distribution with $\mathbf{m} = \mathbf{0}$ and $\mathbf{V} = c^{-1}\mathbf{I}$, then the posterior mean of $\boldsymbol{\beta}$ is

$$\mathbf{m}^\star = (\mathbf{X}'\mathbf{X} + c\mathbf{I})^{-1}\mathbf{X}'\mathbf{y}, \tag{9.57}$$

which is the ridge regression estimator of A**19.12**. Therefore this classical estimator corresponds to a rather special kind of prior information about $\boldsymbol{\beta}$. The elements of $\boldsymbol{\beta}$ are *a priori* independent and identically distributed, with zero mean. The zero prior means cause the posterior estimates (9.57) to be generally shrunk towards zero, relative to the classical least squares estimates $\hat{\boldsymbol{\beta}}$. The larger c is, the smaller is the prior variance, causing the prior to have more influence, and so the shrinkage towards the origin is greater. This is the general behaviour of the ridge regression estimate. It is used in classical statistics when the data are deficient in the sense that $\mathbf{X}'\mathbf{X}$ is nearly singular. This is a situation in which we would expect prior information to be important. Instead of then adopting the simplistic prior $\mathbf{m} = \mathbf{0}$, $\mathbf{V} = c^{-1}\mathbf{I}$ of the ridge regression estimator, it would be appropriate to give careful consideration to Your genuine prior beliefs about $\boldsymbol{\beta}$.

9.41 The first part of this chapter has comprised a thorough study of the normal linear model with the conjugate normal-inverse-gamma prior. This represents the simplest formulation of proper prior information, and that analysis is therefore also simple enough to provide important insights into how the data might generally modify Your prior beliefs. However, in practice these prior distributions are not realistic. The normal-inverse-gamma family suffers from the usual restrictions of conjugate families for many parameters. In particular, it imposes a specific form of relationship between $\boldsymbol{\beta}$ and σ^2. The conditional distribution of $\boldsymbol{\beta}$ given σ^2 is $N(\mathbf{m}, \sigma^2\mathbf{V})$, so the conditional variance must be proportional to σ^2. This means that if You were to learn that the true value of σ^2 is small then this would lead You to a small variance for $\boldsymbol{\beta}$, and a correspondingly strong belief that $\boldsymbol{\beta}$ should be close to its prior mean $\mathbf{m}$. But if You learnt that the true value of σ^2 is very large, then You would instead be very unsure of the value of $\boldsymbol{\beta}$, and would believe it likely to be far from $\mathbf{m}$. Conversely, the prior mean of σ^2 given $\boldsymbol{\beta}$ is $(d-2)^{-1}\{a+(\boldsymbol{\beta}-\mathbf{m})\mathbf{V}^{-1}(\boldsymbol{\beta}-\mathbf{m})\}$

and if You learnt that $\boldsymbol{\beta}$ is close to **m** You would expect σ^2 to be small, or if You learnt that $\boldsymbol{\beta}$ is far from **m** You would then expect σ^2 to be large.

Such a specific relationship between $\boldsymbol{\beta}$ and σ^2 will not reflect Your actual prior beliefs in very many practical situations. There is therefore a need to consider other forms of prior distribution. Almost all of the remainder of this chapter consists of an exploration of alternative prior formulations.

Known variance

9.42 It is sometimes reasonable to suppose that the residual error variance σ^2 is known. In that case the difficulty with the conjugate prior distribution disappears. We now regard σ^2 as fixed in the likelihood (9.2), and denote it by $f(\mathbf{y}\mid\boldsymbol{\beta})$. Only $\boldsymbol{\beta}$ is unknown, and we require a suitable prior distribution $f(\boldsymbol{\beta})$. Using also (9.4), the likelihood simplifies to

$$f(\mathbf{y}\mid\boldsymbol{\beta}) \propto \exp\{-(\boldsymbol{\beta}-\hat{\boldsymbol{\beta}})'\mathbf{X}'\mathbf{X}(\boldsymbol{\beta}-\hat{\boldsymbol{\beta}})/(2\sigma^2)\}, \tag{9.58}$$

and therefore the natural conjugate prior family is the family of normal distributions. Suppose therefore that $\boldsymbol{\beta}$ has the $N(\mathbf{m},\mathbf{W})$ prior distribution,

$$f(\boldsymbol{\beta}) \propto \exp\{-(\boldsymbol{\beta}-\mathbf{m})'\mathbf{W}^{-1}(\boldsymbol{\beta}-\mathbf{m})/2\}. \tag{9.59}$$

Then $f(\boldsymbol{\beta}\mid\mathbf{y}) \propto f(\mathbf{y}\mid\boldsymbol{\beta})f(\boldsymbol{\beta}) \propto \exp(-Q/2)$, where

$$\begin{aligned} Q &= \sigma^{-2}(\boldsymbol{\beta}-\hat{\boldsymbol{\beta}})'\mathbf{X}'\mathbf{X}(\boldsymbol{\beta}-\hat{\boldsymbol{\beta}}) + (\boldsymbol{\beta}-\mathbf{m})'\mathbf{W}^{-1}(\boldsymbol{\beta}-\mathbf{m}) \\ &= \boldsymbol{\beta}'(\mathbf{W}^{-1}+\sigma^{-2}\mathbf{X}'\mathbf{X})\boldsymbol{\beta} + \boldsymbol{\beta}'(\mathbf{W}^{-1}\mathbf{m}+\sigma^{-2}\mathbf{X}'\mathbf{y}) + (\mathbf{W}^{-1}\mathbf{m}+\sigma^{-2}\mathbf{X}'\mathbf{y})'\boldsymbol{\beta} + R_1 \\ &= (\boldsymbol{\beta}-\mathbf{m}^\star)'(\mathbf{W}^\star)^{-1}(\boldsymbol{\beta}-\mathbf{m}^\star) + R_2, \end{aligned}$$

and where

$$\mathbf{m}^\star = (\mathbf{W}^{-1}+\sigma^{-2}\mathbf{X}'\mathbf{X})^{-1}(\mathbf{W}^{-1}\mathbf{m}+\sigma^{-2}\mathbf{X}'\mathbf{y}), \tag{9.60}$$

$$\mathbf{W}^\star = (\mathbf{W}^{-1}+\sigma^{-2}\mathbf{X}'\mathbf{X})^{-1} \tag{9.61}$$

and R_1, R_2 are constants. Therefore,

$$f(\boldsymbol{\beta}\mid\mathbf{y}) \propto \exp\{-(\boldsymbol{\beta}-\mathbf{m}^\star)'(\mathbf{W}^\star)^{-1}(\boldsymbol{\beta}-\mathbf{m}^\star)/2\},$$

i.e. the posterior distribution of $\boldsymbol{\beta}$ is $N(\mathbf{m}^\star,\mathbf{W}^\star)$.

9.43 The analysis is very similar to the case of unknown σ^2 in several respects. In particular, if we let $\mathbf{W}=\sigma^2\mathbf{V}$, then $\mathbf{m}^\star$ in (9.60) is exactly the same as (9.17) and $\mathbf{W}^\star=\sigma^2\mathbf{V}^\star$, with $\mathbf{V}^\star$ as in (9.16). The explanation of this agreement is that if $\mathbf{W}=\sigma^2\mathbf{V}$ the prior distribution $N(\mathbf{m},\mathbf{W})$ for $\boldsymbol{\beta}$ is the same as the conditional prior distribution $f(\boldsymbol{\beta}\mid\sigma^2)$ in the case of unknown σ^2. Then the posterior distribution $N(\mathbf{m}^\star,\mathbf{W}^\star)$ is the same as the conditional posterior distribution $f(\boldsymbol{\beta}\mid\mathbf{y},\sigma^2)$ in the unknown σ^2 case, because of Bayes' theorem

$$f(\boldsymbol{\beta}\mid\mathbf{y},\sigma^2) = f(\boldsymbol{\beta}\mid\sigma^2)f(\mathbf{y}\mid\boldsymbol{\beta},\sigma^2)/f(\mathbf{y}\mid\sigma^2).$$

Knowing σ^2 is the same as conditioning on σ^2, and in particular after observing the data $\mathbf{y}$ our information is $(\mathbf{y},\sigma^2)$, and so the relevant distribution of $\boldsymbol{\beta}$ is $f(\boldsymbol{\beta}\mid\mathbf{y},\sigma^2)$.

9.44 The same analysis can be obtained from a very different perspective by using the Bayes Linear Estimator (see **6.49**). Since only first and second order moments are required in this approach, it is not necessary to assume normality. Suppose then that the model (9.1) simply implies that $E(\mathbf{y} \mid \boldsymbol{\beta}) = \mathbf{X}\boldsymbol{\beta}$ and $\operatorname{var}(\mathbf{y} \mid \boldsymbol{\beta}) = \sigma^2\mathbf{I}$. Again assume that σ^2 is known. Normality is not needed for the prior distribution, either, and we simply have $E(\boldsymbol{\beta}) = \mathbf{m}$, $\operatorname{var}(\boldsymbol{\beta}) = \mathbf{W}$. The Bayes linear estimator (6.29) for a scalar parameter is generalized to the case of estimating a vector parameter in Exercise 6.9. It may then be shown (see Exercise 9.4) that the Bayes linear estimator of $\boldsymbol{\beta}$ based on data $\mathbf{y}$ is $\mathbf{m}^\star$, equation (9.60). Furthermore, $\mathbf{W}^\star$ is the corresponding dispersion matrix (6.33).

If $\mathbf{W} = \sigma^2\mathbf{V}$, $\mathbf{m}^\star$ will still be the Bayes linear estimator in the case of unknown σ^2, but now the dispersion matrix involves σ^2. It is not reasonable to use linear functions of $\mathbf{y}$ to estimate σ^2, and therefore the Bayes linear estimation approach becomes much more complex when σ^2 is unknown.

Conditional conjugate analysis

9.45 Returning to the normal linear model with unknown σ^2, one way to escape from the difficulies inherent in the conjugate family is to give independent prior distributions to $\boldsymbol{\beta}$ and σ^2,

$$f(\boldsymbol{\beta}, \sigma^2) = f(\boldsymbol{\beta})f(\sigma^2)$$

Such a prior represents a situation in which learning about σ^2 would not change Your beliefs about $\boldsymbol{\beta}$, and vice versa. If You assigned a $N(\mathbf{m}, \mathbf{W})$ prior distribution to $\boldsymbol{\beta}$, as in **9.42**, and an $IG(a, d)$ distribution for σ^2, then

$$f(\boldsymbol{\beta}, \sigma^2) \propto \exp\{-(\boldsymbol{\beta} - \mathbf{m})'\mathbf{W}^{-1}(\boldsymbol{\beta} - \mathbf{m})/2\}(\sigma^2)^{-(d+2)/2}\exp\{-a/(2\sigma^2)\}, \qquad (9.62).$$

The posterior distribution is then proportional to the product of (9.62) and the likelihood (9.2). Notice that for fixed σ^2 we can proceed exactly as in **9.42**, and note that the posterior conditional distribution $f(\boldsymbol{\beta} \mid \sigma^2, \mathbf{y})$ is $N(\mathbf{m}^\star, \mathbf{W}^\star)$, with $\mathbf{m}^\star$ and $\mathbf{W}^\star$ given by (9.60) and (9.61). We can therefore integrate with respect to $\boldsymbol{\beta}$ to obtain the marginal posterior distribution for σ^2. However, this $f(\sigma^2 \mid \mathbf{y})$ will not be an inverse-gamma distribution, and it will not be practical to obtain summaries analytically. Nevertheless, summarizing a univariate distribution numerically is a straightforward computational task. Rather than pursue this analysis here, however, we will first generalize to a much larger conditional conjugate family.

9.46 Consider a prior distribution of the form

$$f(\boldsymbol{\beta}, \sigma^2) \propto p(\sigma^2)\exp[-\{\boldsymbol{\beta} - \mathbf{m}(\sigma^2)\}'\mathbf{W}(\sigma^2)^{-1}\{\boldsymbol{\beta} - \mathbf{m}(\sigma^2)\}/2], \qquad (9.63)$$

when $p(\sigma^2)$, $\mathbf{m}(\sigma^2)$ and $\mathbf{W}(\sigma^2)$ are arbitrary functions of σ^2, subject only to conditions that $p(\sigma^2)$ is positive for all σ^2 and $\mathbf{W}(\sigma^2)$ is a positive definite $p \times p$ matrix for all σ^2. The conditional prior distribution of $\boldsymbol{\beta}$ given σ^2 is $N(\mathbf{m}(\sigma^2), \mathbf{W}(\sigma^2))$ and the marginal prior distribution of σ^2 is given by

$$f(\sigma^2) \propto p(\sigma^2)|\mathbf{W}(\sigma^2)|^{1/2}. \qquad (9.64)$$

Further conditions on $p(\sigma^2)$ and $\mathbf{W}(\sigma^2)$ will be required to ensure that this is a proper distribution. The family of distributions (9.63) is a general conditional conjugate family, since we have seen in **9.42** that the normal family of distributions for $\boldsymbol{\beta}$ is conjugate to the likelihood for given σ^2.

9.47 Arguing as in **9.42** we obtain the posterior joint density

$$f(\boldsymbol{\beta}, \sigma^2 \mid \mathbf{y}) \propto p^\star(\sigma^2)\exp[-\{\boldsymbol{\beta} - \mathbf{m}^\star(\sigma^2)\}'\mathbf{W}^\star(\sigma^2)^{-1}\{\boldsymbol{\beta} - \mathbf{m}^\star(\sigma^2)\}/2],$$

where

$$p^\star(\sigma^2) = p(\sigma^2)\sigma^{-n}\exp\{-R_2(\sigma^2)/2\}, \tag{9.65}$$

$$R_2(\sigma^2) = \{\hat{\boldsymbol{\beta}} - \mathbf{m}(\sigma^2)\}'\{\mathbf{W}(\sigma^2) + \sigma^2(\mathbf{X}'\mathbf{X})^{-1}\}^{-1}\{\hat{\boldsymbol{\beta}} - \mathbf{m}(\sigma^2)\} + \sigma^{-2}S,$$

$$\mathbf{m}^\star(\sigma^2) = \{\mathbf{W}(\sigma^2)^{-1} + \sigma^{-2}\mathbf{X}'\mathbf{X}\}^{-1}\{\mathbf{W}(\sigma^2)^{-1}\mathbf{m} + \sigma^{-2}\mathbf{X}'\mathbf{y}\}. \tag{9.66}$$

$$\mathbf{W}^\star(\sigma^2) = \{\mathbf{W}(\sigma^2)^{-1} + \sigma^{-2}\mathbf{X}'\mathbf{X}\}^{-1}, \tag{9.67}$$

and where S is the classical residual sum of squares $(\mathbf{y} - \mathbf{X}\hat{\boldsymbol{\beta}})'(\mathbf{y} - \mathbf{X}\hat{\boldsymbol{\beta}})$. This is clearly a member of the same family, with the 'hyperparameters' $p(\sigma^2)$, $\mathbf{m}(\sigma^2)$ and $\mathbf{W}(\sigma^2)$ replaced by (9.65) to (9.67).

9.48 To summarize the prior or posterior distribution we exploit the conditional conjugacy property to reduce the dimensionality. That is, we use the normal conditional distribution of $\boldsymbol{\beta}$ given σ^2 to reduce computations to operations on the one-dimensional marginal distribution (9.64) or its posterior counterpart. For instance,

$$E(\boldsymbol{\beta}) = E(\mathbf{m}(\sigma^2))$$

requires the calculation of p expectations with respect to $f(\sigma^2)$, which can be done with a single one-dimensional numerical integration exercise. var $(\boldsymbol{\beta})$ can also be obtained from the p^2 expectations

$$E(\boldsymbol{\beta}\boldsymbol{\beta}') = E(\mathbf{W}(\sigma^2) + \mathbf{m}(\sigma^2)\mathbf{m}(\sigma^2)')$$

and var $(\boldsymbol{\beta}) = E(\boldsymbol{\beta}\boldsymbol{\beta}') - E(\boldsymbol{\beta})E(\boldsymbol{\beta})'$. The mode of the joint density $f(\boldsymbol{\beta}, \sigma^2)$ is $(\mathbf{m}(\hat{\sigma}^2), \hat{\sigma}^2)$ where $\hat{\sigma}^2$ maximizes $p(\sigma^2)$.

This dimensionality reduction technique does not, however, give shape summaries of the marginal distribution of $\boldsymbol{\beta}$, or of any single element of $\boldsymbol{\beta}$. This limits the value of being able to compute $E(\boldsymbol{\beta})$ and var $(\boldsymbol{\beta})$ simply.

9.49 The conditional conjugacy facilitates efficient use of Gibbs sampling. Each iteration can be implemented in just two steps. First, using the current σ^2 a new $\boldsymbol{\beta}$ is generated from the multivariate normal distribution $N(\mathbf{m}(\sigma^2), \mathbf{V}(\sigma^2))$ for which efficient algorithms exist (see references in Chapter 8). Then for this $\boldsymbol{\beta}$ a new σ^2 is generated from (9.63) regarded as a function of σ^2 alone, using a method such as the ratio of uniforms, **8.63**. The values of moments like $E(\boldsymbol{\beta})$, var $(\boldsymbol{\beta})$, $E(\sigma^2)$ or var (σ^2), which can all be obtained accurately by one-dimensional quadrature, provide a further check on convergence of the Gibbs sampler.

Although a sample from $f(\boldsymbol{\beta}, \sigma^2)$ does not provide accurate shape summaries, it will certainly assist in interpreting moments. It will, for instance, be adequate to identify possible multimodality or marked skewness.

Generalized error variance

9.50 A small generalization of the linear model is achieved by allowing the random errors comprising $\boldsymbol{\epsilon}$ to be correlated, but assuming that the correlation structure is known. That is, we replace the variance matrix $\sigma^2\mathbf{I}$ by $\sigma^2\mathbf{D}$, where $\mathbf{D}$ is a known positive definite matrix. If the prior distribution of $(\boldsymbol{\beta}, \sigma^2)$ is $NIG(a, d, \mathbf{m}, \mathbf{V})$ as before, then the posterior distribution is also normal-inverse-gamma and we denote it as before by $NIG(a^\star, d^\star, \mathbf{m}^\star, \mathbf{V}^\star)$. We still have $d^\star = d + n$ but formulae for the other posterior parameters become

$$\mathbf{V}^\star = (\mathbf{V}^{-1} + \mathbf{X}'\mathbf{D}^{-1}\mathbf{X})^{-1},$$

$$\mathbf{m}^\star = (\mathbf{V}^{-1} + \mathbf{X}'\mathbf{D}^{-1}\mathbf{X})^{-1}(\mathbf{V}^{-1}\mathbf{m} + \mathbf{X}'\mathbf{D}^{-1}\mathbf{y}),$$

$$a^\star = a + \mathbf{m}'\mathbf{V}^{-1}\mathbf{m} + \mathbf{y}'\mathbf{D}^{-1}\mathbf{y} - (\mathbf{m}^\star)'(\mathbf{V}^\star)^{-1}\mathbf{m}^\star.$$

Equation (9.19) can be derived, expressing the posterior mean as a matrix-weighted average of the prior mean and the classical estimator $\hat{\boldsymbol{\beta}}$, but now $\hat{\boldsymbol{\beta}} = (\mathbf{X}'\mathbf{D}^{-1}\mathbf{X})^{-1}\mathbf{X}'\mathbf{D}^{-1}\mathbf{y}$ (the generalized Least Squares estimator of A**19.19**) and the weight matrix is $\mathbf{A} = (\mathbf{V}^{-1} + \mathbf{X}'\mathbf{D}^{-1}\mathbf{X}')^{-1}\mathbf{X}'\mathbf{D}^{-1}\mathbf{X}$. It is straightforward to generalize other results to this case.

9.51 We can now combine these results with those of **9.42**, to consider the case of known but arbitrary error variance. Let the variance matrix of the random errors $\boldsymbol{\epsilon}$ be $\mathbf{C}$ and known. This corresponds to $\mathbf{C} = \sigma^2\mathbf{D}$ above, but σ^2 is now supposed known as in **9.42**. Therefore the distribution of the data $\mathbf{y}$ given the parameters $\boldsymbol{\beta}$ becomes $N(\mathbf{X}\boldsymbol{\beta}, \mathbf{C})$, with likelihood

$$f(\mathbf{y} \mid \boldsymbol{\beta}) \propto \exp\{-(\mathbf{y} - \mathbf{X}\boldsymbol{\beta})'\mathbf{C}^{-1}(\mathbf{y} - \mathbf{X}\boldsymbol{\beta})/2\}.$$

Let the prior distribution of $\boldsymbol{\beta}$ be $N(\mathbf{m}, \mathbf{W})$ as in **9.42**. Then simple algebra shows that the posterior distribution of $\boldsymbol{\beta}$ is $N(\mathbf{m}^\star, \mathbf{W}^\star)$, where

$$\mathbf{m}^\star = (\mathbf{W}^{-1} + \mathbf{X}'\mathbf{C}^{-1}\mathbf{X})^{-1}(\mathbf{W}^{-1}\mathbf{m} + \mathbf{X}'\mathbf{C}^{-1}\mathbf{y}), \tag{9.68}$$

$$\mathbf{W}^\star = (\mathbf{W}^{-1} + \mathbf{X}'\mathbf{C}^{-1}\mathbf{X})^{-1}. \tag{9.69}$$

The posterior mean $\mathbf{m}^\star$ is a matrix-weighted average of the prior mean $\mathbf{m}$ and the generalized least squares $\hat{\boldsymbol{\beta}} = (\mathbf{X}'\mathbf{C}^{-1}\mathbf{X})^{-1}\mathbf{X}'\mathbf{C}^{-1}\mathbf{y} = (\mathbf{X}'\mathbf{D}^{-1}\mathbf{X})^{-1}\mathbf{X}'\mathbf{D}^{-1}\mathbf{y}$.

Hierarchical linear models

9.52 The method of hierarchical modelling was introduced in **6.39**. As in **9.51**, we suppose that σ^2 is known, but allow a general correlation structure. Suppose, then, that the distribution of $\mathbf{y}$ given $\boldsymbol{\beta}$ is $N(\mathbf{X}\boldsymbol{\beta}, \mathbf{C})$, where $\mathbf{C}$ is a known $n \times n$ positive definite matrix. A hierarchical prior distribution is now proposed for $\boldsymbol{\beta}$. The prior distribution of β is expressed conditional on hyper-parameters γ as $N(\mathbf{X}_1\gamma, \mathbf{C}_1)$. The prior distribution of γ is finally given as $N(\mathbf{m}_2, \mathbf{C}_2)$.

9.53 The posterior distribution is

$$f(\boldsymbol{\beta},\boldsymbol{\gamma}\mid\mathbf{y}) \propto f(\mathbf{y}\mid\boldsymbol{\beta},\boldsymbol{\gamma})f(\boldsymbol{\beta}\mid\boldsymbol{\gamma})f(\boldsymbol{\gamma}) \propto \exp(-Q/2),$$

where

$$Q = (\mathbf{y}-\mathbf{X}\boldsymbol{\beta})'\mathbf{C}^{-1}(\mathbf{y}-\mathbf{X}\boldsymbol{\beta}) + (\boldsymbol{\beta}-\mathbf{X}_1\boldsymbol{\gamma})'\mathbf{C}_1^{-1}(\boldsymbol{\beta}-\mathbf{X}_1\boldsymbol{\gamma}) + (\boldsymbol{\gamma}-\mathbf{m}_2)'\mathbf{C}_2^{-1}(\boldsymbol{\gamma}-\mathbf{m}_2). \qquad (9.70)$$

Since Q is a quadratic expression in $\boldsymbol{\beta}$ and $\boldsymbol{\gamma}$, their joint posterior distribution is clearly normal. Let $\boldsymbol{\theta} = (\boldsymbol{\beta}', \boldsymbol{\gamma}')'$ and write $Q = \boldsymbol{\theta}'\mathbf{V}\boldsymbol{\theta} - \boldsymbol{\theta}'\mathbf{z} - \mathbf{z}'\boldsymbol{\theta} + R$, where

$$\mathbf{V} = \begin{pmatrix} \mathbf{C}_1^{-1} + \mathbf{X}'\mathbf{C}^{-1}\mathbf{X} & -\mathbf{C}_1^{-1}\mathbf{X}_1 \\ -\mathbf{X}_1'\mathbf{C}_1^{-1} & \mathbf{C}_2^{-1} + \mathbf{X}_1'\mathbf{C}_1^{-1}\mathbf{X}_1 \end{pmatrix}, \qquad \mathbf{z} = \begin{pmatrix} \mathbf{X}'\mathbf{C}^{-1}\mathbf{y} \\ \mathbf{C}_2^{-1}\mathbf{m}_2 \end{pmatrix}, \qquad (9.71)$$

$$R = \mathbf{y}'\mathbf{C}^{-1}\mathbf{y} + \mathbf{m}_2'\mathbf{C}_2^{-1}\mathbf{m}_2.$$

Then the posterior distribution of $\boldsymbol{\theta}$ is $N(\mathbf{V}^{-1}\mathbf{z}, \mathbf{V}^{-1})$.

9.54 This gives the full joint posterior distribution of $\boldsymbol{\beta}$ and $\boldsymbol{\gamma}$, and from it we can derive marginal posterior distributions for $\boldsymbol{\beta}$ and $\boldsymbol{\gamma}$ separately. It is possible to invert $\mathbf{V}$ symbolically in a number of different ways, and so obtain a variety of formulae for the posterior means of $\boldsymbol{\beta}$ and $\boldsymbol{\gamma}$. One simple approach does this indirectly by collapsing the hierarchy in two different ways.

First note that we can express the prior distribution $N(\mathbf{X}_1\boldsymbol{\gamma}, \mathbf{C}_1)$ of $\boldsymbol{\beta}$ given $\boldsymbol{\gamma}$ by writing $\boldsymbol{\beta} = \mathbf{X}_1\boldsymbol{\gamma} + \boldsymbol{\delta}$, where $\boldsymbol{\delta}$ is distributed as $N(\mathbf{0}, \mathbf{C}_1)$ independently of $\boldsymbol{\gamma}$. Then since $\boldsymbol{\gamma}$ has the $N(\mathbf{m}_2, \mathbf{C}_2)$ distribution, we can immediately deduce that the marginal prior distribution of $\boldsymbol{\beta}$ is $N(\mathbf{m}, \mathbf{W})$, where

$$\mathbf{m} = \mathbf{X}_1\mathbf{m}_2, \qquad \mathbf{W} = \mathbf{C}_1 + \mathbf{X}_1\mathbf{C}_2\mathbf{X}_1'. \qquad (9.72)$$

Now apply the theory of **9.51**, so that the posterior distribution of $\boldsymbol{\beta}$ is $N(\mathbf{m}^\star, \mathbf{W}^\star)$, where $\mathbf{m}^\star$ and $\mathbf{W}^\star$ are given by (9.68) and (9.69) after inserting (9.72). The posterior mean $\mathbf{m}^\star$ is thereby expressed as a matrix-weighted average of the prior mean $\mathbf{X}_1\mathbf{m}_2$ and the generalized least squares $\hat{\boldsymbol{\beta}}$.

The marginal posterior distribution of $\boldsymbol{\gamma}$ may be found similarly by noting that the original model is $\mathbf{y} = \mathbf{X}\boldsymbol{\beta} + \boldsymbol{\epsilon}$ where $\boldsymbol{\epsilon}$ is distributed as $N(\mathbf{0}, \mathbf{C})$ and the distribution of $\boldsymbol{\beta}$ given $\boldsymbol{\gamma}$ is $N(\mathbf{X}_1\boldsymbol{\gamma}, \mathbf{C}_1)$. Therefore the distribution of $\mathbf{y}$ given $\boldsymbol{\gamma}$ is $N(\mathbf{X}\mathbf{X}_1\boldsymbol{\gamma}, \mathbf{C} + \mathbf{X}\mathbf{C}_1\mathbf{X}')$. This corresponds to another linear model in which $\mathbf{X}$ is replaced by $\mathbf{X}\mathbf{X}_1$ and $\mathbf{C}$ by $\mathbf{C} + \mathbf{X}\mathbf{C}_1\mathbf{X}'$. We again use the theory of **9.51** to obtain a posterior distribution $N(\mathbf{m}_2^\star, \mathbf{C}_2^\star)$ for $\boldsymbol{\gamma}$, where

$$\begin{aligned} \mathbf{C}_2^\star &= (\mathbf{C}_2^{-1} + \mathbf{X}_1'\mathbf{X}'(\mathbf{C} + \mathbf{X}\mathbf{C}_1\mathbf{X}')^{-1}\mathbf{X}\mathbf{X}_1)^{-1} \\ &= \mathbf{C}_2 - \mathbf{C}_2\mathbf{X}_1'\mathbf{X}'(\mathbf{C} + \mathbf{X}\mathbf{C}_1\mathbf{X}' + \mathbf{X}\mathbf{X}_1\mathbf{C}_2\mathbf{X}_1'\mathbf{X}')^{-1}\mathbf{X}\mathbf{X}_1\mathbf{C}_2, \end{aligned}$$

$$\mathbf{m}_2^\star = \mathbf{C}_2^\star(\mathbf{C}_2^{-1}\mathbf{m}_2 + \mathbf{X}_1'\mathbf{X}'(\mathbf{C} + \mathbf{X}\mathbf{C}_1\mathbf{X}')^{-1}\mathbf{y}). \qquad (9.73)$$

This expresses the posterior mean of $\boldsymbol{\gamma}$ as a matrix-weighted average of its prior mean $\mathbf{m}_2$ and a corresponding generalized least-squares estimator

$$\hat{\boldsymbol{\gamma}} = \{\mathbf{X}_1'\mathbf{X}'(\mathbf{C} + \mathbf{X}\mathbf{C}_1\mathbf{X}')^{-1}\mathbf{X}\mathbf{X}_1\}^{-1}\mathbf{X}_1'\mathbf{X}'(\mathbf{C} + \mathbf{X}\mathbf{C}_1\mathbf{X}')^{-1}\mathbf{y}. \qquad (9.74)$$

9.55 An alternative derivation by directly inverting **V**, Exercise 9.7, yields the formulae

$$\mathbf{m}^\star = \hat{\boldsymbol{\beta}} - (\mathbf{X}'\mathbf{C}^{-1}\mathbf{X})^{-1}\mathbf{D}^{-1}(\hat{\boldsymbol{\beta}} - \mathbf{X}_1\mathbf{m}_2), \tag{9.75}$$

$$\mathbf{m}_2^\star = \mathbf{m}_2 - \mathbf{C}_2\mathbf{X}_1'\mathbf{D}^{-1}(\mathbf{X}_1\mathbf{m}_2 - \hat{\boldsymbol{\beta}}), \tag{9.76}$$

where

$$\mathbf{D} = \mathbf{C}_1 + \mathbf{X}_1\mathbf{C}_2\mathbf{X}_1' + (\mathbf{X}'\mathbf{C}^{-1}\mathbf{X})^{-1}.$$

(9.75) is a straightforward restatement of the value for $\mathbf{m}^\star$ given by (9.68) and (9.72). (9.76), however, takes a very different form from (9.74), showing more clearly that the inference depends on the data $\mathbf{y}$ only through the sufficient statistic $\hat{\boldsymbol{\beta}}$.

9.56 Yet another approach is to use the conditional posterior distributions. We note from (9.70) (or from (9.68) with $\mathbf{m} = \mathbf{X}_1\boldsymbol{\gamma}$ and $\mathbf{W} = \mathbf{C}_1$) that the conditional posterior mean of $\boldsymbol{\beta}$ given $\boldsymbol{\gamma}$ is

$$\mathbf{m}^\star(\boldsymbol{\gamma}) = (\mathbf{C}_1^{-1} + \mathbf{X}'\mathbf{C}^{-1}\mathbf{X})^{-1}(\mathbf{C}_1^{-1}\mathbf{X}_1\boldsymbol{\gamma} + \mathbf{X}'\mathbf{C}^{-1}\mathbf{y}).$$

Therefore

$$\mathbf{m}^\star = E(\mathbf{m}^\star(\boldsymbol{\gamma})) = (\mathbf{C}_1^{-1} + \mathbf{X}'\mathbf{C}^{-1}\mathbf{X})^{-1}(\mathbf{C}_1^{-1}\mathbf{X}_1\mathbf{m}_2^\star + \mathbf{X}'\mathbf{C}^{-1}\mathbf{y}). \tag{9.77}$$

Similarly, we obtain an equation for $\mathbf{m}_2^\star$ in terms of $\mathbf{m}^\star$,

$$\mathbf{m}_2^\star = (\mathbf{C}_2^{-1} + \mathbf{X}_1'\mathbf{C}_1^{-1}\mathbf{X}_1)^{-1}(\mathbf{C}_2^{-1}\mathbf{m}_2 + \mathbf{X}_1'\mathbf{C}_1^{-1}\mathbf{m}^\star). \tag{9.78}$$

Uniform second stage prior distribution

9.57 An important special case arises if the prior distribution for $\boldsymbol{\gamma}$, in the final stage of the hierarchy, is an improper uniform distribution representing weak prior information about $\boldsymbol{\gamma}$. This is achieved by letting $\mathbf{C}_2^{-1} \to \mathbf{0}$. (9.73) shows that then $\mathbf{m}_2^\star = \hat{\boldsymbol{\gamma}}$, and from (9.78) $\mathbf{m}_2^\star = (\mathbf{X}_1'\mathbf{C}_1^{-1}\mathbf{X}_1)^{-1}\mathbf{X}_1'\mathbf{C}_1^{-1}\mathbf{m}^\star$. Inserting the first of these into (9.77) expresses $\mathbf{m}^\star$ as a matrix-weighted average of $\mathbf{X}_1\hat{\boldsymbol{\gamma}}$ and $\hat{\boldsymbol{\beta}}$. Inserting the second into (9.77) and solving for $\mathbf{m}^\star$ yields

$$\mathbf{m}^\star = (\mathbf{C}_1^{-1} - \mathbf{C}_1^{-1}\mathbf{X}_1(\mathbf{X}_1'\mathbf{C}_1^{-1}\mathbf{X}_1)^{-1}\mathbf{X}_1'\mathbf{C}_1^{-1} + \mathbf{X}'\mathbf{C}^{-1}\mathbf{X})^{-1}\mathbf{X}'\mathbf{C}^{-1}\mathbf{y}. \tag{9.79}$$

The same result is obtained by inverting $\mathbf{W}$ in (9.72) to give $\mathbf{C}_1^{-1} - \mathbf{C}_1^{-1}\mathbf{X}_1(\mathbf{C}_2^{-1} + \mathbf{X}_1'\mathbf{C}_1^{-1}\mathbf{X}_1)^{-1}\mathbf{X}_1'\mathbf{C}_1^{-1}$, setting $\mathbf{C}_2^{-1} = \mathbf{0}$ and substituting in (9.68).

Example 9.9

Independent samples of m measurements are made on each of p subjects. Denote by y_{ij} the jth measurement on subject i ($i = 1, 2, \ldots, p$, $j = 1, 2, \ldots, m$). We propose the model $y_{ij} = \mu_i + \epsilon_{ij}$, where μ_i is the true mean measurement for subject i and then ϵ_{ij}s are independent measurement errors. We can write this as a linear model $\mathbf{y} = \mathbf{X}\boldsymbol{\beta} + \boldsymbol{\epsilon}$ by letting $\mathbf{y}$ be a single vector of all $n = mp$ observations y_{ij}, $\boldsymbol{\beta} = (\mu_1, \mu_2, \ldots, \mu_k)'$. The $n \times p$ matrix $\mathbf{X}$ has elements zero and one arranged as

$$\mathbf{X} = \begin{pmatrix} \mathbf{1}_m & \mathbf{0}_m & \ldots & \mathbf{0}_m \\ \mathbf{0}_m & \mathbf{1}_m & \ldots & \mathbf{0}_m \\ \vdots & \vdots & \ddots & \vdots \\ \mathbf{0}_m & \mathbf{0}_m & \ldots & \mathbf{1}_m \end{pmatrix},$$

where each $\mathbf{0}_m$ is an $m \times 1$ vector of zeros and each $\mathbf{1}_m$ is an $m \times 1$ vector of ones. Assume that the variance matrix of the errors $\boldsymbol{\epsilon}$ is $\mathbf{C} = \sigma^2\mathbf{I}_n$, and suppose also that σ^2 is known. Simple calculations yield $\mathbf{X}'\mathbf{C}^{-1}\mathbf{X} = \sigma^{-2}\mathbf{X}'\mathbf{X} = m\sigma^{-2}\mathbf{I}_k$, $\hat{\boldsymbol{\beta}} = (\bar{y}_1, \bar{y}_2, \ldots, \bar{y}_p)'$, where $\bar{y}_i = m^{-1}\sum_{j=1}^{m} y_{ij}$, or $\hat{\mu}_i = \bar{y}_i$.

For the prior distribution write $\mu_i = \xi + \delta_i$, where ξ is an overall true average observation for all subjects and the δ_is are independent discrepancies of an individual subject's mean μ_i from that overall mean. We can write this as $\boldsymbol{\beta} = \mathbf{X}_1\boldsymbol{\gamma} + \boldsymbol{\delta}$, where $\boldsymbol{\gamma} = (\xi)$ is a scalar hyperparameter and $\mathbf{X}_1 = \mathbf{1}_p$ is a $p \times 1$ vector of ones. Let the variance matrix of $\boldsymbol{\delta}$ be $\mathbf{C}_1 = \tau^2\mathbf{I}_p$ with τ^2 assumed known. Now $\mathbf{X}\mathbf{X}_1 = \mathbf{1}_n$ and

$$\begin{aligned}(\mathbf{C} + \mathbf{X}\mathbf{C}_1\mathbf{X}')^{-1} &= \mathbf{C}^{-1} - \mathbf{C}^{-1}\mathbf{X}(\mathbf{C}_1^{-1} + \mathbf{X}'\mathbf{C}^{-1}\mathbf{X})^{-1}\mathbf{X}'\mathbf{C}^{-1} \\ &= \sigma^{-2}\mathbf{I}_n - \sigma^{-4}\mathbf{X}\{(\tau^{-2} + m\sigma^{-2})\mathbf{I}_p\}^{-1}\mathbf{X}' \\ &= \sigma^{-2}\mathbf{I}_n - \sigma^{-4}(\tau^2 - m\sigma^{-2})^{-1}\mathbf{X}\mathbf{X}'.\end{aligned}$$

Therefore

$$\begin{aligned}\mathbf{X}_1'\mathbf{X}'(\mathbf{C} + \mathbf{X}\mathbf{C}_1\mathbf{X}')^{-1} &= \sigma^{-2}\mathbf{1}_n' - \sigma^{-4}(\tau^2 - m\sigma^{-2})^{-1}m\mathbf{1}_n' \\ &= (\sigma^2 + m\tau^2)^{-1}\mathbf{1}_n'.\end{aligned}$$

Hence $\hat{\gamma} = \hat{\xi} = \bar{y} = k^{-1}\sum_{i=1}^{p}\bar{y}_i$.

If we now assume weak prior information about ξ, its posterior mean will be $\bar{y}$. Its posterior variance $\mathbf{C}_2^\star$ is $\{(\sigma^2 + m\tau^2)^{-1}\mathbf{1}_n'\mathbf{1}_n\}^{-1} = (\sigma^2 + m\tau^2)/n$. Now to find the posterior mean of $\boldsymbol{\beta}$ we apply (9.77) and need first to derive

$$(\mathbf{C}_1^{-1} + \mathbf{X}'\mathbf{C}^{-1}\mathbf{X})^{-1} = (\tau^{-2} + m\sigma^{-2})^{-1}\mathbf{I}_p.$$

Then

$$\mathbf{m}^\star = a\bar{y}\mathbf{1} + (1 - a)\hat{\boldsymbol{\beta}},$$

where

$$a = \tau^{-2}(\tau^{-2} + m\sigma^{-2})^{-1} = \sigma^2/(\sigma^2 + m\tau^2).$$

The posterior mean of each μ_i is $a\bar{y} + (1 - a)\bar{y}_i$, a weighted average of the mean of the observations on subject i and the mean of all observations. This is a *shrinkage* estimator, discussed in **6.42-43**. Indeed, this example is basically Example 6.16 rephrased in linear model terms. The degree of shrinkage is governed by the weight a attached to $\bar{y}$. a will be small if the data provide good information about each subject, either through σ^2 being small or through having a large number m of observations on each subject. Conversely, a will be large if the information on each subject is not strong and the variability τ^2 between subjects is small.

9.58 An interesting feature of the hierarchical model with uniform prior distribution for γ is that the prior distribution for $\boldsymbol{\beta}$ also becomes improper. It was noted in **9.54** that the marginal prior distribution for $\boldsymbol{\beta}$ is $N(\mathbf{X}_1\mathbf{m}_2, \mathbf{C}_1 + \mathbf{X}_1\mathbf{C}_2\mathbf{X}_1')$. Consider a scalar linear function of $\boldsymbol{\beta}$, $\alpha = \mathbf{b}'\boldsymbol{\beta}$. Its prior distribution is therefore $N(\mathbf{b}'\mathbf{X}_1\mathbf{m}_2, \mathbf{b}'\mathbf{C}_1\mathbf{b} + \mathbf{b}'\mathbf{X}_1\mathbf{C}_2\mathbf{X}_1'\mathbf{b})$. Now as $\mathbf{C}_2^{-1}$ goes to $\mathbf{0}$, $\mathbf{b}'\mathbf{X}_1\mathbf{C}_2\mathbf{X}_1'\mathbf{b}$ will go to infinity unless $\mathbf{b}'\mathbf{X}_1 = \mathbf{0}$, a zero vector. If $\mathbf{b}'\mathbf{X}_1 = \mathbf{0}$, the prior distribution of α is $N(0, \mathbf{b}'\mathbf{C}_1\mathbf{b})$. This results directly from the first stage of the hierarchical prior distribution which asserts that α has this prior distribution independently of γ. With a uniform prior distribution on γ, every other linear function $\alpha = \mathbf{b}'\boldsymbol{\beta}$ also has an improper uniform prior distribution.

In this case, therefore, we can interpret the hierarchical model as providing 'structural' information about $\boldsymbol{\beta}$ by giving zero prior mean and finite variance to those linear functions $\mathbf{b}'\boldsymbol{\beta}$ with $\mathbf{b}'\mathbf{X}_1 = \mathbf{0}$, but providing no other information about $\boldsymbol{\beta}$.

Example 9.10

In Example 9.9, $\mathbf{X}_1 = \mathbf{1}$. The structural prior information is that every linear contrast $\sum_{i=1}^p b_i\mu_i$ with $\sum_{i=1}^p b_i = 0$ has zero prior expectation, but no prior information is given about any other functions of the μ_is. In particular, each μ_i alone has a uniform prior distribution.

Hierarchical models with unknown variances

9.59 The assumption of known variances in a hierarchical model is mathematically convenient but will rarely be realistic in practice. As in **9.50**, we might let $\mathbf{C} = \sigma^2\mathbf{D}$, where $\mathbf{D}$ is known but σ^2 is unknown. The case $\mathbf{D} = \mathbf{I}$ corresponds to the usual linear model formulation. Since the first stage of the hierarchical prior distribution is also formulated as a linear model, it is also useful to let $\mathbf{C}_1 = \tau^2\mathbf{D}_1$ with $\mathbf{D}_1$ known but τ^2 unknown. Now all the preceding theory applies in the sense that it gives the conditional posterior distributions of $\boldsymbol{\beta}$ and $\boldsymbol{\delta}$ given σ^2 and τ^2, but we need to find also the posterior distributions of σ^2 and τ^2.

The joint posterior distribution of all the parameters is

$$\begin{aligned} f(\boldsymbol{\beta}, \gamma, \sigma^2, \tau^2 \mid \mathbf{y}) &\propto f(\mathbf{y} \mid \boldsymbol{\beta}, \gamma, \sigma^2\tau^2) f(\boldsymbol{\beta} \mid \gamma, \sigma^2, \tau^2) f(\gamma \mid \sigma^2, \tau^2) f(\sigma^2, \tau^2) \\ &\propto |\mathbf{C}|^{-1/2} |\mathbf{C}_1|^{-1/2} \exp(-Q/2) f(\sigma^2, \tau^2), \end{aligned} \tag{9.80}$$

where Q is given by (9.70), but now involves σ^2 and τ^2 through $\mathbf{C}$ and $\mathbf{C}_1$. As before, the conditional posterior distribution of $\boldsymbol{\beta}$ and γ given σ^2 and τ^2 is $N(\mathbf{V}^{-1}\mathbf{z}, \mathbf{V}^{-1})$ with $\mathbf{V}$ and $\mathbf{z}$ as in (9.71). We can therefore integrate out $\boldsymbol{\beta}$ and γ to yield the marginal posterior distribution of σ^2 and τ^2,

$$f(\sigma^2, \tau^2 \mid \mathbf{y}) \propto (\sigma^2)^{-n/2} (\tau^2)^{-p/2} |\mathbf{V}|^{-1/2} \exp(-T/2) f(\sigma^2, \tau^2), \tag{9.81}$$

where

$$T = \mathbf{y}'\mathbf{C}^{-1}\mathbf{y} + \mathbf{m}_2'\mathbf{C}_2^{-1}\mathbf{m}_2 - \mathbf{z}'\mathbf{V}^{-1}\mathbf{z}. \tag{9.82}$$

Both T and $\mathbf{V}$ are typically complex functions of σ^2 and τ^2. As a result, (9.81) will generally be mathematically intractable. Nevertheless, the underlying linear model structure allows some of the computational methods of Chapter 8 to be applied very efficiently.

9.60 By integrating out $\boldsymbol{\beta}$ and γ we have reduced dimensionality to the two-dimensional marginal distribution (9.81). By simple quadrature over this distribution we can obtain many summaries of interest. For instance, the posterior means of $\boldsymbol{\beta}$ and γ are expectations of $\mathbf{m}^*$ and $\mathbf{m}_2^*$ (which are both now functions of σ^2 and τ^2) with respect to (9.81). Obviously posterior inference about σ^2 and τ^2 is obtainable directly from (9.81).

Gibbs sampling can be implemented simply and efficiently to provide inferences not obtainable through the dimensionality reduction device. $\boldsymbol{\beta}$ and γ can be sampled in a single step using their joint multivariate normal distribution given σ^2 and τ^2. It remains to sample σ^2 and τ^2 from their conditional distributions, using (9.81) regarded as a function of σ^2 or τ^2 alone. For a quite general prior distribution $f(\sigma^2, \tau^2)$ this may be done by rejection or 'ratio-of-uniforms' methods, but if σ^2 and τ^2 are given independent inverse-gamma prior distributions we have a kind of double conditional conjugacy. Then it is easy to find the conditional posterior distributions of σ^2 and τ^2 given $\boldsymbol{\beta}$ and γ, which are also independent inverse-gamma distributions. Gibbs sampling is then even easier to implement.

9.61 It is tempting now to represent weak prior information about σ^2 and τ^2 by adopting the improper prior formulation $f(\sigma^2, \tau^2) \propto \sigma^{-2}\tau^{-2}$. Unfortunately this leads to an improper posterior distribution. To see why this is the case, it is instructive to consider why no such problem arises in general if we set $f(\sigma^2) \propto \sigma^{-2}$. The posterior distribution of σ^2 will be given by

$$f(\sigma^2 \mid \mathbf{y}) \propto f(\mathbf{y} \mid \sigma^2) f(\sigma^2),$$

where $f(\mathbf{y} \mid \sigma^2)$ is the appropriate integrated likelihood function. Now the prior distribution $f(\sigma^2)$ is improper in both tails, by which we mean that $\int_a^b f(\sigma^2) d\sigma^2$ diverges as either $b \to \infty$ or $a \to 0$. The function σ^{-2} tends to infinity too fast as $\sigma^2 \to 0$, and does not tend to zero fast enough as $\sigma^2 \to \infty$, for the prior distribution to be proper. The posterior distribution will only be proper if multiplying by $f(\mathbf{y} \mid \sigma^2)$ remedies these defects. This does indeed happen because the data provide information that σ^2 is neither zero nor infinite so $f(\mathbf{y} \mid \sigma^2) \to 0$ as $\sigma^2 \to 0$ or $\sigma^2 \to \infty$. In particular, as long as the residual sum of squares $S = (\mathbf{y} - \mathbf{X}\hat{\boldsymbol{\beta}})'(\mathbf{y} - \mathbf{X}\hat{\boldsymbol{\beta}})$ is positive it is clear that σ^2 cannot be zero. This manifests itself generally through the appearance of a term like $\exp\{-S/(2\sigma^2)\}$ in $f(\mathbf{y} \mid \sigma^2)$, which tends rapidly to zero as $\sigma^2 \to 0$.

This will happen also in the hierarchical linear model, and no difficulty arises there when we set $f(\sigma^2) \propto \sigma^{-2}$. Consider, however

$$f(\tau^2 \mid \mathbf{y}) \propto f(\mathbf{y} \mid \tau^2) f(\tau^2). \tag{9.83}$$

Now the data $\mathbf{y}$ do not deny the possibility that $\tau^2 = 0$. If $\tau^2 = 0$ then $\boldsymbol{\beta}$ must equal $\mathbf{X}_1\gamma$ for some γ and the data cannot refute this entirely. The data may suggest an estimate $\hat{\boldsymbol{\beta}}$ that is very far from $\mathbf{X}_1\gamma$ for any γ, but this only makes $\tau^2 = 0$ highly improbable rather than actually impossible. As a result, it can be proved that in general $f(\mathbf{y} \mid \tau^2 = 0)$ is positive. When this is multiplied by $f(\tau^2) \propto \tau^{-2}$ in (9.83) the result is an improper posterior distribution.

9.62 Hierarchical linear models were first introduced by Lindley and Smith (1972). They have been applied and extended in many different ways. See Smith (1973a, 1973b), Fearn (1975), Young (1977), Haitovsky (1987), Lee (1987), Albert (1988a, 1988b), Pericchi and Nazaret (1988), Polasek (1988), Datta and Ghosh (1991), Stroud (1991) and Lange et al. (1992) for various early and more recent examples.

Hierarchical models and prior beliefs

9.63 It is interesting to examine the hierarchical linear model in terms of the extent to which it allows a wider range of prior beliefs to be expressed, in the light of the criticism in **9.41** of the natural conjugate family for the linear model. First, note that with known variances the hierarchical model produces a multivariate normal prior distribution for $\boldsymbol{\beta}$, which is a member of the natural conjugate family of **9.42**. In this case the hierarchical modelling does not offer any greater variety of prior distributions. It does, however, provide a framework for thinking about a prior distribution for $\boldsymbol{\beta}$. In Example 9.9, for instance, prior beliefs about the μ_is would certainly include correlation between them, since if You learn that one subject has a high measurement You will tend to expect a higher measurement of others. It is generally hard to think about correlations, and the hierarchical model simplifies this process by introducing a common mean hyperparameter ξ, conditional on which the μ_is are independent. The primary strength of hierarchical models is in this structuring of possibly complex prior beliefs in terms of simpler constructs.

If we now consider the case of unknown variances it is clear that the hierarchical model with two unknown variances σ^2 and τ^2, as in **9.59**, is distinct from the simple linear model. The way those variances enter into the conditional distribution of $\boldsymbol{\beta}$ and γ given σ^2 and τ^2 is more rigid than in the most general conditional conjugate family (9.63), but it would be simple to generalize in the same way.

Heavy-tailed models

9.64 Another way to broaden the class of prior distributions is to use heavy-tailed priors. The use of heavy-tailed distributions in the context of robustness is discussed in **7.26** to **7.29**. It is certainly convenient, having expressed a prior mean and standard deviation for a parameter θ, to complete the prior specification by assuming a normal prior distribution, but often this gives unrealistically thin tails. With a normal prior distribution, the prior probability that θ will lie more than, say, two and a half prior standard deviations from its prior mean is very small (0.0124), and in practice we wish to allow somewhat more probability to the event of the prior mean being far from the true value. In other words, prior beliefs are often better represented by a heavier-tailed distribution than the normal.

A useful family of heavy-tailed distributions is the t family. Under the natural conjugate normal-inverse-gamma family, the marginal distribution of $\boldsymbol{\beta}$ is a t distribution, so it seems that the natural conjugate distributions already incorporate hevy tails. However, this is a feature of the relationship between $\boldsymbol{\beta}$ and σ^2 in the natural conjugate family, which is criticized in **9.41**. The distribution of $\boldsymbol{\beta}$ given σ^2 is normal, and when in **9.45** we removed the dependence between $\boldsymbol{\beta}$ and σ^2 we proposed instead a normal marginal distribution for $\boldsymbol{\beta}$, (9.62). The more general conditional conjugate family (9.63) still does not admit a heavy-tailed marginal distribution for $\boldsymbol{\beta}$ except by inducing the same form of correlation between $\boldsymbol{\beta}$ and σ^2 as is imposed by the natural conjugate family.

9.65 Let us instead assign a $t_d(\mathbf{m}, \mathbf{W})$ distribution for $\boldsymbol{\beta}$ independently of σ^2, so that the prior distribution is

$$f(\boldsymbol{\beta}, \sigma^2) \propto \{1 + (\boldsymbol{\beta} - \mathbf{m})'\mathbf{W}^{-1}(\boldsymbol{\beta} - \mathbf{m})\}^{-(d+p)/2} f(\sigma^2). \tag{9.84}$$

This is not a member of the natural conjugate or conditional conjugate families. Multiplying by the likelihood function yields an intractable posterior distribution, in the sense that the conditional distribution of $\boldsymbol{\beta}$ given σ^2 is not normal and we cannot integrate analytically with respect to $\boldsymbol{\beta}$. Nevertheless, we can achieve a relatively tractable analysis by a simple device that recognizes the derivation of a t distribution as a marginal distribution in a normal-inverse-gamma joint distribution.

Consider a hierarchical model in which we introduce a hyperparameter τ^2 by letting the conditional prior distribution of $\boldsymbol{\beta}$ be $N(\mathbf{m}, \tau^2\mathbf{V})$. Then at the next stage of the hierarchy we give τ^2 an $IG(a, d)$ distribution. Then

$$f(\boldsymbol{\beta}, \sigma^2, \tau^2) \propto (\tau^2)^{-p/2} \exp\{-(\boldsymbol{\beta} - \mathbf{m})'\mathbf{V}^{-1}(\boldsymbol{\beta} - \mathbf{m})/(2\tau^2)\}(\tau^2)^{-(d+2)/2} \exp\{-a/(2\tau^2)\} f(\sigma^2). \tag{9.85}$$

Integrating τ^2 out of (9.85) yields (9.84) with $\mathbf{W} = a\mathbf{V}$. The joint distribution of $\boldsymbol{\beta}$ and τ^2 is of course $NIG(a, d, \mathbf{m}, \mathbf{V})$.

Using the representation (9.85), multiplying by the likelihood (9.2) yields the posterior distribution $f(\boldsymbol{\beta}, \sigma^2, \tau^2 \mid \mathbf{y})$. The conditional posterior distribution of $\boldsymbol{\beta}$ given σ^2 and τ^2 is $N(\mathbf{m}^\star, \mathbf{W}^\star)$, where $\mathbf{m}^\star$ and $\mathbf{W}^\star$ are given by (9.68) and (9.69) respectively, with $\mathbf{W} = \tau^2\mathbf{V}$ and $\mathbf{C} = \sigma^2\mathbf{I}$. We can then integrate with respect to $\boldsymbol{\beta}$ to obtain

$$f(\sigma^2, \tau^2 \mid \mathbf{y}) \propto |\mathbf{W}^\star|^{-1/2}(\sigma^2)^{-n/2}(\tau^2)^{-(d+p+2)/2} \exp(-T/2) \exp\{-a/(2\tau^2)\} f(\sigma^2), \tag{9.86}$$

where

$$T = \sigma^{-2}\mathbf{y}'\mathbf{y} + \tau^{-2}\mathbf{m}'\mathbf{V}^{-1}\mathbf{m} - (\mathbf{m}^\star)'(\mathbf{W}^\star)^{-1}\mathbf{m}^\star.$$

As with the hierarchical linear model with unknown variances, we have reduced the dimensionality to this two-dimensional posterior distribution for σ^2 and τ^2. Many inferences about $\boldsymbol{\beta}$ can be made via this representation, such as calculating $E(\boldsymbol{\beta} \mid \mathbf{y})$ as the expectation of $\mathbf{m}^\star$ (a function of σ^2 and τ^2) with respect to (9.86). Gibbs sampling is also very easy to implement, particularly if $f(\sigma^2)$ is an inverse gamma distribution.

9.66 This device, of introducing an extra unknown variance, is very generally applicable. The new variance parameter is given an inverse-gamma distribution, and the original parameters have a normal distribution conditional on the unknown variance but a heavy-tailed t distribution unconditionally. However many unknown variances we introduce in this way tractable normal posterior distributions are obtained conditional on all the unknown variances. We can integrate down to the marginal posterior distribution of the unknown variances, which is generally intractable but will be amenable to numerical integration if sufficiently low dimensional. Gibbs sampling is always straightforward, and only requires sampling from normal and inverse-gamma distributions. See references in **7.26**.

Example 9.11
The hierarchical linear model with unknown variances, analysed in **9.59**, is an example of this technique. We could also give γ a t distribution by letting $\mathbf{C}_2 = \omega^2\mathbf{V}_2$, say, and assuming an inverse-gamma distribution for ω^2.

Example 9.12
Instead of an error variance matrix $\sigma^2\mathbf{I}$ in the linear model, where each ϵ_i has the same variance σ^2, we could let each ϵ_i have its own variance σ_i^2. We now have n unknown error variances, equivalent to assuming a heavy-tailed error distribution. Lindley (1971) presents a hierarchical prior distribution for such a set of error variances.

Generalizations of the linear model

9.67 The analysis presented in this chapter can be generalized in a variety of ways. Multivariate linear models, in which each observation of the response variable y is a vector random variable, are introduced in **10.28**, in the simplest case of a multivariate normal sample. Non-normal structures for the error can be considered within a class of generalized linear models, for which Bayesian analysis is given by Albert (1988b), Ibrahim and Laud (1991) and Eaves and Chang (1992).

Dynamic linear models allow the parameter vector $\boldsymbol{\beta}$ to evolve over time. These and other models with variable parameters are considered in **10.42** to **10.47**, and **10.50** to **10.52**. For other variations on the structure and assumptions of the linear model, see for example Reilly and Patino-Leal (1981), Buonaccorsi and Gatsonis (1988), Bagchi et al. (1990), Chib and Tiwari (1991) and Lee (1992). Other specialized forms of linear model are important in econometrics; see Morales (1971), Zellner (1971), Ilmakunnan (1985), Tsurumi (1985), Steel (1991), Steel and Richard (1991), Percy (1992) and Chib (1993).

A good source of theory concerning linear models generally, and covering several generalizations, is Broemeling, (1985).

EXERCISES

9.1 Consider the simple regression model of Example 9.2 with conjugate prior distribution. Inference is required for $\xi = -\alpha/\beta$, the intercept of the regression line with the x-axis. Prove that $E(\xi \mid \mathbf{y})$ does not exist.

In the case of known σ^2, with a normal prior distribution as in **9.42**, show that $f(\xi \mid \mathbf{y})$ can be obtained explicitly in terms of the standard normal d.f. Φ by differentiating

$$P(\xi \leqslant t \mid \mathbf{y}) = P(\alpha - \beta t \leqslant 0, \beta \geqslant 0) + P(\alpha - \beta t \geqslant 0, \beta \leqslant 0).$$

9.2 As an alternative to the determinant and trace criteria for experimental design mentioned in **9.14**, one might follow the approach of scoring rules and choose an experiment to minimize the entropy (2.54) of the appropriate posterior distribution.

Prove first that the entropy of the $NIG(a, d, \mathbf{m}, \mathbf{V})$ distribution is

$$[d + p + (d + p + 2)\{\log(a/2) - \psi(d/2)\}]/2,$$

where $\psi(t)$ is the digamma function $\mathrm{d}\log\Gamma(t)/\mathrm{d}t$. Deduce that in the case of a conjugate $NIG(a, d, \mathbf{m}, \mathbf{V})$ distribution the entropy of the full joint posterior distribution is not a useful criterion for experimental design.

Prove also that in the case of known variance presented in (9.42), minimizing the entropy of the posterior density of $\boldsymbol{\beta}$ produces the determinant criterion of maximizing $|\mathbf{W}^{-1} + \sigma^{-2}\mathbf{X}'\mathbf{X}|$.

9.3 Observations y_{ij} $(i = 1, 2,\ j = 1, 2, \ldots, n_i)$ are independently distributed as $N(\mu_i, \sigma^2)$ given $\boldsymbol{\theta} = (\mu_1, \mu_2, \sigma^2)$. Write this as a linear model and consider constructing a Bayes factor for this model against the alternative that $\mu_2 = 0$, under weak prior information. Using the approach of **9.35** a minimal experiment must either have $n_1 = 1$, $n_2 = 2$ or $n_1 = 2$, $n_2 = 1$. Show that the Bayes factors in these two cases differ by a factor of $\sqrt{2}$.

9.4 Consider a linear model with known variance as in **9.42**. However, normality is not assumed, and so the model simply states that $E(\mathbf{y} \mid \boldsymbol{\beta}) = \mathbf{X}\boldsymbol{\beta}$, $\mathrm{var}(\mathbf{y} \mid \boldsymbol{\beta}) = \sigma^2\mathbf{I}$. Similarly, the only assertions of prior information are $E(\boldsymbol{\beta}) = \mathbf{m}$, $\mathrm{var}(\boldsymbol{\beta}) = \mathbf{W}$. Prove that the Bayes linear estimator (6.34) for $\boldsymbol{\beta}$ as a linear function of $\mathbf{y}$ is (9.60), and that the corresponding dispersion matrix (6.33) is (9.61).

9.5 Express the hierarchical model of **9.52** as a linear model

$$\mathbf{y} = \mathbf{Z}\begin{pmatrix}\boldsymbol{\beta}\\ \boldsymbol{\gamma}\end{pmatrix} + \mathbf{e}$$

for appropriate matrix $\mathbf{Z}$ and a non-hierarchical prior distribution for $\boldsymbol{\beta}, \boldsymbol{\gamma}$ and σ^2. Verify that the generalized posterior mean and variance (9.68) and (9.69) for $\boldsymbol{\beta}$ and $\boldsymbol{\gamma}$ are as given in **9.53**.

9.6 With $\hat{\boldsymbol{\beta}}$ and $\hat{\boldsymbol{\gamma}}$ defined as in **9.54**, prove that an alternative expression for $\hat{\boldsymbol{\gamma}}$ is

$$\hat{\boldsymbol{\gamma}} = (\mathbf{X}_1'\mathbf{P}^{-1}\mathbf{X}_1)^{-1}\mathbf{X}_1'\mathbf{P}^{-1}\hat{\boldsymbol{\beta}},$$

where $\mathbf{P} = \mathbf{C}_1 + (\mathbf{X}'\mathbf{C}^{-1}\mathbf{X})^{-1}$.

9.7 Prove the results (9.75) and (9.76) by inverting the matrix $\mathbf{V}$ in (9.71).

CHAPTER 10

OTHER STANDARD MODELS

10.1 Certain kinds of prior formulation arise repeatedly in applications. This is partly because parameters with similar structures may arise in a variety of contexts, and partly because certain prior distributions have been found to offer more flexibility than others, and so have been adapted to different applications. This chapter begins with a good example of this duality. Wherever observations are classified into a finite number of categories a natural model is the multinomial distribution, whose parameters are the probabilities of a single observation falling in the different classes. We are therefore led to look for distributions for a set of parameters which must be positive and sum to one. The natural conjugate distributions form the Dirichlet family. Prior distributions using Dirichlet distributions in various ways can be useful in a variety of categorical data problems.

However, like natural conjugates generally, the Dirichlet family is rather limited in its ability to represent a useful range of prior beliefs. The multivariate normal distribution is much more flexible, and by applying a suitable transformation it can be adapted to these problems.

After extending these ideas to an infinite number of categories, we consider inference for samples from a finite population. Analysis of data following a multivariate normal distribution is examined next, followed by variations on those problems. Priors representing beliefs about data in the form of expert judgements are then considered. The chapter ends with a brief discussion of graphical modelling.

The Dirichlet family

10.2 Suppose that $x_1, x_2, \ldots, x_n$ are identically distributed discrete random variables taking k possible values with probabilities $\theta_1, \theta_2, \ldots, \theta_k$. Inference is required about these k parameters, although notice that there are effectively only $k-1$ parameters because $\sum_{j=1}^{k} \theta_j = 1$. Assuming that the x_is are independent given the θ_js the likelihood is

$$f(\mathbf{x} \mid \boldsymbol{\theta}) = \prod_{j=1}^{k} \theta_j^{n_j}, \tag{10.1}$$

where n_j is the number of x_is observed to take the jth possible value, $j = 1, 2, \ldots, k$. The n_js are sufficient statistics. If we only observed the n_js (i.e. we observe how many x_is fall into each category but not which ones) the likelihood would instead be

$$f(\mathbf{n} \mid \boldsymbol{\theta}) = \binom{n}{n_1, n_2, \ldots, n_k} \prod_{j=1}^{k} \theta_j^{n_j}, \tag{10.2}$$

where

$$\binom{n}{n_1, n_2, \ldots, n_k} = \frac{n!}{n_1!\,n_2!\ldots n_k!}$$

is a generalization of the usual binomial coefficient $\binom{n}{r} = \binom{n}{r, n-r}$. (10.2) is derivable from (10.1) by showing that there are $\binom{n}{n_1, n_2, \ldots, n_k}$ ways of allocating the x_is to the various categories subject to the fixed n_js. (10.2) is known as a *multinomial* distribution, generalizing the binomial distribution which is the case of two categories. (See also A**5.49** and A**30.4**.)

The two likelihoods are proportional and therefore equivalent (corresponding to the n_js being sufficient). We refer to the simpler form (10.1) as the multinomial likelihood.

10.3 The multinomial likelihood is a member of an exponential family, and from the theory of **6.16** the corresponding natural conjugate family has the form

$$f(\boldsymbol{\theta}) \propto \prod_{j=1}^{k} \theta_j^{a_j}. \tag{10.3}$$

The proportionality constant is required to make $\int f(\theta)\mathrm{d}\theta = 1$ when integrated over the simplex region defined by $\theta_j \geqslant 0$ $(j = 1, 2, \ldots, k)$ and $\sum_{j=1}^{k} \theta_j = 1$. We can do this integration as a series of beta integrals and find that

$$f(\boldsymbol{\theta}) = B(\mathbf{a})^{-1} \prod_{j=1}^{k} \theta_j^{a_j - 1}, \tag{10.4}$$

where the switch from powers a_j in (10.3) to $a_j - 1$ helps to show this as a generalization of the beta distribution, and where

$$B(\mathbf{a}) = B(a_1, a_2, \ldots, a_k) = \{\textstyle\prod_{j=1}^{k} \Gamma(a_j)\}/\Gamma\{\textstyle\sum_{j=1}^{k} a_j\} \tag{10.5}$$

is a generalization of the beta function.

(10.4) is known as a Dirichlet distribution, and we have seen that the Dirichlet distributions form the natural conjugate family to the multinomial likelihood. When $\boldsymbol{\theta}$ has the density (10.4) we say that it has the Dirichlet distribution with parameter $\mathbf{a}$, and we denote this distribution by $D(\mathbf{a})$.

10.4 When $\boldsymbol{\theta}$ has the $D(\mathbf{a})$ distribution, marginal distributions are easily found. First define $\boldsymbol{\phi}$ by $\phi_j = \theta_j$, $j = 1, 2, \ldots, k-2$ and $\phi_{k-1} = \theta_{k-1} + \theta_k$. Then the joint distribution of $\boldsymbol{\phi}$ and θ_k is obtained from (10.4) by this linear transformation, as

$$f(\boldsymbol{\phi}, \theta_k) = B(\mathbf{a})^{-1}\{\textstyle\prod_{j=1}^{k-2} \phi_j^{a_j - 1}\}(\phi_{k-1} - \theta_k)^{a_{k-1} - 1}\theta_k^{a_k - 1}.$$

It is now simple to integrate out θ_k over its possible range $[0, \phi_{k-1}]$ to deduce that $\boldsymbol{\phi}$ has the $D(a_1, a_2, \ldots, a_{k-2}, a_{k-1} + a_k)$ distribution. Indeed, it is repeated use of this integration that establishes the correctness of the normalizing constant in (10.4). It also proves as its last step that θ_1 has the beta distribution

$$f(\theta_1) = B(a_1, a - a_1)\theta_1^{a_1 - 1}(1 - \theta_1)^{a - a_1 - 1}, \tag{10.6}$$

where $a = \sum_{j=1}^{k} a_j$. In general, the marginal distribution of any θ_j is a beta distribution with parameters a_j and $a - a_j$.

10.5 Moments can also be obtained simply by

$$E(\textstyle\prod_{j=1}^{k} \theta_j^{m_j}) = B(\mathbf{a})^{-1} \displaystyle\int \prod_{j=1}^{k} \theta_j^{a_j+m_j-1}\, \mathrm{d}\boldsymbol{\theta},$$

$$= B(\mathbf{a}+\mathbf{m})/B(\mathbf{a})$$

which can be further simplified using (10.5). In particular

$$E(\theta_j) = \frac{\Gamma(a_j+1)\Gamma(a)}{\Gamma(a_j)\Gamma(a+1)} = a_j/a. \tag{10.7}$$

Similar expressions for $E(\theta_j^2)$ and $E(\theta_j\theta_l)$ produce

$$\operatorname{var}(\theta_j) = a_j(a-a_j)/\{a^2(a+1)\}, \tag{10.8}$$

$$\operatorname{cov}(\theta_j, \theta_l) = -a_j a_l/\{a^2(a+1)\}. \tag{10.9}$$

10.6 We now see the usual difficulty of natural conjugate prior families, that they can represent only a rather limited range of prior beliefs. (10.7) shows that the *relative* magnitudes of the a_js specify the prior means of the θ_js. We then have essentially one hyperparameter left. We can think of the overall magnitude $a = \sum_{j=1}^{k} a_j$ of the hyperparameters as representing the strength of prior information, because by increasing a (keeping the relative magnitudes a_j/a fixed) we decrease the prior variances (10.8). We can therefore specify a prior Dirichlet distribution by specifying prior means for each θ_j plus a single, overall measure of strength of information. Albert and Gupta (1982) and Leonard and Novick (1986) provide some generalizations of Dirichlet distributions by hierarchical modelling. A generalization appropriate to censored data problems is considered by Dickey (1983).

Example 10.1

With $k = 3$, suppose that Your prior means for the θ_is are $E(\theta_1) = 0.1$, $E(\theta_2) = 0.3$, $E(\theta_3) = 0.6$. Then the corresponding Dirichlet hyperparameters have the form $a_1 = 0.1a$, $a_2 = 0.3a$, $a_3 = 0.6a$, and $\operatorname{var}(\theta_1) = 0.09/(a+1)$, $\operatorname{var}(\theta_2) = 0.21/(a+1)$, $\operatorname{var}(\theta_3) = 0.24/(a+1)$, $\operatorname{cov}(\theta_1, \theta_2) = -0.03/(a+1)$, $\operatorname{cov}(\theta_1, \theta_3) = -0.06/(a+1)$, $\operatorname{cov}(\theta_2, \theta_3) = -0.18/(a+1)$. Although by choosing a You can now fix the overall magnitude of variances and covariances, the general structure of the covariance matrix is forced upon You by the Dirichlet family. For instance, the correlation between θ_2 and θ_3 is now fixed at -0.802 independent of a.

10.7 Given the multinomial likelihood (10.1) and Dirichlet prior $D(\mathbf{a})$, we immediately find that the posterior distribution is $D(\mathbf{a}+\mathbf{n})$. Posterior summaries are now easily derived, such as

$$E(\theta_j \mid \mathbf{x}) = (a_j + n_j)/(a+n). \tag{10.10}$$

So the posterior mean of θ_j is a weighted average of the prior mean a_j/a and the classical maximum likelihood estimate n_j/n with weights a and n respectively. The weights reflect strength of information, since n is the quantity of data and we have seen that a determines the strength of prior information. The strength of posterior information is now represented by $\sum_{j=1}^{k}(a_j + n_j) = a + n$. Modes of the posterior distribution are found in Example 1.7 and Exercise 1.7.

Log contrasts

10.8 In order to compare two θ_js we might consider the posterior distribution of $\theta_i - \theta_j = \phi$. Moments of ϕ are simple to obtain. No simple expression can be given for the marginal distribution of ϕ, but using the approach of **10.4** we first find that the joint posterior distribution of $(\theta_i, \theta_j, 1-\theta_i-\theta_j)$ is $D(a_i+n_i, a_j+n_j, a+n-a_i-a_j-n_i-n_j)$, and the distribution of ϕ may then be found numerically using methods described in Chapter 8. Alternatively, if all the $(a_j + n_j)$s are sufficiently large the Dirichlet distribution is well approximated by a multivariate normal distribution with the same first and second order moments, hence the distribution of ϕ or any other linear combination of θ_js will then be approximately normal. However this approximation suffers from the fact that linear combinations of θ_js are bounded, e.g. $-1 \leqslant \theta_i - \theta_j \leqslant 1$, whereas the normal distribution is unbounded. So unless the (a_j+n_j)s are genuinely large the approximation will necessarily be poor in the tails.

10.9 A better normal approximation for moderate values of the hyperparameters is found as follows. Suppose that $\boldsymbol{\theta}$ has the $D(\mathbf{a})$ distribution and let $\phi = \sum_{j=1}^{k} c_j \log\theta_j$ be a *log contrast* in the θ_js, where $\sum_{j=1}^{k} c_j = 0$. Then the moment generating function of ϕ is easily found.

$$M(t) = E(e^{t\phi}) = E\{\textstyle\prod_{j=1}^{k}\theta_j^{tc_j}\} = B(\mathbf{a} + t\mathbf{c})/B(\mathbf{a})$$
$$= \prod_{j=1}^{k}\Gamma(a_j + tc_j)/\Gamma(a_j) \qquad (10.11)$$

using (10.5) and the fact that $\sum_{j=1}^{k} c_j = 0$. Moments of ϕ may be found by expanding (10.11) as a power series in t. In fact, using Stirling's series for the log gamma function (see A**3.24**),

$$\log M(t) = t\{\textstyle\sum_{j=1}^{k} c_j \log a_j - \sum_{j=1}^{k} c_j/(2a_j)\} + t^2\sum_{j=1}^{k} c_j^2/(2a_j) + R, \qquad (10.12)$$

where the remainder term R is of order a_j^{-2}.

If R is zero, $M(t)$ is the moment generating function of a normal random variable, so that if all a_js are large enough the distribution of ϕ is approximately normal with mean

$$E(\textstyle\sum_{j=1}^{k} c_j \log\theta_j) \approx \sum_{j=1}^{k} c_j \log a_j - \sum_{j=1}^{k} c_j/(2a_j) \qquad (10.13)$$

and variance

$$\mathrm{var}\,(\textstyle\sum_{j=1}^{k} c_j \log\theta_j) \approx \sum_{j=1}^{k} c_j^2/a_j. \qquad (10.14)$$

If we consider two log contrasts and find their joint moment generating function, then to the same level of approximation

$$\mathrm{cov}\left(\sum_{j=1}^{k} c_j \log \theta_j, \sum_{j=1}^{k} d_j \log \theta_j\right) \approx \sum_{j=1}^{k} c_j d_j / a_j. \tag{10.15}$$

Example 10.2
Let $\boldsymbol{\theta}$ be distributed as $D(a_0, a_0, \ldots, a_0)$ so that all the a_js are equal, and suppose that a_0 is large. Let $\phi_1, \phi_2, \ldots, \phi_{k-1}$ be a set of orthonormal log contrasts, i.e. $\phi_i = \sum_{j=1}^{k} c_{ij} \log \theta_j$, $\sum_{j=1}^{k} c_{ij} = 0$, $\sum_{j=1}^{k} c_{ij}^2 = 1$ and $\sum_{j=1}^{k} c_{ij} c_{hj} = 0$ for all i and $h \neq i$. Then the ϕ_is are approximately independent $N(0, a_0^{-1})$ random variables.

10.10 Another way to compare θ_i and θ_j is therefore to consider the posterior distribution of $\log(\theta_i/\theta_j)$, which is approximately normal with

$$E\{\log(\theta_i/\theta_j) \mid \mathbf{x}\} \approx \log\{(a_i + n_i)/(a_j + n_j)\} + \{(a_j + n_j)^{-1} - (a_i + n_i)^{-1}\}/2, \tag{10.16}$$

$$\mathrm{var}\{\log(\theta_i/\theta_j) \mid \mathbf{x}\} \approx (a_i + n_i)^{-1} + (a_j + n_j)^{-1}. \tag{10.17}$$

The second term in the mean can be seen as a correction for skewness. The approximations ignore terms of order $(a_j+n_j)^{-2}$ and so are accurate provided these are small. In particular, they will be good if none of the observed frequencies n_j is small. Also of interest is the fact that the *exact* distribution of $\log(\theta_i/\theta_j)$, as given by (10.11), depends only on $a_i + n_i$ and $a_j + n_j$, and not on any other of the posterior hyperparameters.

Alternative approximations to (10.13) to (10.15) are given by Lindley (1964) and Bloch and Watson (1967). Lindley suggests

$$E\left(\sum_{j=1}^{k} c_j \log \theta_j\right) \approx \sum_{j=1}^{k} c_j \log(a_j - 0.5), \tag{10.18}$$

$$\mathrm{var}\left(\sum_{j=1}^{k} c_j \log \theta_j\right) \approx \sum_{j=1}^{k} c_j^2/(a_j - 0.5). \tag{10.19}$$

$$\mathrm{cov}\left(\sum_{j=1}^{k} c_j \log \theta_j, \sum_{j=1}^{k} d_j \log \theta_j\right) \approx \sum_{j=1}^{k} c_j d_j/(a_j - 0.5). \tag{10.20}$$

These approximations are also correct to order a_j^{-2}, but (10.18) is slightly simpler than (10.13). Bloch and Watson give a further improvement, replacing $a_j - 0.5$ in (10.18) to (10.20) by $a_j - 0.5 - 1/(24a_j)$. (10.16) and (10.17) can obviously be modified in accordance with either of these.

Prior distributions based on normality

10.11 The multivariate normal approximation to the Dirichlet distribution itself or to the distribution of a set of log contrasts, suggests a more flexible family of prior distributions. Whereas the Dirichlet family is limited in its capacity to represent actual prior beliefs because of not having enough hyperparameters, the multivariate normal family is much richer. It has enough hyperparameters to specify separately all means, variances and covariances.

The simplest approach would be to assume that $\boldsymbol{\theta}$ is distributed as $N(\mathbf{m}, \mathbf{V})$. The fact that $\sum_{j=1}^{k} \theta_j = 1$ means that $\mathbf{m}$ and $\mathbf{V}$ must be chosen so that $\sum_{j=1}^{k} m_j = 1$ and

$\sum_{i=1}^{k}\sum_{j=1}^{k} v_{ij} = 0$. Then **V** has rank $k-1$ and the multivariate normal distribution is essentially $(k-1)$-dimensional as required. However, this can only ever be an approximate prior specification because the normal distribution cannot respect the inequality constraints $\theta_j \geqslant 0$ which also apply. The approximation must be poor unless the prior variances expressed in **V** are small, which corresponds to quite strong prior information. This formulation is therefore inappropriate in many practical situations.

10.12 An alternative is to assign a multivariate normal prior distribution to a set of log contrasts. We can define $\phi_i = \sum_{j=1}^{k} c_{ij} \log \theta_j$ for $i = 1, 2, \ldots, k-1$, which we can write as $\boldsymbol{\phi} = \mathbf{C} \log \boldsymbol{\theta}$, where the $(k-1) \times k$ matrix $\mathbf{C} = (c_{ij})$ has rank $k-1$. Together with the condition $\sum_{j=1}^{k} \theta_j = 1$, this defines a one-to-one transformation from $\boldsymbol{\phi}$ to $\boldsymbol{\theta}$ which automatically satisfies the constraints that $\theta_j \geqslant 0$. Therefore we are free to assign any $(k-1)$-dimensional normal prior distribution to $\boldsymbol{\phi}$. Only in the case that the means, variances and covariances in this distribution approximate to (10.13) to (10.15) for some **a** will the implied prior distribution on $\boldsymbol{\theta}$ approximate to a Dirichlet. Aitchison (1982) shows that these transformed normal distributions are indeed a rich class of distributions on the simplex, offering important distributional shapes that cannot be achieved within the Dirichlet family. Notice that by assigning normal distributions to different sets of log contrasts, i.e. with different **C** matrices, we obtain different classes of transformed normal distributions for $\boldsymbol{\theta}$.

10.13 Formally, if the prior distribution of $\boldsymbol{\phi}$ is $N(\mathbf{m}, \mathbf{V})$ then the posterior distribution is

$$f(\boldsymbol{\theta} \mid \mathbf{x}) \propto J\{\textstyle\prod_{j=1}^{k} \theta_j^{n_j}\} \exp\{-(\mathbf{C} \log \boldsymbol{\theta} - \mathbf{m})' \mathbf{V}^{-1} (\mathbf{C} \log \boldsymbol{\theta} - \mathbf{m})/2\}, \tag{10.21}$$

where J is the Jacobian of the transformation from $\boldsymbol{\theta}$ to $\boldsymbol{\phi}$. J can be derived as follows. Remembering that $\boldsymbol{\theta}$ is really $(k-1)$-dimensional, let $\boldsymbol{\theta}_- = (\theta_1, \theta_2, \ldots, \theta_{k-1})$ and consider the transformation from $\boldsymbol{\theta}_-$ to $\boldsymbol{\phi}$ defined by $\boldsymbol{\phi} = \mathbf{C} \log \boldsymbol{\theta}$ and the constraint $\theta_k = 1 - \theta_1 - \theta_2 - \ldots - \theta_{k-1} = 1 - \mathbf{1}'\boldsymbol{\theta}_-$, where $\mathbf{1}$ is a $(k-1) \times 1$ vector of ones. That is,

$$\boldsymbol{\phi} = \mathbf{C}_- \log \boldsymbol{\theta}_- + \mathbf{c}_k \log(1 - \mathbf{1}'\boldsymbol{\theta}_-),$$

where $\mathbf{C} = (\mathbf{C}_-, \mathbf{c}_k)$ partitions $\mathbf{C}$ into the $(k-1) \times (k-1)$ matrix $\mathbf{C}_-$ and the $(k-1) \times 1$ vector $\mathbf{c}_k$. Notice that since $\boldsymbol{\phi}$ is a vector of log contrasts, the rows of $\mathbf{C}$ all sum to zero, so that

$$\mathbf{C}_-\mathbf{1} + \mathbf{c}_k = \mathbf{0},$$

$$\therefore \boldsymbol{\phi} = \mathbf{C}_-\{\log \boldsymbol{\theta}_- - \mathbf{1} \log(1 - \mathbf{1}'\boldsymbol{\theta}_-)\}.$$

The matrix of partial derivatives is therefore

$$\partial\boldsymbol{\phi}/\partial\boldsymbol{\theta}_- = \mathbf{C}_-(\mathbf{D} + \mathbf{1}\mathbf{1}'d),$$

where $\mathbf{D}$ is the diagonal matrix with diagonal elements $\theta_1^{-1}, \theta_2^{-1}, \ldots, \theta_{k-1}^{-1}$ and $d = \theta_k^{-1} = (1 - \mathbf{1}'\boldsymbol{\theta}_-)^{-1}$. Then the Jacobian is the determinant of this matrix,

$$J = |\partial\boldsymbol{\phi}/\partial\boldsymbol{\theta}_-| = |\mathbf{C}_-||\mathbf{D} + \mathbf{1}\mathbf{1}'d| = |\mathbf{C}_-| \prod_{j=1}^{k} \theta_j^{-1}. \tag{10.22}$$

The last step is proved using the general identity $|\mathbf{A}+\mathbf{UBV}| = |\mathbf{A}||\mathbf{B}||\mathbf{B}^{-1}+\mathbf{VA}^{-1}\mathbf{U}|$, setting $\mathbf{A}=\mathbf{D}$, $\mathbf{B}=d$ and $\mathbf{U}=\mathbf{V}'=\mathbf{1}$.

10.14 Substituting (10.22) into (10.21), the posterior density of $\boldsymbol{\theta}$ is

$$f(\boldsymbol{\theta}\,|\,\mathbf{x}) \propto \{\textstyle\prod_{j=1}^{k}\theta_j^{n_j-1}\}\exp\{-(\mathbf{C}\log\boldsymbol{\theta}-\mathbf{m})'\mathbf{V}^{-1}(\mathbf{C}\log\boldsymbol{\theta}-\mathbf{m})/2\}. \tag{10.23}$$

Distributions of this general form are defined and studied by Aitchison (1985). We can obtain a representation of weak prior information by giving $\boldsymbol{\phi}$ a uniform prior distribution so that the second term in (10.18) vanishes (corresponding to letting $\mathbf{V}^{-1}$ be a zero matrix). Then in $\boldsymbol{\theta}$ space, this is the improper prior distribution

$$f(\boldsymbol{\theta}) \propto \prod_{j=1}^{k}\theta_j^{-1}. \tag{10.24}$$

Equation (10.24) generalizes the prior distribution $f(\theta) \propto \theta^{-1}(1-\theta)^{-1}$ often used to express weak prior information about the parameter θ of a binomial observation, and has been used by Villegas (1976). It is also a limiting Dirichlet distribution (10.4) in which $a_j = 0$ for all j, so that the sum of the a_js is also zero, corresponding to the weakest possible prior information (see **10.6**). Other forms of weak prior distribution are discussed by Dyer and Pierce (1993).

10.15 We cannot integrate (10.23) analytically over $\boldsymbol{\theta}$ space to find the normalizing constant or posterior moments. We can, however, differentiate to obtain an equation for the posterior mode. Maximizing (10.23) subject to the constraint that $\sum\theta_j = 1$ using the Lagrange multiplier method, gives the system of equations

$$\mathbf{C}'\mathbf{V}^{-1}(\mathbf{C}\log\boldsymbol{\theta}-\mathbf{m}) = \mathbf{n}-\mathbf{1}+\lambda\boldsymbol{\theta}, \tag{10.25}$$

where λ is the Lagrange multiplier. Summing the k equations (10.25) results in $0 = n-k+\lambda$, since the rows of $\mathbf{C}$ sum to zero and $\sum\theta_j = 1$. Therefore $\lambda = k-n$, so that the mode of (10.23) is a solution to

$$\mathbf{C}'\mathbf{V}^{-1}(\mathbf{C}\log\boldsymbol{\theta}-\mathbf{m}) = \mathbf{n}-\mathbf{1}-(n-k)\boldsymbol{\theta}. \tag{10.26}$$

In the case of the weak prior distribution (10.24), the left hand side is zero and the mode is given by $\theta_j = (n_j-1)/(n-k)$. Otherwise, (10.26) can be solved by appropriate numerical methods.

Example 10.3

If the data form a histogram, i.e. n_i is the number of times that some variable is observed to take values in the ith range, then it is usually reasonable to have a prior belief that the underlying probabilities θ_i form a smooth distribution. We can represent this very simply by a prior belief that $\log\theta_i$ is close to $\log\theta_{i+1}$. So defining $\phi_i = \log\theta_i - \log\theta_{i+1}$ for $i = 1, 2, \ldots, k-1$, we let this prior belief be represented by a $N(\mathbf{0}, v\mathbf{I})$ prior distribution for

Table 10.1 Apple diameters data, Example 10.3

x_i	2.05	2.15	2.25	2.35	2.45	2.55	2.65	2.75	2.85	2.95
n_i	5	4	10	18	15	20	25	27	12	6
$n\hat{\theta}_i$	4.92	5.52	9.49	15.72	16.14	20.28	24.71	24.51	12.78	7.94

$\boldsymbol{\phi}$, where the prior variance v determines the strength of prior belief in the smoothness of the histogram. Then $\mathbf{m} = \mathbf{0}$ and

$$\mathbf{C}'\mathbf{V}^{-1}\mathbf{C}\log\boldsymbol{\theta} = v^{-1}(\phi_1, \phi_2 - \phi_1, \phi_3 - \phi_2, \ldots, \phi_{k-1} - \phi_{k-2}, -\phi_{k-1})',$$

forming the left hand side of (10.26).

To see the effect of this analysis consider some real data. The values n_i in Table 10.1 are observed frequencies of sizes of apples reported in Leonard (1973). The values x_i are the midpoints of the intervals (in inches) of apple diameters. Using the above prior distribution, set $v = 0.2$, so that ϕ_i is generally expected to lie in the range $\pm\sqrt{0.2} = \pm 0.447$, and therefore θ_i/θ_{i+1} is expected to lie in the range 0.64 to 1.56. Solving (10.26) and multiplying the resulting θ_is by $n = 142$ gives the smoothed histogram also shown in Table 10.1.

10.16 The posterior distribution of $\boldsymbol{\phi}$ is

$$f(\boldsymbol{\phi} \mid \mathbf{x}) \propto \{\textstyle\prod_{j=1}^{k} t_j(\boldsymbol{\phi})^{n_j}\} \exp\{-(\boldsymbol{\phi} - \mathbf{m})'\mathbf{V}^{-1}(\boldsymbol{\phi} - \mathbf{m})/2\}, \tag{10.27}$$

where $t_j(\boldsymbol{\phi})$ is θ_j expressed as a function of $\boldsymbol{\phi}$. (10.27) is essentially the same as (10.21) except that $n_j - 1$ is changed to n_j (removing the Jacobean (10.22)). In particular, its mode is found by making corresponding changes to (10.26), i.e. by solving

$$\mathbf{C}'\mathbf{V}^{-1}(\mathbf{C}\log\boldsymbol{\theta} - \mathbf{m}) = \mathbf{n} - n\boldsymbol{\theta}$$

for $\boldsymbol{\theta}$ and then setting $\boldsymbol{\phi} = \mathbf{C}\log\boldsymbol{\theta}$. However, neither (10.21) nor (10.27) is easy to summarize fully, since no analytical results can be obtained. (Numerical techniques are needed even to calculate the mode.) An alternative is to use the approximations of **10.9** to replace the first term in (10.27) by a multivariate normal expression. Using (10.18) to (10.20) we find the approximation

$$f(\boldsymbol{\phi} \mid \mathbf{x}) \propto \exp\{-(\boldsymbol{\phi} - \mathbf{C}\mathbf{n}^{\star})'(\mathbf{C}\mathbf{N}\mathbf{C}')^{-1}(\boldsymbol{\phi} - \mathbf{C}\mathbf{n}^{\star})/2\} \exp\{-(\boldsymbol{\phi} - \mathbf{m})'\mathbf{V}^{-1}(\boldsymbol{\phi} - \mathbf{m})/2\} \tag{10.28}$$

where $\mathbf{n}^{\star}$ has jth element $\log(n_j - 0.5)$ and $\mathbf{N}$ is a diagonal matrix with jth diagonal element $(n_j - 0.5)^{-1}$. Combining the two normal terms in (10.28), the posterior distribution of $\boldsymbol{\phi}$ is approximately normal with mean vector

$$\mathbf{V}^{\star}\{\mathbf{V}^{-1}\mathbf{m} + (\mathbf{C}\mathbf{N}\mathbf{C}')^{-1}\mathbf{C}\mathbf{n}^{\star}\}$$

and covariance matrix

$$\mathbf{V}^{\star} = \{\mathbf{V}^{-1} + (\mathbf{C}\mathbf{N}\mathbf{C}')^{-1}\}^{-1}.$$

10.17 Aitchison (1982) gives several other possible transformations (from $\boldsymbol{\theta}$ constrained to the simplex to an unconstrained $\boldsymbol{\phi}$) other than the log contrast approach. Leonard (1982) points out that it is possible to set a multivariate normal distribution directly on $\log\boldsymbol{\theta}$. Although $\log\theta_j$s are then constrained by $\sum_j \theta_j = 1$ the implied distribution of any set of log contrasts is independent of this constraint. Leonard's approach is rather more complex, because of the constraint, but he argues that it is more natural. It is applied in Leonard (1973, 1975, 1977), and a similar formulation is used by Lindley (1985) to model expert opinion. Leonard (1978) extends it to estimating a continuous density function. Aitchison (1986) gives a thorough treatment of these distributions.

Much of previous work has been applied to contingency tables. Leonard (1975) deals with two-way tables and this is extended by Nazaret (1987), Tsutakawa (1988) and Knuiman and Speed (1988) to three-way and higher order tables.

The Dirichlet process

10.18 The preceding theory offers a variety of ways of analysing discrete data. We began **10.2** by letting $x_1, x_2, \ldots, x_n$ be discrete random variables independently and identically distributed with probabilities $\theta_1, \theta_2, \ldots, \theta_k$ of taking each of k possible discrete values. If we now let k become very large, we can hope to develop corresponding analysis for continuous data. Let $y_1, y_2, \ldots, y_n$ be independently and identically distributed with d.f. $F(y)$. By grouping the possible values of y_i into classes, we can create corresponding discrete random variables. Thus, let x_i take the value j if $b_{j-1} < y_i \leqslant b_j$, for $-\infty = b_0 < b_1 < b_2 < \ldots < b_k = \infty$. Then the x_is are independent and identically distributed discrete random variables with $\theta_j = F(b_j) - F(b_{j-1})$. Let the prior distribution for $\theta_1, \theta_2, \ldots, \theta_k$ be the Dirichlet distribution (10.4), i.e. $D(\mathbf{a})$.

10.19 Now the property **10.4** of the Dirichlet distribution means that if we merge some of the classes into which y_i is grouped to define x_i, then the prior distribution of the new class probabilities $\boldsymbol{\phi}$ is also Dirichlet. If, for instance, we remove the class boundary b_k and set $b_{k-1} = \infty$ to produce $k-1$ classes, then x_i is a discrete random variable with probabilities $\phi_1 = \theta_1, \phi_2 = \theta_2, \ldots, \phi_{k-2} = \theta_{k-2}, \phi_{k-1} = \theta_{k-1} + \theta_k$ for its $k-1$ possible values. **10.4** shows that $\boldsymbol{\phi}$ has a Dirichlet distribution with hyperparameters $a_1, a_2, \ldots, a_{k-2}, a_{k-1} + a_k$. When we merge classes we simply add the corresponding a_js to obtain the new Dirichlet distribution. Similarly, we can split into more than k classes by introducing more class boundaries. We can assign a Dirichlet prior distribution to the new class probabilities $\boldsymbol{\phi}$, and it will be consistent with the Dirichlet distribution for $\boldsymbol{\theta}$ as long as its hyperparameters return to $a_1, a_2, \ldots, a_k$ when we remerge the newly split classes. For instance, if we split the last class by introducing a new boundary $b_k > b_{k-1}$ and let $b_{k+1} = \infty$, then the new class probabilities are $\phi_1 = \theta_1$, $\phi_2 = \theta_2, \ldots, \phi_{k-1} = \theta_{k-1}$, ϕ_k and ϕ_{k+1} such that $\phi_k + \phi_{k+1} = \theta_k$. A Dirichlet prior distribution $D(\mathbf{a}^\star)$ for $\boldsymbol{\phi}$ will be consistent with a distribution $D(\mathbf{a})$ for $\boldsymbol{\theta}$ if $a_1^\star = a_1$, $a_2^\star = a_2, \ldots, a_{k-1}^\star = a_{k-1}$ and $a_k^\star + a_{k+1}^\star = a_k$.

Now consider splitting classes by introducing more and more new boundaries until there is an infinite number of boundaries and the width $b_j - b_{j-1}$ of any class tends to zero. The result of splitting the Dirichlet like this until it effectively becomes an infinite-dimensional distribution is called a Dirichlet process. Formally, the Dirichlet process is defined as follows.

The unknown distribution function $F(y)$ has the Dirichlet process distribution with parameter $G(y)$, written $DP(G)$, if the distribution of $\boldsymbol{\theta}$ is always Dirichlet, however we define the class boundaries. That is, for any boundaries $-\infty = b_0 < b_1 < b_2 < \ldots < b_{k-1} < b_k = \infty$, the random variables $\theta_1 = F(b_1)$, $\theta_2 = F(b_2) - F(b_1), \ldots, \theta_{k-1} = F(b_{k-1}) - F(b_{k-2})$ and $\theta_k = 1 - F(b_{k-1})$ have a Dirichlet distribution $D(\mathbf{a})$ with hyperparameters $a_1 = G(b_1)$, $a_2 = G(b_2) - G(b_1), \ldots, a_{k-1} = G(b_{k-1}) - G(b_{k-2})$, $a_k = g - G(b_{k-1})$. $G(y)$ is a non-decreasing function with $\lim_{y\to-\infty} G(y) = 0$ and $\lim_{y\to\infty} G(y) = g$. It is clear from our discussion of splitting and merging classes that it is possible to define a distribution like this so that $\boldsymbol{\theta}$ has a Dirichlet distribution for any set of class boundaries.

Moments of the Dirichlet process

10.20 The Dirichlet process allows us to define a prior distribution for the entire d.f. $F(y)$ of a continuous random variable. Some of its properties are very easy to deduce. For instance, introducing only the boundary b_1 with $k = 2$, the distribution of $F(b_1)$ is beta with parameters $G(b_1)$ and $g - G(b_1)$. Therefore the expectation of $F(b_1)$ is $g^{-1}G(b_1)$. Or in general the expectation of the function F is the function $g^{-1}G$:

$$E(F(y)) = g^{-1}G(y). \tag{10.29}$$

Similarly, we obtain the covariance between $F(b_1)$ and $F(b_2)$ (with $b_2 > b_1$) by looking at the Dirichlet distribution of $F(b_1), F(b_2) - F(b_1)$ and $1 - F(b_2)$. From (10.8) and (10.9),

$$\text{var}\,(F(y)) = G(y)\{g - G(y)\}/\{g^2(g+1)\}, \tag{10.30}$$

$$\text{cov}\,(F(y), F(y') - F(y)) = -G(y)\{G(y') - G(y)\}/\{g^2(g+1)\},$$

and hence

$$\text{cov}\,(F(y), F(y')) = G(y)\{g - G(y')\}/\{g^2(g+1)\} \tag{10.31}$$

for $y' > y$. (Obviously (10.30) can be obtained directly from (10.31) by letting $y' \to y$.)

10.21 We see the same difficulty with the Dirichlet process as a family of prior distributions as we found in **10.6** for the Dirichlet distribution. Apart from the scaling factor g, the entire hyperparameter function $G(y)$ serves simply to define the prior expectation of $F(y)$. We then only have g with which to specify prior variances and covariances alike, leaving correlations unchanged. So g serves as a single measure of the strength of prior information.

10.22 Suppose that F has the Dirichlet process distribution $DP(G)$ and consider the (predictive) distribution of a future observation y. Conditional on F, the probability that y lies in any interval $[y', y'']$ is $F(y'') - F(y')$. Therefore unconditionally,

$$\begin{aligned} P(y \in [y', y'']) &= E\{P(y \in [y', y''] \mid F)\} \\ &= E\{F(y'') - F(y')\} \\ &= g^{-1}\{G(y'') - G(y')\}. \end{aligned} \tag{10.32}$$

Therefore $g^{-1}G$ is also the predictive distribution function of a future observation. We can think of $G(y)$ as specifying this predictive distribution, leaving g as a single hyperparameter to describe association between two or more future observations.

The posterior Dirichlet process

10.23 Suppose that $F(y)$ has a Dirichlet process prior distribution. We now observe $y_1, y_2, \ldots, y_n$ and require the posterior distribution of $F(y)$. This is easy to determine by considering the posterior distribution of $\boldsymbol{\theta}$ based on any given set of class boundaries. $\boldsymbol{\theta}$ has a Dirichlet posterior distribution with hyperparameters $a_1 + n_1, a_2 + n_2, \ldots, a_k + n_k$, where $n_1, n_2, \ldots, n_k$ are counts of y_is falling between the class boundaries. By considering the limit of this distribution as we split the classes as before, the posterior distribution of $F(y)$ is a Dirichlet process with hyperparameter function $G(y) + G_n(y)$, where for any y the value of $G_n(y)$ is the number of observed y_is which are less than or equal to y. Therefore in particular the posterior mean is

$$E(F(y) \mid \mathbf{y}) = (g + n)^{-1}\{G(y) + G_n(y)\}, \tag{10.33}$$

which is a weighted average of the prior mean $g^{-1}G(y)$ and the data estimate $n^{-1}G_n(y)$, with weights g and n. The data estimate $F_n(y) = n^{-1}G_n(y)$ is called the *empirical distribution function.* (10.33) also confirms our interpretation of g as measuring the quantity of prior information, on the same scale as n measures the quantity of data.

Example 10.4

Various kinds of posterior inference can now be derived from this Dirichlet process distribution. Consider for instance the mean μ of the distribution $F(y)$, given by

$$\mu = \int_{-\infty}^{\infty} y \, \mathrm{d}F(y).$$

In the same way as $E(aX + bY) = aE(X) + bE(Y)$, the posterior expectation of μ is

$$\begin{aligned} E(\mu \mid \mathbf{y}) &= \int_{-\infty}^{\infty} y \, \mathrm{d}E(F(y) \mid \mathbf{y}) \\ &= (g + n)^{-1} \left\{ \int_{-\infty}^{\infty} y \, \mathrm{d}G(y) + \int_{-\infty}^{\infty} y \, \mathrm{d}G_n(y) \right\} \\ &= (g + n)^{-1}(gE(\mu) + n\bar{y}), \end{aligned}$$

i.e. a weighted average of the prior mean $E(\mu)$ and the sample mean $\bar{y}$, with weights g and n.

Density estimation

10.24 This result, however, brings out another difficulty with the Dirichlet process. $G_n(y)$ is a step function which jumps from $k - 1$ to k when y equals the kth sample order statistic $y_{(k)}$. Therefore the posterior mean (10.33) also has a jump at each $y_{(k)}$. Because of (10.32), the posterior predictive distribution of a future observation given the observations $y_1, y_2, \ldots, y_n$ is also (10.33). Letting $y'' = y_{(k)}$ and y' tend to $y_{(k)}$ from below, shows there is a probability $(g + n)^{-1}$ that a future observation will exactly equal $y_{(k)}$. In fact, there is a probability $n/(g+n)$ that the next observation will exactly equal one of the n values already observed. If we believe that y is a continuous random variable then it would be highly unrealistic to give a non-zero probability to a future observation exactly equalling any past observation. Since assigning a prior Dirichlet process has precisely this implication, the Dirichlet process is an unrealistic representation of beliefs about a continuous d.f.

The reason for this problem is a fundamental property of a Dirichlet process, that if F is distributed as $DP(G)$ for any G then with probability one F is discrete. This result is proved by Ferguson (1973) and Blackwell (1973), and very clearly makes the Dirichlet process inappropriate for continuous F. It also implies that posterior inference from a Dirichlet process prior may not be consistent in the sense of **3.20**; see Diaconis and Freedman (1986). Several alternatives have been suggested, all of them rather more complicated than the Dirichlet process. Lavine (1992) describes the Polya tree distribution, which is a generalization of the Dirichlet process based on the same idea of successively splitting the range of y into smaller and smaller classes as in **10.19**. However, this is done by a series of conditional Dirichlet or beta distributions such that the full joint distribution of any partition is no longer Dirichlet. Another class of distributions are often known as mixtures of Dirichlets, but the term *Dirichlet mixtures* would be more appropriate. A genuine mixture of Dirichlet distributions would arise if we supposed that F was distributed as $DP(G_\phi)$ conditional on a further hyperparameter ϕ. Then giving ϕ a distribution (discrete or continuous) results in the unconditional distribution of F being a mixture of the Dirichlet processes $DP(G_\phi)$. Such a distribution would still be discrete with probability one. A Dirichlet mixture distribution reverses the conditioning by supposing that

$$F(y) = \int F_1(y-y')\,\mathrm{d}F_2(y'), \qquad (10.34)$$

where F_2 is distributed as $DP(G)$ and F_1 is a fixed, known continuous distribution function known as the *kernel*. With probability one F_2 will be discrete and (10.34) will reduce to the discrete mixture of kernels

$$F(y) = \sum_j f_j F_1(y - y_j),$$

where the discrete realization of F_2 has given probabilities f_j to the points y_j. This is of course continuous, so with probability one the Dirichlet mixture distribution produces a continuous F. The expectation of F is the continuous convolution

$$E\{F(y)\} = g^{-1}\int F_1(y-y')\,\mathrm{d}G(y').$$

Examples of the use of Dirichlet mixtures are given by Antoniak (1974), Ferguson (1983), West (1991), Escobar and West (1991).

Leonard (1978) adapts the idea of assuming a normal distribution for log contrasts to the case of a continuous distribution. Some theory of general distributions for continuous F is given by Ferguson (1974). Inference about a continuous d.f. F is often phrased in terms of inference about its density function, and is generally called *density estimation.* The use of Dirichlet mixtures can be seen as a Bayesian formulation of kernel density estimates, which are considered instead as an *ad hoc* procedure by Silverman (1986).

Finite populations

10.25 The relationship between classical statistical inference from finite populations and Bayesian theory generally was discussed in **5.21** to **5.24**. We now consider a fully Bayesian analysis. Consider a population of N units, and suppose that the value of a characteristic of interest for unit i is ξ_i, $i = 1, 2, \ldots, N$. You wish to make inference about the total $\tau = \sum_{i=1}^{N} \xi_i$. (Inference about the average τ/N would then follow trivially.) Data consist of observing the value of the characteristic for each of a sample of n units. For simplicity of notation suppose that these are the first n units in the population. The posterior distribution of τ is therefore $f(\tau \mid \xi_1, \xi_2, \ldots, \xi_n)$.

A number of points should be noticed about this solution. In classical inference, it is essential that a sample be chosen in some random way. The data are $x_1, x_2, \ldots, x_n$, where each x_j is an observation of a random ξ_i. As noted in **5.22**, classical inference is not possible if the data include knowledge of which units have been sampled, for then, since parameters are not random in classical inference, the x_js cease to be random variables. There is no such difficulty in Bayesian inference, which is based on the posterior distribution of τ given the observed values of the characteristics of n particular units. Whether those units were chosen at random is not generally relevant since the randomization mechanism itself is ancillary. See Godambe (1966), Ericson (1969a), Scott and Smith (1973), Scott (1977), Sugden and Smith (1984) and Smith (1991) for a fuller discussion of this matter, and for consideration of exceptional circumstances when the sampling rule might be informative. Notice also that the observations are assumed to be made without error, so that they can be thought of as observations of the true values of the parameters $\xi_i, \xi_2, \ldots, \xi_n$. It would of course be straightforward to model the possibility of observation error.

To obtain an explicit posterior distribution $f(\tau \mid \xi_1, \xi_2, \ldots, \xi_n)$, we begin with the joint prior distribution $f(\xi_1, \xi_2, \ldots, \xi_N)$ for all N units of the population. Then the posterior distribution of the unobserved units is just the conditional distribution $f(\xi_{n+1}, \xi_{n+2}, \ldots, \xi_N \mid \xi_1, \xi_2, \ldots, \xi_n)$ obtained directly from the prior. Since τ is equal to the sum $\sum_{i=1}^{n} \xi_i$ of the observations plus the sum $\sum_{i=n+1}^{N} \xi_i$ of the unobserved units, its posterior distribution may be deduced from the previous step. In particular,

$$E(\tau \mid \xi_1, \xi_2, \ldots, \xi_n) = \sum_{i=1}^{n} \xi_i + \sum_{i=n+1}^{N} E(\xi_i \mid \xi_1, \xi_2, \ldots, \xi_n),$$

$$\operatorname{var}(\tau \mid \xi_1, \xi_2, \ldots, \xi_n) = \sum_{i=n+1}^{N} \sum_{j=n+1}^{N} \operatorname{cov}(\xi_i, \xi_j \mid \xi_1, \xi_2, \ldots, \xi_n).$$

10.26 The key step in this solution is to establish the joint prior distribution $f(\xi_1, \xi_2, \ldots, \xi_N)$ for the whole population. N may be very large, and then the specification of such a high-dimensional joint distribution is potentially an enormous task. Nevertheless, prior knowledge of the population can often be modelled in simple ways. The simplest useful structure is to suppose that the ξ_is are independent and identically distributed with common distribution $g(\xi \mid \boldsymbol{\theta})$, conditional on unknown hyperparameters

$\boldsymbol{\theta}$. This hierarchical model is then completed by a prior distribution $f(\boldsymbol{\theta})$ for $\boldsymbol{\theta}$. Then

$$f(\xi_1, \xi_2, \ldots, \xi_N) = \int \prod_{i=1}^{N} g(\xi_i \mid \boldsymbol{\theta}) f(\boldsymbol{\theta})\, d\boldsymbol{\theta}.$$

The next step of conditioning on the observations results in

$$f(\xi_{n+1}, \xi_{n+2}, \ldots, \xi_N \mid \xi_1, \xi_2, \ldots, \xi_n) = \int \prod_{i=n+1}^{N} g(\xi_i \mid \boldsymbol{\theta}) f(\boldsymbol{\theta} \mid \xi_1, \xi_2, \ldots, \xi_n)\, d\boldsymbol{\theta}. \qquad (10.35)$$

In order to implement this approach, therefore, we need to derive $f(\boldsymbol{\theta} \mid \xi_1, \xi_2, \ldots, \xi_n)$. This is simply the posterior distribution of $\boldsymbol{\theta}$ in a conventional model of iid observations. Then (10.35) is just the posterior predictive distribution of $N - n$ further observations.

Example 10.5

Let the ξ_is be independently distributed as $N(\theta, v)$ and let the prior distribution of θ be $N(m, w)$. Then the posterior distribution of θ given $\xi_1, \xi_2, \ldots, \xi_n$ is $N(m_1, w_1)$, where $m_1 = (nw\bar{\xi} + vm)/(nw + v)$, $w_1 = vw/(nw + v)$ and $\bar{\xi} = n^{-1}\sum_{i=1}^{n} \xi_i$. The predictive distribution of each unobserved ξ_i is then $N(m_1, v + w_1)$ (see Exercise 3.8). For inference about τ, the sum of $N - n$ unobserved ξ_is has a $N((N-n)\theta, (N-n)v)$ distribution given θ, and its posterior predictive distribution is thereby found to be $N((N-n)m_1, (N-n)v + (N-n)^2 w_1)$. Finally, the posterior distribution of τ is therefore normal with mean

$$E(\tau \mid \xi_1, \xi_2, \ldots, \xi_n) = n\bar{\xi} + (N-n)m_1 = N\{a\bar{\xi} + (1-a)m\}, \qquad (10.36)$$

where $a = (Nnw + nv)/(Nnw + Nv)$, and with variance

$$\operatorname{var}(\tau \mid \xi_1, \xi_2, \ldots, \xi_n) = (N-n)v + (N-n)^2 w_1 = (N-n)v(Nw + v)/(nw + v). \qquad (10.37)$$

10.27 The model assumed in this case for the ξ_is is one of exchangeability. Specifically, it corresponds to assuming that the ξ_is can be thought of as N members of an infinite exchangeable sequence $\xi_1, \xi_2, \ldots$, and that in this infinite superpopulation the ξ_is make up a normal distribution with variance v but unknown mean θ. It therefore corresponds directly with the classical use of superpopulations described in **5.23**. Indeed, if $w \to \infty$ in Example 10.5, to represent weak prior information about θ, the results (10.36) and (10.37) correspond exactly with the classical example in **5.23**. The analysis given there does not assume normality, but a Bayesian analogue giving the same mean and variance is obtained using the Bayes linear estimator, see Smouse (1984). Ericson (1969a) introduces the Bayesian approach of **10.26**. Further development is given by Scott and Smith (1969), Thompson (1978), Lo (1986), Bolfarine and Rodrigues (1990), Meeden and Vardeman (1991), Binder (1992) and Nandram and Sedransk (1993).

Priors for a multivariate normal sample

10.28 Suppose that $x_1, x_2, \ldots, x_n$ are independently and identically distributed as $N(\mu, \sigma^2)$. This case of a sample from a normal distribution yields quite elementary analysis with conjugate prior distributions. In Example 1.5 we assumed σ^2 known and a normal prior distribution for μ. Chapter 9 greatly extends that analysis since a single normal sample is the simplest case of a normal linear model. An alternative generalization is to a sample from a multivariate normal distribution $N(\boldsymbol{\mu}, \boldsymbol{\Sigma})$. The likelihood function is

$$f(\mathbf{x}_1, \mathbf{x}_2, \ldots, \mathbf{x}_n \mid \boldsymbol{\mu}, \boldsymbol{\Sigma}) \propto \prod_{i=1}^{n} |\boldsymbol{\Sigma}|^{-1/2} \exp\{-(\mathbf{x}_i - \boldsymbol{\mu})' \boldsymbol{\Sigma}^{-1} (\mathbf{x}_i - \boldsymbol{\mu})/2\}$$
$$= |\boldsymbol{\Sigma}|^{-n/2} \exp\{-\textstyle\sum_{i=1}^{n} (\mathbf{x}_i - \boldsymbol{\mu})' \boldsymbol{\Sigma}^{-1} (\mathbf{x}_i - \boldsymbol{\mu})/2\}. \quad (10.38)$$

Now expanding the quadratic forms we have

$$\sum_i (\mathbf{x}_i - \boldsymbol{\mu})' \boldsymbol{\Sigma}^{-1} (\mathbf{x}_i - \boldsymbol{\mu}) = n\boldsymbol{\mu}' \boldsymbol{\Sigma}^{-1} \boldsymbol{\mu} - n\bar{\mathbf{x}}' \boldsymbol{\Sigma}^{-1} \boldsymbol{\mu} - n\boldsymbol{\mu}' \boldsymbol{\Sigma}^{-1} \bar{\mathbf{x}} + \sum_i \mathbf{x}_i' \boldsymbol{\Sigma}^{-1} \mathbf{x}_i$$
$$= n(\boldsymbol{\mu} - \bar{\mathbf{x}})' \boldsymbol{\Sigma}^{-1} (\boldsymbol{\mu} - \bar{\mathbf{x}}) + r,$$

where $\bar{\mathbf{x}} = n^{-1} \sum_i \mathbf{x}_i$ is the sample mean vector,

$$r = \sum_i \mathbf{x}_i' \boldsymbol{\Sigma}^{-1} \mathbf{x}_i - n\bar{\mathbf{x}}' \boldsymbol{\Sigma}^{-1} \bar{\mathbf{x}}$$
$$= \sum_i (\mathbf{x}_i - \bar{\mathbf{x}})' \boldsymbol{\Sigma}^{-1} (\mathbf{x}_i - \bar{\mathbf{x}})$$
$$= \sum_i tr \boldsymbol{\Sigma}^{-1} (\mathbf{x}_i - \bar{\mathbf{x}})(\mathbf{x}_i - \bar{\mathbf{x}})'$$
$$= n\, tr \boldsymbol{\Sigma}^{-1} \mathbf{S}$$

and $\mathbf{S} = n^{-1} \sum_i (\mathbf{x}_i - \bar{\mathbf{x}})(\mathbf{x}_i - \bar{\mathbf{x}})'$ is the sample covariance matrix. Therefore

$$f(\mathbf{x}_1, \mathbf{x}_2, \ldots, \mathbf{x}_n \mid \boldsymbol{\mu}, \boldsymbol{\Sigma}) \propto |\boldsymbol{\Sigma}|^{-n/2} \exp\{-n(\boldsymbol{\mu} - \bar{\mathbf{x}})' \boldsymbol{\Sigma}^{-1} (\boldsymbol{\mu} - \bar{\mathbf{x}})/2 - n\, tr \boldsymbol{\Sigma}^{-1} \mathbf{S}/2\}. \quad (10.39)$$

10.29 Now if $\boldsymbol{\Sigma}$ is known, two terms can be dropped from (10.39) leaving

$$f(\mathbf{x}_1, \mathbf{x}_2, \ldots, \mathbf{x}_n \mid \boldsymbol{\mu}) \propto \exp\{-n(\boldsymbol{\mu} - \bar{\mathbf{x}})' \boldsymbol{\Sigma}^{-1} (\boldsymbol{\mu} - \bar{\mathbf{x}})/2\}. \quad (10.40)$$

Now applying the ideas of **6.16** and **6.17** to (10.40) shows that the strict natural conjugate family of distributions for $\boldsymbol{\mu}$ is $N(\mathbf{a}, d\boldsymbol{\Sigma})$, with prior mean $\mathbf{a}$ and prior covariance matrix a multiple of $\boldsymbol{\Sigma}$. This would suffer from the usual difficulty with natural conjugates of having hyperparameters $\mathbf{a}$ to determine the prior mean and a single hyperparameter d to determine all the variances and covariances. However, it is very easy to work with the more general family of all multivariate normal distributions. For if we let the prior distribution of $\boldsymbol{\mu}$ be $N(\mathbf{a}, \mathbf{B})$ we find

$$f(\boldsymbol{\mu} \mid \mathbf{x}_1, \mathbf{x}_2, \ldots, \mathbf{x}_n) \propto f(\boldsymbol{\mu}) f(\mathbf{x}_1, \mathbf{x}_2, \ldots, \mathbf{x}_n \mid \boldsymbol{\mu})$$
$$\propto \exp\{-(\boldsymbol{\mu} - \mathbf{a})' \mathbf{B}^{-1} (\boldsymbol{\mu} - \mathbf{a})/2 - n(\boldsymbol{\mu} - \bar{\mathbf{x}})' \boldsymbol{\Sigma}^{-1} (\boldsymbol{\mu} - \bar{\mathbf{x}})/2\}.$$

Now complete the square by

$$(\boldsymbol{\mu} - \mathbf{a})' \mathbf{B}^{-1} (\boldsymbol{\mu} - \mathbf{a}) + n(\boldsymbol{\mu} - \bar{\mathbf{x}})' \boldsymbol{\Sigma}^{-1} (\boldsymbol{\mu} - \bar{\mathbf{x}})$$
$$= \boldsymbol{\mu}' (\mathbf{B}^{-1} + n\boldsymbol{\Sigma}^{-1}) \boldsymbol{\mu} - (\mathbf{a}' \mathbf{B}^{-1} + n\bar{\mathbf{x}}' \boldsymbol{\Sigma}^{-1}) \boldsymbol{\mu} - \boldsymbol{\mu}' (\mathbf{B}^{-1} \mathbf{a} + n\boldsymbol{\Sigma}^{-1} \bar{\mathbf{x}})$$
$$+ (\mathbf{a}' \mathbf{B}^{-1} \mathbf{a} + n\bar{\mathbf{x}}' \boldsymbol{\Sigma}^{-1} \bar{\mathbf{x}})$$
$$= (\boldsymbol{\mu} - \mathbf{a}^{\star})' (\mathbf{B}^{\star})^{-1} (\boldsymbol{\mu} - \mathbf{a}^{\star}) + R, \quad (10.41)$$

where

$$\mathbf{B}^\star = (\mathbf{B}^{-1} + n\boldsymbol{\Sigma}^{-1})^{-1},$$

$$\mathbf{a}^\star = \mathbf{B}^\star(\mathbf{B}^{-1}\mathbf{a} + n\boldsymbol{\Sigma}^{-1}\bar{\mathbf{x}}),$$

and R is a constant. The posterior distribution of $\boldsymbol{\mu}$ is therefore $N(\mathbf{a}^\star, \mathbf{B}^\star)$. The posterior mean is a matrix-weighted average of the prior mean $\mathbf{a}$ and the sample mean $\bar{\mathbf{x}}$. The weights are respectively the prior *precision matrix* $\mathbf{B}^{-1}$ and the analogous data precision $n\boldsymbol{\Sigma}^{-1}$ (being the inverse of the covariance matrix of $\bar{\mathbf{x}}$). The posterior precision matrix is the sum of $\mathbf{B}^{-1}$ and $n\boldsymbol{\Sigma}^{-1}$, and the posterior covariance matrix is smaller than either the prior covariance matrix $\mathbf{B}$ or the data analogue $n^{-1}\boldsymbol{\Sigma}$, in the sense that both $\mathbf{B} - \mathbf{B}^\star$ and $n^{-1}\boldsymbol{\Sigma} - \mathbf{B}^\star$ are positive definite.

10.30 The ability to break out of the confines of the natural conjugate family to the general normal prior $N(\mathbf{a}, \mathbf{B})$ is a powerful tool. The general normal family has enough hyperparameters to represent any prior means, variances and covariances for the elements of $\boldsymbol{\mu}$. It is this versatility of the normal family that makes transformations to normality in other problems, such as in **10.12** to **10.17**, attractive.

Unknown variance components

10.31 Now suppose that the mean $\boldsymbol{\mu}$ is known but the covariance matrix $\boldsymbol{\Sigma}$ is unknown. Then the likelihood (10.38) becomes

$$f(\mathbf{x}_1, \mathbf{x}_2, \ldots, \mathbf{x}_n \mid \boldsymbol{\Sigma}) \propto |\boldsymbol{\Sigma}|^{-n/2} \exp(-n\, tr \boldsymbol{\Sigma}^{-1}\mathbf{S}_\mu/2), \tag{10.42}$$

where $\mathbf{S}_\mu = n^{-1}\sum_i(\mathbf{x}_i - \boldsymbol{\mu})(\mathbf{x}_i - \boldsymbol{\mu})'$. In the one-dimensional case, $\boldsymbol{\Sigma}$ and $\mathbf{S}_\mu$ are scalars σ^2 and s_μ^2, and (10.42) becomes

$$f(\mathbf{x} \mid \sigma^2) \propto \sigma^{-n} \exp(-n\sigma^{-2}s_\mu^2/2). \tag{10.43}$$

The natural conjugate family of prior distributions for σ^2 is the inverse-gamma family. If σ^2 has the $IG(a, d)$ prior distribution (9.8) then the posterior distribution is $IG(a{+}ns_\mu^2, d{+}n)$.

Therefore in particular

$$E(\sigma^2 \mid \mathbf{x}) = (a + ns_\mu^2)/(d + n - 2),$$

$$\operatorname{var}(\sigma^2 \mid \mathbf{x}) = 2(a + ns_\mu^2)^2/\{(d + n - 2)^2(d + n - 4)\}$$

and the posterior mode is $(a + ns_\mu^2)/(d + n + 2)$. The posterior mean is a weighted average of the prior mean and s_μ^2 with weights $d - 2$ and n. The skewness of $f(\sigma^2 \mid \mathbf{x})$ decreases as the hyperparameter d increases, so the posterior distribution will be more symmetric than the prior.

10.32 In the more general p-dimensional case, the covariance matrix $\boldsymbol{\Sigma}$ could be entirely unknown or we might have knowledge about its structure. The strongest assumption is that $\boldsymbol{\Sigma}$ is known up to a single unknown scaling parameter, i.e. $\boldsymbol{\Sigma} = \sigma^2\mathbf{C}$, where $\mathbf{C}$ is known. Then (10.42) becomes

$$f(\mathbf{x}_1, \mathbf{x}_2, \ldots, \mathbf{x}_n \mid \sigma^2) \propto (\sigma^2)^{-np/2} \exp(-n\sigma^{-2}c/2), \tag{10.44}$$

where $c = tr(\mathbf{C}^{-1}\mathbf{S}_\mu) = n^{-1}\sum_i(\mathbf{x}_i - \boldsymbol{\mu})'\mathbf{C}^{-1}(\mathbf{x}_i - \boldsymbol{\mu})$ and we have used the fact that $|\sigma^2\mathbf{C}| = \sigma^{2p}|\mathbf{C}|$. The inverse gamma prior is again the natural conjugate for this problem, and the posterior distribution is $IG(a + nc, d + np)$. Its mean and variance are

$$E(\sigma^2 \mid \mathbf{x}_1, \mathbf{x}_2, \ldots, \mathbf{x}_n) = (a + nc)/(d + np - 2),$$

$$\text{var}\,(\sigma^2 \mid \mathbf{x}_1, \mathbf{x}_2, \ldots, \mathbf{x}_n) = 2(a + nc)^2/\{(d + np - 2)^2(d + np - 4)\}.$$

The posterior mode is $(a + nc)/(d + np + 2)$.

Example 10.6

Many other special structures might be assumed for $\boldsymbol{\Sigma}$. One is to suppose that all its diagonal elements are equal and all its off-diagonal elements are equal. Then we can write $\boldsymbol{\Sigma} = \sigma^2(\mathbf{I}_p - p^{-1}\mathbf{J}_p) + \tau^2 p^{-1}\mathbf{J}_p$, where $\mathbf{I}_p$ is the $p \times p$ identity matrix and $\mathbf{J}_p$ the $p \times p$ matrix of ones. $\boldsymbol{\Sigma}$ is positive definite for all $\sigma^2 > 0$ and $\tau^2 > 0$. Then since $|\boldsymbol{\Sigma}| = \tau^2(\sigma^2)^{p-1}$ and $\boldsymbol{\Sigma}^{-1} = \sigma^{-2}(\mathbf{I}_p - p^{-1}\mathbf{J}_p) + \tau^{-2}p^{-1}\mathbf{J}_p$, (10.42) now becomes

$$f(\mathbf{x}_1, \mathbf{x}_2, \ldots, \mathbf{x}_n \mid \sigma^2, \tau^2) \propto (\sigma^2)^{-n(p-1)/2}(\tau^2)^{-n/2} \exp(-n\sigma^{-2}c/2 - n\tau^{-2}d/2), \tag{10.45}$$

where $c = tr\{(\mathbf{I}_p - p^{-1}\mathbf{J}_p)\mathbf{S}_\mu\} = n^{-1}\sum_i(\mathbf{x}_i - \boldsymbol{\mu})'(\mathbf{x}_i - \boldsymbol{\mu}) - d$, $d = tr\,(p^{-1}\mathbf{J}_p\mathbf{S}_\mu) = n^{-1}p\sum_i(\bar{x}_i - \bar{\mu})^2$ and $\bar{x}_i$ and $\bar{\mu}$ are the averages of the p elements of $\mathbf{x}_i$ and $\boldsymbol{\mu}$.

Now the natural conjugate prior gives σ^2 and τ^2 independent inverse-gamma distributions. They then have independent inverse-gamma posterior distributions and posterior summarization is straightforward. The independence between σ^2 and τ^2 is an obvious restriction imposed by the natural conjugate family. Exercise 10.6 concerns a generalization of this example.

Inverse-Wishart distribution

10.33 If $\boldsymbol{\Sigma}$ is unknown with no special structure assumed, then the likelihood is (10.42) and the prior distribution for $\boldsymbol{\Sigma}$ must be a distribution over the space of all positive definite symmetric matrices. A family of such distributions is the Wishart family, studied in classical statistics as the distribution of the sample covariance matrix for a sample from a multivariate normal distribution; see Anderson (1958). The natural conjugate family

$$f(\boldsymbol{\Sigma}) \propto |\boldsymbol{\Sigma}|^{-(d+p+1)/2} \exp\{-(tr\boldsymbol{\Sigma}^{-1}\mathbf{A})/2\} \tag{10.46}$$

is in fact the family of *inverse-Wishart* distributions, since $\boldsymbol{\Sigma}^{-1}$ has a Wishart distribution. We denote the distribution with this density by $IW(\mathbf{A}, d)$. Its hyperparameters are a positive definite symmetric matrix $\mathbf{A}$ and a scalar $d > p$. Mardia et al. (1979) give the normalizing constant, so that the $IW(\mathbf{A}, d)$ density function is actually

$$f(\boldsymbol{\Sigma}) = k^{-1}|\mathbf{A}|^{d/2}|\boldsymbol{\Sigma}|^{-(d+p+1)/2} \exp\{-(tr\boldsymbol{\Sigma}^{-1}\mathbf{A})/2\}, \tag{10.47}$$

where

$$k = 2^{dp/2}\pi^{p(p-1)/4}\prod_{i=1}^{p}\Gamma\{(d+1-i)/2\}.$$

Combining (10.42) and (10.46), the posterior distribution is $IW(\mathbf{A}+n\mathbf{S}_\mu, d+n)$.

10.34 To derive summaries of the inverse-Wishart distribution we begin with the moment generating function of $\boldsymbol{\Sigma}^{-1}$. From (10.47),

$$E\{\exp(tr\boldsymbol{\Sigma}^{-1}\mathbf{T})\} = |\mathbf{A}|^{d/2}|\mathbf{A}-2\mathbf{T}|^{-d/2}. \tag{10.48}$$

Now note some standard results of differentiating with respect to the elements of a matrix $\mathbf{X}$,

$$\mathrm{d}(tr\mathbf{XA})/\mathrm{d}\mathbf{X} = \mathbf{A}', \tag{10.49}$$

$$\mathrm{d}|\mathbf{X}|/\mathrm{d}\mathbf{X} = |\mathbf{X}|\mathbf{X}^{-1}. \tag{10.50}$$

So differentiating both sides of (10.48) with respect to $\mathbf{T}$ gives

$$E\{\boldsymbol{\Sigma}^{-1}\exp(tr\boldsymbol{\Sigma}^{-1}\mathbf{T})\} = d|\mathbf{A}|^{d/2}|\mathbf{A}-2\mathbf{T}|^{-d/2}(\mathbf{A}-2\mathbf{T})^{-1},$$

and letting $\mathbf{T}$ go to a zero matrix gives

$$E(\boldsymbol{\Sigma}^{-1}) = d\mathbf{A}^{-1}.$$

This is a standard result for the mean of a Wishart distribution. We cannot find $E(\boldsymbol{\Sigma})$ so easily because we cannot obtain $E\{\exp(tr\boldsymbol{\Sigma}\mathbf{T})\}$. In fact this expectation does not exist because not all moments of $\boldsymbol{\Sigma}$ exist (in the same way as not all moments of the inverse-gamma distribution exist). However, it can be shown that

$$E(\boldsymbol{\Sigma}) = (d-p-1)^{-1}\mathbf{A} \tag{10.51}$$

provided $d > p+1$, see Mardia et al. (1979, p85), and Siskind (1972) derives the variances and covariances of elements of $\boldsymbol{\Sigma}$. Since the hyperparameter $\mathbf{A}$ can be thought of as determining the prior mean of $\boldsymbol{\Sigma}$, there is only the scalar d left to determine all these variances and covariances. The situation is analogous to the Dirichlet distribution, and is another instance of the restrictive nature of natural conjugate distributions in multi-parameter models. The inverse Wishart family cannot represent most forms of practical prior information.

10.35 To find the mode of $\boldsymbol{\Sigma}$ we need only differentiate (10.46) or, more simply, its logarithm. Again using (10.49) and (10.50),

$$\mathrm{d}\log f(\boldsymbol{\Sigma})/\mathrm{d}\boldsymbol{\Sigma}^{-1} = (d+p+1)\boldsymbol{\Sigma} - \mathbf{A},$$

using the fact that $\mathbf{A}$ is symmetric. Equating to zero gives

$$\boldsymbol{\Sigma} = (d+p+1)^{-1}\mathbf{A}. \tag{10.52}$$

That this is the mode of (10.46) follows by noting first that $\log f(\boldsymbol{\Sigma})$ is maximized for the same $\boldsymbol{\Sigma}$ as $f(\boldsymbol{\Sigma})$. Second, we can find this maximum by regarding $f(\boldsymbol{\Sigma})$ as a function of $\boldsymbol{\Sigma}^{-1}$, differentiating with respect to Σ^{-1} instead of Σ. Third, we should really be maximizing over the set of all positive definite symmetric matrices, but since the unconstrained solution (10.52) is positive definite symmetric we do not need to apply the constraint. Fourth, since there are no other turning points of (10.46), this is the unique mode.

Unknown mean and variance

10.36 In the general case of a multivariate normal sample with both $\boldsymbol{\mu}$ and Σ unknown the likelihood is (10.39). The conjugate prior distribution

$$f(\boldsymbol{\mu}, \Sigma) \propto |\Sigma|^{-(d+p+2)/2} \exp\{-c(\boldsymbol{\mu}-\mathbf{a})'\Sigma^{-1}(\boldsymbol{\mu}-\mathbf{a})/2 - (tr\Sigma^{-1}\mathbf{A})/2\} \tag{10.53}$$

is known as the *normal-inverse-Wishart* distribution $NIW(\mathbf{A}, d, \mathbf{a}, c)$. Its hyperparameters are positive scalars d and c, a positive definite symmetric matrix $\mathbf{A}$ and a vector $\mathbf{a}$. Then combining this with (10.39) we have a posterior distribution

$$f(\boldsymbol{\mu}, \Sigma \mid \mathbf{x}_1, \mathbf{x}_2, \ldots, \mathbf{x}_n) \propto |\Sigma|^{-(d+p+2+n)/2} \exp(-Q/2),$$

where

$$Q = tr\,\Sigma^{-1}(\mathbf{A} + n\mathbf{S}) + c(\boldsymbol{\mu}-\mathbf{a})'\Sigma^{-1}(\boldsymbol{\mu}-\mathbf{a}) + n(\boldsymbol{\mu}-\bar{\mathbf{x}})'\Sigma^{-1}(\boldsymbol{\mu}-\bar{\mathbf{x}}).$$

Expanding the two quadratic forms in $\boldsymbol{\mu}$ and collecting terms as in (10.41) gives

$$Q = tr\Sigma^{-1}\mathbf{A}^\star + c^\star(\boldsymbol{\mu}-\mathbf{a}^\star)'\Sigma^{-1}(\boldsymbol{\mu}-\mathbf{a}^\star),$$

where

$$\mathbf{a}^\star = (c+n)^{-1}(c\mathbf{a} + n\bar{\mathbf{x}}), \tag{10.54}$$

$$c^\star = c + n, \tag{10.55}$$

$$\mathbf{A}^\star = \mathbf{A} + n\mathbf{S} + cn(c+n)^{-1}(\mathbf{a}-\bar{\mathbf{x}})(\mathbf{a}-\bar{\mathbf{x}})'. \tag{10.56}$$

Therefore the posterior distribution is $NIW(\mathbf{A}^\star, d^\star, \mathbf{a}^\star, c^\star)$, where $d^\star = d + n$.

Example 10.7

Example 1.6 deals with the case $p = 1$. Then $\boldsymbol{\mu}$, Σ, $\mathbf{a}$ and $\mathbf{A}$ all reduce to scalars and we obtain a family of bivariate distributions with four hyperparameters.

10.37 To summarize the normal-inverse-Wishart distributions we can easily find marginal and conditional distributions. First regarding (10.53) as a function of $\boldsymbol{\mu}$ for fixed Σ,

$$f(\boldsymbol{\mu} \mid \Sigma) \propto \exp\{-c(\boldsymbol{\mu}-\mathbf{a})'\Sigma^{-1}(\boldsymbol{\mu}-\mathbf{a})\},$$

so that the conditional distribution of $\boldsymbol{\mu}$ given Σ is $N(\mathbf{a}, c^{-1}\Sigma)$.

Integrating (10.53) with respect to $\boldsymbol{\mu}$ involves simply integrating out this normal distribution, and the result is just (10.46). Therefore the marginal distribution of Σ is $IW(\mathbf{A}, d)$. Then the posterior mean of Σ is

$$E(\Sigma \mid \mathbf{x}_1, \mathbf{x}_2, \ldots, \mathbf{x}_n) = (d^\star - p - 1)^{-1}\mathbf{A}^\star,$$

which is a weighted average of three components – the prior mean (10.51) with weight $(d-p-1)$, the classical unbiased estimator $(n-1)^{-1}\mathbf{S}$ with weight $n-1$ and a third estimator

$$\hat{\Sigma} = cn(c+n)^{-1}(\mathbf{a}-\bar{\mathbf{x}})(\mathbf{a}-\bar{\mathbf{x}})' \tag{10.57}$$

with weight 1. Since the ditribution of $\bar{\mathbf{x}}$ given $\boldsymbol{\mu}$ and Σ is $N(\boldsymbol{\mu}, n^{-1}\Sigma)$ and the distribution of $\boldsymbol{\mu}$ given Σ is $N(\mathbf{a}, c^{-1}\Sigma)$, the distribution of $\bar{\mathbf{x}}$ given just Σ is $N(\mathbf{a}, (c^{-1}+n^{-1})\Sigma)$. It follows that the expectation of $(\bar{\mathbf{x}}-\mathbf{a})(\bar{\mathbf{x}}-\mathbf{a})'$ given Σ is $(c^{-1}+n^{-1})\Sigma$, and hence

$$E(\hat{\Sigma} \mid \Sigma) = \Sigma.$$

Therefore $\hat{\Sigma}$ is also a kind of unbiased estimator of Σ, but instead of having expectation Σ given $\boldsymbol{\mu}$ and Σ for all $\boldsymbol{\mu}$ (which is classical unbiasedness) its expectation is Σ given only Σ after integrating out $\boldsymbol{\mu}$ using its prior conditional distribution. Thus, information about Σ is coming from the prior marginal distribution of Σ, from the data alone, and from a combination of the data and the prior conditional distribution of $\boldsymbol{\mu}$ given Σ. The situation is analogous to that discussed in **9.15** for the normal-inverse-gamma distribution.

10.38 Now regarding (10.53) as a function of Σ for fixed $\boldsymbol{\mu}$ write

$$f(\Sigma \mid \boldsymbol{\mu}) \propto |\Sigma|^{-(d+p+2)/2} \exp\{-(tr\Sigma^{-1}\mathbf{A}_\mu)/2\},$$

where $\mathbf{A}_\mu = \mathbf{A} + c(\boldsymbol{\mu}-\mathbf{a})(\boldsymbol{\mu}-\mathbf{a})'$. Therefore the conditional distribution of Σ given $\boldsymbol{\mu}$ is $IW(\mathbf{A}_\mu, d+1)$.

Then integrating (10.53) with respect to Σ implies integrating over this inverse-Wishart distribution, and from (10.47)

$$\begin{aligned} f(\boldsymbol{\mu}) &\propto |\mathbf{A}_\mu|^{-(d+1)/2} = |\mathbf{A} + c(\boldsymbol{\mu}-\mathbf{a})(\boldsymbol{\mu}-\mathbf{a})'|^{-(d+1)/2} \\ &\propto \{c^{-1} + (\boldsymbol{\mu}-\mathbf{a})'\mathbf{A}^{-1}(\boldsymbol{\mu}-\mathbf{a})\}^{-(d+1)/2} \end{aligned} \tag{10.58}$$

using the same determinant identity as for (10.22). This is the $t_{d+1-p}(\mathbf{a}, c^{-1}\mathbf{A})$ distribution. Its mean and variance are therefore $E(\boldsymbol{\mu}) = \mathbf{a}$ and $\text{var}(\boldsymbol{\mu}) = (d-p-1)^{-1}c^{-1}\mathbf{A}$. The posterior mean of $\boldsymbol{\mu}$ is $\mathbf{a}^*$, given by (10.54). It is a simple weighted average of the prior mean $\mathbf{a}$ and the sample mean $\bar{\mathbf{x}}$.

10.39 The natural conjugate family, as before, is too restrictive to represent most realistic prior information. The conditional distributions establish a specific form of association between $\boldsymbol{\mu}$ and Σ, so that the conditional variance of $\boldsymbol{\mu}$ is a multiple of Σ. Therefore acquiring knowledge about Σ affects the *strength* of information about $\boldsymbol{\mu}$. Conversely, learning about how far $\boldsymbol{\mu}$ is from its prior mean $\mathbf{a}$ affects the estimate of Σ. Another aspect of this constraint is seen in the marginal distributions. The prior covariance matrix of $\boldsymbol{\mu}$ must be a multiple of the prior expectation of Σ. The prior distribution of Σ, being inverse-Wishart also suffers from the same restrictions as we identified in **10.34** for the case of known $\boldsymbol{\mu}$. Essentially, only the two scalars c and d exist to determine the entire variance-covariance structure of $\boldsymbol{\mu}$ and Σ.

Alternative prior distributions

10.40 In order to analyse a multivariate normal sample with more realistic prior information, it is necessary to find more flexible families of prior distributions than the inverse-Wishart and normal-inverse-Wishart. Some alternatives to the inverse-Wishart family are considered by Dickey et al. (1985), Leonard and Hsu (1992) and Brown et al. (1993), but in general it is difficult to construct distributions over the space of positive definite symmetric matrices. It is possible to generalize the normal-inverse-Wishart family to a conditional conjugate family in the same way as the normal-inverse-gamma family was generalized in **9.46**. We can then integrate the posterior density with respect to $\boldsymbol{\mu}$, leaving the posterior distribution $f(\boldsymbol{\Sigma} \mid \mathbf{x}_1, \mathbf{x}_2, \ldots, \mathbf{x}_n)$. This, however, must be summarized numerically, which is a major computational task unless p is small.

Multivariate linear model

10.41 The distributions we have used in considering a multivariate normal sample, and much of the algebra, are similar to the analysis of the normal linear model in Chapter 9. We can in fact generalize both models to define a multivariate linear model. Some details and related models are given by Tiao and Zellner (1964), Zellner and Chetty (1965), Box and Tiao (1973) and Richard and Steel (1988).

Dynamic linear models

10.42 Another generalization is the dynamic linear model, introduced by Harrison and Stevens (1976) to model time series data. Consider a series of observations $\mathbf{y}_1, \mathbf{y}_2, \mathbf{y}_3, \ldots$, such that the observation $\mathbf{y}_t$ at time t follows a linear model, known as the *observation equation*,

$$\mathbf{y}_t = \mathbf{X}_t\boldsymbol{\beta}_t + \boldsymbol{\epsilon}_t \tag{10.59}$$

with $\boldsymbol{\epsilon}_t$ distributed as $N(\mathbf{0}, \mathbf{C}_t)$ and $\mathbf{X}_t$ a known matrix which may be different at each time point. The model relates the parameters $\boldsymbol{\beta}_t$ at successive time points through a *system equation*

$$\boldsymbol{\beta}_t = \mathbf{G}\boldsymbol{\beta}_{t-1} + \boldsymbol{\omega}_t, \tag{10.60}$$

where $\mathbf{G}$ is a known constant matrix and $\boldsymbol{\omega}_t$ is another 'error' vector distributed as $N(\mathbf{0}, \mathbf{W}_t)$. The formulation is completed by a prior distribution $N(\mathbf{m}_0, \mathbf{V}_0)$ for $\boldsymbol{\beta}_0$. The interpretation of the model is that there is an underlying parameter vector $\boldsymbol{\beta}$ which is evolving through time, so that its value at time t is $\boldsymbol{\beta}_t$. The system equation defines the evolution of $\boldsymbol{\beta}$ as a Markov chain (i.e. $\boldsymbol{\beta}_t$ depends only on the last value $\boldsymbol{\beta}_{t-1}$). We do not, however, observe the underlying system parameters, but the observation $\mathbf{y}_t$ at time t is related to the system parameter $\boldsymbol{\beta}_t$ through the observation equation. Suppose first that all variances are known.

10.43 Given the prior $N(\mathbf{m}_0, \mathbf{V}_0)$ for $\boldsymbol{\beta}_0$, we can use the system equation (10.60) to find the distribution of $\boldsymbol{\beta}_1$, i.e. $N(\mathbf{G}\mathbf{m}_0, \mathbf{G}\mathbf{V}_0\mathbf{G}' + \mathbf{W}_0)$. This is the distribution of $\boldsymbol{\beta}_1$ given the prior information, which is the information available at time 0. We now observe $\mathbf{y}_1$, and through the observation equation (10.59) we can deduce the posterior distribution of

$\boldsymbol{\beta}_1$ using the linear model theory of **9.42**. This is the distribution of $\boldsymbol{\beta}_1$ given information available at time 1, and is again a normal distribution. We can clearly continue to work through in this way from one time interval to the next. The general results are as follows.

10.44 Let the information available at time t be denoted by $\mathbf{y}^t = (\mathbf{y}_1, \mathbf{y}_2, \ldots, \mathbf{y}_t)$. Let the distribution of $\boldsymbol{\beta}_{t-1}$ given $\mathbf{y}^{t-1}$ be $N(\mathbf{m}_{t-1}, \mathbf{V}_{t-1})$. Then the distribution of $\boldsymbol{\beta}_t$ given $\mathbf{y}^{t-1}$ is $N(\mathbf{G}\mathbf{m}_{t-1}, \mathbf{R}_t)$, where

$$\mathbf{R}_t = \mathbf{G}\mathbf{V}_{t-1}\mathbf{G}' + \mathbf{W}_t. \tag{10.61}$$

Then we observe $\mathbf{y}_t$, so that the information becomes $\mathbf{y}^t$. Using (9.60) and (9.61), the posterior distribution of $\boldsymbol{\beta}_t$, i.e. its distribution given $\mathbf{y}^t$, is $N(\mathbf{m}_t, \mathbf{V}_t)$, where

$$\mathbf{m}_t = \mathbf{V}_t(\mathbf{R}_t^{-1}\mathbf{G}\mathbf{m}_{t-1} + \mathbf{X}_t'\mathbf{C}_t^{-1}\mathbf{y}_t), \tag{10.62}$$

$$\mathbf{V}_t = (\mathbf{R}_t^{-1} + \mathbf{X}_t'\mathbf{C}_t^{-1}\mathbf{X}_t)^{-1}. \tag{10.63}$$

Equations (10.61) to (10.63) define a recursive procedure for deriving the distribution of the process parameter at time t, i.e. $\boldsymbol{\beta}_t$, given all the information $\mathbf{y}^t$ available at time t. It is this distribution which is most important in the context where we wish to forecast the future evolution of the system parameters, or to forecast future observations.

10.45 Forecasting is easily achieved using the observation and system equations. Given $\mathbf{y}^t$ the distribution of $\boldsymbol{\beta}_t$ is $N(\mathbf{m}_t, \mathbf{V}_t)$, and we can then use the system equation (10.60) k times to obtain the distribution of $\boldsymbol{\beta}_{t+k}$ given $\mathbf{y}^t$. To forecast a future observation $\mathbf{y}_{t+k}$ we simply apply the observation equation once to derive the distribution of $\mathbf{y}_{t+k}$ given $\mathbf{y}^t$ from that of $\boldsymbol{\beta}_{t+k}$ given $\mathbf{y}^t$. Extensive applications and variations on this basic system may be found in West and Harrison (1989) and Queen and Smith (1993). Gamerman and Migon (1993) present a model combining the features of dynamic linear models and hierarchical models.

10.46 In the forecasting context, the recursive nature of the basic equations is also important. At time t, the full set of parameters relating to the data $\mathbf{y}^t$ is $\boldsymbol{\beta}^t = (\boldsymbol{\beta}_1, \boldsymbol{\beta}_2, \ldots, \boldsymbol{\beta}_t)$. In a long series of data, the dimensionality of the parameter space for $\boldsymbol{\beta}^t$ would rapidly become unmanageable. It is therefore important that we only need to retain information on $\boldsymbol{\beta}_t$ given $\mathbf{y}_t$ in order both to forecast and to apply Bayes' theorem to new data. In order to make the recursive equations as computationally efficient as possible, it is usual to convert (10.63) to

$$\mathbf{V}_t = \mathbf{R}_t - \mathbf{A}_t\mathbf{H}_t\mathbf{A}_t', \tag{10.64}$$

where

$$\mathbf{H}_t = \mathbf{C}_t + \mathbf{X}_t\mathbf{R}_t\mathbf{X}_t', \tag{10.65}$$

$$\mathbf{A}_t = \mathbf{R}_t\mathbf{X}_t'\mathbf{H}_t^{-1}. \tag{10.66}$$

Then applying (10.64) to (10.62) we have

$$\mathbf{m}_t = \mathbf{G}\mathbf{m}_{t-1} + \mathbf{A}_t(\mathbf{y}_t - \mathbf{X}_t\mathbf{G}\mathbf{m}_{t-1}). \tag{10.67}$$

The recursion then consists of (10.61), (10.65), (10.66), (10.67) and (10.64), requiring only one matrix inversion in (10.66). This system is known to engineers as the *Kalman filter*, originally attributed to Kalman (1963).

Example 10.8

The 'steady' model relates a scalar observation y_t to a scalar parameter β_t by $y_t = \beta_t + \epsilon_t$. The process evolves randomly through the system equation $\beta_t = \beta_{t-1} + \omega_t$. Let the error variances $\mathbf{C}_t$ and $\mathbf{W}_t$ be scalars c and w, independent of t. The prior distribution of β_0 is $N(m_0, v_0)$. Then $R_t = v_{t-1}+w$, $H_t = c+R_t = c+v_{t-1}+w$, $A_t = R_t/H_t = (v_{t-1}+w)/(c+v_{t-1}+w)$,

$$m_t = m_{t-1} + A_t(y_t - m_{t-1})$$
$$= \frac{(v_{t-1}+w)y_t + cm_{t-1}}{v_{t-1}+w+c},$$

$$v_t = R_t - R_t^2/H_t = \frac{(v_{t-1}+w)c}{v_{t-1}+w+c}. \tag{10.68}$$

Repeated use of (10.68) produces $v_1, v_2, v_3, \ldots$, starting from v_0. It is clear that v_t converges to the positive solution v of the quadratic equation $v^2 + vw - wc = 0$. Then m_t becomes a weighted average of y_t and m_{t-1} with constant weights for large t. We can then expand to obtain

$$m_t = (1-a)y_t + am_{t-1} = (1-a)y_t + a(1-a)y_{t-1} + a^2 m_{t-2}$$
$$= (1-a)(y_t + ay_{t-1} + a^2 y_{t-2} + a^3 y_{t-3} + \ldots),$$

where $a = c/(v+w+c)$. This form of estimate is known in the study of time series as an exponentially weighted moving average; see for instance Cryer (1986). The fact that the posterior variance converges to a positive constant is in apparent contradiction to the usual Bayesian result that if we obtain an infinite amount of data posterior variances will tend to zero (**3.21**). However, for each t we are concerned with the posterior distribution of a new parameter β_t. We never obtain an infinite amount of information about the current β_t. In the equilibrium situation for large t, the gain in information from the new observation y_t is exactly balanced by the loss of information as we pass from β_{t-1} to β_t through the system equation.

10.47 If variances are not known, we might assume them known up to scalar multiples, as in $\mathbf{C}_t = \sigma^2 \mathbf{D}_t$ and $\mathbf{W}_t = \tau^2 \mathbf{Q}_t$. Now given data $\mathbf{y}^t$, we can regard the above analysis as yielding the conditional posterior distribution of $\boldsymbol{\beta}_t$ given σ^2 and τ^2, i.e. $f(\boldsymbol{\beta}_t \mid \sigma^2, \tau^2, \mathbf{y}^t)$. We then need also to derive the marginal posterior distribution $f(\sigma^2, \tau^2 \mid \mathbf{y}^t)$. That would allow inferences to be made about σ^2 and τ^2, and about $\boldsymbol{\beta}_t$ through its normal conditional distribution. For instance, $E(\boldsymbol{\beta}_t \mid \mathbf{y}^t) = E(\mathbf{m}_t \mid \sigma^2, \tau^2, \mathbf{y}^t)$, where $\mathbf{m}_t$ is now regarded as a function of σ^2 and τ^2. However, this is difficult to put into practice and certainly implies a substantial loss of the benefits of recursive computation offered by the Kalman filter. There are two senses in which this is true. First, the Kalman filter does not allow us to obtain $\mathbf{m}_t$ recursively as a function of σ^2 and τ^2. Second, we cannot use the Kalman filter to derive $f(\sigma^2, \tau^2 \mid \mathbf{y}^t)$, and in fact this is a complex calculation to perform.

Harrison and Stevens (1976) approach this problem by allowing σ^2 and τ^2 to take a small number of possible values, as an example of a general procedure they call multiprocess modelling. They show how it is then possible to make inference recursively by running parallel versions of the Kalman filter, one for each combination of possible σ^2 and τ^2 values. These methods are further refined in West and Harrison (1989).

Inference about functions

10.48 In **10.19** we constructed a distribution, the Dirichlet process, for an unknown distribution function $F(x)$. A distribution function is a special kind of function, constrained to be increasing (or non-decreasing) from 0 to 1 as x goes from $-\infty$ to ∞, and the Dirichlet process is appropriate to this case. We now consider formulating prior beliefs about a general unknown function $z(x)$. Let the prior expectation of $z(x)$ be $m(x)$, for all x, so that the function m becomes the prior expectation of the unknown function z. Next let the prior covariance between $z(x)$ and $z(x^\star)$ be $w(x, x^\star)$, for all x and $x^\star$. Then in particular $\text{var}\,(z(x)) = w(x, x)$. Now let the marginal distribution of $z(x)$ be normal for every x, let the joint distribution of $z(x)$ and $z(x^\star)$ be bivariate normal for every x and $x^\star$, and in general let the joint distribution of z at any p specific points be a p-dimensional multivariate normal distribution.

There is no inconsistency in assuming joint normality in this way for all sets of p points on the function and for arbitrarily large p. It can be shown that this procedure formally defines a distribution for the entire unknown function z. We call this distribution a Gaussian process, and denote the distribution of z by $GP(m, w)$. The 'hyperparameters' of the distribution are the mean function m and the covariance function w. Gaussian processes have been studied quite extensively, and occur in many applications of probability theory; see for instance Doob (1953), Grimmett and Stirzaker (1982).

10.49 We then observe the function z at n points $x_1, x_2, \ldots, x_n$. If these observations are $\mathbf{y} = (y_1, y_2, \ldots, y_n)'$, define $\mathbf{z} = (z(x_1), z(x_2), \ldots, z(x_n))'$, then let $\mathbf{y} = \mathbf{z} + \boldsymbol{\epsilon}$, where $\boldsymbol{\epsilon}$ is a vector of observation errors distributed as $N(\mathbf{0}, \mathbf{C})$. We wish to derive the posterior distribution of the function z after observing $\mathbf{y}$. Now the joint distribution of $\mathbf{y}$ and $z(x)$ is normal with

$$E\begin{pmatrix} \mathbf{y} \\ z(x) \end{pmatrix} = \begin{pmatrix} \mathbf{m} \\ m(x) \end{pmatrix}, \qquad \text{var}\begin{pmatrix} \mathbf{y} \\ z(x) \end{pmatrix} = \begin{pmatrix} \mathbf{W} + \mathbf{C} & \mathbf{w}(x)' \\ \mathbf{w}(x) & w(x,x) \end{pmatrix},$$

where

$$\mathbf{m} = \begin{pmatrix} m(x_1) \\ m(x_2) \\ \vdots \\ m(x_n) \end{pmatrix}, \qquad \mathbf{w}(x) = \begin{pmatrix} w(x_1, x) \\ w(x_2, x) \\ \vdots \\ w(x_n, x) \end{pmatrix},$$

$$\mathbf{W} = \begin{pmatrix} w(x_1,x_1) & w(x_1,x_2) & \ldots & w(x_1,x_n) \\ w(x_2,x_1) & w(x_2,x_2) & \ldots & w(x_2,x_n) \\ \vdots & \vdots & & \vdots \\ w(x_n,x_1) & w(x_n,x_2) & \ldots & w(x_n,x_n) \end{pmatrix}.$$

Therefore the conditional distribution of $z(x)$ given $\mathbf{y}$ is normal with mean

$$m^\star(x) = m(x) + \mathbf{w}(x)'(\mathbf{W}+\mathbf{C})^{-1}(\mathbf{y}-\mathbf{m}) \tag{10.69}$$

and variance

$$w^\star(x,x) = w(x,x) - \mathbf{w}(x)'(\mathbf{W}+\mathbf{C})^{-1}\mathbf{w}(x). \tag{10.70}$$

This is the posterior distribution of a single arbitrary point $z(x)$ on the function z. The argument generalizes trivially to show that the posterior distribution of any p points on the function is multivariate normal. Hence the posterior distribution of z is $GP(m^\star, w^\star)$, where the posterior mean function $m^\star$ is given by (10.69) and the posterior covariance function is obtained by generalizing (10.70) to

$$w^\star(x,x^\star) = w(x,x^\star) - \mathbf{w}(x)'(\mathbf{W}+\mathbf{C})^{-1}\mathbf{w}(x^\star). \tag{10.71}$$

Nonparametric regression

10.50 This result forms the basis of Bayesian analyses of a variety of problems with unknown functions. One application is to regression. The simple linear regression model $y_i = \alpha + \beta x_i + \epsilon_i$ implies an underlying linear relationship that we might denote by $z(x) = \alpha + \beta x$. A general linear model similarly implies a specific form of regression relationship between the response variable and the regressor variables. By allowing instead an arbitrary relationship and assigning a Gaussian process prior distribution to the regression function z, we avoid assuming any specific form for the regression.

It may, however, be reasonable to suppose *a priori* that z will at least approximate to a simple form such as the linear regression $z(x) = \alpha + \beta x$. We could represent this by giving z a GP prior distribution but with prior expectation $m(x) = \alpha + \beta x$. This leads to a hierarchical model. In general, we can give $m(x)$ a linear model form

$$m(x) = \mathbf{h}(x)'\boldsymbol{\beta}, \tag{10.72}$$

where $\mathbf{h}(x)$ is known and defines the expected form of the regression, and $\boldsymbol{\beta}$ is a vector of unknown parameters. (For instance, $\mathbf{h}(x) = (1,x)'$ would correspond to a linear expectation $m(x) = \alpha + \beta x$ with unknown α and β.) Then a prior distribution, such as $N(\mathbf{b},\mathbf{B})$, for $\boldsymbol{\beta}$ completes the hierarchical model.

10.51 In fact $\boldsymbol{\beta}$ can now be removed to obtain the implied marginal prior distribution for z, which is a Gaussian process with mean function $\mathbf{h}(x)'\mathbf{b}$ and covariance function $w(x,x^\star) + \mathbf{h}(x)'\mathbf{B}\mathbf{h}(x^\star)$. We can therefore apply (10.69) and (10.71) directly with these functions replacing $m(x)$ and $w(x,x^\star)$, to obtain the posterior distribution of z. In particular we find $\mathbf{W}$ is replaced by $\mathbf{W}+\mathbf{H}'\mathbf{B}\mathbf{H}$ where $\mathbf{H} = (\mathbf{h}(x_1), \mathbf{h}(x_2), \ldots, \mathbf{h}(x_n))$. Then $(\mathbf{W}+\mathbf{C})^{-1}$ becomes

$$(\mathbf{W}+\mathbf{C}+\mathbf{H}'\mathbf{B}\mathbf{H})^{-1} = (\mathbf{W}+\mathbf{C})^{-1} - (\mathbf{W}+\mathbf{C})^{-1}\mathbf{H}'\{\mathbf{B}^{-1}+\mathbf{H}(\mathbf{W}+\mathbf{C})^{-1}\mathbf{H}'\}^{-1}\mathbf{H}(\mathbf{W}+\mathbf{C})^{-1}. \tag{10.73}$$

Similarly, $\mathbf{m}$ is replaced by $\mathbf{H}'\mathbf{b}$ and $\mathbf{w}(x)$ becomes $\mathbf{w}(x) + \mathbf{H}'\mathbf{B}\mathbf{h}(x)$.

Now let $\mathbf{B}^{-1} = \mathbf{0}$ to represent weak prior information about $\boldsymbol{\beta}$. Then the posterior distribution of z is a Gaussian process whose mean function can be expressed in the form

$$\hat{z}(x) = \mathbf{h}(x)'\hat{\boldsymbol{\beta}} + \mathbf{w}(x)'(\mathbf{W}+\mathbf{C})^{-1}(\mathbf{y}-\mathbf{H}'\hat{\boldsymbol{\beta}}), \tag{10.74}$$

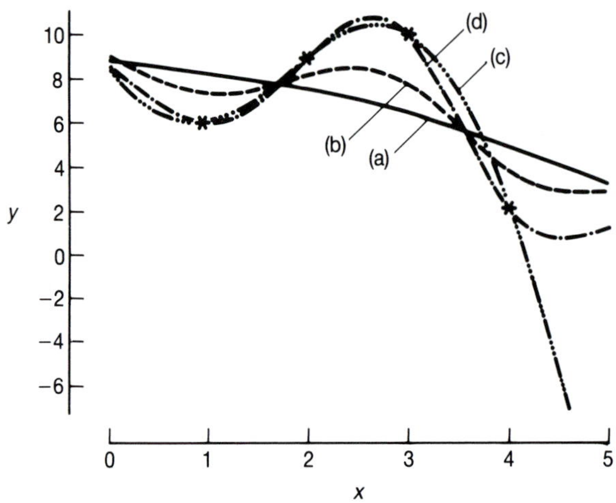

Fig. 10.1 Nonparametric regression estimators. (a) $r = 0.5$, $b = 0.1$; (b) $r = 0.5$, $b = 1$; (c) $r = 0$, $b = 0.1$; (d) $r = 0$, $b = 1$

where

$$\hat{\boldsymbol{\beta}} = \{\mathbf{H}(\mathbf{W}+\mathbf{C})^{-1}\mathbf{H}'\}^{-1}\mathbf{H}(\mathbf{W}+\mathbf{C})^{-1}\mathbf{y}. \tag{10.75}$$

The mean function (10.74) is a non-parametric regression estimator, comprising a fitted regression term $\mathbf{h}(x)'\hat{\boldsymbol{\beta}}$ in which the parameter vector $\boldsymbol{\beta}$ is estimated by the generalized least squares form (10.75), plus a term which estimates the departure of z from the strict linear model form represented by (10.72), based on the residuals $\mathbf{y} - \mathbf{H}'\hat{\boldsymbol{\beta}}$.

Example 10.9

In Figure 10.1, four observations $\mathbf{y} = (6, 9, 10, 2)$ are shown at equally spaced points on the function z. The underlying regression defining the prior mean is linear, given by $\mathbf{h}(x) = (1, x)'$. The covariance function of the departures from the linear regression is $w(x, x^\star) = \tau^2 \exp\{-b(x-x^\star)^2\}$ and the observation error variance is $\mathbf{C} = \sigma^2\mathbf{I}_4$. The estimator (10.74) depends on σ^2 and τ^2 only through their ratio $r = \sigma^2/\tau^2$, which determines how closely $\hat{z}(x_i)$ comes to the actual observations y_i. Large σ^2 implies large observation error, and the estimator $\hat{z}(x)$ lies close to the fitted line $\mathbf{h}(x)'\hat{\boldsymbol{\beta}}$ for all x. At the other estreme, if σ^2 goes to zero then we observe $z(x_i)$ without error and the estimator $\hat{z}(x)$ will pass exactly through the observations.

The parameter b determines how smooth z is believed to be. For small b, the correlation between $z(x)$ and $z(x^\star)$ will be high, even when x and $x^\star$ are far apart, so that the function will have a high degree of smoothness. Figure 10.1 shows $\hat{z}(x)$ for four combinations of b and σ^2/τ^2.

10.52 Bayesian nonparametric regression models based on Gaussian processes are presented by Blight and Ott (1975), O'Hagan (1978) and Angers and Delampady (1992).

Silverman (1985) employs an alternative Bayesian approach based on spline functions. The work of Wahba (1978,1983) links both Gaussian processes and splines.

Bayesian numerical analysis

10.53 The case $\mathbf{C} = \mathbf{0}$, where the function is observed without error, is useful in developing Bayesian methods for a range of problems traditionally considered in the field of numerical analysis. The two estimators $\hat{z}(x)$ in Figure 10.1 corresponding to $r = 0$ are examples of Bayesian interpolation. In general, when $\mathbf{C} = \mathbf{0}$, the posterior mean function m^* will pass exactly through all the observed points on the curve, providing interpolation between those points. Another example is a Bayesian approach to quadrature developed in O'Hagan (1991b). The function z is then considered to be the integrand in an integral such as $\int_a^b z(x)\mathrm{d}x$. The observations $y_1, y_2, \ldots, y_n$ are the evaluations $z(x_1), z(x_2), \ldots, z(x_n)$ of the integrand at the design points $x_1, x_2, \ldots, x_n$, as would be used in conventional quadrature rules. The posterior expectation of the integral can be shown to be $\int_a^b m^*(x)dx$. Bayesian approaches to numerical analysis generally are considered in Diaconis (1988) and O'Hagan (1992).

Expert opinion

10.54 Suppose that an expert is asked to consider whether an event A will occur, and asserts his or her probability p for that event. In **2.54** we considered the use of scoring rules as loss functions for the expert's task of assigning that probability. We now consider the use of the expert's opinion as data to assist You, who have an interest in whether A will occur. We therefore wish to obtain Your posterior probability $P(A \mid p)$, based on the received expert's opinion p. From Bayes' theorem,

$$P(A \mid p) = P(A)f(p \mid A)/f(p), \tag{10.76}$$

where $P(A)$ is Your prior probability for A, $f(p \mid A)$ is Your probability distribution for the 'observation' p, and $f(p)$ is the marginal distribution for p given by

$$f(p) = f(p \mid A)P(A) + f(p \mid A^c)P(A^c). \tag{10.77}$$

In addition to Your prior probability $P(A) = 1 - P(A^c)$, application of (10.76) and (10.77) requires Your beliefs about the expert's opinion, expressed in the form of distributions $f(p \mid A)$ and $f(p \mid A^c)$.

Expressed in terms of odds, (10.76) becomes

$$\frac{P(A \mid p)}{P(A^c \mid p)} = \frac{P(A)}{P(A^c)}.B, \tag{10.78}$$

where $B = f(p \mid A)/f(p \mid A^c)$ is a Bayes factor. B determines how Your beliefs about A are influenced by the expert's opinion. One extreme case would be if You believed the expert to be noninformative. This would be expressed by $B = 1$, so that Your posterior probability for A equals Your prior probability and You ignore p. In this case $f(p \mid A) = f(p \mid A^c)$ for all p, and hence p and A are independent.

More realistically, the expert's opinion will have a positive influence on Your beliefs, such that $P(A \mid p)$ is an increasing function of p. This is equivalent to B being an increasing function of p.

10.55 An obvious generalization of the problem is to consider a set of mutually exclusive and exhaustive events $A_1, A_2, \ldots, A_k$. (A single event A then corresponds to $k = 2$, with $A_1 = A$ and $A_2 = A^c$.) You have prior probabilities $P(A_j)$ such that $\sum_{j=1}^k P(A_j) = 1$. An expert then gives probabilities p_j such that $\sum_{j=1}^k p_j = 1$ and You regard $\mathbf{p} = (p_1, p_2, \ldots, p_k)$ as data to modify Your prior probabilities. Your beliefs about the expert's expertise are expressed in distributions $f(\mathbf{p} \mid A_j)$ for $\mathbf{p}$. Your posterior probabilities are then $P(A_j \mid \mathbf{p}) = P(A_j)f(\mathbf{p} \mid A_j)/f(\mathbf{p})$, where now

$$f(\mathbf{p}) = \sum_{j=1}^{k} P(A_j)f(\mathbf{p} \mid \mathbf{A}_j)$$

replaces (10.77).

The case of an uninformative expert is easy to generalize. Your posterior probabilities will equal Your prior probabilities if $f(\mathbf{p} \mid A_j)$ is the same function of $\mathbf{p}$ for every A_j, $j = 1, 2, \ldots, k$. Then $\mathbf{p}$ and $(A_1, A_2, \ldots, A_k)$ are independent.

Accepting the expert's opinion

10.56 The opposite position to ignoring the expert is to accept the expert's probabilities as Your own, i.e. $P(A_j \mid \mathbf{p}) = p_j$ for all j and all $\mathbf{p}$. This certainly entails a positive response to the expert's opinion. Intuitively, it is a reasonable response if You have faith in the expert and if Your own knowledge about which A_j will occur is weak. Consider the set of all problems comprising k mutually exclusive and exhaustive events. If You believe the expert's probability judgements to be *calibrated* over that set (see **6.4**), then if $(E_1, E_2, \ldots, E_k)$ is a randomly chosen problem from that set, $P(E_j \mid \mathbf{p}) = p_j$.

Although $P(E_j \mid \mathbf{p}) = p_j$ will hold if You believe the expert to be calibrated, it could equally well represent the weaker position that You do not have reason to believe otherwise. That is, You have no specific knowledge about the expert's probability judgements that would lead You to assign either a larger or smaller value than p_j to $P(E_j \mid \mathbf{p})$.

Now if You consider Your own knowledge about $(A_1, A_2, \ldots, A_k)$ to be weak, and in particular to be much weaker than the expert's, You may decide that You cannot distinguish $(A_1, A_2, \ldots, A_k)$ from the randomly chosen problem $(E_1, E_2, \ldots, E_k)$ given that the expert has stated probabilities $\mathbf{p}$. Then $P(A_j \mid \mathbf{p}) = P(E_j \mid \mathbf{p})$. Therefore You accept the expert's opinion, and have posterior probabilities $P(A_j \mid \mathbf{p}) = p_j$, if You believe the expert to be calibrated (or have no reason to believe otherwise) and if Your own knowledge about the A_js is much weaker than the expert's.

10.57 Consider further the view that $P(E_j \mid \mathbf{p}) = p_j$ for a randomly chosen problem $(E_1, E_2, \ldots, E_k)$ from the set of all k-probability problems, expressing a belief that the expert's probability judgements are calibrated over that set. Another reasonable assumption, when you do not have particular knowledge about the expert's probability judgements in general, is that Your distributions $f(\mathbf{p} \mid A_j)$ are not specific to the particular problem $(A_1, A_2, \ldots, A_k)$ but would apply to any problem in that set. So we denote $f(\mathbf{p} \mid A_j)$ by $f_j(\mathbf{p})$

as representing Your beliefs about the expert's probabilities $\mathbf{p}$ in any case where the jth event actually occurs. Then

$$P(E_j \mid \mathbf{p}) = \frac{P(E_j)f_j(\mathbf{p})}{\sum_{l=1}^{k} P(E_l)f_l(\mathbf{p})},$$

but we can argue that $P(E_1) = P(E_2) = \ldots = P(E_k) = k^{-1}$ since the set of problems concerning k mutually exclusive and exhaustive events includes permutations of any such problem. Therefore a belief that the expert is calibrated implies

$$p_j = f_j(\mathbf{p})/\textstyle\sum_{l=1}^{k} f_l(\mathbf{p}),$$

or, for any $l \neq j$, $f_j(\mathbf{p})/f_l(\mathbf{p}) = p_j/p_l$. Returning to the original problem, we can conclude that a reasonable representation of a belief that the expert is calibrated might be

$$f(\mathbf{p} \mid A_j)/f(\mathbf{p} \mid A_l) = p_j/p_l \tag{10.79}$$

for all $j \neq l$ and all $\mathbf{p}$.

An immediate consequence of (10.79) is that Your posterior probabilities are

$$P(A_j \mid \mathbf{p}) = \frac{p_j P(A_j)}{\sum_{l=1}^{k} p_l P(A_l)}. \tag{10.80}$$

If, in addition, Your prior information about the A_js is represented by equal probabilities $P(A_1) = P(A_2) = \ldots = P(A_k) = k^{-1}$ then (10.80) reduces to $P(A_j \mid \mathbf{p}) = p_j$ and You will therefore accept the expert's opinions. This will arise in particular if Your prior information is weak.

Models for expert opinion

10.58 The calibration argument leading to (10.79) makes some strong assumptions, and thereby completely determines the likelihood function. The likelihood is $f(\mathbf{p} \mid A_j)$ regarded as a function of j, which (10.79) implies is $f(\mathbf{p} \mid A_j) \propto p_j$. If that argument is not appropriate in a given context, a variety of alternative models are possible. An obvious family of distributions is the Dirichlet family, so we could let $f(\mathbf{p} \mid A_j)$ be the $D(\mathbf{a}_j)$ distribution. The condition (10.79) implies that all the elements of $\mathbf{a}_j$ are equal except the jth. Also, if the common value of the other elements is a, the jth element must be $a+1$, so that

$$f(\mathbf{p} \mid A_j) = \frac{\Gamma(a+1)\{\Gamma(a)\}^{k-1}}{\Gamma(ka+1)} p_j \prod_{l=1}^{k} p_l^{a-1}. \tag{10.81}$$

The following example illustrates the kind of inferences $P(A_j \mid \mathbf{p})$ that are possible using other Dirichlet distributions.

Example 10.10

Let $f(\mathbf{p} \mid A_j)$ be the $D(\mathbf{a}_j)$ distribution in which the jth element of $\mathbf{a}_j$ is $a+t$ and all other elements are a. With $t = 1$ this is (10.81). Then instead of (10.80) we have generally

$$P(A_j \mid \mathbf{p}) = \frac{p_j^t P(A_j)}{\sum_{l=1}^{k} p_l^t P(A_l)}.$$

Table 10.2 Posterior probabilities under two Dirichlet models

j	1	2	3	4
p_j	0.1	0.2	0.3	0.4
$P(A_j \mid \mathbf{p})$, $t=2$	0.033	0.133	0.3	0.533
$P(A_j \mid \mathbf{p})$, $t=0.5$	0.163	0.230	0.282	0.325

If Your prior information is given by $P(A_j) = k^{-1}$ for all j, this reduces to $P(A_j \mid \mathbf{p}) = p_j^t / \sum p_l^t$. As we have seen, $t = 1$ leads to You accepting the expert's opinions. If $t > 1$, Your response to the expert's stated $\mathbf{p}$ is to give the A_js posterior probabilities which vary more than the p_js, whereas if $t < 1$ Your posterior probabilites will vary less than the expert's. Table 10.2 gives an example with $k = 4$.

10.59 An alternative to the Dirichlet family is to define multivariate normal distributions on sets of log constrasts, as in **10.12**. Let $\mathbf{q} = \mathbf{C} \log \mathbf{p}$ be a vector of $k-1$ log contrasts in the p_js, defined by a $(k-1) \times k$ matrix $\mathbf{C}$ whose rows all sum to zero. Given A_j, we now assume that Your beliefs about the expert are represented by a $N(\mathbf{m}_j, \mathbf{V}_j)$ distribution for $\mathbf{q}$. Transforming from $\mathbf{q}$ to $\mathbf{p}$ using the Jacobian (10.22),

$$f(\mathbf{p} \mid A_j) = k \exp\{-(\mathbf{C} \log \mathbf{p} - \mathbf{m}_j)\mathbf{V}_j^{-1}(\mathbf{C} \log \mathbf{p} - \mathbf{m}_j)/2\} \prod_{l=1}^{k} p_l^{-1}, \tag{10.82}$$

where k is a constant. An interesting special case arises if $\mathbf{V}_j = \mathbf{V}$, the same variance matrix for all j. Then (10.82) reduces to

$$f(\mathbf{p} \mid A_j) = k^{\star} c_j \prod_{l=1}^{k} p_l^{b_{jl}-1}, \tag{10.83}$$

where $(b_{j1}, b_{j2}, \ldots, b_{jk}) = \mathbf{m}_j' \mathbf{V}^{-1} \mathbf{C}$,

$$c_j = \exp(-\mathbf{m}_j' \mathbf{V}_j^{-1} \mathbf{m}_j / 2)$$

and $k^{\star}$ does not depend on j. Therefore $k^{\star}$ cancels out when we apply Bayes' theorem. The result then is the same as from a Dirichlet model with $\mathbf{a}_j = \mathbf{C}'\mathbf{V}^{-1}\mathbf{m}_j$, except that c_j will not in general be proportional to the normalizing constant of the $D(\mathbf{a}_j)$ distribution. Related models based on normal distributions are developed by French (1980, 1981) and Lindley (1982).

10.60 From the expert giving a discrete set of probabilities $p_1, p_2, \ldots, p_k$, we can generalize to the expert giving his or her personal probability distribution $p(x)$ for a, possibly continuous, random variable x. The generalization of (10.80) for continuous x is now

$$f(x \mid p) = f(x)p(x) / \int f(z)p(z)\,dz,$$

where $f(x)$ is Your prior probability distribution for x. West (1988) develops a general Dirichlet model for this problem. Lindley (1988) obtains some results generalizing the multivariate normal model.

An alternative approach, based on limited specification of Your beliefs about the expert, is followed by Genest and Schervish (1985) and West and Crosse (1992).

Opinion pooling

10.61 If several experts give You their opinions, You need to state Your beliefs about the experts in terms of joint distributions for their opinions conditional on each A_j. Lindley (1985) generalizes the multivariate normal model to this case. A very different, and less formally Bayesian, approach is to consider directly a rule for *pooling* the opinions of the different experts. Let $P_i(A_j)$ be the probability for A_j given by expert i, $i = 1, 2, \ldots, r$. We require to give a pooled probability $P_0(A_j)$ as a function of these opinions. The previous problem of You receiving a single expert's opinion could be seen as pooling the views of *two* experts. A generalization of (10.80) of the form

$$P_0(A_j) = \frac{\prod_{i=1}^{r} P_i(A_j)}{\sum_{l=1}^{k} \prod_{i=1}^{r} P_i(A_l)}$$

might be generalized further by giving a weight w_i to expert i,

$$P_0(A_j) = \frac{\prod_{i=1}^{r} \{P_i(A_j)\}^{w_i}}{\sum_{l=1}^{k} \prod_{i=1}^{r} \{P_i(A_l)\}^{w_i}}.$$

This is known as the *logarithmic opinion pool.* This and other forms of opinion pool have been proposed by Stone (1961), Weerahandi and Zidek (1981) and Winkler (1968). McConway (1981) and French (1985) discuss various properties of opinion pools and their relationship to formal Bayesian analysis of expert opinions.

Graphical models

10.62 It is possible to build very complex probabilistic structures from quite simple components. Lauritzen and Spiegelhalter (1988) describe what they refer to as a simple example of modelling used in computerized 'expert systems'. Conditional on one of a number of underlying muscle diseases, probabilities are given for a range of pathophysiological disorders, and conditional on any of those there are probabilities for an even larger range of symptoms. Observations may be made of symptoms, or diagnostic tests might identify particular pathophysiological disorders, or ultimately a definite diagnosis may be made of a patient's disease. Wherever information enters the system, it is desired to obtain posterior joint and marginal probabilities for the other unknown quantities. Such models may be thought of as graphs, in which random quantities are *nodes*, and correlated quantities are connected by *arcs*. Typically, there are relatively few arcs, showing that, conditional on the values of nodes to which a given quantity is connected, that quantity is independent of all other nodes in the graph. The concept of conditional independence is crucial to the development of such models. It facilitates building graphs whose structure is very complex, but where each node is connected to, and therefore potentially dependent on, only a few others. Probabilistic rules concerning conditional independence (see, for example, Dawid (1979)) convert into rules for manipulating graphs and the probability distributions associated with arcs. Lauritzen and Spiegelhalter (1988) develop rules which can be applied mechanistically to make posterior computations on very large systems feasible. Many examples of such graphical models and related theory can be found in Oliver and Smith (1989).

EXERCISES

10.1 Let $x_1, x_2, \ldots, x_k$ be independently distributed with gamma distributions, $f(x_i) = a^{b_i} x_i^{b_i - 1} \exp(-a x_i)/\Gamma(b_i)$. Let $y_i = x_i / \sum_{j=1}^{k} x_j$, $i = 1, 2, \ldots, k$. Prove that $\mathbf{y}$ has the $D(\mathbf{b})$ distribution. Use this representation to give an alternative derivation of results in **10.4**.

10.2 Observations n_{11}, n_{12}, n_{21} and n_{22} form a 2×2 contingency table. That is, n_{ij} is the observed count of cases falling into category i in one classification and category j in a second classification. Let θ_{ij} be the probability that each case falls into the (i, j) category, and assume cases are independent so that the likelihood is

$$f(\mathbf{n} \mid \boldsymbol{\theta}) \propto \prod_{i=1}^{2} \prod_{j=1}^{2} \theta_{ij}^{n_{ij}}.$$

(We assume also that only the total number of cases $n = n_{11} + n_{12} + n_{21} + n_{22}$ is fixed. A different likelihood arises if one of the margins of the 2×2 table is fixed.) Your prior belief is that the degree of association between the two classifications is likely to be small, which You express by giving a $N(0, v)$ prior distribution to the log cross-ratio

$$\phi = \log \frac{\theta_{12}\theta_{21}}{\theta_{11}\theta_{22}}.$$

Your prior distribution is completed by giving uniform prior distributions to any two other log contrasts in the θ_{ij}s (so that Your information is otherwise weak). Show that the posterior mode $(\hat{\theta}_{11}, \hat{\theta}_{12}, \hat{\theta}_{21}, \hat{\theta}_{22})$, obtained by solving (10.26), has the property that

$$(n-4)(\hat{\theta}_{i1} + \hat{\theta}_{i2}) = n_{i1} + n_{i2} - 2, \qquad (n-4)(\hat{\theta}_{1j} + \hat{\theta}_{2j}) = n_{1j} + n_{2j} - 2$$

for $i = 1, 2$ and $j = 1, 2$, regardless of v.

10.3 Let F have the Dirichlet process prior distribution $DP(G)$, and let ϕ be the median of F, i.e. $F(\phi) = 1/2$. Observations $y_1, y_2, \ldots, y_n$ are drawn from the distribution with d.f. F, so that the posterior distribution of F is $DP(G + G_n)$, where $n^{-1} G_n$ is the empirical distribution function as in **10.23**. Prove that the posterior median f of ϕ is the median of the posterior mean of F, i.e. $E\{F(f) \mid \mathbf{y}\} = 1/2$, and hence that f lies between the prior median of ϕ and the median of the y_is.

10.4 Following Example 10.4, define $\sigma^2 = \int_{-\infty}^{\infty} y^2 \, \mathrm{d}F(y) - \mu^2$. Prove that

$$E(\sigma^2 \mid \mathbf{y}) + \operatorname{var}(\mu \mid \mathbf{y}) = (g + n)^{-1}[g\{E(\sigma^2) + \operatorname{var}(\mu)\} + nt],$$

where

$$t = n^{-1} \sum (y_i - \bar{y})^2 + (g + n)^{-1} g\{E(\mu) - \bar{y}\}^2.$$

10.5 Generalizing Example 10.5, suppose that the prior distribution for the units $\boldsymbol{\xi} = (\xi_1, \xi_2, \ldots, \xi_n)'$ in a population of size N is formulated hierarchically as a linear model. First $\boldsymbol{\xi}$ is distributed as $N(\mathbf{A}\boldsymbol{\beta}, \sigma^2 \mathbf{I})$ given hyperparameters $(\boldsymbol{\beta}, \sigma^2)$, then $(\boldsymbol{\beta}, \sigma^2)$ has the $NIG(a, d, \mathbf{m}, \mathbf{V})$ prior distribution. Now units $\mathbf{y} = (\xi_1, \xi_2, \ldots, \xi_n)'$ are observed as a sample from the population. Let $\mathbf{X}$ comprise the first n rows of $\mathbf{A}$ and $\mathbf{z}$ be the sum of the remaining $N - n$ rows of $\mathbf{A}$. Prove that the posterior distribution of $\tau = \sum_{i=1}^{N} \xi_i$ is $t_{d+n}(n\bar{\xi} + \mathbf{z}'\mathbf{m}^*, a^*(N - n + \mathbf{z}'\mathbf{V}^*\mathbf{z}))$, where $\bar{\xi} = n^{-1} \sum_{i=1}^{n} \xi_i$ is the sample mean and where $\mathbf{V}^*$, $\mathbf{m}^*$ and a^* are given by (9.16), (9.17) and (9.18).

10.6 Generalizing Example 10.6, suppose that $\boldsymbol{\Sigma} = \sum_{i=1}^{k} \alpha_i \mathbf{A}_i$, where the $\mathbf{A}_i$s are known symmetric non-negative definite matrices satisfying $\mathbf{A}_i \mathbf{A}_i = \mathbf{I}$ for $i = 1, 2, \ldots, k$ and $\mathbf{A}_i \mathbf{A}_j = \mathbf{0}$ for all $i \neq j$, and where the α_is are unknown variance components. Prove that if the α_is have independent inverse-gamma prior distributions then they also have independent inverse-gamma posterior distributions. (O'Hagan, 1977.)

10.7 Prove that in the model of **10.51** the posterior covariance function of z is given by

$$w^{\star}(x, x^{\star}) = w(x, x^{\star}) - \mathbf{w}(x)'(\mathbf{W} + \mathbf{C})^{-1}\mathbf{w}(x^{\star}) + \{\mathbf{h}(x) - \mathbf{H}(\mathbf{W} + \mathbf{C})^{-1}\mathbf{w}(x)\}'\{\mathbf{H}(\mathbf{W} + \mathbf{C})^{-1}\mathbf{H}'\}^{-1}\{\mathbf{h}(x^{\star}) - \mathbf{H}(\mathbf{W} + \mathbf{C})^{-1}\mathbf{w}(x^{\star})\}.$$

REFERENCES

ABRAHAM, B. and BOX, G. E. P. (1978). Linear models and spurious observations. *Appl. Statist.*, **27,** 120–130.

ABRAMOWITZ, M. and STEGUN, I. A. (eds) (1965). *Handbook of Mathematical Functions.* National Bureau of Standards: Washington DC.

ACHCAR, J. A. and SMITH, A. F. M. (1990). Aspects of reparameterization in approximate Bayesian inference. In: *Essays in Honour of G. A. Barnard,* J. Hodges (ed.). North-Holland: Amsterdam.

AITCHISON, J. (1982). The statistical analysis of compositional data (with discussion). *J. R. Statist. Soc.,* **B, 44,** 139–177.

AITCHISON, J. (1985). A general class of distributions on the simplex. *J. R. Statist. Soc.,* **B, 47,** 136–146.

AITCHISON, J. (1986). *The Statistical Analysis of Compositional Data.* Chapman and Hall: London.

AITCHISON, J. and DUNSMORE, I. R. (1975). *Statistical Prediction Analysis.* Cambridge University Press.

AITKIN, M. (1991). Posterior Bayes factors (with discussion). *J. R. Statist. Soc.,* **B, 53,** 111–142.

AKAIKE, H. (1973). Information theory and an extension of the maximum likelihood principle. In: *2nd International Symposium on Information Theory,* 267–281. Akademia Kiado: Budapest.

AKAIKE, H. (1978). A new look at the Bayes procedure. *Biometrika,* **65,** 53–59.

ALBERT, J. H. (1988a). Bayesian estimation of Poisson means using a heirarchical log-linear model. In: *Bayesian Statistics 3.* J. M. Bernardo et al. (ed.), 519–531. Oxford University Press.

ALBERT, J. H. (1988b). Computational methods using a Bayesian heirarchical generalised linear model. *J. Amer. Statist. Ass.,* **83,** 1037–1044.

ALBERT, J. H. and GUPTA, A. K. (1982). Mixtures of Dirichlet distributions and estimation in contingency tables. *Ann. Statist.,* **10,** 1251–1268.

ALPERT, M. and RAIFFA, H. (1982). A progress report on the training of probability assessors. In: *Judgement Under Uncertainty; Heuristics and Biases,* Kahneman, D. et al. (ed.). Cambridge University Press.

ANDERSON, T. W. (1958). *An Introduction to Multivariate Statistical Analysis.* Wiley: New York.

ANGERS, J. F. and BERGER, J. (1991). Robust hierarchical Bayes estimation of exchangeable means. *Canad. J. Statist,* **19,** 39–56.

ANGERS, J. F. and DELAMPADY, M. (1992). Hierarchical Bayesian curve fitting and smoothing. *Canad. J. Statist.,* **20,** 35–49.

ANTONIAK, C. E. (1974). Mixtures of Dirichlet processes with applications to nonparametric problems. *Ann. Statist.,* **2,** 1152–1174.

ATKINSON, A. C. (1978). Posterior probabilities for choosing a regression model. *Biometrika,* **65,** 39–48.

BAGCHI, P. and KADANE, J. B. (1991). Laplace approximations to posterior moments and marginal distributions on circles, spheres and cylinders. *Canad. J. Statist.,* **19,** 67–77.

BAGCHI, P., DRAPER, N. and GUTTMAN, I. (1990). Bayesian assessment of assumptions of regression analysis. *Comm. Statist. Th. Meth.,* **19,** 921–934.

BARNETT, V. (1991). *Sample Survey Principles and Methods. 2nd ed.* Edward Arnold: London.

BARTLETT, M. S. (1957). A comment on D.V. Lindley's statistical paradox. *Biometrika,* **44,** 533–534.

BASU, D. (1969). Role of the sufficiency and likelihood principles in sample surveys. *Sankhyā* **A, 31,** 441–454.

BASU, D. (1980). Randomization analysis of experimental data: the Fisher randomization test (with discussion). *J. Amer. Statist. Ass.*, **75,** 575–595.

BATES, D. M. and WATTS, D. G. (1988). *Nonlinear Regression Analysis and its Applications.* Wiley: New York.

BAYARRI, M. J. and BERGER, J. O. (1992). Robust Bayesian bounds for outlier detection. Technical Report 92-43C, Department of Statistics, Purdue University.

BELLMAN, R. E. (1957). *Dynamic Programming.* Princeton University Press.

BERGER, J. O. (1985). *Statistical Decision Theory and Bayesian Analysis. 2nd ed.* Springer-Verlag, N.Y.

BERGER, J. (1986). *Multivariate Estimation: Bayes, Empirical Bayes and Stein Approaches.* SIAM: Philadelphia.

BERGER, J. O. (1990). Robust Bayesian analysis – sensitivity to the prior. *J. Statist. Plan. Inf.*, **25,** 303–328.

BERGER, J. and BERLINER, M. (1986). Robust Bayes and empirical Bayes analysis with ϵ-contaminated priors. *Ann. Statist.*, **14,** 461–486.

BERGER, J. O. and BERNARDO, J. M. (1992). On development of the reference prior method. In: *Bayesian Statistics 4*, J. M. Bernardo et al. (ed.), 35–60. Oxford University Press.

BERGER, J. O. and DELAMPADY, M. (1987). Testing precise hypotheses (with discussion). *Statist. Sci.*, **2,** 317–352.

BERGER, J. O. and MORTERA, J. (1991). Interpreting the stars in precise hypothesis testing. *Int. Statist. Rev.*, **59,** 337–353.

BERGER, J. and O'HAGAN, A. (1986). Ranges of posterior probabilities for unimodal priors with specified quantiles. In: *Bayesian Statistics 3*, J. M. Bernardo et al. (ed.), 45–65. Oxford University Press.

BERGER, J. O. and PERICCHI, L. R. (1993). The intrinsic Bayes factor for model selection and prediction. Technical Report 93–43 C, Department of Statistics, Purdue University.

BERGER, J. O. and SELLKE, T. (1987). Testing of a point null hypothesis: the irreconcilability of significance levels and evidence (with discussion). *J. Amer. Statist. Ass.*, **82,** 112–139.

BERGER, J. O. and WOLPERT, R. L. (1988). *The Likelihood Principle. 2nd ed.* Institute of Mathematical Statistics: Hayward, California.

BERLINER, L. M. and GOEL, P. (1986). Incorporating partial prior information: ranges of posterior probabilities. Technical Report 357, Department of Statistics, Ohio State University.

BERNARDO, J. M. (1979a). Reference posterior distributions for Bayesian inference (with discussion). *J. R. Statist. Soc.*, **B, 41,** 113–147.

BERNARDO, J. M. (1979b). Expected information as expected utility. *Ann. Statist.*, **7,** 686–690.

BESAG, J. and GREEN, P. J. (1993). Spatial statistics and Bayesian computation. *J. R. Statist. Soc.*, **B, 55,** 25–37.

BINDER, D. A. (1992). Non-parametric Bayesian models for samples from finite populations. *J. R. Statist. Soc.*, **B, 44,** 388–393.

BIRNBAUM, A. (1962). On the foundations of statistical inference. *J. Amer. Statist. Ass.*, **65,** 402.

BLACKWELL, D. (1973). Discreteness of Ferguson selections. *Ann. Statist.*, **1,** 356–358.

BLIGHT, B. J. N. and OTT, L. (1975). A Bayesian approach to model inadequacy for polynomial regression. *Biometrika*, **62,** 79–88.

BLOCH, D. A. and WATSON, G. S. (1967). A Bayesian study of the multinomial distribution. *Ann. Math. Statist.*, **38,** 1423–1435.

BLYTH, C. R. (1972). On Simpson's paradox and the sure-thing principle (with discussion). *J. Amer. Statist. Ass.*, **67,** 364–381.

BOLFARINE, H. and RODRIGUES, J. (1990). Finite population prediction under a linear function superpopulation model: a Bayesian perspective. *Comm. Statist. Th. Meth.*, **19,** 2577–2594.

BOX, G. E. P. (1980). Sampling and Bayes inference in scientific modelling and robustness (with discussion), *J. R. Statist. Soc.*, **A, 143,** 383–430.

BOX, G. E. P. and TIAO, G.C (1962). A further look at robustness via Bayes's theorem. *Biometrika*, **49,** 419.

Box, G. E. P. and Tiao, G. C. (1964). A note on criterion robustness and inference robustness. *Biometrika*, **51,** 169.

Box, G. E. P. and Tiao, G. C. (1968). A Bayesian approach to some outlier problems. *Biometrika*, **55,** 119–129.

Box, G. E. P. and Tiao, G. C. (1973). *Bayesian Inference in Statistical Analysis.* Reading, Mass: Addison-Wesley.

Broemeling, L. D. (1985). *Bayesian Analysis of Linear Models.* Marcel Dekker: New York.

Brooks, R. J. (1972). A decision theory approach to optimal regression designs. *Biometrika*, **59,** 563–571.

Brooks, R. J. (1976). Optimal regression designs for prediction when prior knowledge is available. *Metrika*, **23,** 217–221.

Brooks, R. J. (1980). On the relative efficiency of two paired-data experiments. *J. R. Statist. Soc.*, **B, 42,** 186–191.

Brown, P. J., Le, N. D. and Zidek, J. V. (1993). Inference for a covariance matrix. Technical Report 121, Department of Statistics, University of British Columbia.

Buehler, R. J. and Fedderson, A. P. (1963). Note on a conditional property of Student's *t*. *Ann. Math. Statist.*, **34**, 1098–1100.

Buonaccorsi, J. P. and Gatsonis, C. A. (1988). Bayesian inference for ratios of coefficients in a linear model. *Biometrics*, **44**, 87–101.

Carlin, B. P. and Polson, N. G. (1991). Inference for nonconjugate Bayesian models using the Gibbs sampler. *Canad. J. Statist.*, **19,** 399–405.

Cassel, C-M., Särndal, C. E. and Wretman, J. H. (1977). *Foundations of Inference in Survey Sampling*. New York: Wiley.

Chaloner, K. (1984). Optimal Bayesian experimental designs for linear models. *Ann. Statist.*, **12,** 283–300.

Chaloner, K. (1989). Bayesian designs for estimating the turning point of a quadratic regression. *Comm. Statist. The. Meth.*, **18,** 1385–1400.

Chaloner, K. and Brant, R. (1988). A Bayesian approach to outlier detection and residual analysis. *Biometrika*, **75,** 651–659.

Chib, S. (1993). Bayes regression with autoregressive errors — a Gibbs sampling approach. *J. Econometrics*, **58**, 275–294.

Chib, S. and Tiwari, R. C. (1991). Robust Bayes analysis in normal linear regression with an improper mixture prior. *Comm. Statist. Th. Meth.*, **20**, 807–829.

Clarke, B. and Barron A (1990). Information-theoretic asymptotics of Bayes methods. *IEEE Trans. Inform. Theory*, **36,** 453–471.

Clarke, B. and Wasserman, L. (1992). Information tradeoff. Technical Report 558, Department of Statistics, Carnegie Mellon University.

Copas, J. B. (1983). Regression, prediction and shrinkage. *J. R. Statist. Soc.*, **B, 45,** 311.

Copas, J. B. (1988). Binary regression models for contaminated data (with discussion). *J. R. Statist. Soc.*, **B, 50,** 225–265.

Cryer, J. D. (1986). *Time Series Analysis*. PWS: Boston, Mass.

Cuevas, A. and Sanz, P. (1988). On differentiability properties of Bayes operators. In: *Bayesian Statistics 3*, J. M. Bernardo et al. (ed.), 569–577. Oxford University Press.

Dalal, S. and Hall, W. J. (1983). Approximating priors by mixtures of natural conjugate priors. *J. R. Statist. Soc.*, **B, 45,** 278–286.

Datta, G. S. and Ghosh, M. (1991). Bayesian prediction in linear models: applications to small-area estimation. *Ann. Statist.*, **19**, 1748–1770.

Davis, P. J. and Rabinowitz, P. (1984). *Methods of Numerical Integration. 2nd ed.* Academic Press: Orlando, Florida.

Dawid, A. P. (1973). Posterior expectations for large observations. *Biometrika*, **60,** 664–667.

DAWID, A. P. (1977). Spherical matrix distributions and a multivariate model. *J. R. Statist. Soc.*, **B**, **39**, 254–261.

DAWID, A. P. (1979). Conditional independence in statistical theory. *J. R. Statist. Soc.*, **B**, **41**, 1–31.

DAWID, A. P. (1980). A Bayesian look at nuisance parameters. In: *Bayesian Statistics*, J. M. Bernardo et al. (ed.), 167–184. University Press: Valencia.

DAWID, A. P., STONE, M. and ZIDEK, J. V. (1973). Marginalisation paradoxes in Bayesian and structural inference (with discussion). *J. R. Statist. Soc.*, **B**, **35**, 189–233.

DE FINETTI, B. (1961). The Bayesian approach to the rejection of outliers. In: *Proceedings of the Fourth Berkeley Symposium on Probability and Statistics, Volume 1*, (ed.), 199–210. University Press: Berkeley.

DE FINETTI, B. (1974). *Theory of Probability*, 2 vols. Translated by A. Machi and A. F. M. Smith. Wiley: London.

DE GROOT, M. H. (1970). *Optimal Statistical Decisions*. McGraw-Hill: N.Y.

DE GROOT, M. H. and FIENBERG, S. E. (1983). The comparison and evaluation of forecasters. *The Statistician*, **32**, 12–22.

DELAMPADY, M. (1989). Lower bounds on Bayes factors for interval null hypotheses. *J. Amer. Statis. Ass.*, **84**, 120–124.

DEMPSTER, A. P. (1968). A generalisation of Bayesian inference (with discussion). *J. Roy. Statist. Soc.*, **B**, **30**, 205–247.

DEMPSTER, A. P. (1985). Probability, evidence and judgement. In: *Bayesian Statistics 2*, J. M. Bernardo et al. (eds), 119–131. North-Holland: Amsterdam.

DE ROBERTIS, L. and HARTIGAN, J. (1981). Bayesian inference using intervals of measures. *Ann. Statist.*, **9**, 235–244.

DEVROYE, L. (1986). *Non-uniform Random Variate Generation*. Springer-Verlag: New York.

DIACONIS, P. (1988). Bayesian numerical analysis. In: *Statistical Decision Theory and Related Topics IV, Volume 1*, S. S. Gupta and J. Berger (ed.), 163–175. Wiley: New York.

DIACONIS, P. and FREEDMAN, D. (1986). On the consistency of Bayes estimates (with discussion). *Ann. Statist.*, **14**, 1–67.

DIACONIS, P. and YLVISAKER, D. (1979). Conjugate priors for exponential families. *Ann. Statist.*, **7**, 269–281.

DIACONIS, P. and YLVISAKER, D. (1985). Quantifying prior opinion. In: *Bayesian Statistics 2*, J. M. Bernardo et al. (ed.), 133–156. North-Holland: Amsterdam.

DICKEY, J. M. (1983). Multiple hypergeometric functions: probabilistic interpretations and statistical uses. *J. Amer. Statist. Ass.*, **78**, 628–637.

DICKEY, J. M. and CHEN, C. M. (1985). Direct subjective-probability modelling using ellipsoidal distributions. In: *Bayesian Statistics 2*, J. M. Bernardo et al. (ed.), 157–182. Amsterdam: North-Holland.

DOOB, J. L. (1953). *Stochastic Processes*. Wiley: New York.

DRAPER, N. R. and HUNTER, W. G. (1967). The use of prior distributions in the design of experiments. *Biometrika*, **54**, 147–153.

DYER, D. and PIERCE, R. L. (1993). On the choice of prior distribution in hypergeometric sampling. *Comm. Statist. Th. Meth.*, **22**, 2125–2146.

EATWELL, J., MILGATE, M. and NEWMAN, P. eds (1990). *The New Palgrave: Utility and Probability*. Macmillan Press: London.

EAVES, D. M. and CHANG, T. (1992). Posterior mode estimation for the generalized linear model. *Ann. Inst. Stat. Math.*, **44**, 417–434.

EDWARDS, W. and PHILLIPS, L. D. (1964). Man as a transducer for probabilities in Bayesian command and control systems. In: *Human Judgements and Optimality*, Bryan, G. L. and Shelley, M. W (ed.). Wiley: New York.

EFRON, B. and MORRIS, C. (1973). Combining possibly related estimation problems (with discussion). *J. R. Statist. Soc.*, **B**, **35**, 379–422.

Ericson, W. A. (1969a). Subjective Bayesian models in sampling finite populations (with discussion). *J. R. Statist. Soc.*, **B**, **31**, 195–233.

Ericson, W. A. (1969b). A note on the posterior mean of a population mean. *J. R. Statist. Soc.*, **B**, **31**, 332–334.

Escobar, M. and West, M. (1991). Bayesian density estimation and inference using mixtures. Technical Report 533, Department of Statistics, Carnegie Mellon University.

Fan, T. H. and Berger, J. O. (1992). Behaviour of the posterior distribution and inferences for a normal mean with t prior distributions. *Statist. Decis.*, **10**, 99–120.

Fang, K. T., Kotz, S. and Ng, K. W. (1990). *Symmetric Multivariate and Related Distributions.* Chapman and Hall: London.

Farrow, M. and Goldstein, M. (1993). Bayes linear methods for grouped multivariate repeated measurement studies, with application to crossover trials. *Biometrika*, **80**, 39–59.

Fearn, T. (1975). A Bayesian approach to growth curves. *Biometrika*, **62**, 89–100.

Fedorov, V. V. (1972). *Theory of Optimal Experiments.* Translated by W. J. Studden and E. M. Klimko. Academic Press: New York.

Feller, W. (1968). *An Introduction to Probability Theory and its Applications, Volume 1*, 3rd edition. Wiley: New York.

Ferguson, T. S. (1967). *Mathematical Statistics, A Decision Theoretic Approach.* Academic Press: New York.

Ferguson, T. S. (1973). A Bayesian analysis of some nonparametric problems. *Ann. Statist.*, **1**, 209–230.

Ferguson, T. S. (1974). Prior distributions on spaces of probability measures. *Ann. Statist.*, **2**, 615–629.

Ferguson, T. S. (1983). Bayesian density estimation by mixtures of normal distributions. In: *Recent Advances in Statistics*, H. Rizvi and J. Rustagi (ed.), 287–302. New York: Academic Press.

Fine, T. L. (1973). *Theories of Probability.* Academic Press: New York.

Freeman, P. R. (1970). Optimal Bayesian sequential estimation of the median effective dose. *Biometrika*, **57**, 79–89.

French, S. (1980). Updating of belief in the light of someone else's opinion. *J. R. Statist. Soc.*, **A**, **143**, 43–48.

French, S. (1981). Consensus of opinion. *Eur. J. Opl. Res.*, **7**, 332–340.

French, S. (1985). Group consensus probability distributions: a critical survey. In: *Bayesian Statistics 2*, J. M. Bernardo et al. (ed.), 183–202. North-Holland: Amsterdam.

French, S. (1988). *Decision Theory: An Introduction to the Mathematics of Rationality.* Ellis Horwood: Chichester, Sussex.

Gamerman, D. and Migon, H. S. (1993). Dynamic hierarchical models. *J. R. Statist. Soc.*, **B**, **55**, 629–642.

Garthwaite, P. H. (1989). Fractile assessments for a linear regression model: an experimental study. *Org. Behaviour and Human Dec. Proc.*, **43**, 188–206.

Garthwaite, P. H. (1992). Preposterior expected loss as a scoring rule for prior distributions. *Comm. Statist. Th. Meth.*, **21**, 3601–3619.

Garthwaite, P. H. and Dickey, J. M. (1988). Quantifying expert opinion in linear regression problems. *J. Roy. Statist. Soc.*, **B**, **50**, 462–474.

Garthwaite, P. H. and Dickey, J. M. (1992). Elicitation of prior distributions for variable selection problems in regression. *Ann. Statist.*, **20**, 1697–1719.

Geisser, S. (1971). The inferential use of predictive distributions (with discussion). In: *Foundations of Statistical Inference*, V P. Godambe and D. A. Sprott (ed.), 456–469. Holt, Rinehart and Winston: Toronto.

Geisser, S. (1980). A predictivistic primer. In: *Bayesian Analysis in Econometrics and Statistics: Essays in Honour of Harold Jeffreys*, A. Zellner (ed.), 363–381. North-Holland, Amsterdam.

Geisser, S. (1985). On predicting observables: a selective update. In: *Bayesian Statistics 2*, J. M. Bernardo et al. (ed.), 203–230. North-Holland: Amsterdam.

GEISSER, S. (1987). Influential observations, diagnostics and discordancy tests. *Appl. Statist.*, **14**, 133–142.

GEISSER, S. (1989). Predictive discordancy testing for exponential observations. *Canad. J. Statist.*, **17**, 19–26.

GEISSER, S. and EDDY, W. (1979). A predictive approach to model selection. *J. Amer. Statist. Ass.*, **74**, 153–160.

GELFAND, A. E. and SMITH, A. F. M. (1990). Sampling-based approaches to calculating marginal densities. *J. Amer. Statist. Ass.*, **85**, 398–409.

GELMAN, A. and RUBIN, D. R. (1992). A single series from the Gibbs sampler provides a false sense of security. In: *Bayesian Statistics 4*, J. M. Bernardo et al. (ed.), 625–631. Oxford University Press.

GENEST, C. and SCHERVISH, M. J. (1985). Modelling expert judgements for Bayesian updating. *Ann. Statist.*, **13**, 1198–1212.

GEWEKE, J. (1988). Antithetic acceleration of Monte Carlo integration in Bayesian inference. *J. Econometrics*, **38**, 73–90.

GEWEKE, J. (1989). Bayesian inference in econometric models using Monte Carlo integration. *Econometrika*, **57**, 1317–1339.

GEWFKE, J. (1992a). Evaluating the accuracy of sampling based approaches to the calculation of posterior moments. In: *Bayesian Statistics 4*, J. M. Bernardo et al. (ed.), 169–193. Oxford University Press.

GEWEKE, J. (1992b). Priors for macroeconomic time series and their applications. Discussion Paper 44, Institute for Empirical Macroeconomics, Federal Reserve Bank of Minneapolis.

GILKS, W. R. and WILD, P. (1992). Adaptive rejection sampling for Gibbs sampling. *Appl. Statist.*, **41**, 337–348.

GILKS, W. R., CLAYTON, D. G., SPIEGELHALTER, D. J., BEST, N. G., MCNEIL, A. J., SHARPLES, L. D. and KIRBY, A. J. (1993). Modelling complexity: application of Gibbs sampling in medicine. *J. R. Statist. Soc.*, **B**, **55**, 39–52.

GIOVAGNOLI, A. and VERDINELLI, I. (1985). Optimal block designs under a hierarchical linear model. In: *Bayesian Statistics 2*, J. M. Bernardo et al. (ed.), 655–661. North-Holland: Amsterdam.

GIRÓN, F. J., MARTÍNEZ, M. L. and MORCILLO, C. (1992). A Bayesian justification for the analysis of residuals and influence measures. In: *Bayesian Statistics 4*, J. M. Bernardo et al. (ed.), 651–660. Oxford University Press.

GITFINS, J. C. (1989). *Multi-armed Bandit Allocation Indices*. Wiley: Chichester.

GLAZEBROOK, K. D. (1978). On the optimal allocation of two or more treatments in a controlled clinical trial. *Biometrika*, **65**, 335–340.

GODAMBE, V. R. (1966). A new approach to sampling from finite populations 1: sufficiency and linear estimation. *J. R. Statist. Soc.*, **B**, **28**, 310–319.

GOEL, P. K. and DE GROOT, M. H. (1980). Only normal distributions have linear posterior expectations in linear regression. *J. Amer. Statist. Ass.*, **75**, 895–900.

GOLDSTEIN, M. (1975a). Approximate Bayes solutions to some nonparametric problems. *Ann. Statist.*, **3**, 512–517.

GOLDSTEIN, M. (1975b). A note on some Bayesian non-parametric estimates. *Ann. Statist.*, **3**, 736–740.

GOLDSTEIN, M. (1977). On contractions of Bayes estimators for exponential family distributions. *Ann. Statist.*, **5**, 1235–1239.

GOLDSTEIN, M. (1979). The variance modified linear Bayes estimator. *J. R. Statist. Soc.*, **B**, **41**, 96–100.

GOLDSTEIN, M. (1981). Revising provisions: a geometric interpretation. *J. R. Statist. Soc.*, **B**, **43**, 105–130.

GOLDSTEIN, M. (1982). Contamination distributions. *Ann. Statist.*, **10**, 174–183.

GOLDSTEIN, M. (1985). Temporal coherence. In: *Bayesian Statistics 2*, J. M. Bernardo et al. (ed.), 231–248. North-Holland: Amsterdam.

GOLDSTEIN, M. (1986). Exchangeable belief structures. *J. Amer. Statist. Ass.*, **81**, 971–976.

GOLDSTEIN, M. (1988a). Adjusting belief structures. *J. R. Statist. Soc.*, **B**, **50**, 133–154.

GOLDSTEIN, M. (1988b). The data trajectory (with discussion). In: *Bayesian Statistics 3*, J. M. Bernardo et al. (ed.), 189–209. Oxford University Press.

GOLDSTEIN, M. (1990). Influence and belief adjustment. In: *Influence Diagrams, Belief Nets and Decision Analysis*, R. Oliver and J. Q. Smith (ed.). Wiley: Chichester.

GOLDSTEIN, M. (1991). Belief transforms and the comparison of hypotheses. *Ann. Statist.*, **13**, 2067–2089.

GOOD, I. J. (1959). Kinds of probability. *Science*, **129**, 443–447.

GOOD, I. J. (1983). *Good Thinking*. University of Minnesota Press.

GOUTIS, C. (1993). Ranges of posterior measures for some classes of priors with specified moments. *Int. Statist. Rev.*

GRIMMETT, G. R. and STIRZAKER, D. R. (1982). *Probability and Random Processes*. Clarendon Press: Oxford.

GUSTAFSON, P. and WASSERMAN, L. (1993). *Local sensitivity diagnostics for Bayesian inference*. Technical Report 574, Department of Statistics, Carnegie Mellon University.

GUTTMAN, L. and PEÑA, D. (1988). Outliers and influence: evaluation by posteriors of parameters in the linear model. In: *Bayesian Statistics 3*, J. M. Bernardo et al. (ed.), 631–640. Oxford University Press.

GUTTMAN, L., DUTTER, R. and FREEMAN, P. R. (1978). Care and handling of univariate outliers in the general linear model to detect spuriosity – a Bayesian approach. *Technometrics*, **20**, 187–193.

HAITOVSKY, Y. (1987). On multivariate ridge regression. *Biometrika*, **74**, 563–570.

HAMMERSLEY, J. M. and HANDSCOMB, D. C. (1964). *Monte Carlo Methods*. Chapman and Hall: London.

HARRISON, P. J. and STEVENS, C. F. (1976). Bayesian Forecasting (with discussion). *J. R. Statist. Soc.*, **B**, **38**, 205–247.

HARTIGAN, J. A. (1969). Linear Bayes methods. *J. R. Statist. Soc.*, **B**, **31**, 446–454.

HARTIGAN, J. A. (1983). *Bayes Theory*. Springer-Verlag: New York.

HARTLEY, H. O. and RAO, J. N. K. (1968). A new estimation theory for sample surveys. *Biometrika*, **55**, 547–557.

HASTINGS, W. K. (1970). Monte Carlo sampling methods using Markov chains and their applications. *Biometrika*, **57**, 97–109.

HEWITT, E. and SAVAGE, L. (1955). Symmetric measures on cartesian products. *Trans. Amer. Math. Soc.*, **80**, 470–501.

HEYDE, C. C. and JOHNSTONE, I. M. (1979). On asymptotic posterior normality for stochastic processes. *J. R. Statist. Soc.*, **B**, **41**, 184–189.

HILL, B. M. (1974). On coherence, inadmissibility and inference about many parameters in the theory of least squares. In: *Studies in Bayesian Econometrics and Statistics*, S. E. Fienberg and A. Zellner (ed.), 555–586. North-Holland: Amsterdam.

HILL, B. and LANE, D. (1984). Conglomerability and countable additivity. In: *Bayesian Inference and Decision Techniques with Applications*, P. K. Goel and A. Zellner (ed.). North-Holland: Amsterdam.

HILLS, S. E. and SMITH, A. F. M. (1992). Parameterization issues in Bayesian inference. In: *Bayesian Statistics 4*, J. M. Bernardo et al. (ed.), 227–246. Oxford University Press.

HILLS, S. E. and SMITH, A. F. M. (1993). Diagnostic plots for improved parameterization in Bayesian inference. *Biometrika*, **80**, 61–74.

HOUSEHOLDER, A. S. (1964). *The Theory of Matrices in Numerical Analysis*. Blaisdell: New York.

HOWSON, C. and URBACH, P. (1989). *Scientific Reasoning: The Bayesian Approach*. Open Court: La Salle, Illinois.

IBRAHIM, J. G. and LAND, P. W. (1991). On Bayesian analysis of generalized linear models using Jeffreys prior, *J. Amer. Statist. Ass.*, **86**, 981–986.

ILMAKUNNAS, P. (1985). Bayesian estimation of cost functions with stochastic or exact constraints on parameters. *Int. Econ. Rev.*, **26**, 111–134.

JAMES, W. and STEIN, C. (1961). Estimation with quadratic loss. *Fourth Berkeley Symposium Math. Statist. and Prob.* **1**, 361, U. of California Press, Berkeley.

JAYNES, E. T. (1968). Prior probabilities. *IEEE Trans. Syst. Sci. Cyber.*, **4**, 227–241.

JAYNES, E. T. (1981). Marginalization and prior probabilities. In: *Bayesian Analysis in Econometrics and Statistics*, A. Zellner (ed.). Amsterdam: North-Holland.

JAYNES, E. T. (1983). *Papers on Probability, Statistics and Statistical Physics*, R. D. Rosenkrantz (ed.). Reidel: Dordrecht.

JEFFREYS, H. (1967). *Theory of Probability*. Oxford University Press.

JOHNSON, N. L. and KOTZ, S. (1969). *Distributions in Statistics: Discrete Distributions*. Wiley: New York.

JOHNSON, N. L. and KOTZ, S. (1970a). *Distributions in Statistics: Continuous Univariate Distributions – 1*. Wiley: New York.

JOHNSON, N. L. and KOTZ, S. (1970b). *Distributions in Statistics: Continuous Univariate Distributions – 2*. Wiley: New York.

JOHNSON, N. L. and KOTZ, S. (1972). *Distributions in Statistics: Continuous Multivariate Distributions*. Wiley: New York.

JOHNSON, R. A. (1967). An asymptotic expansion for posterior distributions. *Ann. Math. Statist.*, **38**, 1899–1906.

JOHNSON, R. A. (1970). Asymptotic expansions associated with posterior distributions. *Ann. Math. Statist.*, **41**, 851–864.

KADANE, J. B. (1980). Predictive and structural methods for eliciting prior distributions. In: *Bayesian Analysis in Econometrics and Statistics: Essays in Honour of Harold Jeffreys*, A. Zellner (ed.). 89–93. Amsterdam: North-Holland.

KADANE, J. B. and CHUANG, D. T. (1978). Stable decision problems. *Ann. Statist.*, **6**, 1095–1110.

KADANE, J. B. and O'HAGAN, A. (1993). *Modelling using finitely additive probability: uniform distributions on the natural numbers.* Statistics research Report 93-5, University of Nottingham.

KADANE, J. B. and SEIDENFELD, J. (1990). Randomization in a Bayesian perspective. *J. Statist. Plan. Inf.*, **25**, 329–345.

KADANE, J. B., DICKEY, J. M., WINKLER, R. L., SMITH, W. S. and PETERS, S. C. (1980). Interactive elicitation of opinion for a normal linear model. *J. Amer. Statist. Assoc.*, **75**, 845–854.

KALMAN, R. E. (1963). New methods in Wiener filtering theory. In: *Proceedings of the First Symposium on Engineering Applications of Random Function Theory and Probability*, J. L. Bogdanoff and F. Kozin (ed.). Wiley: New York.

KASS, R. E. and SLATE, E. H. (1992). Reparameterization and diagnostics of non-normality (with discussion). In: *Bayesian Statistics 4*, J. M. Bernardo et al. (ed.), 289–305. Oxford University Press.

KASS, R. E., TIERNEY, L. and KADANE, J. B. (1988). Asymptotics in Bayesian computation. In: *Bayesian Statistics 3*, J. M. Bernardo et al. (ed.), 261–278. Oxford University Press.

KENNEDY, W. J. and GENTLE, J. E. (1980). *Statistical Computing*. Marcel Dekker: New York.

KINDERMAN, A. J. and MONAHAN, J. F. (1977). Computer generation of random variables using the ratio of random deviates. *ACM Trans. in Math. Software*, **3**, 257–260.

KLOEK, T. and VAN DIJK, H. K. (1978). Bayesian estimates of equation system parameters: an application of integration by Monte Carlo. *Econometrika*, **46**, 1–19.

KNUIMAN, M. W. and SPEED, T. P. (1988). Incorporating prior information into the analysis of contingency tables. *Biometrics*, **44**, 1061–1071.

LANGE, N., CARLIN, B. P. and GELFAND, A. E. (1992). Hierarchical Bayes models for the progression of HIV infection using longitudinal CD4 T-cell numbers. *J. Amer. Statist. Ass.*, **87**, 615–626.

LAURITZEN, S. L. and SPIEGELHALTER, D. J. (1988). Local computations with probabilities on graphical structures and their application to expert systems (with discussion). *J. R. Statist. Soc.*, **B**, **50**, 157–224.

LAVINE, M. (1991a). An approach to robust Bayesian analysis for multidimensional parameter spaces. *J. Amer. Statist. Ass.*, **86**, 400–403.

LAVINE, M. (1991b). Sensitivity in Bayesian Statistics: the prior and the likelihood. *J. Amer. Statist. Ass.*, **86**, 396–399.

LAVINE, M. (1992). Some aspects of Polya tree distributions for statistical modelling. *Ann. Statist*, **20**, 1222–1235.

LAVINE, M., WASSERMAN, L. and WOLPERT, R. L. (1991). Bayesian inference with specified prior marginals. *J. Amer. Statist. Ass.*, **86**, 964–971.

LE, H. and O'HAGAN, A. (1993). A class of bivariate heavy-tailed distributions. Research Report, Nottingham University Statistics Group.

LEE, S. Y. (1992). Bayesian analysis of stochastic constraints in structural equation models. *Brit. J. Math. Stat. Psy.*, **45**, 93–107.

LEE, T. D. (1987). Assessment of inter- and intra-laboratory variances: a Bayesian alternative to BS5497. *The Statistician*, **36**, 161–170.

LEMPERS, F. B. (1971). *Posterior Probabilities of Alternative Linear Models*. Rotterdam University Press.

LEONARD, T. (1973). A Bayesian method for histograms. *Biometrika*, **60**, 297–308.

LEONARD, T. (1975). Bayesian estimation methods for two-way contingency tables. *J. Roy. Statist. Soc.*, **B**, **37**, 23–37.

LEONARD, T. (1977). An alternative Bayesian approach to the Bradley–Terry model for paired comparisons. *Biometrika*, **63**, 69–75.

LEONARD, T. (1978). Density estimation, stochastic processes and prior information (with discussion). *J. R. Statist. Soc.*, **B**, **40**, 113–146.

LEONARD, T. (1982). Discussion of 'The statistical analysis of compositional data'. *J. R. Statist. Soc.*, **B**, **44**, 167–168.

LEONARD, T. and HSU, J. S. J. (1992). Bayesian inference for a covariance matrix. *Ann. Statist.*, **20**, 1669–1696.

LEONARD, T. and NOVICK, M. R. (1986). Bayesian full rank marginalisation for two-way contingency tables. *J. Educ. Statist.*, **11**, 33–56.

LICHTENSTEIN, S., FISCHHOFF, B. and PHILLIPS, L. D. (1982). Calibration of probabilities: the state of the art to 1980. In: *Judgement Under Uncertainty: Heuristics and Biases*, Kahneman, D. et al. (ed.). Cambridge University Press.

LINDLEY, D. V. (1956). On a measure of the information provided by an experiment. *Ann. Math. Statist.*, **27**, 986–1005.

LINDLEY, D. V. (1957). A statistical paradox. *Biometrika*, **44**, 187–192.

LINDLEY, D. V. (1960). Dynamic programming and decision theory. *Appl. Statist.*, **10**, 39–51.

LINDLEY, D. V. (1964). The Bayesian analysis of contingency tables. *Ann. Math. Statist.*, **35**, 1622–1643.

LINDLEY, D. V. (1971). The estimation of many parameters. In: *Foundations of Statistical Inference*, V. P. Godambe and D. A. Sprott (ed.), 435–455. Holt, Rinehart and Winston: Toronto.

LINDLEY, D. V. (1980). Approximate Bayesian methods (with discussion). In: *Bayesian Statistics*, J. M. Bernardo et al. (ed.), 223–245. Valencia University Press.

LINDLEY, D. V. (1982). The improvement of probability judgments. *J. R. Statist. Soc.*, **A**, **148**, 117–126.

LINDLEY, D. V. (1985). Reconciliation of discrete probability distributions. In: *Bayesian Statistics 2*, J. M. Bernardo et al. (ed.), 375–390. North-Holland: Amsterdam.

LINDLEY, D. V. (1988). The use of probability statements. In: *Accelerated Life Tests and Expert Opinion in Reliability*, C. A. Clarotti and D. V. Lindley (ed.), 25–57. North-Holland: Amsterdam.

LINDLEY, D. V. and NOVICK, M. R. (1981). The role of exchangeability in inference. *Ann. Statist.*, **9**, 45–58.

LO, A. Y. (1986). Bayesian statistical inference for sampling a finite population. *Ann. Statist.*, **14**, 1226–1233.

MCCONWAY, K. J. (1981). Marginalisation and linear opinion pools. *J. Amer. Statist. Ass.*, **76**, 410–414.

MALEC, D. and SEDRANSK, J. (1985). Bayesian inference for finite population parameters in multistage cluster sampling. *J. Amer. Statist. Ass.*, **80**, 897–902.

MARDIA, K. V. (1970). Measures of multivariate skewness and kurtosis with applications. *Biometrika*, **57**, 519–530.

MARDIA, K. V., KENT, J. T. and BIBBY, J. M. (1979). *Multivariate Analysis*. Academic Press: London.

MEEDEN, G. and VARDEMAN, S. (1985). Bayes and admissible set estimation. *J. Amer. Statist. Ass.*, **80**, 465–471.

MEEDEN, G. and VAPDEMAN, S. (1991). A noninformative Bayesian approach to interval estimation in finite population sampling. *J. Amer. Statist. Ass.*, **86**, 972–980.

MEINHOLD, R. J. and SINGPURWALLA, N. D. (1989). Robustification of Kalman filter models. *J. Amer. Statist. Ass.*, **84**, 479–486.

METROPOLIS, N., ROSENBLUTH, A. W., ROSENBLUTH, M. N., TELLER, A. H. and TELLER, E. (1953). Equations of state calculations by fast computing machine. *Jour. Chem. Phys.*, **21**, 1087–1091.

MORALES, J.-A. (1971). *Bayesian Full Information Structural Analysis*. Lecture Notes in Operations Research and Mathematical Systems 43, Springer-Verlag: Berlin.

MORENO, E. and CANO, J. A. (1991). Robust Bayesian analysis with ϵ-contaminations partially known. *J. R. Statist. Soc.*, **B**, **53**, 143–155.

MORRIS, C. (1983). Parametric empirical Bayes inference: theory and applications. *J. Amer. Statist. Ass.*, **78**, 47–65.

MOSTELLER, F. and WALLACE, D. L. (1984). *Applied Bayesian and Classical Inference: The Case of the Federalist Papers*. Springer-Verlag: New York. (Second edition *of Inference and Disputed Authorship: The Federalist*, Addison-Wesley, 1964).

NANDRAM, B. and SEDRANSK, J. (1993). Bayesian predictive inference for a finite population proportion: two-stage cluster sampling. *J. R. Statist. Soc.*, **B**, **55**, 399–408.

NAYLOR, J. C. and SMITH, A. F. M. (1982). Application of a method for the efficient computation of posterior distributions. *Appl. Statist.*, **31**, 214–225.

NAZARET, W. A. (1987). Bayesian log linear estimates for three-way contingency tables. *Biometrika*, **74**, 401–410.

NOVICK, M. R. and JACKSON, P. H. (1974). *Statistical Methods for Educational and Psychological Research*, McGraw-Hill: New York.

O'HAGAN, A. (1977). A general structure for inference about variances and covariances. In: *Recent Developments in Statistics*, J. R. Barra et al. (eds), 545–549. North-Holland: Amsterdam.

O'HAGAN, A. (1978). On curve fitting and optimal design for regression (with discussion). *J. R. Statist. Soc.*, **B**, **40**, 1–42.

O'HAGAN, A. (1979). On outlier rejection phenomena in Bayes inference. *J. R. Statist. Soc.*, **B**, **41**, 358–367.

O'HAGAN, A. (1981). A moment of indecision. *Biometrika*, **68**, 329–330.

O'HAGAN, A. (1987a). Monte Carlo is fundamentally unsound. *The Statistician*, **36**, 247–249.

O'HAGAN, A. (1987b). Exploring a high-dimensional posterior density. *Comp. Statist. Q*, **3**, 85–96.

O'HAGAN, A. (1988). *Probability; Methods and Measurement*. Chapman and Hall: London.

O'HAGAN, A. (1990). On outliers and credence for location parameter inference. *J. Amer. Statist. Assoc.*, **85**, 172–176.

O'HAGAN, A. (1991a). Discussion of Aitkin (1991). *J. R. Statist. Soc.*, **B**, **53**, 137.

O'HAGAN, A. (1991b). Bayes-Hermite quadrature. *J. Statist. Plan. Inf.*, **29**, 245–260.

O'HAGAN, A. (1992). Some Bayesian numerical analysis (with discussion). In: *Bayesian Statistics 4*, J. M. Bernardo et al. (ed.), 345–363. Oxford University Press.

O'HAGAN, A. (1993). Fractional Bayes factors for model comparison. Statistical Research Report 93–6, University of Nottingham.

O'HAGAN, A. and LE, H. (1993). Conflicting information and a class of bivariate heavy-tailed distributions.

O'HAGAN, A. and LEONARD, T. (1976). Bayes estimation subject to uncertainty about parameter constraints. *Biometrika*, **63**, 201–203.

O'HAGAN, A. and WELLS, F. S. (1993). Use of prior information to estimate costs in a sewerage operation. In: *Case Studies in Bayesian Statistics*, C. Gatsonis, J. S. Hodges and R. E. Kass (eds), 118–162. Springer-Verlag: New York.

OLIVER, R. M. and SMITH, J. Q. eds. (1989). *Influence Diagrams, Belief Nets and Decision Analysis.* Wiley: Chichester.

OWEN, R. J. (1970). The optimum design of a two-factor experiment using prior information. *Ann. Math. Statist.*, **41**, 1917–1934.

OWEN, R. J. (1975). A Bayesian sequential procedure for quantal response in the context of adaptive mental testing. *J. Amer. Statist. Ass.*, **70**, 351–356.

PEÑA, D. and TIAO, G. C. (1992). Bayesian robustness functions for linear models. In: *Bayesian Statistics 4*, J. M. Bernardo et al. (ed.), 365–368. Oxford University Press.

PERCY, D. F. (1992). Prediction for seemingly unrelated regressions. *J. R. Statist. Soc.*, **B**, **54**, 243–252.

PERICCHI, L. R. (1984). An alternative to the standard Bayesian procedure for discrimination between normal linear models. *Biometrika*, **71**, 515–586.

PERICCHI, L. R. and NAZARET, W. A. (1988). On being imprecise at the higher levels of a hierarchical linear model. In: *Bayesian Statistics 3*, J. M. Bernardo et al. (ed.), 361–375. North-Holland: Amsterdam.

PERICCHI, L. R. and SMITH, A. F. M. (1992). Exact and approximate posterior moments for a normal location parameter. *J. R. Statist. Soc.*, **B**, **54**, 793–804.

PETTIT, L. I. (1988). Bayes methods for outliers in exponential samples. *J. R. Statist. Soc.*, **B**, **50**, 371–380.

PETTIT, L. I. (1990). The conditional predictive ordinate for the normal distribution. *J. R. Statist. Soc.*, **B**, **52**, 175–184.

PETTIT, L. I. and SMITH, A. F. M. (1985). Outliers and influential observations in linear models. In: *Bayesian Statistics 2*, J. M. Bernardo et al. (ed.), 473–494. North-Holland: Amsterdam.

PILZ, J. (1983). *Bayesian Estimation and Experimental Design in Linear Regression Models.* Teubner-Texte zur Mathematik, DDR. (New edition, 1991, published by Wiley).

POINCARÉ, H. (1905). *Science and Hypothesis.* Reprinted 1952. Dover: New York.

POLASEK, W. and POTZELBERGER, K. (1988). Robust Bayesian analysis in hierarchial models. In: *Bayesian Statistics 3*, J. M. Bernardo et al. (ed.), 377–394. Oxford University Press.

POSKITT, D. S. (1987). Precision, complexity and Bayesian model determination. *J. R. Statist. Soc.*, **B**, **49**, 199–208.

QUEEN, C. M. and SMITH, J. Q. (1993). Multiregression dynamic models. *J. R. Statist. Soc.*, **B**, **55**, 849–870.

RAFTERY, A. E. and LEWIS, S. M. (1992). How many iterations in the Gibbs sampler? In: *Bayesian Statistics 4*, J. M. Bernardo et al. (ed.), 763–773. Oxford University Press.

RALESCU, D. and RALESCU, S. (1981). A class of nonlinear admissible estimators in the one-parameter exponential family. *Ann. Statist.*, **9**, 177–183.

REILLY, P. M. and PATINO-LEAL, H. (1981). A Bayesian study of the error-in-variables model. *Technometrics*, **23**, 221–231.

RICHARD, J.-F. and STEEL, M. F. J. (1988). Bayesian analysis of seemingly unrelated regression equations under a recursive extended natural conjugate prior density. *J. Econometrics*, **38**, 7–37.

RIPLEY, B. D. (1987). *Stochastic Simulation.* Wiley: New York.

ROBBINS, H. (1955). An empirical Bayes approach to statistics. In: *Proceedings of Third Berkeley Symposium on Mathematical Statistics and Probability, Volume 1*, (ed.), 151–164. Univ. of Calif. Press, Berkeley.

ROBERTS, G. O. (1992). Convergence diagnostics of the Gibbs sampler. In: *Bayesian Statistics 4*, J. M. Bernardo et al. (ed.), 775–782. Oxford University Press.

ROBERTS, G. O. and SMITH, A. F. M. (1993). Some convergence theory for Markov Chain Monte Carlo.

ROBINSON, G. K. (1975). Some counterexamples to the theory of confidence intervals. *Biometrika*, **62**, 155–161.

ROBINSON, G. K. (1979). Conditional properties of statistical procedures. *Ann. Statist.*, **7**, 742–755.

ROYALL, R. M. (1968). An old approach to finite population sampling theory. *J. Amer. Statist. Ass.*, **63**, 1269–1279.

ROYALL, R. M. (1970a). Finite population sampling – on labels in estimation. *Ann. Math. Statist.*, **41**, 1774–1779.

ROYALL, R. M. (1970b). On finite population sampling under certain linear regression models. *Biometrika*, **57**, 377–387.

RUBIN, D. B. (1978). Bayesian inference for causal effects: the role of randomization. *Ann. Statist.*, **6**, 34–58.

RUBIN, D. B. (1987). *Multiple Imputation for Nonresponse in Surveys*. Wiley: New York.

RUGGERI, F. (1990). Posterior ranges of functions of parameters under priors with specified quantiles. *Comm. Statist. The. Meth.*, **19**, 127–144.

RUGGERI, F. and WASSERMAN, L. (1993). Infinitessimal sensitivity of posterior distributions. *Canad. J. Statist.*, **22**, 195–203.

SALINETTI, G. (1992). Stability of Bayesian decisions. Paper presented at Bayesian Robustness conference, Milan, May 1992.

SALZER, H. E., ZUCKER, R. and CAPUANO, R. (1952). Tables of the zeroes and weight factors of the first twenty Hermite polynomials. *J. Res. Nat. Bur. Stand.*, **48**, 111–116.

SAN MARTINI, A. and SPEZZAFERRI, F. (1984). A predictive model selection criterion. *J. R. Statist. Soc.*, **B**, **46**, 296–303.

SAVAGE, L. J. (1972). *The Foundations of Statistics*, 2nd ed. Dover: New York.

SAVAGE, L. J. (1975). Elicitation of personal probabilities and expectations. In: *Studies in Bayesian Econometrics and Statistics*, S. E. Fienberg and A. Zellner (ed.), 111–156. North-Holland: Amsterdam.

SCHERVISH, M. J. and CARLIN, B. P. (1992). On the convergence of successive substitution sampling. *J. Comp. Graph. Statist.*, **1**, 111–127.

SCHWARZ, G. (1978). Estimating the dimension of a model. *Ann. Statist.*, **6**, 461–464.

SCOTT, A. J. (1977). On the problem of randomization in survey sampling. *Sankhyā*, **39c**, 1–9.

SCOTT, A. and SMITH, T. M. F. (1969). Estimation in multi-stage surveys. *J. Amer. Statist. Ass.*, **64**, 830–840.

SCOTT, A. and SMITH, T. M. F. (1973). Survey design, symmetry and posterior distributions. *J. R. Statist. Soc.*, **B**, **35**, 57–60.

SEIDENFELD, T. and WASSERMAN, L. (1991). Dilation for sets of probabilities. Technical Report 537, Department of Statistics, Carnegie Mellon University.

SHAFER, G. (1976). *A Mathematical Theory of Evidence*. Princeton University Press.

SHAFER, G. (1981). Constructive probability. *Synthese*, **48**, 1–60.

SHARPLES, L. D. (1990). Identification and accommodation of outliers in general hierarchical models. *Biometrika*, **77**, 445–453.

SHAW, J. E. H. (1988). Aspects of numerical integration and summarisation (with discussion). In: *Bayesian Statistics 3*, J. M. Bernardo et al. (ed.), 411–428. Oxford University Press.

SILVERMAN, B. W. (1985). Some aspects of the spline smoothing approach to non-parametric regression curve fitting (with discussion). *J. R. Statist. Soc.*, **B**, **47**, 1–52.

SILVERMAN, B. W. (1986). *Density Estimation for Statistics and Data Analysis*. Chapman and Hall: London.

SIMPSON, E. H. (1951). The interpretation of interaction in contingency tables. *J. R. Statist. Soc.*, **B**, **13**, 238–241.

SISKIND, V. (1972). Second moments of inverse Wishart-matrix elements. *Biometrika*, **59**, 690–691.

SIVAGANESAN, S. (1991). Sensitivity of some posterior summaries when the prior is unimodal with specified quantiles. *Canad. J. Statist.*, **19**, 57–65.

SIVAGANESAN, S. (1993). Robust Bayesian diagnostics. *J. Statist. Plan. Inf.*, **35**, 171–188.

SIVAGANESAN, S. and BERGER, J. (1989). Ranges of posterior measures for priors with unimodal contaminations. *Ann. Statist.*, **17**, 868–889.

SMITH, A. F. M. (1973a). Bayes estimates in one-way and two-way models. *Biometrika*, **60**, 319–330.

SMITH, A. F. M. (1973b). A general Bayesian linear model. *J. R. Statist. Soc.*, **B**, **35**, 67–75.

SMITH, A. F. M. and ROBERTS, G. O. (1993). Bayesian computation via the Gibbs samples and related Markov Chain Monte Carlo methods. *J. R. Statist. Soc.*, **B**, **55**, 3–23.

SMITH, A. F. M. and SPIEGELHALTER, D. J. (1980). Bayes factors and choice criteria for linear models. *J. R. Statist. Soc.*, **B**, **42**, 213–220.

SMITH, A. F. M., SKENE, A. M., SHAW, J. E. H., NAYLOR, J. C. and DRANSFIELD, M. (1985). The implementation of the Bayesian paradigm. *Commun. Statist. A*, **14**, 1079–1102.

SMITH, J. Q. (1988). *Decision Analysis – A Bayesian Approach*. Chapman and Hall: London.

SMITH, T. M. F. (1991). Post-stratification. *The Statistician*, **40**, 315–323.

SMOUSE, E. P. (1984). A note on Bayesian least squares inference for finite population models. *J. Amer. Statist. Ass.*, **79**, 390–392.

SPEZZAFERRI, F. (1988). Nonsequential designs for model discrimination and parameter estimation. In: *Bayesian Statistics 3*, J. M. Bernardo et al. (ed.), 777–783. Oxford University Press.

SPIEGELHALTER, D. J. and SMITH, A. F. M. (1982). Bayes factors for linear and loglinear models with vague prior information. *J. R. Statist. Soc.*, **B**, **44**, 377–387.

STEEL, M. F. J. (1991). A Bayesian analysis of simultaneous equation models by combining recursive analytical and numerical approaches. *J. Econometrics*, **48**, 83–117.

STEEL, M. F. J. and RICHARD, J.-F. (1991). Bayesian multivariate exogeneity analysis – an application to a UK money demand equation. *J. Econometrics*, **49**, 239–274.

STEIN, C. (1956). Inadmissibility of the usual estimator for the mean of a multivariate normal distribution. In: *Proc. Third Berkeley Symp. Math. Statist. and Prob.*, **I**, 137. U. California Press, Berkeley.

STEINBERG, D. M. and HUNTER, W. G. (1984). Experimental design – review and comment. *Technometrics*, **26**, 71–89.

STEWART, L. (1983). Bayesian analysis using Monte Carlo integration – a powerful methodology for handling some difficult problems. *The Statistician*, **32**, 195–200.

STIGLER, S. M. (1990). The 1990 Neyman memorial lecture: a Galtonian perspective on shrinkage estimators. *Statist. Sci*, **5**, 147.

STONE, M. (1959). Application of a measure of information to the design and comparison of regression experiments. *J. R. Statist. Soc.*, **B**, **21**, 55–70.

STONE, M. (1961). The linear opinion pool. *Ann. Math. Statist.*, **32**, 1339–1342.

STONE, M. (1963). Robustness of non-ideal decision procedures. *J. Amer. Statist. Ass.*, **58**, 480–486.

STONE, M. (1979). Review and analysis of some inconsistencies related to improper priors and finite additivity. In: *Proceedings of the Sixth International Congress on Logic, Methodology and Philosophy of Science*, L. J. Cohen et al. (ed.). North-Holland: Amsterdam.

STROUD, T. W. F. (1991). Hierarchical Bayes predictive means and variances with application to sample survey inference. *Commun. Statist. Th. Meth.*, **20**, 13–36.

SUGDEN, R. A. and SMITH, T. M. F. (1984). Ignorable and informative designs in survey sampling inference. *Biometrika*, **71**, 495–506.

TANNER, M. and WONG, W. (1987). The calculation of posterior distributions by data augmentation (with discussion). *J. Amer. Statist. Ass.*, **82**, 528–550.

THISTED, R. A. (1986). *Elements of Statistical Computing: Numerical computation*. Chapman and Hall: London.

THOMPSON, M. E. (1978). Stratified sampling with exchangeable prior distributions. *Ann. Statist.*, **6**, 1168–1169.

TIAO, G. C. and ZELLNER, A. (1964). On the Bayesian estimation of multivariate regression. *J. R. Statist. Soc.*, **B**, **26**, 277–285.

TIERNEY, L. and KADANE, J. B. (1986). Accurate approximations for posterior moments and marginal densities. *J. Amer. Statist. Ass.*, **81**, 82–86.

TIERNEY, L., KASS, R. E. and KADANE, J. B. (1989). Approximate marginal densities of nonlinear functions. *Biometrika*, **76**, 425–433.

TSURUMI, H. (1985). Limited information Bayesian analysis of a structural coefficient in a simultaneous equations system. *Comm. Statist. Th. Meth.*, **14**, 1103–1120.

TSUTAKAWA, R. K. (1988). Estimation of cancer mortality rates: a Bayesian analysis of small frequencies. *Biometrika*, **41**, 69–79.

TVERSKY, A. (1974). Assessing uncertainty (with discussion). *J. R. Statist. Soc.*, **B**, **36**, 148–159.

TVERSKY, A. and KAHNEMAN, D. (1974). Judgement under uncertainty: Heuristics and biases. *Science*, **185**, 1124–1131.

VAN DIJK, H. K. and KLOEK, T. (1983). Monte Carlo analysis of skew posterior distributions: an illustrative economic example. *The Statistician*, **32**, 216–223.

VAN DIJK, H. K. and KLOEK, T. (1985). Experiments with some alternatives for simple importance sampling in Monte Carlo integration. In: *Bayesian Statistics 2*, J. M. Bernardo et al. (ed.), 511–530. North-Holland: Amsterdam.

VERDINELLI, I. (1983). Computing Bayes D- and A-optimal designs for a two-way model. *The Statistician*, **32**, 161–167.

VERDINELLI, I. (1992). Advances in Bayesian experimental design. In: *Bayesian Statistics 4*, J. M. Bernardo et al. (ed.), 467–481. Oxford University Press.

VERDINELLI, L. and WASSERMAN, L. (1991). Bayesian analysis of outlier problems using the Gibbs sampler. *Statist. and Comput*, **1**, 105–117.

VILLEGAS, C. (1977a). Inner statistical inference. *J. Amer. Statist. Ass.*, **72**, 453–458.

VILLEGAS, C. (1977b). On the representation of ignorance. *J. Amer. Statist. Ass.*, **72**, 651–654.

VILLEGAS, C. (1981). Inner statistical inference II. *Ann. Statist.*, **9**, 768–776.

WAHBA, G. (1978). Improper priors, spline smoothing and the problem of guarding against model errors in regression. *J. R. Statist. Soc.*, **B**, **40**, 364–372.

WAHBA, G. (1983). Bayesian 'confidence intervals' for the cross-validated smoothing spline. *J. R. Statist. Soc.*, **B**, **45**, 133–150.

WALKER, A. M. (1969). On the asymptotic behaviour of posterior distributions. *J. R. Statist. Soc.*, **B**, **31**, 80–88.

WALLEY, P. (1991). *Statistical Reasoning with Imprecise Probabilities*. Wiley: New York.

WASSERMAN, L. (1990). Belief functions and statistical inference. *Canad. J. Statist.*, **18**, 183–196.

WASSERMAN, L. (1992a). The conflict between improper priors and robustness. Technical Report 559, Department of Statistics, Carnegie Mellon University.

WASSERMAN, L. (1992b). Recent methodological advances in robust Bayesian inference (with discussion). In: *Bayesian Statistics 4*, J. M. Bernardo et al. (ed.), 483–502. Oxford University Press.

WASSERMAN, L. and KADANE, J. B. (1990). Bayes' theorem for Choquet capacities. *Ann. Statist.*, **18**, 1328–1339.

WASSERMAN, L. and KADANE, J. B. (1992). Computing bounds on expectations. *J. Amer. Statist. Ass.*, **87**, 512.

WEERAHANDI, S. and ZIDEK, J. V. (1981). Multi-Bayesian statistical decision theory. *J. R. Statist. Soc.*, **A**, **144**, 85–93.

WEST, M. (1984). Outlier models and prior distributions in Bayesian linear regression. *J. R. Statist. Soc.*, **B**, **46**, 431–439.

WEST, M. (1985). Generalised linear models: scale parameters, outlier accommodation and prior distribution. In: *Bayesian Statistics 2*, J. M. Bernardo et al. (ed.), 531–558. North-Holland: Amsterdam.

WEST, M. (1988). Modelling expert opinion. In: *Bayesian Statistics 3*, J. M. Bernardo et al. (ed.), 493–508. Oxford University Press.

WEST, M. (1991). Kernel density estimation and marginalization consistency. *Biometrika*, **78**, 421–425.

WEST, M. and CROSSE, J. (1992). Modelling probabilistic agent opinion. *J. R. Statist. Soc.*, **B**, **54**, 285–299.

WEST, M. and HARRISON, P. J. (1989). *Bayesian Forecasting and Dynamic Models.* Springer: New York.

WHITTLE, P. (1958). On the smoothing of probability density functions. *J. R. Statist. Soc.*, **B**, **20**, 334–343.

WHITTLE, P. (1983). *Optimisation Over Time (2 Volumes).* Wiley: Chichester.

WILD, P. and GILKS, W. R. (1993). AS 287. Adaptive rejection sampling from log-concave density functions. *Appl. Statist.*, **42**, 701–709.

WINKLER, R. L. (1968). The consensus of subjective probability distributions. *Mgmt. Sci*, **15**, B61–B75.

WINKLER, R. L. (1969). Scoring rules 1 and the evaluation of probability assessors. *J. Amer. Statist. Ass.*, **64**, 1073–1078.

WINKLER, R. L. (1980). Prior information, predictive distributions and Bayesian model-building. In: *Bayesian Analysis in Econometrics and Statistics: Essays in Honour of Harold Jeffreys*, A. Zellner (ed.). North-Holland: Amsterdam.

YOUNG, A. S. (1977). A Bayesian approach to prediction using polynomials. *Biometrika*, **64**, 309–317.

ZELLNER, A. (1971). *An Introduction to Bayesian Inference in Econometrics.* Wiley: New York.

ZELLNER, A. and CHETTY, V. K. (1965). Prediction and decision problems in regression models from the Bayesian point of view. *J. Amer. Statist. Ass.*, **60**, 608–616.

INDEX OF EXAMPLES IN TEXT

Chapter	1	2	3	4	5	6	7	8	9	10
Examples										
.1	**.5**	**.3**	**.4**	**.2**	**.4**	**.7**	**.4**	**.4**	**.12**	**.6**
.2	**.5**	**.3**	**.4**	**.18**	**.9**	**.13**	**.5**	**.5**	**.12**	**.9**
.3	**.6**	**.3**	**.6**	**.32**	**.16**	**.13**	**.5**	**.13**	**.23**	**.15**
.4	**.10**	**.4**	**.8**	**.33**	**.25**	**.13**	**.5**	**.14**	**.23**	**.23**
.5	**.11**	**.5**	**.9**	**.34**	**.30**	**.17**	**.12**	**.24**	**.35**	**.26**
.6	**.12**	**.6**	**.9**	**.42**	**.35**	**.17**	**.13**	**.32**	**.35**	**.32**
.7	**.12**	**.13**	**.9**	**.42**	**.35**	**.28**	**.13**	**.45**	**.36**	**.36**
.8	**.28**	**.14**	**.12**	**.43**	**.35**	**.30**	**.16**	**.53**	**.39**	**.46**
.9		**.17**	**.17**	**.43**	**.40**	**.33**	**.16**	**.57**	**.57**	**.51**
.10		**.17**	**.31**	**.45**	**.41**	**.35**	**.17**	**.59**	**.58**	**.58**
.11		**.18**	**.32**	**.45**	**.42**	**.36**	**.22**	**.65**	**.66**	
.12		**.36**	**.41**		**.42**	**.37**	**.24**	**.65**	**.66**	
.13		**.49**	**.48**		**.42**	**.37**	**.24**	**.75**		
.14		**.51**	**.48**			**.37**	**.39**			
.15		**.51**	**.58**			**.42**	**.42**			
.16		**.54**				**.42**	**.43**			
.17		**.59**				**.45**	**.45**			
.18						**.46**	**.46**			
.19						**.46**	**.56**			
.20						**.52**	**.56**			

Each entry in the table gives the section number that preceeds the Example numbered in the left margin, in the chapter at the head of the column. Thus the entry **.43**, with co-ordinates (.16,7) means that Example 7.16 appears in section **7.43**.

INDEX

(References are to chapter-sections, displayed at the tops of pages. Examples in the text may be located using the *Index of Examples in Text* on page 325. Exercises appear at the ends of chapters.)